Lecture Notebook

to accompany

EIGHTH EDITION **LIFE** The Science of Biology

Sadava • Heller • Orians • Purves • Hillis

 Sinauer Associates, Inc.

 W. H. Freeman and Company

Cover photograph © Max Billder.

Lecture Notebook to accompany *Life: The Science of Biology,* **Eighth Edition**

Address editorial correspondence to:
Sinauer Associates, Inc.
23 Plumtree Road
Sunderland, MA 01375 U.S.A.
Fax: 413-549-1118
Internet: www.sinauer.com; publish@sinauer.com

Address orders to:
W.H. Freeman and Company
VHPS/W.H. Freeman & Co. Order Department
16365 James Madison Highway, U.S. Route 15
Gordonsville, VA 22942 U.S.A.

ISBN 978-0-7167-7894-3
Printed in U.S.A.

Contents

1 Studying Life

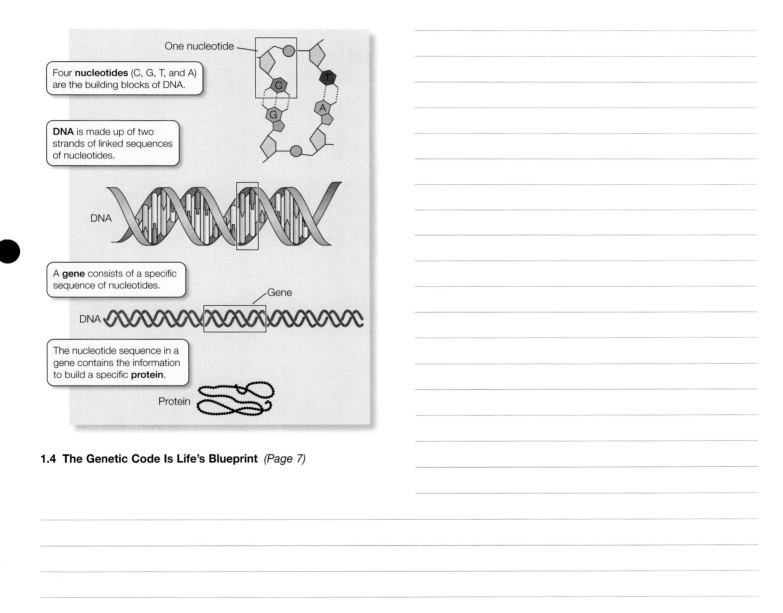

Four **nucleotides** (C, G, T, and A) are the building blocks of DNA.

One nucleotide

DNA is made up of two strands of linked sequences of nucleotides.

DNA

A **gene** consists of a specific sequence of nucleotides.

Gene

DNA

The nucleotide sequence in a gene contains the information to build a specific **protein**.

Protein

1.4 The Genetic Code Is Life's Blueprint *(Page 7)*

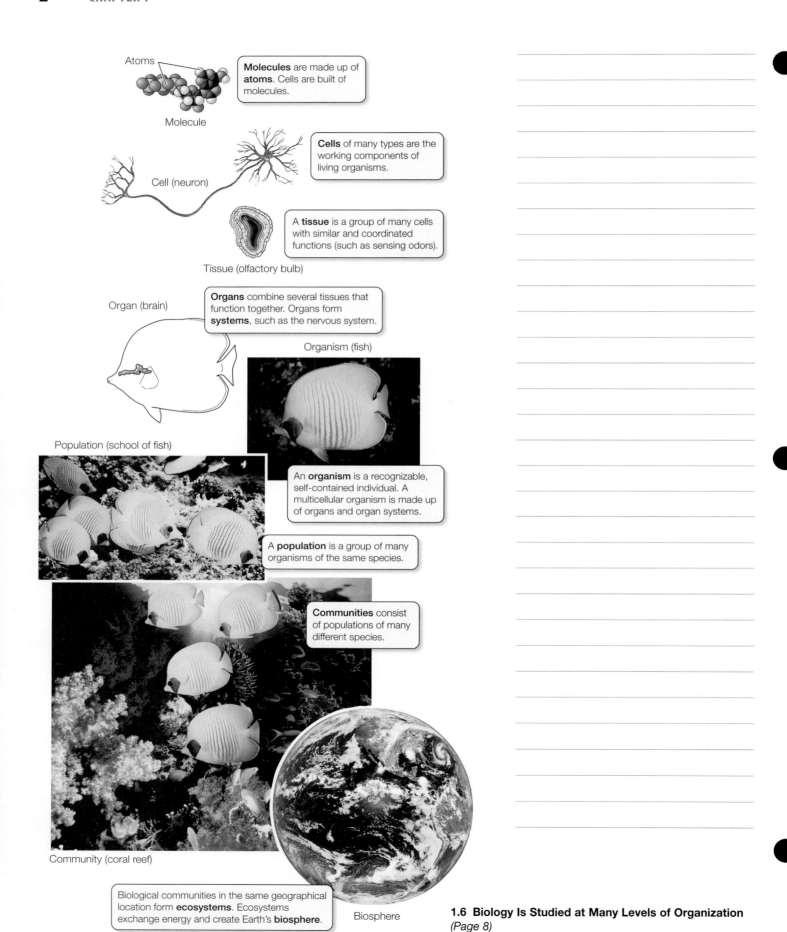

Atoms

Molecule

Molecules are made up of **atoms**. Cells are built of molecules.

Cell (neuron)

Cells of many types are the working components of living organisms.

Tissue (olfactory bulb)

A **tissue** is a group of many cells with similar and coordinated functions (such as sensing odors).

Organ (brain)

Organs combine several tissues that function together. Organs form **systems**, such as the nervous system.

Organism (fish)

An **organism** is a recognizable, self-contained individual. A multicellular organism is made up of organs and organ systems.

Population (school of fish)

A **population** is a group of many organisms of the same species.

Communities consist of populations of many different species.

Community (coral reef)

Biological communities in the same geographical location form **ecosystems**. Ecosystems exchange energy and create Earth's **biosphere**.

Biosphere

1.6 Biology Is Studied at Many Levels of Organization
(Page 8)

Each "day" represents about 150 million years.

Life probably arose more than 4 billion years ago.

		1 Earth forms	2	3	4	5 Origin of life
6	7 Oldest fossils	8	9	10	11	12
13	14 Photo-synthesis evolves	15	16	17	18	19
20 Eukaryotic cells evolve	21	22	23	24 Multi-cellular organisms	25	26
27	28	29	30			

27 Aquatic life — Abundant fossils

28 First land plants — First land animals

29 Coal-forming forests — Insects — First mammals — Dinosaurs dominant

30 First birds — First flowering plants — Rise of mammals

← First hominids
← Homo sapiens

Homo sapiens (modern humans) appeared in the last 10 minutes of day 30.

Recorded history fills the last few seconds of day 30.

1.9 Life's Calendar *(Page 10)*

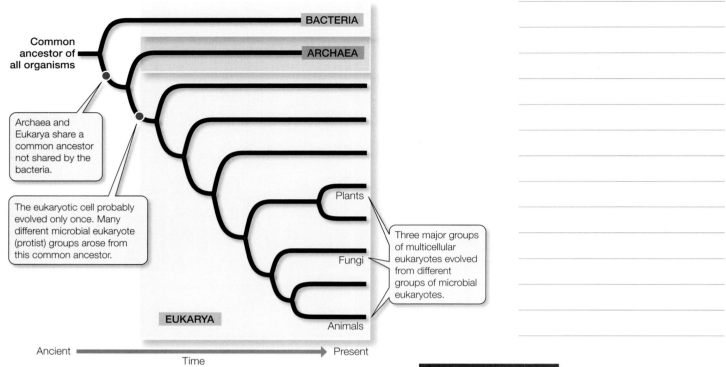

Common ancestor of all organisms

BACTERIA

ARCHAEA

Archaea and Eukarya share a common ancestor not shared by the bacteria.

The eukaryotic cell probably evolved only once. Many different microbial eukaryote (protist) groups arose from this common ancestor.

Plants

Three major groups of multicellular eukaryotes evolved from different groups of microbial eukaryotes.

Fungi

EUKARYA

Animals

Ancient ——————————> Present
Time

1.11 The Tree of Life *(Page 12)*

EXPERIMENT

HYPOTHESIS: Something in the environment is causing developmental limb abnormalities in Pacific tree frogs (*Hyla regilla*).

METHOD

1. Identify a test area of small ponds in an area where abnormal tree frogs have been found (agricultural land in Santa Clara County, California).
2. Collect and analyze water samples from the ponds.
3. Census the organisms in the ponds.
4. Look for correlations between the presence of frog abnormalities and the characteristics of the ponds.

Deformed hind leg

RESULTS

Pacific tree frogs were found in 13 of 35 ponds. Frogs with limb abnormalities were found in 4 of these 13 ponds. Water and census analyses of the 13 ponds containing frogs revealed no difference in water pollution, but did reveal the presence of snails infested with parasitic flatworms of the genus *Ribeiroia* in the 4 ponds with abnormal frogs.

Santa Clara County, CA

	Pesticide residues in water?	Heavy metals in water?	Industrial chemicals in water?	Snails in water?	*Ribeiroia* in water?	*Ribeiroia* larvae in frogs?
Ponds with normal frogs	No	No	No	No	No	No
Ponds with abnormal frogs	No	No	No	Yes	Yes	Yes

CONCLUSION: Infection by parasitic *Ribeiroia* may cause abnormalities in the limb development of Pacific tree frogs.

EXPERIMENT

HYPOTHESIS: Infection of Pacific tree frog tadpoles by the parasite *Ribeiroia* causes developmental limb abnormalities.

METHOD

1. Collect *Hyla regilla* eggs from a site with no record of abnormal frogs.
2. Allow eggs to hatch in laboratory aquaria. Randomly divide equal numbers of the resulting tadpoles into control and experimental groups.
3. Allow the control group to develop normally. Subject the experimental groups to infection with *Ribeiroia*, a different parasite (*Alaria*), and a combination of both parasites.
4. Follow tadpole development. Count and assess the resulting adult frogs.

Control (no parasites)   Experiment 1 (with *Alaria*)   Experiment 2 (with *Ribeiroia*)   Experiment 3 (with *Alaria* and *Ribeiroia*)

RESULTS

■ Survivorship (percent of tadpoles reaching adulthood)

■ Abnormality rate (percent of adults with limb abnormalities)

CONCLUSION: *Ribeiroia* causes developmental limb abnormalities in Pacific tree frogs.

1.14 Controlled Experiments Manipulate a Variable *(Page 15)*

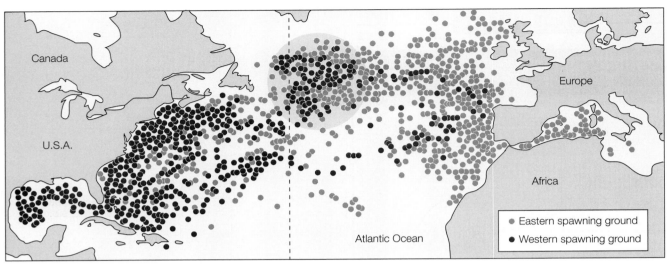

1.15 Bluefin Tuna Do Not Recognize the Lines Drawn on Maps by International Commissions *(Page 17)*

CHAPTER 2 The Chemistry of Life

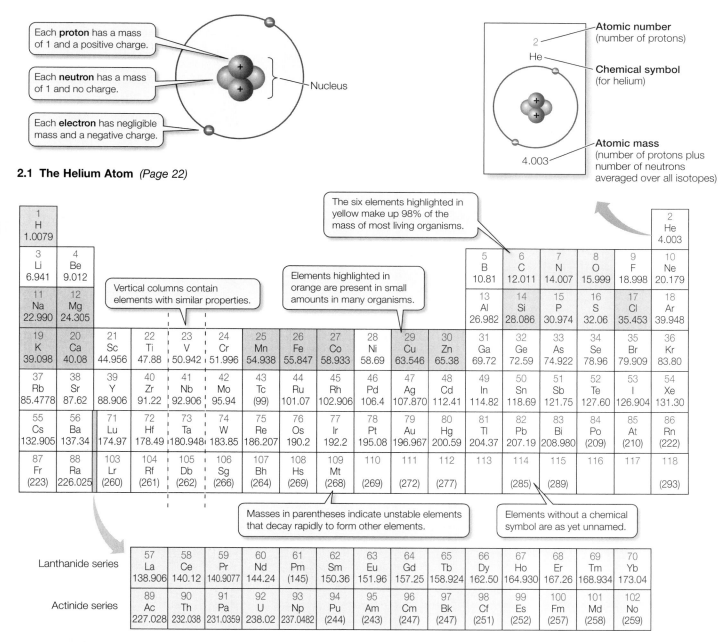

Each **proton** has a mass of 1 and a positive charge.

Each **neutron** has a mass of 1 and no charge.

Each **electron** has negligible mass and a negative charge.

Nucleus

Atomic number
(number of protons)

2

He

Chemical symbol
(for helium)

4.003

Atomic mass
(number of protons plus
number of neutrons
averaged over all isotopes)

2.1 The Helium Atom *(Page 22)*

The six elements highlighted in yellow make up 98% of the mass of most living organisms.

Elements highlighted in orange are present in small amounts in many organisms.

Vertical columns contain elements with similar properties.

1 H 1.0079																	2 He 4.003
3 Li 6.941	4 Be 9.012											5 B 10.81	6 C 12.011	7 N 14.007	8 O 15.999	9 F 18.998	10 Ne 20.179
11 Na 22.990	12 Mg 24.305											13 Al 26.982	14 Si 28.086	15 P 30.974	16 S 32.06	17 Cl 35.453	18 Ar 39.948
19 K 39.098	20 Ca 40.08	21 Sc 44.956	22 Ti 47.88	23 V 50.942	24 Cr 51.996	25 Mn 54.938	26 Fe 55.847	27 Co 58.933	28 Ni 58.69	29 Cu 63.546	30 Zn 65.38	31 Ga 69.72	32 Ge 72.59	33 As 74.922	34 Se 78.96	35 Br 79.909	36 Kr 83.80
37 Rb 85.4778	38 Sr 87.62	39 Y 88.906	40 Zr 91.22	41 Nb 92.906	42 Mo 95.94	43 Tc (99)	44 Ru 101.07	45 Rh 102.906	46 Pd 106.4	47 Ag 107.870	48 Cd 112.41	49 In 114.82	50 Sn 118.69	51 Sb 121.75	52 Te 127.60	53 I 126.904	54 Xe 131.30
55 Cs 132.905	56 Ba 137.34	71 Lu 174.97	72 Hf 178.49	73 Ta 180.948	74 W 183.85	75 Re 186.207	76 Os 190.2	77 Ir 192.2	78 Pt 195.08	79 Au 196.967	80 Hg 200.59	81 Tl 204.37	82 Pb 207.19	83 Bi 208.980	84 Po (209)	85 At (210)	86 Rn (222)
87 Fr (223)	88 Ra 226.025	103 Lr (260)	104 Rf (261)	105 Db (262)	106 Sg (266)	107 Bh (264)	108 Hs (269)	109 Mt (268)	110 (269)	111 (272)	112 (277)	113	114 (285)	115 (289)	116	117	118 (293)

Masses in parentheses indicate unstable elements that decay rapidly to form other elements.

Elements without a chemical symbol are as yet unnamed.

Lanthanide series	57 La 138.906	58 Ce 140.12	59 Pr 140.9077	60 Nd 144.24	61 Pm (145)	62 Sm 150.36	63 Eu 151.96	64 Gd 157.25	65 Tb 158.924	66 Dy 162.50	67 Ho 164.930	68 Er 167.26	69 Tm 168.934	70 Yb 173.04
Actinide series	89 Ac 227.028	90 Th 232.038	91 Pa 231.0359	92 U 238.02	93 Np 237.0482	94 Pu (244)	95 Am (243)	96 Cm (247)	97 Bk (247)	98 Cf (251)	99 Es (252)	100 Fm (257)	101 Md (258)	102 No (259)

2.2 The Periodic Table *(Page 22)*

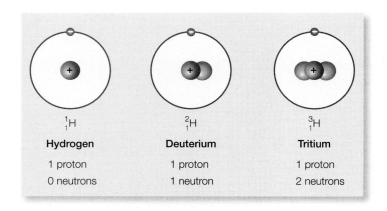

¹₁H   ²₁H   ³₁H

Hydrogen   **Deuterium**   **Tritium**

1 proton   1 proton   1 proton

0 neutrons   1 neutron   2 neutrons

2.3 Isotopes Have Different Numbers of Neutrons *(Page 23)*

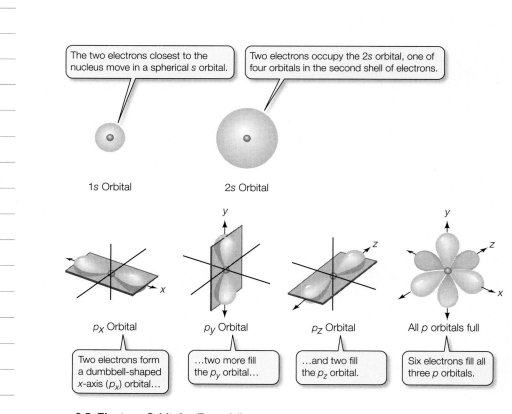

The two electrons closest to the nucleus move in a spherical *s* orbital.

Two electrons occupy the 2*s* orbital, one of four orbitals in the second shell of electrons.

1*s* Orbital   2*s* Orbital

*p*ₓ Orbital   *p*_y Orbital   *p*_z Orbital   All *p* orbitals full

Two electrons form a dumbbell-shaped *x*-axis (*p*ₓ) orbital…

…two more fill the *p*_y orbital…

…and two fill the *p*_z orbital.

Six electrons fill all three *p* orbitals.

2.5 Electron Orbitals *(Page 24)*

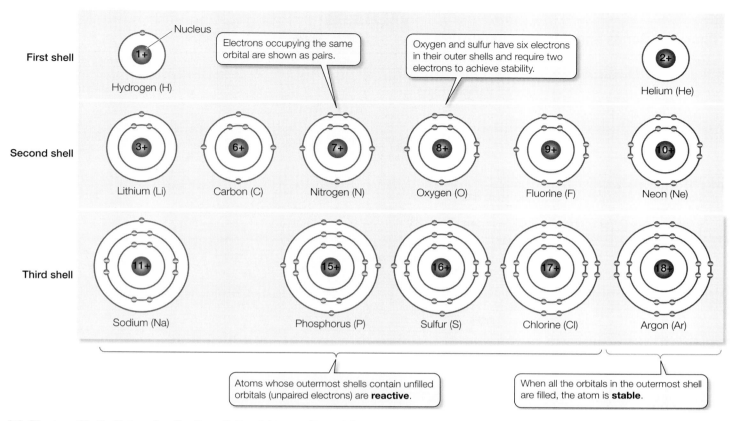

First shell

Nucleus

Hydrogen (H)

Electrons occupying the same orbital are shown as pairs.

Oxygen and sulfur have six electrons in their outer shells and require two electrons to achieve stability.

Helium (He)

Second shell

Lithium (Li)   Carbon (C)   Nitrogen (N)   Oxygen (O)   Fluorine (F)   Neon (Ne)

Third shell

Sodium (Na)   Phosphorus (P)   Sulfur (S)   Chlorine (Cl)   Argon (Ar)

Atoms whose outermost shells contain unfilled orbitals (unpaired electrons) are **reactive**.

When all the orbitals in the outermost shell are filled, the atom is **stable**.

2.6 Electron Shells Determine the Reactivity of Atoms *(Page 24)*

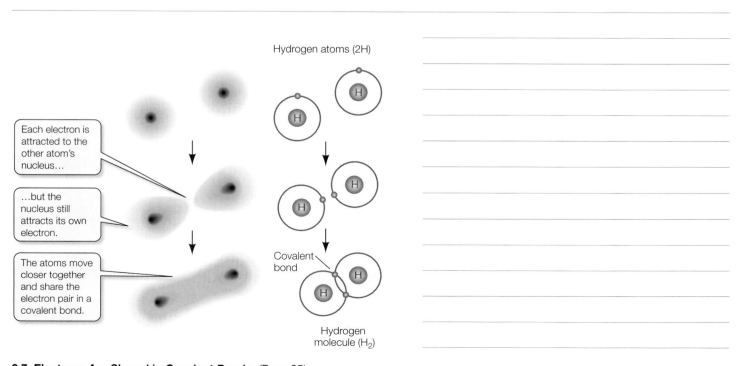

Hydrogen atoms (2H)

Each electron is attracted to the other atom's nucleus...

...but the nucleus still attracts its own electron.

The atoms move closer together and share the electron pair in a covalent bond.

Covalent bond

Hydrogen molecule (H_2)

2.7 Electrons Are Shared in Covalent Bonds *(Page 25)*

TABLE 2.1

Chemical Bonds and Interactions

NAME	BASIS OF INTERACTION	STRUCTURE	BOND ENERGY[a] (KCAL/MOL)
Covalent bond	Sharing of electron pairs		50–110
Ionic bond	Attraction of opposite changes		3–7
Hydrogen bond	Sharing of H atom		3–7
Hydrophobic interaction	Interaction of nonpolar substances in the presence of polar substances (especially water)		1–2
van der Waals interaction	Interaction of electrons of nonpolar substances		1

[a]*Bond energy* is the amount of energy needed to separate two bonded or interacting atoms under physiological conditions.

(Page 26)

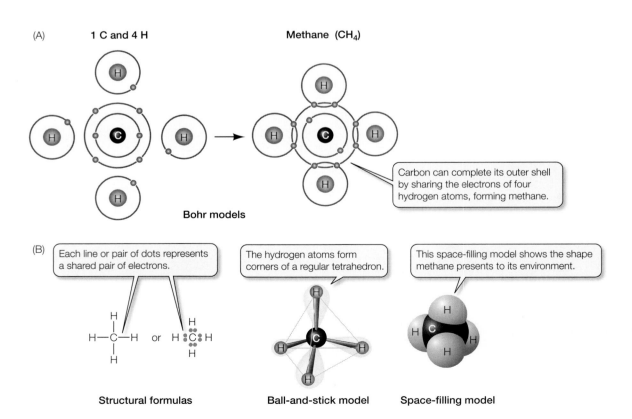

(A)　　1 C and 4 H　　　　　Methane (CH₄)

Carbon can complete its outer shell by sharing the electrons of four hydrogen atoms, forming methane.

Bohr models

(B) Each line or pair of dots represents a shared pair of electrons.

The hydrogen atoms form corners of a regular tetrahedron.

This space-filling model shows the shape methane presents to its environment.

Structural formulas　　　Ball-and-stick model　　Space-filling model

2.8 Covalent Bonding Can Form Compounds *(Page 26)*

TABLE 2.2

Covalent Bonding Capabilities of Some Biologically Important Elements

ELEMENT	USUAL NUMBER OF COVALENT BONDS
Hydrogen (H)	1
Oxygen (O)	2
Sulfur (S)	2
Nitrogen (N)	3
Carbon (C)	4
Phosphorus (P)	5

(Page 27)

TABLE 2.3

Some Electronegativities

ELEMENT	ELECTRONEGATIVITY
Oxygen (O)	3.5
Chlorine (Cl)	3.1
Nitrogen (N)	3.0
Carbon (C)	2.5
Phosphorus (P)	2.1
Hydrogen (H)	2.1
Sodium (Na)	0.9
Potassium (K)	0.8

(Page 27)

Bohr model

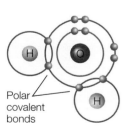

Polar covalent bonds

Ball-and-stick model

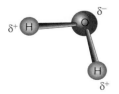

Space-filling model

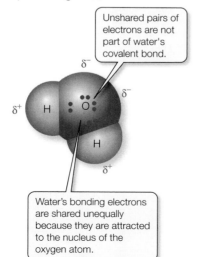

Unshared pairs of electrons are not part of water's covalent bond.

Water's bonding electrons are shared unequally because they are attracted to the nucleus of the oxygen atom.

2.9 Water's Covalent Bonds are Polar *(Page 28)*

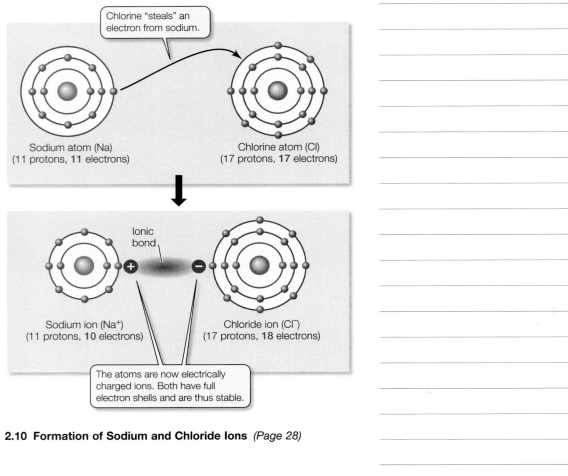

Chlorine "steals" an electron from sodium.

Sodium atom (Na)
(11 protons, **11** electrons)

Chlorine atom (Cl)
(17 protons, **17** electrons)

Ionic bond

Sodium ion (Na⁺)
(11 protons, **10** electrons)

Chloride ion (Cl⁻)
(17 protons, **18** electrons)

The atoms are now electrically charged ions. Both have full electron shells and are thus stable.

2.10 Formation of Sodium and Chloride Ions *(Page 28)*

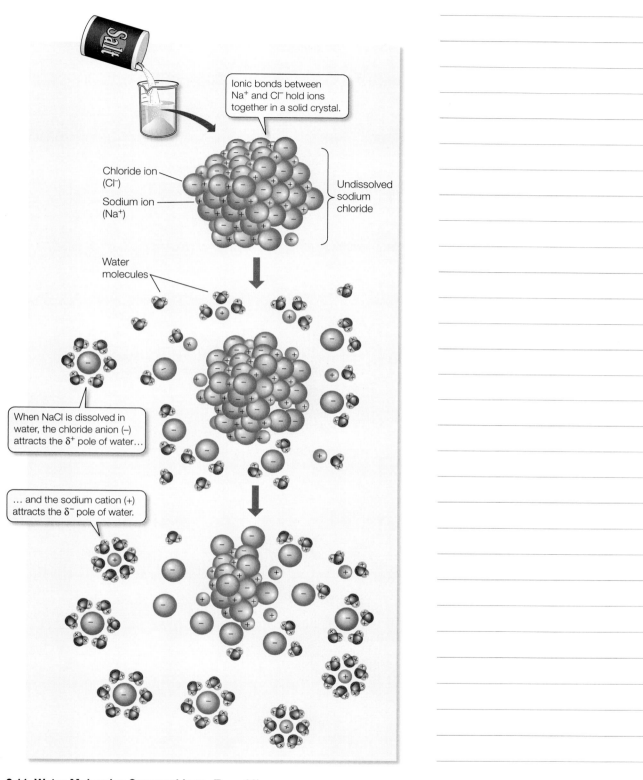

2.11 Water Molecules Surround Ions *(Page 29)*

(A)

(B)

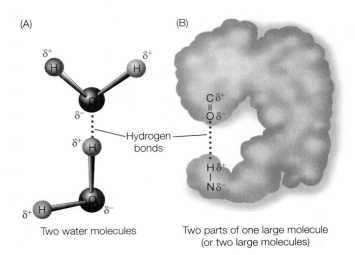

Two water molecules

Two parts of one large molecule
(or two large molecules)

2.12 Hydrogen Bonds Can Form Between or Within Molecules
(Page 29)

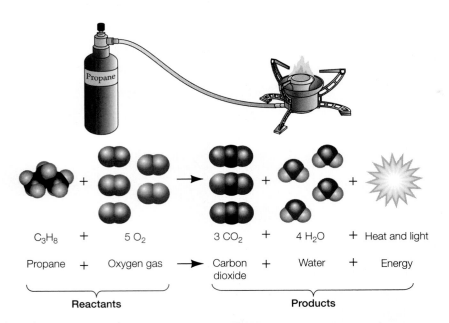

| C_3H_8 | + | $5\ O_2$ | | $3\ CO_2$ | + | $4\ H_2O$ | + | Heat and light |
| Propane | + | Oxygen gas | → | Carbon dioxide | + | Water | + | Energy |

Reactants

Products

2.13 Bonding Partners and Energy May Change in a Chemical Reaction
(Page 30)

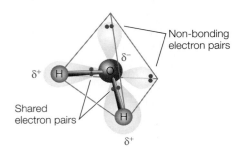

Page 31 In-Text Art

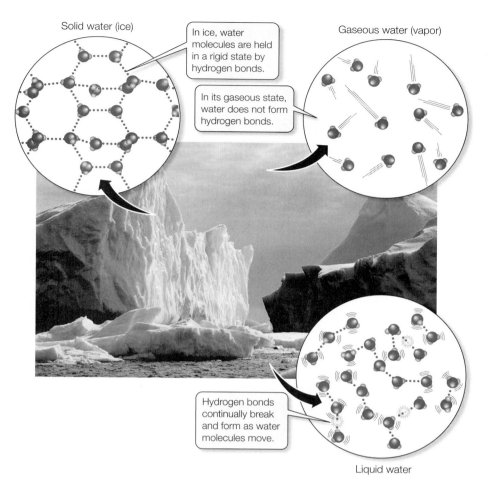

Solid water (ice)

In ice, water molecules are held in a rigid state by hydrogen bonds.

Gaseous water (vapor)

In its gaseous state, water does not form hydrogen bonds.

Hydrogen bonds continually break and form as water molecules move.

Liquid water

2.14 Hydrogen Bonds Hold Water Molecules Together *(Page 31)*

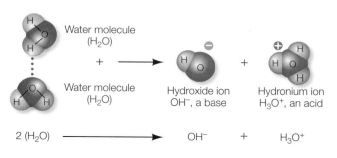

Water molecule
(H_2O)

+

Water molecule
(H_2O)

Hydroxide ion
OH^-, a base

+

Hydronium ion
H_3O^+, an acid

$2\ (H_2O) \longrightarrow OH^- + H_3O^+$

Page 34 In-Text Art

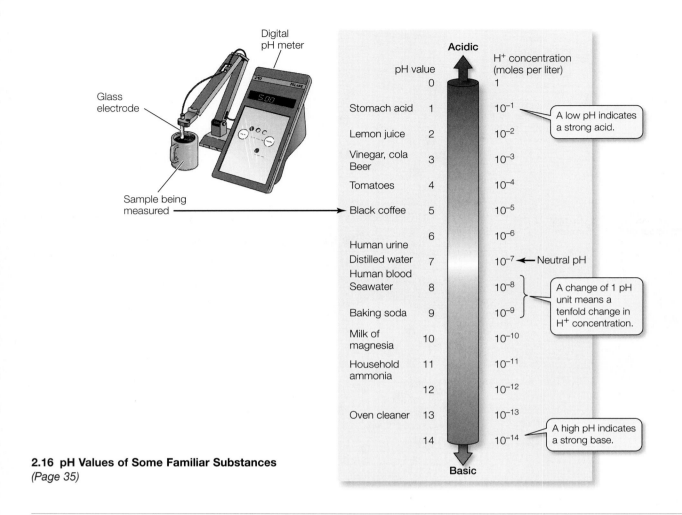

2.16 pH Values of Some Familiar Substances
(Page 35)

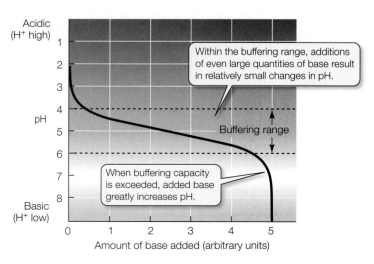

2.17 Buffers Minimize Changes in pH *(Page 35)*

CHAPTER 3 Macromolecules and the Origin of Life

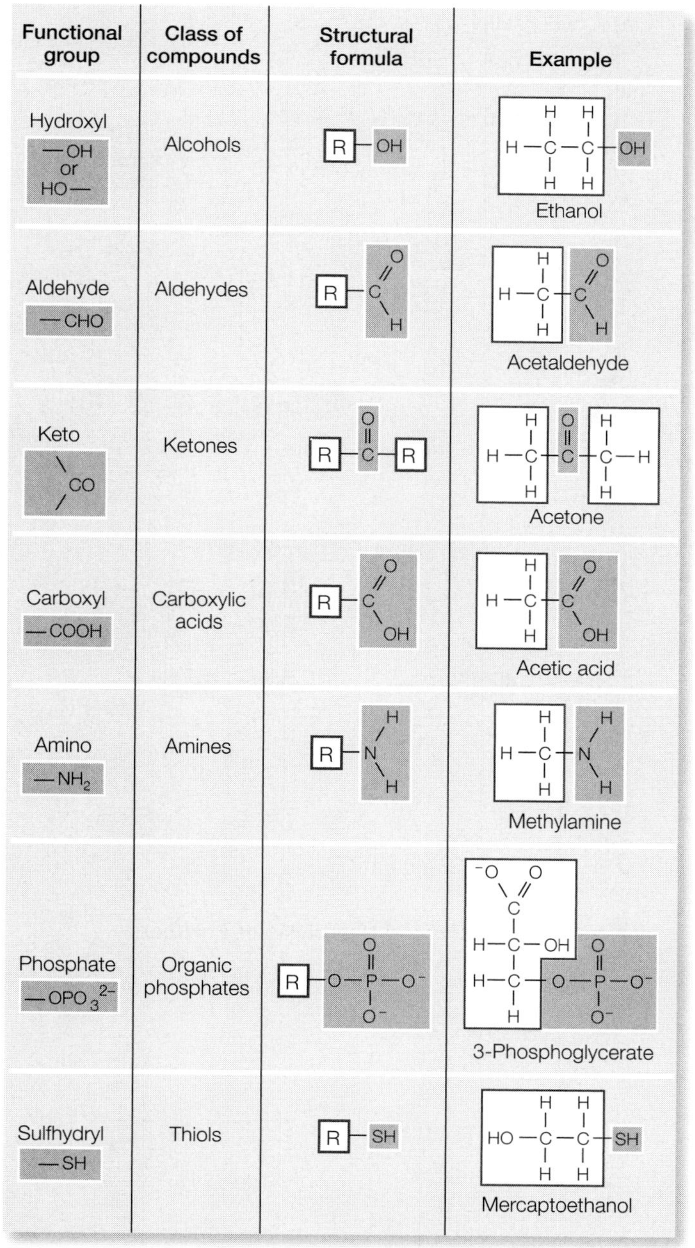

Functional group	Class of compounds	Structural formula	Example
Hydroxyl —OH or HO—	Alcohols	R—OH	Ethanol
Aldehyde —CHO	Aldehydes	R—CHO	Acetaldehyde
Keto \CO	Ketones	R—CO—R	Acetone
Carboxyl —COOH	Carboxylic acids	R—COOH	Acetic acid
Amino —NH₂	Amines	R—NH	Methylamine
Phosphate —OPO₃²⁻	Organic phosphates	R—O—P	3-Phosphoglycerate
Sulfhydryl —SH	Thiols	R—SH	Mercaptoethanol

3.1 Some Functional Groups Important to Living Systems
(Page 40)

TABLE 3.1

The Building Blocks of Organisms

MONOMER	COMPLEX POLYMER (MACROMOLECULE)
Amino acid	Polypeptide (protein)
Monosaccharide (sugar)	Polysaccharide (carbohydrate)
Nucleotide	Nucleic acid

(Page 40)

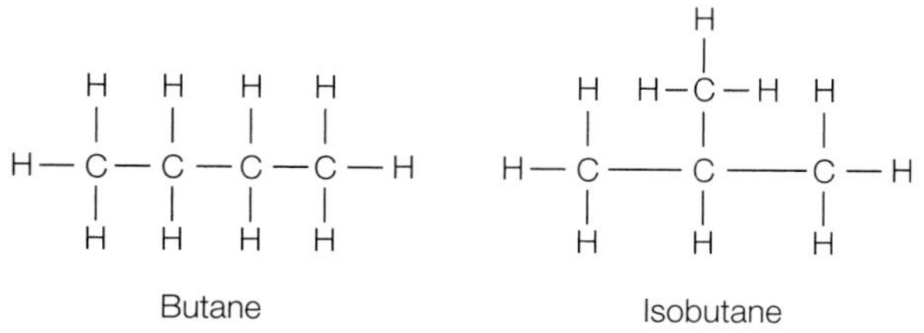

Butane   Isobutane

Page 40 In-Text Art

17

(A)

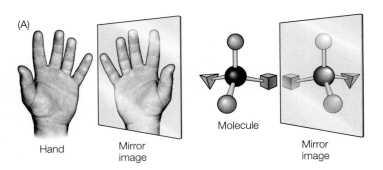

Hand   Mirror image   Molecule   Mirror image

(B)

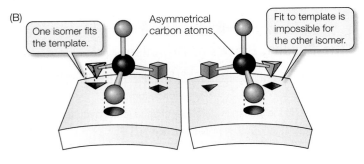

One isomer fits the template.

Asymmetrical carbon atoms

Fit to template is impossible for the other isomer.

3.2 Optical Isomers *(Page 41)*

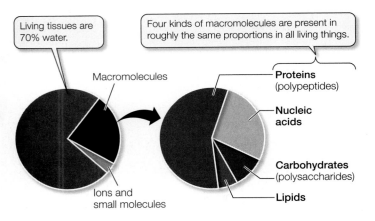

Living tissues are 70% water.

Four kinds of macromolecules are present in roughly the same proportions in all living things.

Macromolecules

Proteins (polypeptides)

Nucleic acids

Carbohydrates (polysaccharides)

Ions and small molecules

Lipids

3.3 Substances Found in Living Tissues *(Page 41)*

(A) Condensation

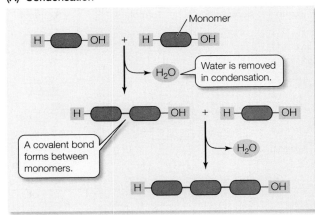

Monomer

Water is removed in condensation.

A covalent bond forms between monomers.

(B) Hydrolysis

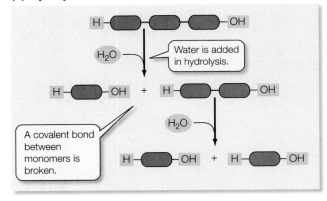

Water is added in hydrolysis.

A covalent bond between monomers is broken.

3.4 Condensation and Hydrolysis of Polymers *(Page 41)*

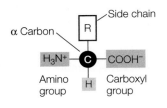

Side chain

α Carbon

R

H_3N^+ C $COOH^-$

Amino group   H   Carboxyl group

Page 42 In-Text Art

TABLE 3.2

The Twenty Amino Acids

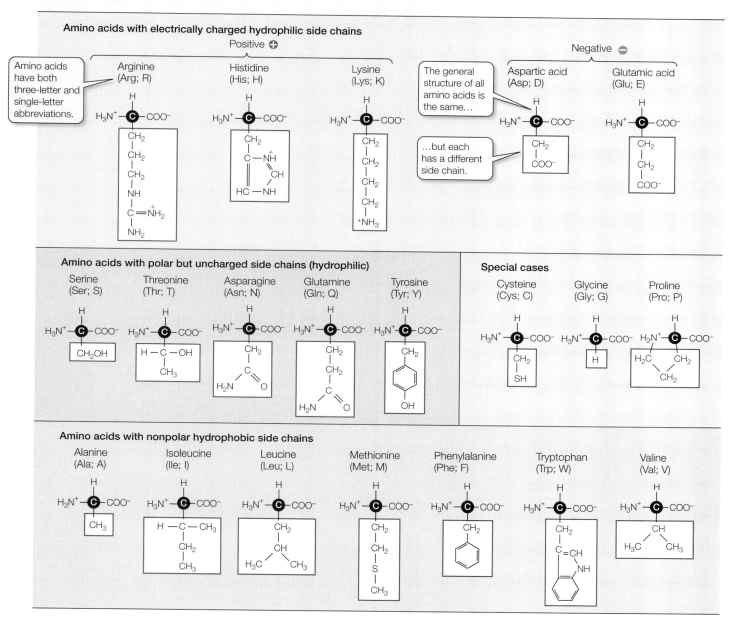

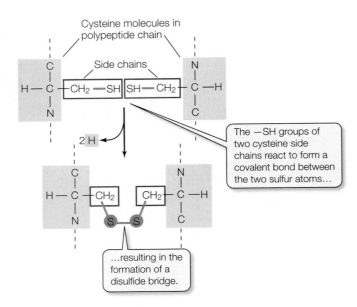

3.5 A Disulfide Bridge *(Page 44)*

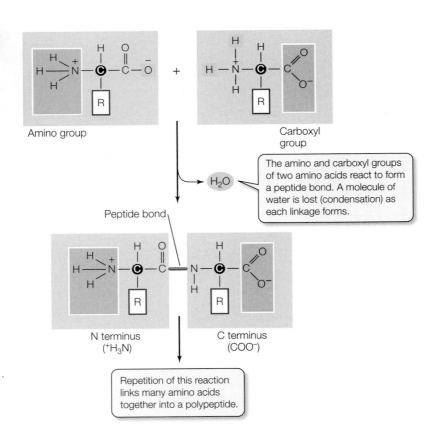

3.6 Formation of Peptide Bonds *(Page 44)*

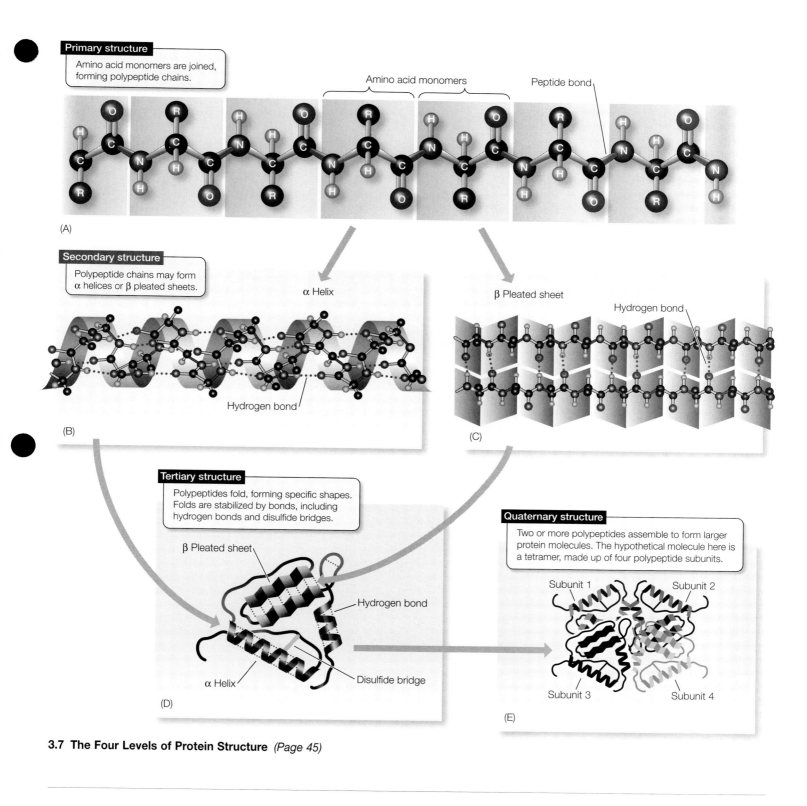

Primary structure

Amino acid monomers are joined, forming polypeptide chains.

Amino acid monomers

Peptide bond

(A)

Secondary structure

Polypeptide chains may form α helices or β pleated sheets.

α Helix

Hydrogen bond

(B)

β Pleated sheet

Hydrogen bond

(C)

Tertiary structure

Polypeptides fold, forming specific shapes. Folds are stabilized by bonds, including hydrogen bonds and disulfide bridges.

β Pleated sheet

Hydrogen bond

α Helix

Disulfide bridge

(D)

Quaternary structure

Two or more polypeptides assemble to form larger protein molecules. The hypothetical molecule here is a tetramer, made up of four polypeptide subunits.

Subunit 1

Subunit 2

Subunit 3

Subunit 4

(E)

3.7 The Four Levels of Protein Structure *(Page 45)*

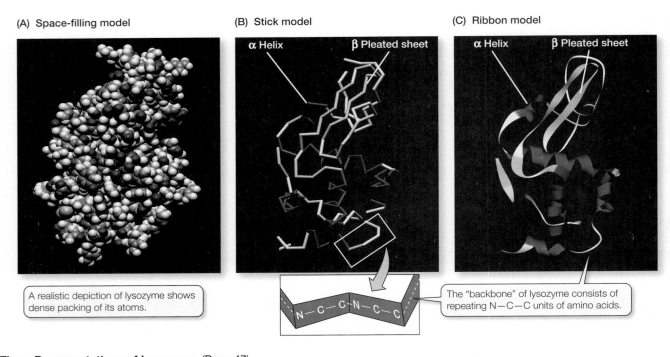

(A) Space-filling model

A realistic depiction of lysozyme shows dense packing of its atoms.

(B) Stick model

α Helix   β Pleated sheet

N—C—C—N—C—C

The "backbone" of lysozyme consists of repeating N—C—C units of amino acids.

(C) Ribbon model

α Helix   β Pleated sheet

3.8 Three Representations of Lysozyme *(Page 47)*

(A)

(B)

α Subunits

β Subunits   Heme

3.9 Quaternary Structure of a Protein *(Page 47)*

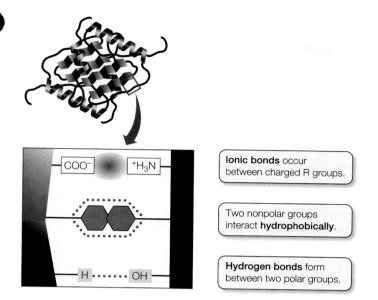

Ionic bonds occur between charged R groups.

Two nonpolar groups interact **hydrophobically**.

Hydrogen bonds form between two polar groups.

3.10 Noncovalent Interactions between Proteins and Other Molecules *(Page 48)*

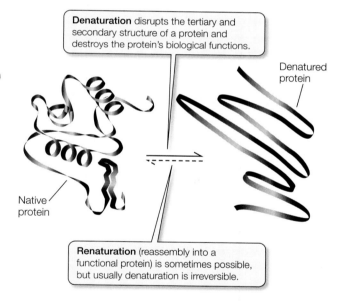

Denaturation disrupts the tertiary and secondary structure of a protein and destroys the protein's biological functions.

Denatured protein

Native protein

Renaturation (reassembly into a functional protein) is sometimes possible, but usually denaturation is irreversible.

3.11 Denaturation Is the Loss of Tertiary Protein Structure and Function *(Page 48)*

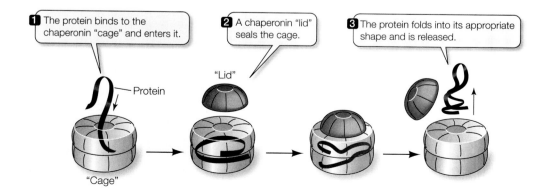

1 The protein binds to the chaperonin "cage" and enters it.

2 A chaperonin "lid" seals the cage.

3 The protein folds into its appropriate shape and is released.

"Lid"

Protein

"Cage"

3.12 Chaperonins Protect Proteins from Inappropriate Binding *(Page 49)*

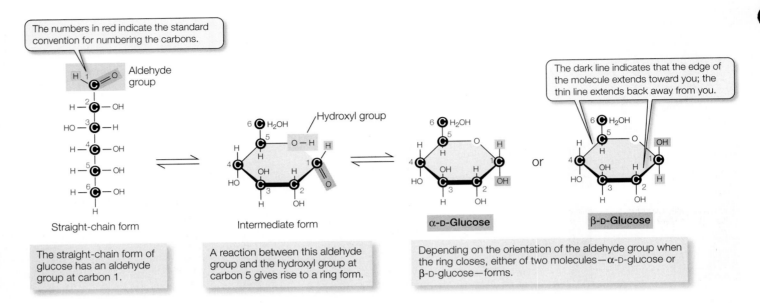

The numbers in red indicate the standard convention for numbering the carbons.

Aldehyde group

The dark line indicates that the edge of the molecule extends toward you; the thin line extends back away from you.

Hydroxyl group

Straight-chain form

Intermediate form

α-D-Glucose

or

β-D-Glucose

The straight-chain form of glucose has an aldehyde group at carbon 1.

A reaction between this aldehyde group and the hydroxyl group at carbon 5 gives rise to a ring form.

Depending on the orientation of the aldehyde group when the ring closes, either of two molecules—α-D-glucose or β-D-glucose—forms.

3.13 Glucose: From One Form to the Other *(Page 50)*

Three-carbon sugar

Glyceraldehyde is the smallest monosaccharide and exists only as the straight-chain form.

Glyceraldehyde

Five-carbon sugars (pentoses)

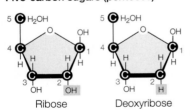

Ribose   Deoxyribose

Ribose and deoxyribose each have five carbons, but very different chemical properties and biological roles.

Six-carbon sugars (hexoses)

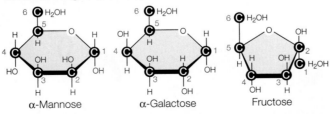

α-Mannose   α-Galactose   Fructose

These hexoses are structural isomers. All have the formula $C_6H_{12}O_6$, but each has distinct biochemical properties.

3.14 Monosaccharides Are Simple Sugars *(Page 50)*

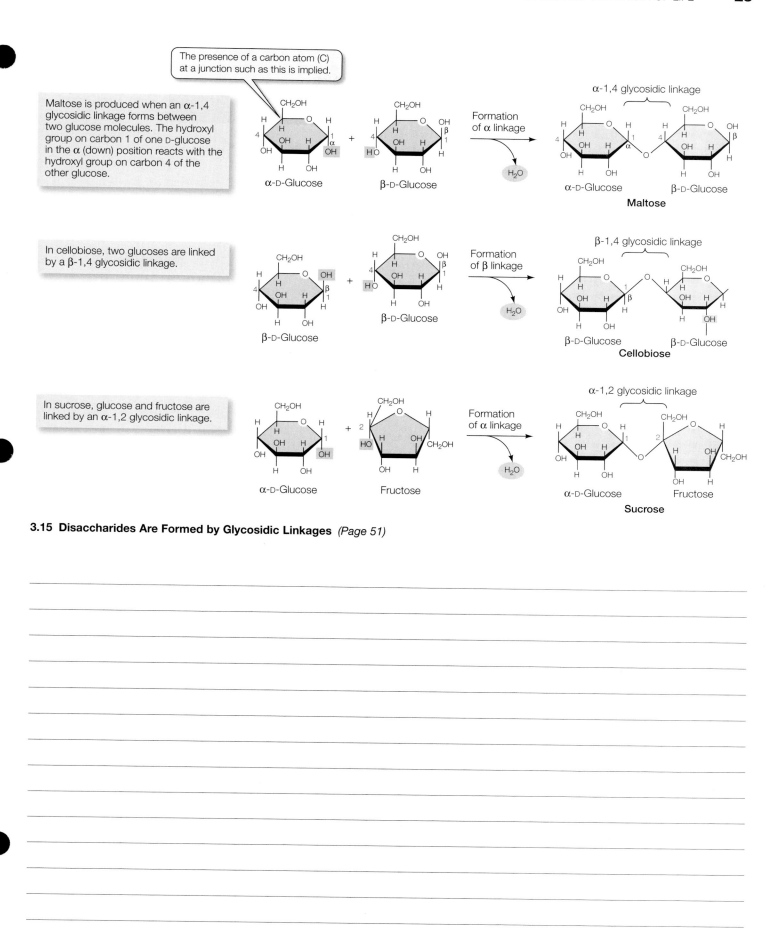

The presence of a carbon atom (C) at a junction such as this is implied.

Maltose is produced when an α-1,4 glycosidic linkage forms between two glucose molecules. The hydroxyl group on carbon 1 of one D-glucose in the α (down) position reacts with the hydroxyl group on carbon 4 of the other glucose.

α-D-Glucose + β-D-Glucose → Formation of α linkage → H₂O → α-1,4 glycosidic linkage → α-D-Glucose + β-D-Glucose → **Maltose**

In cellobiose, two glucoses are linked by a β-1,4 glycosidic linkage.

β-D-Glucose + β-D-Glucose → Formation of β linkage → H₂O → β-1,4 glycosidic linkage → β-D-Glucose + β-D-Glucose → **Cellobiose**

In sucrose, glucose and fructose are linked by an α-1,2 glycosidic linkage.

α-D-Glucose + Fructose → Formation of α linkage → H₂O → α-1,2 glycosidic linkage → α-D-Glucose + Fructose → **Sucrose**

3.15 Disaccharides Are Formed by Glycosidic Linkages *(Page 51)*

(A) Molecular structure

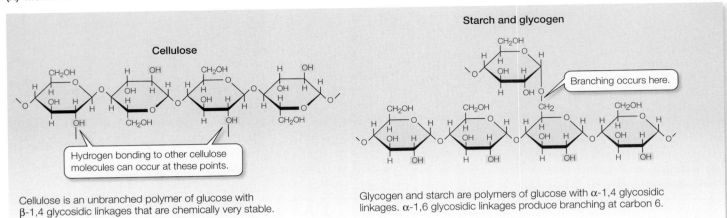

Cellulose

Starch and glycogen

Branching occurs here.

Hydrogen bonding to other cellulose molecules can occur at these points.

Cellulose is an unbranched polymer of glucose with β-1,4 glycosidic linkages that are chemically very stable.

Glycogen and starch are polymers of glucose with α-1,4 glycosidic linkages. α-1,6 glycosidic linkages produce branching at carbon 6.

(B) Macromolecular structure

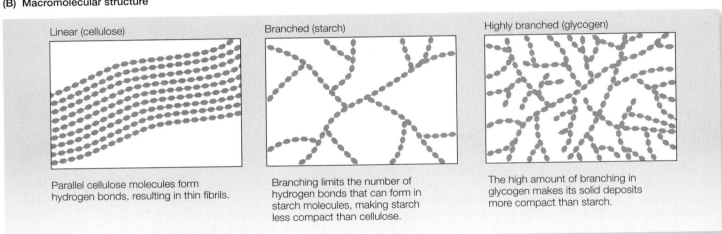

Linear (cellulose)

Branched (starch)

Highly branched (glycogen)

Parallel cellulose molecules form hydrogen bonds, resulting in thin fibrils.

Branching limits the number of hydrogen bonds that can form in starch molecules, making starch less compact than cellulose.

The high amount of branching in glycogen makes its solid deposits more compact than starch.

(C) Polysaccharides in cells

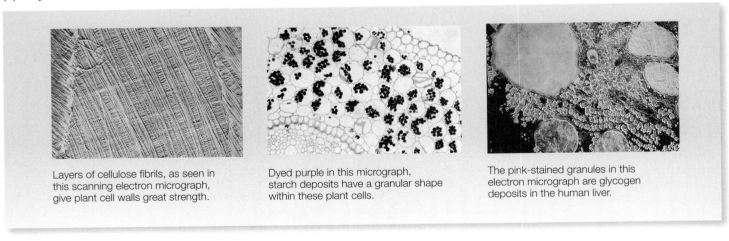

Layers of cellulose fibrils, as seen in this scanning electron micrograph, give plant cell walls great strength.

Dyed purple in this micrograph, starch deposits have a granular shape within these plant cells.

The pink-stained granules in this electron micrograph are glycogen deposits in the human liver.

3.16 Representative Polysaccharides *(Page 52)*

(A) Sugar phosphate

Fructose 1,6 bisphosphate is involved in the reactions that liberate energy from glucose. (The numbers in its name refer to the carbon sites of phosphate bonding; *bis-* indicates that two phosphates are present.)

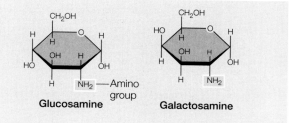

Phosphate groups

Fructose

Fructose 1,6 bisphosphate

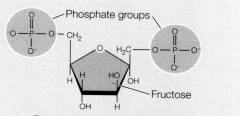

Galactosamine is an important component of cartilage, a connective tissue in vertebrates.

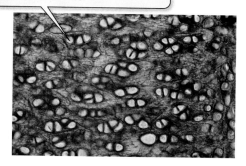

(B) Amino sugars

The monosaccharides glucosamine and galactosamine are amino sugars with an amino group in place of a hydroxyl group.

Amino group

Glucosamine   **Galactosamine**

(C) Chitin

Chitin is a polymer of *N*-acetylglucosamine; *N*-acetyl groups provide additional sites for hydrogen bonding between the polymers.

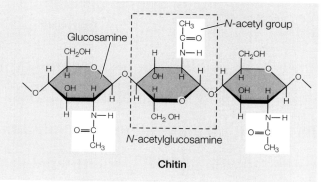

Glucosamine

N-acetyl group

N-acetylglucosamine

Chitin

The external skeletons of insects are made up of chitin.

3.17 Chemically Modified Carbohydrates *(Page 53)*

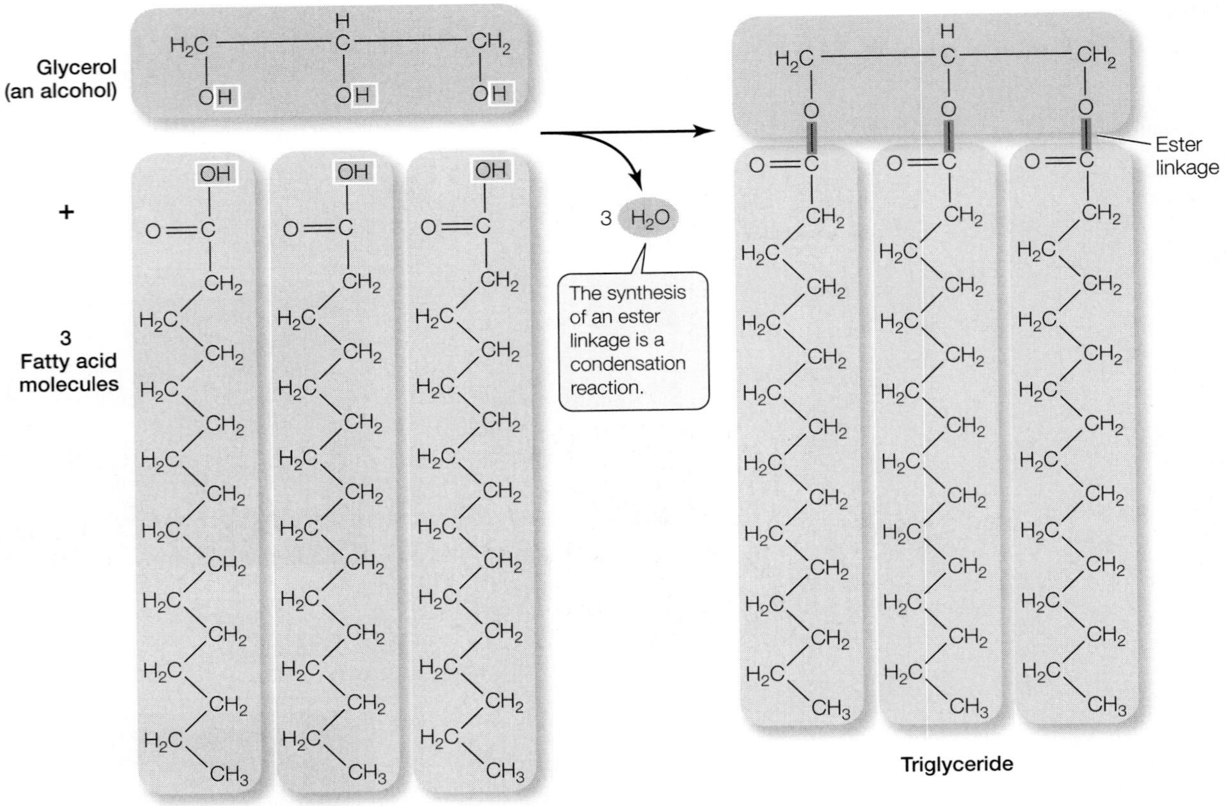

Glycerol
(an alcohol)

+

3
Fatty acid
molecules

3 H₂O

The synthesis
of an ester
linkage is a
condensation
reaction.

Ester
linkage

Triglyceride

3.18 Synthesis of a Triglyceride *(Page 54)*

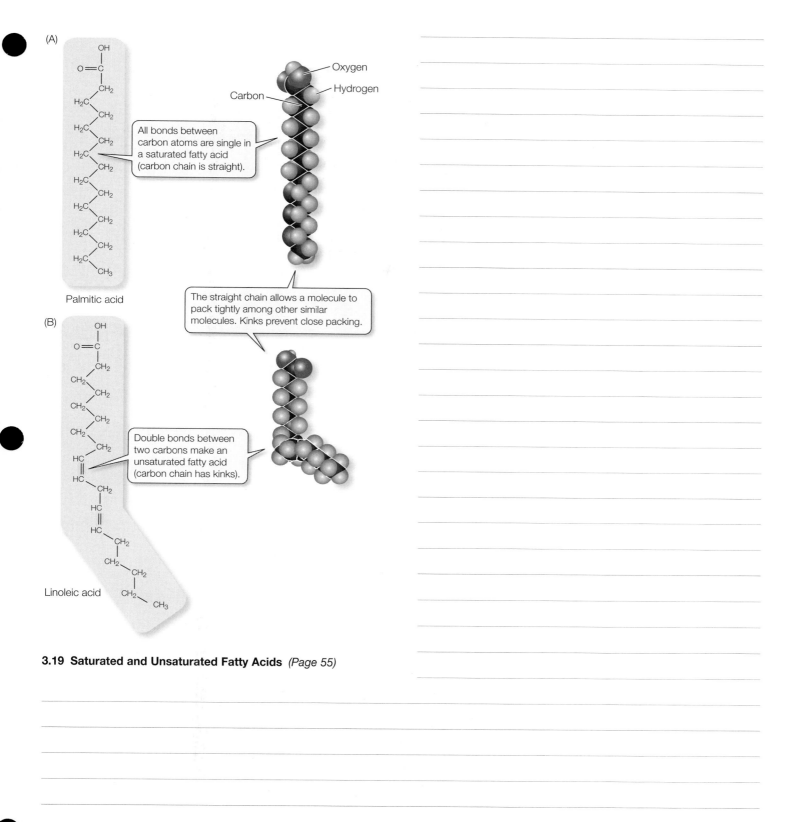

3.19 Saturated and Unsaturated Fatty Acids *(Page 55)*

(A) Phosphatidylcholine

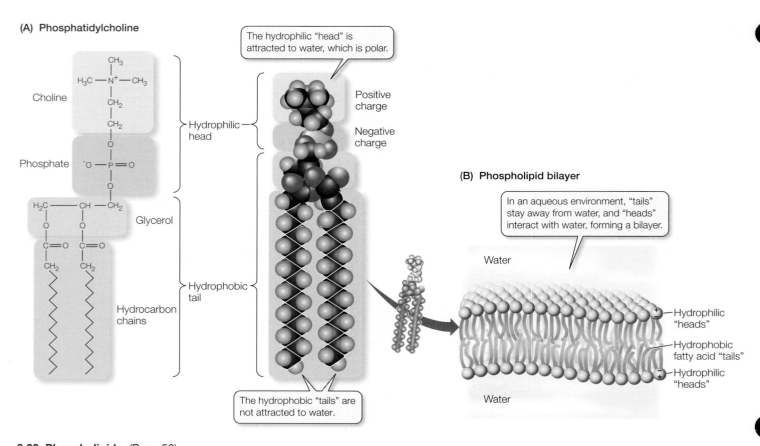

The hydrophilic "head" is attracted to water, which is polar.

Choline

Phosphate

Glycerol

Hydrocarbon chains

Hydrophilic head

Hydrophobic tail

Positive charge

Negative charge

The hydrophobic "tails" are not attracted to water.

(B) Phospholipid bilayer

In an aqueous environment, "tails" stay away from water, and "heads" interact with water, forming a bilayer.

Water

Water

Hydrophilic "heads"

Hydrophobic fatty acid "tails"

Hydrophilic "heads"

3.20 Phospholipids *(Page 56)*

β-Carotene

Vitamin A

Vitamin A

3.21 β-Carotene is the Source of Vitamin A *(Page 56)*

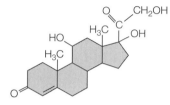

Cholesterol is a constituent of membranes and is the source of steroid hormones.

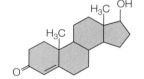

Vitamin D₂ can be produced in the skin by the action of light on a cholesterol derivative.

Cortisol is a hormone secreted by the adrenal glands.

Testosterone is a male sex hormone.

3.22 All Steroids Have the Same Ring Structure *(Page 57)*

$$H_3C - (CH_2)_{14} - \overset{\overset{\displaystyle O}{\|}}{C} - O - CH_2 - (CH_2)_{28} - CH_3$$

Fatty acid · Ester linkage · Alcohol

Page 57 In-Text Art

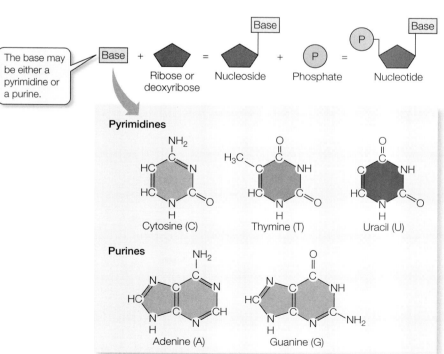

The base may be either a pyrimidine or a purine.

Base + Ribose or deoxyribose = Nucleoside + Phosphate = Nucleotide

Pyrimidines

Cytosine (C) · Thymine (T) · Uracil (U)

Purines

Adenine (A) · Guanine (G)

3.23 Nucleotides Have Three Components *(Page 58)*

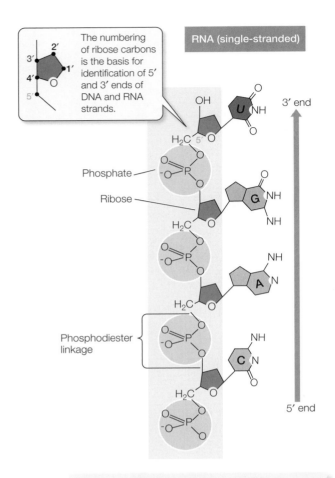

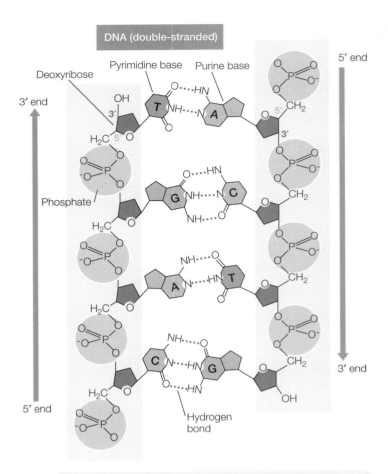

In RNA, the bases are attached to ribose. The bases in RNA are the purines adenine (A) and guanine (G) and the pyrimidines cytosine (C) and uracil (U).

In DNA, the bases are attached to deoxyribose, and the base thymine (T) is found instead of uracil. Hydrogen bonds between purines and pyrimidines hold the two strands of DNA together.

3.24 Distinguishing Characteristics of DNA and RNA *(Page 58)*

TABLE 3.3

Distinguishing RNA from DNA

NUCLEIC ACID	SUGAR	BASES
RNA	Ribose	Adenine
		Cytosine
		Guanine
		Uracil
DNA	Deoxyribose	Adenine
		Cytosine
		Guanine
		Thymine

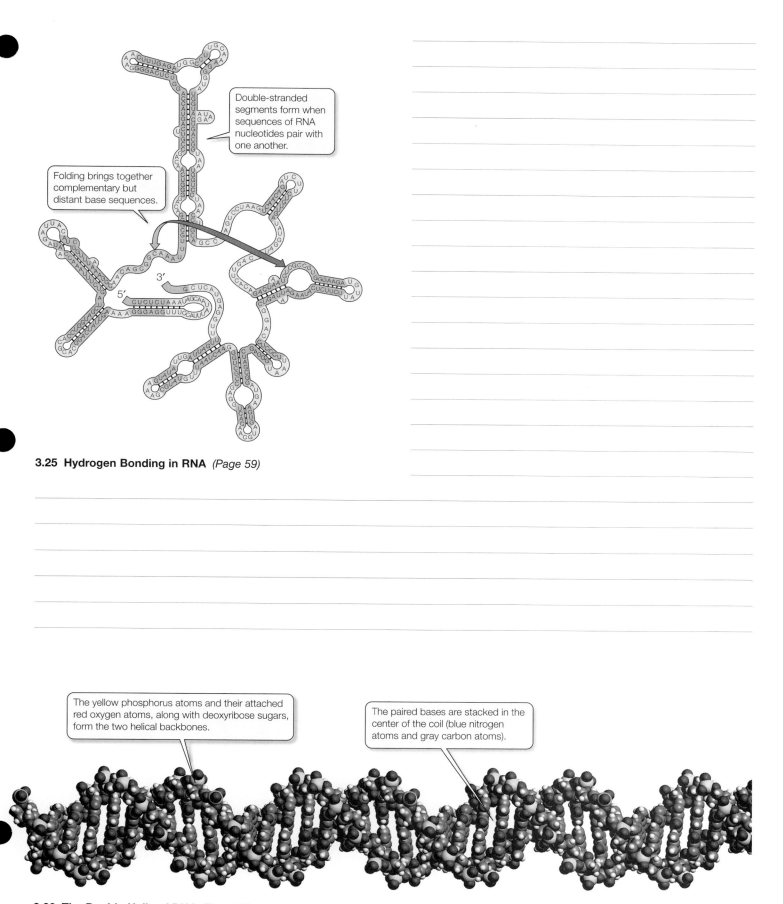

Double-stranded segments form when sequences of RNA nucleotides pair with one another.

Folding brings together complementary but distant base sequences.

3′

5′

3.25 Hydrogen Bonding in RNA *(Page 59)*

The yellow phosphorus atoms and their attached red oxygen atoms, along with deoxyribose sugars, form the two helical backbones.

The paired bases are stacked in the center of the coil (blue nitrogen atoms and gray carbon atoms).

3.26 The Double Helix of DNA *(Page 60)*

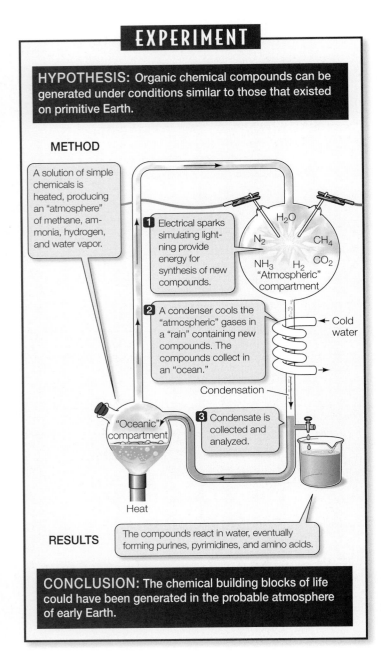

EXPERIMENT

HYPOTHESIS: Organic chemical compounds can be generated under conditions similar to those that existed on primitive Earth.

METHOD

A solution of simple chemicals is heated, producing an "atmosphere" of methane, ammonia, hydrogen, and water vapor.

1 Electrical sparks simulating lightning provide energy for synthesis of new compounds.

H_2O

N_2 CH_4

NH_3 H_2 CO_2
"Atmospheric" compartment

2 A condenser cools the "atmospheric" gases in a "rain" containing new compounds. The compounds collect in an "ocean."

← Cold water

Condensation

3 Condensate is collected and analyzed.

"Oceanic" compartment

Heat

RESULTS

The compounds react in water, eventually forming purines, pyrimidines, and amino acids.

CONCLUSION: The chemical building blocks of life could have been generated in the probable atmosphere of early Earth.

3.28 Synthesis of Prebiotic Molecules in an Experimental Atmosphere *(Page 62)*

1 This folded RNA is a ribozyme and can speed up a reaction.

These short sequences of RNA are complementary to the ribozyme.

5'   3'

2 The short sequences base-pair with the ribozyme.

3'
5'
5'   5'
3'   3'

3 The ribozyme catalyzes the polymerization of the short sequences.

5'   5'
3'   3'

4 The short sequences are now one longer sequence of RNA.

5'   3'

3'   5'

3.29 An Early Catalyst for Life? *(Page 63)*

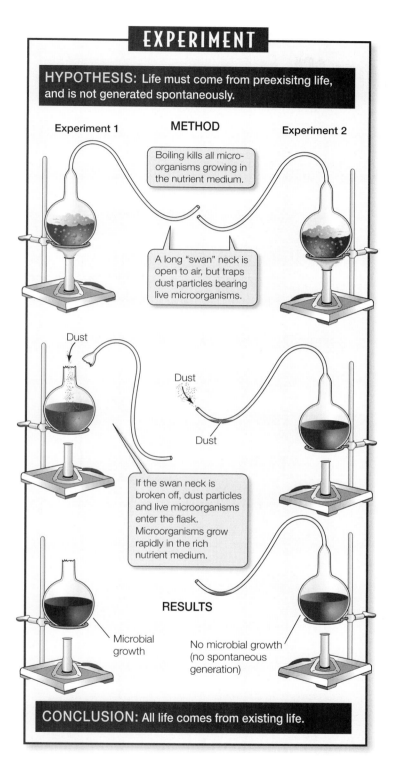

3.30 Disproving the Spontaneous Generation of Life *(Page 64)*

CHAPTER 4 Cells: The Working Units of Life

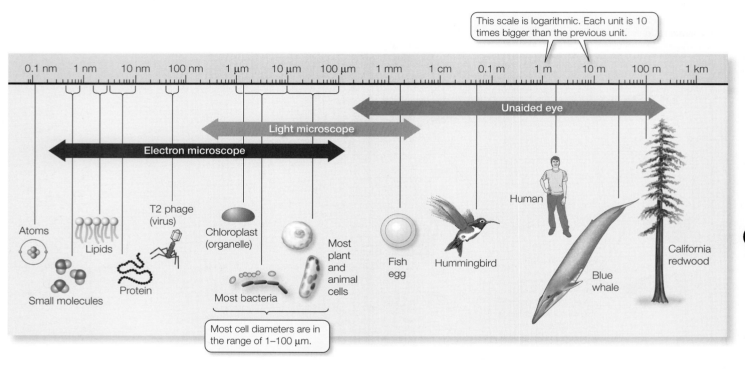

This scale is logarithmic. Each unit is 10 times bigger than the previous unit.

0.1 nm   1 nm   10 nm   100 nm   1 μm   10 μm   100 μm   1 mm   1 cm   0.1 m   1 m   10 m   100 m   1 km

Unaided eye

Light microscope

Electron microscope

Atoms

Lipids

T2 phage (virus)

Protein

Small molecules

Chloroplast (organelle)

Most bacteria

Most plant and animal cells

Fish egg

Hummingbird

Human

Blue whale

California redwood

Most cell diameters are in the range of 1–100 μm.

4.1 The Scale of Life *(Page 70)*

(A) Cubes

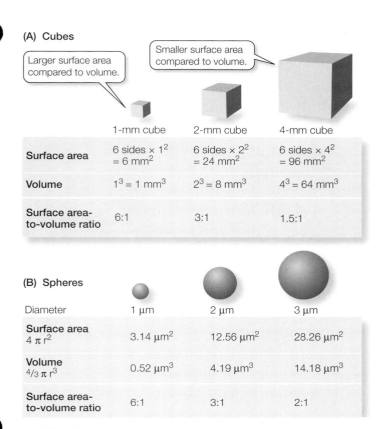

	1-mm cube	2-mm cube	4-mm cube
Surface area	6 sides × 1^2 = 6 mm^2	6 sides × 2^2 = 24 mm^2	6 sides × 4^2 = 96 mm^2
Volume	1^3 = 1 mm^3	2^3 = 8 mm^3	4^3 = 64 mm^3
Surface area-to-volume ratio	6:1	3:1	1.5:1

(B) Spheres

Diameter	1 µm	2 µm	3 µm
Surface area $4 \pi r^2$	3.14 µm^2	12.56 µm^2	28.26 µm^2
Volume $^4/_3 \pi r^3$	0.52 µm^3	4.19 µm^3	14.18 µm^3
Surface area-to-volume ratio	6:1	3:1	2:1

4.2 Why Cells Are Small *(Page 70)*

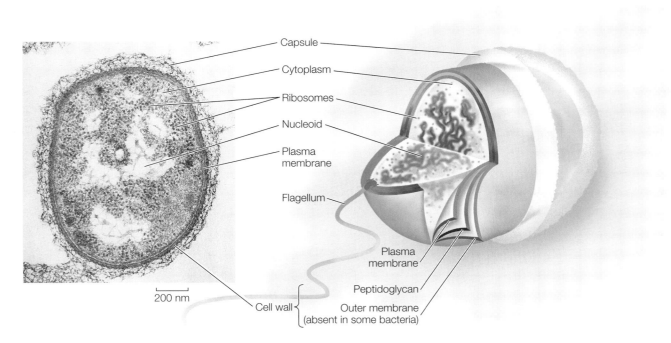

4.4 A Prokaryotic Cell *(Page 73)*

(A)

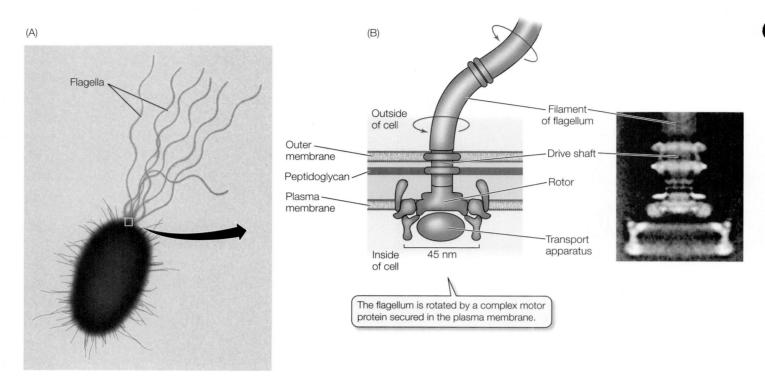

Flagella

(B)

Outside
of cell

Outer
membrane

Peptidoglycan

Plasma
membrane

Inside
of cell

Filament
of flagellum

Drive shaft

Rotor

Transport
apparatus

45 nm

The flagellum is rotated by a complex motor
protein secured in the plasma membrane.

4.5 Prokaryotic Flagella *(Page 74)*

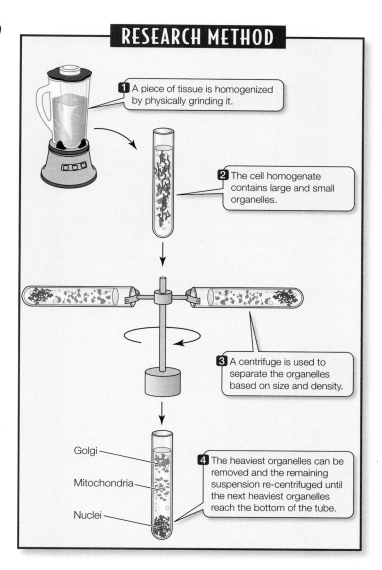

RESEARCH METHOD

1 A piece of tissue is homogenized by physically grinding it.

2 The cell homogenate contains large and small organelles.

3 A centrifuge is used to separate the organelles based on size and density.

Golgi

Mitochondria

Nuclei

4 The heaviest organelles can be removed and the remaining suspension re-centrifuged until the next heaviest organelles reach the bottom of the tube.

4.6 Cell Fractionation *(Page 75)*

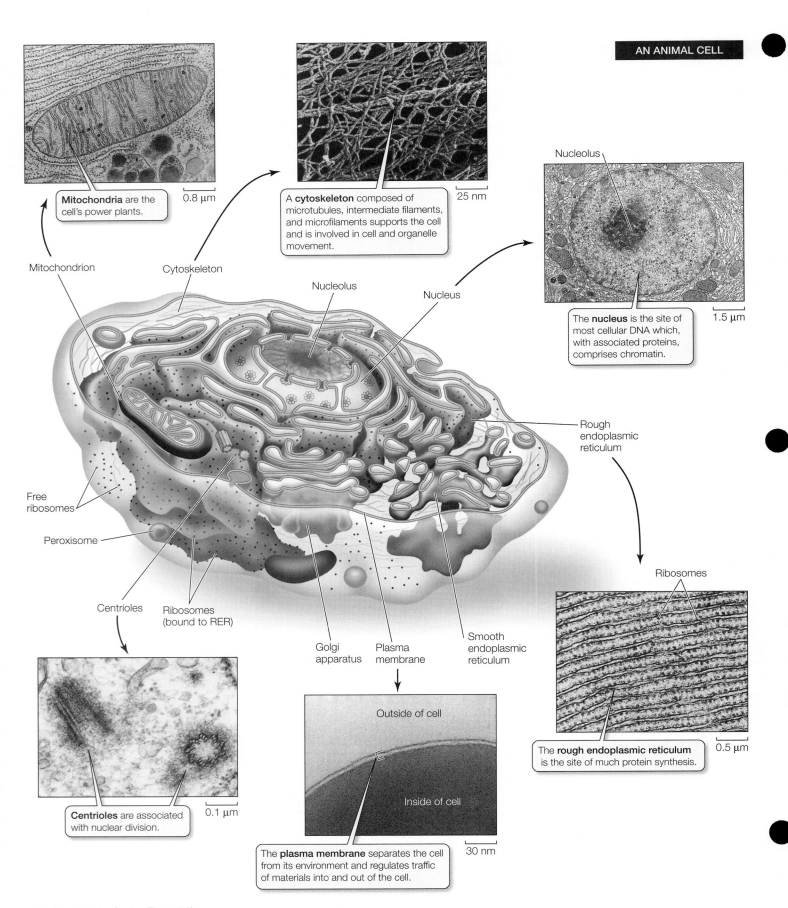

Mitochondria are the cell's power plants.

0.8 μm

A **cytoskeleton** composed of microtubules, intermediate filaments, and microfilaments supports the cell and is involved in cell and organelle movement.

25 nm

Nucleolus

The **nucleus** is the site of most cellular DNA which, with associated proteins, comprises chromatin.

1.5 μm

Mitochondrion

Cytoskeleton

Nucleolus

Nucleus

Rough endoplasmic reticulum

Free ribosomes

Peroxisome

Centrioles

Ribosomes (bound to RER)

Golgi apparatus

Plasma membrane

Smooth endoplasmic reticulum

Ribosomes

Centrioles are associated with nuclear division.

0.1 μm

Outside of cell

Inside of cell

The **plasma membrane** separates the cell from its environment and regulates traffic of materials into and out of the cell.

30 nm

The **rough endoplasmic reticulum** is the site of much protein synthesis.

0.5 μm

4.7 Eukaryotic Cells *(Page 76)*

A PLANT CELL

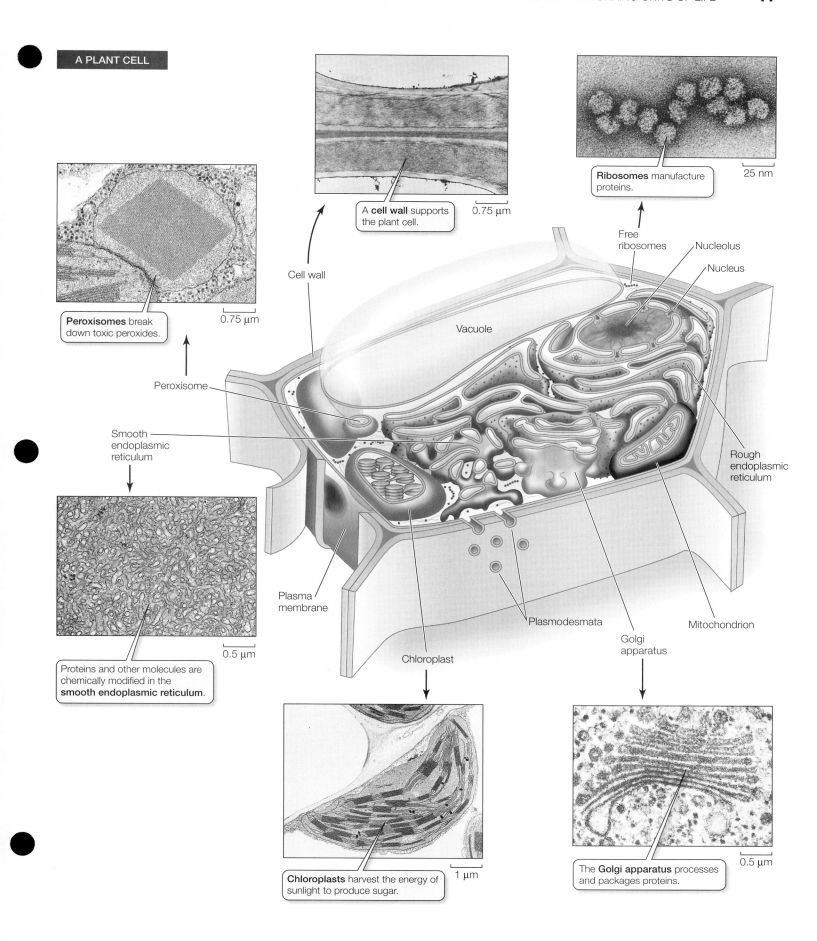

Peroxisome

A **cell wall** supports the plant cell.

0.75 μm

Ribosomes manufacture proteins.

25 nm

Peroxisomes break down toxic peroxides.

0.75 μm

Cell wall

Free ribosomes

Nucleolus

Nucleus

Vacuole

Smooth endoplasmic reticulum

Rough endoplasmic reticulum

Proteins and other molecules are chemically modified in the **smooth endoplasmic reticulum**.

0.5 μm

Plasma membrane

Plasmodesmata

Mitochondrion

Golgi apparatus

Chloroplast

Chloroplasts harvest the energy of sunlight to produce sugar.

1 μm

The **Golgi apparatus** processes and packages proteins.

0.5 μm

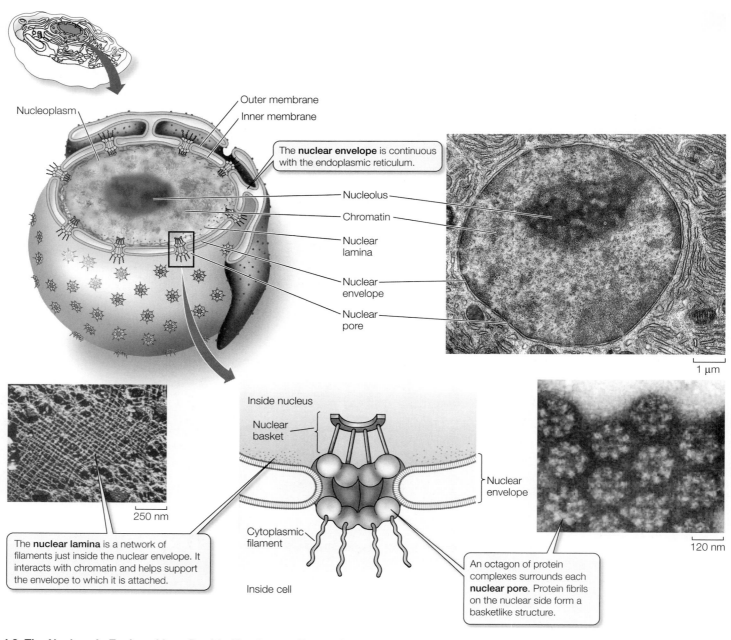

Nucleoplasm

Outer membrane

Inner membrane

The **nuclear envelope** is continuous with the endoplasmic reticulum.

Nucleolus

Chromatin

Nuclear lamina

Nuclear envelope

Nuclear pore

1 μm

250 nm

The **nuclear lamina** is a network of filaments just inside the nuclear envelope. It interacts with chromatin and helps support the envelope to which it is attached.

Inside nucleus

Nuclear basket

Cytoplasmic filament

Inside cell

Nuclear envelope

An octagon of protein complexes surrounds each **nuclear pore**. Protein fibrils on the nuclear side form a basketlike structure.

120 nm

4.8 The Nucleus Is Enclosed by a Double Membrane *(Page 78)*

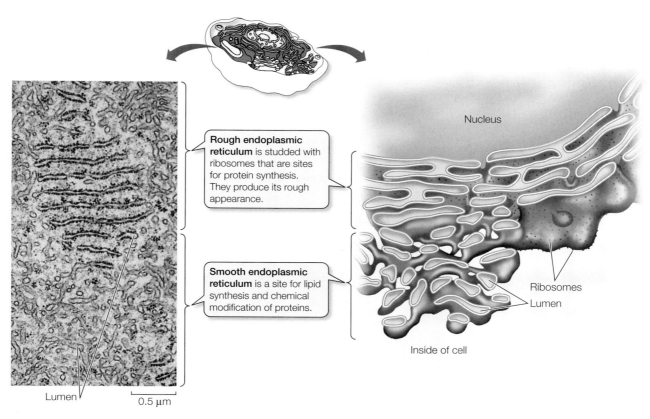

Rough endoplasmic reticulum is studded with ribosomes that are sites for protein synthesis. They produce its rough appearance.

Smooth endoplasmic reticulum is a site for lipid synthesis and chemical modification of proteins.

Lumen

0.5 μm

Nucleus

Ribosomes

Lumen

Inside of cell

4.10 Endoplasmic Reticulum *(Page 80)*

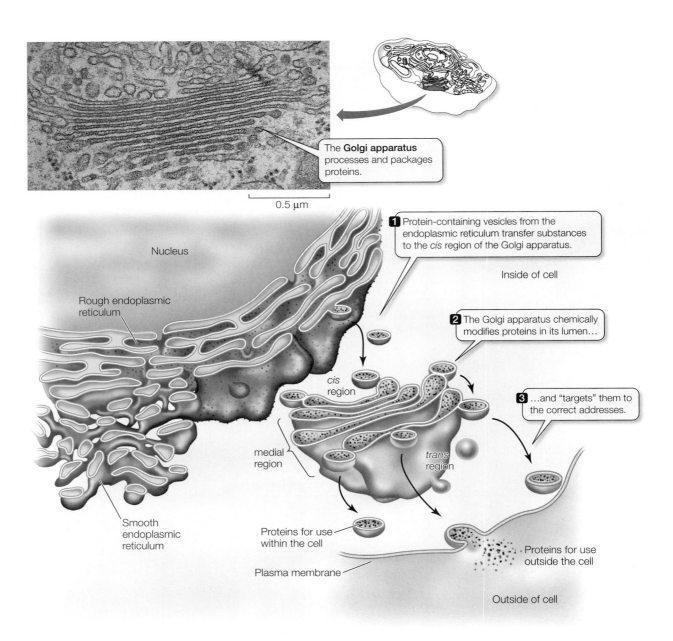

The **Golgi apparatus** processes and packages proteins.

0.5 μm

1 Protein-containing vesicles from the endoplasmic reticulum transfer substances to the *cis* region of the Golgi apparatus.

Inside of cell

2 The Golgi apparatus chemically modifies proteins in its lumen...

3 ...and "targets" them to the correct addresses.

Nucleus

Rough endoplasmic reticulum

Smooth endoplasmic reticulum

cis region

medial region

trans region

Proteins for use within the cell

Proteins for use outside the cell

Plasma membrane

Outside of cell

4.11 The Golgi Apparatus *(Page 81)*

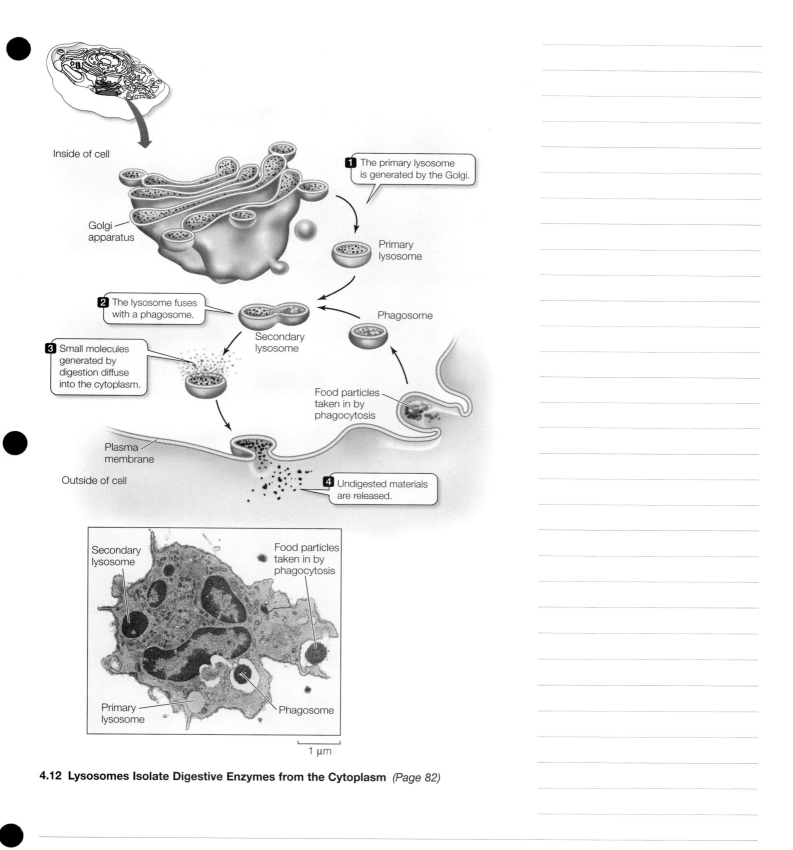

Inside of cell

Golgi apparatus

1 The primary lysosome is generated by the Golgi.

Primary lysosome

2 The lysosome fuses with a phagosome.

Secondary lysosome

Phagosome

3 Small molecules generated by digestion diffuse into the cytoplasm.

Food particles taken in by phagocytosis

Plasma membrane

Outside of cell

4 Undigested materials are released.

Secondary lysosome

Food particles taken in by phagocytosis

Primary lysosome

Phagosome

1 μm

4.12 Lysosomes Isolate Digestive Enzymes from the Cytoplasm *(Page 82)*

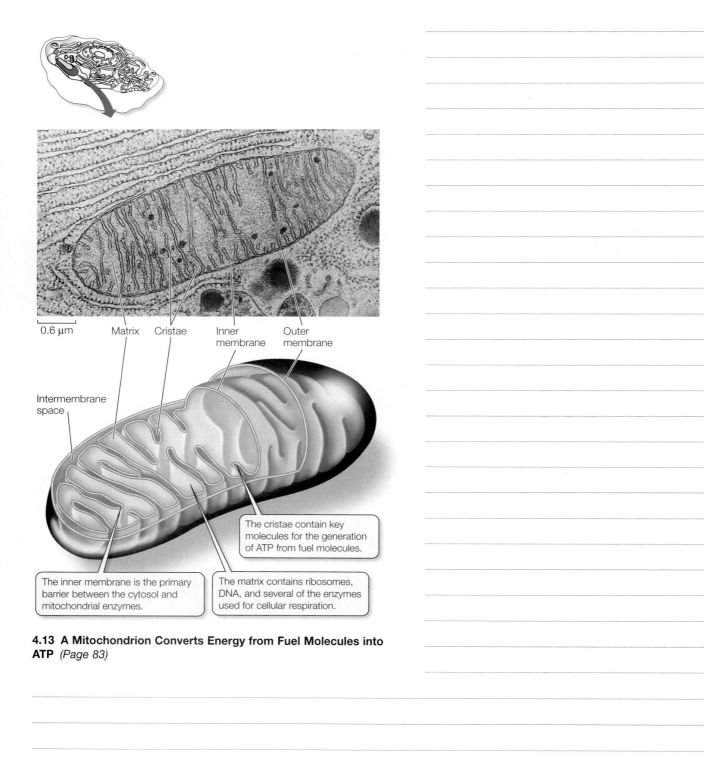

0.6 μm   Matrix   Cristae   Inner   Outer
                           membrane membrane

Intermembrane
space

The cristae contain key
molecules for the generation
of ATP from fuel molecules.

The inner membrane is the primary
barrier between the cytosol and
mitochondrial enzymes.

The matrix contains ribosomes,
DNA, and several of the enzymes
used for cellular respiration.

4.13 A Mitochondrion Converts Energy from Fuel Molecules into ATP *(Page 83)*

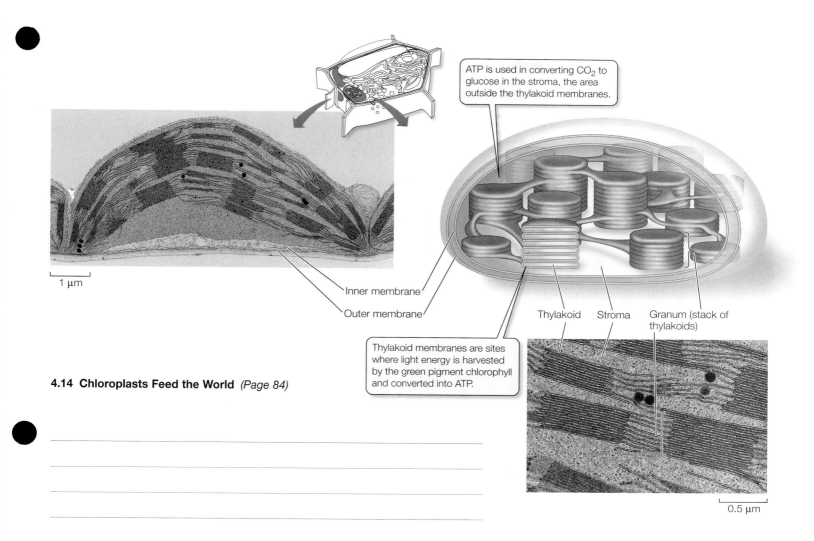

ATP is used in converting CO_2 to glucose in the stroma, the area outside the thylakoid membranes.

Inner membrane

Outer membrane

Thylakoid   Stroma   Granum (stack of thylakoids)

Thylakoid membranes are sites where light energy is harvested by the green pigment chlorophyll and converted into ATP.

1 μm

0.5 μm

4.14 Chloroplasts Feed the World *(Page 84)*

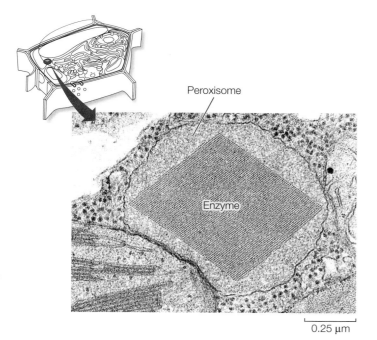

Peroxisome

Enzyme

0.25 μm

4.17 A Peroxisome *(Page 85)*

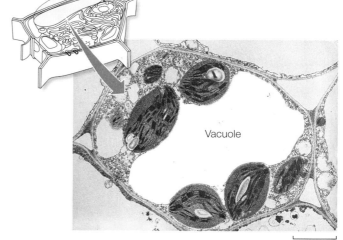

Vacuole

2 μm

4.18 Vacuoles in Plant Cells Are Usually Large *(Page 85)*

EXPERIMENT

HYPOTHESIS: Amoeboid cell movements are caused by the cytoskeleton.

METHOD

Amoeba proteus is a single-celled eukaryote that moves by extending its membrane.

The drug cytochalasin B is a drug that breaks apart microfilaments, part of the cytoskeleton

Amoeba treated with cytochalasin B

Control: Untreated *Amoeba*

RESULTS

Treated *Amoeba* rounds up and does not move

Untreated *Amoeba* continues to move

CONCLUSION: Microfilaments of the cytoskeleton are essential for amoeboid cell movement.

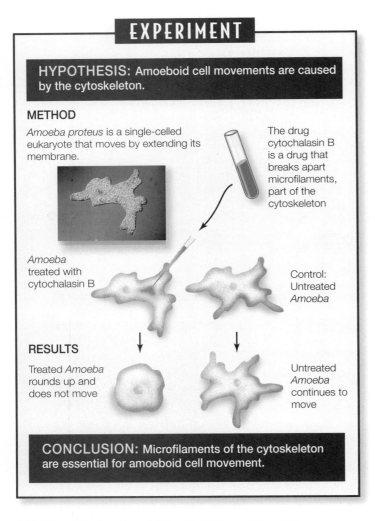

4.19 Showing Cause and Effect in Biology *(Page 86)*

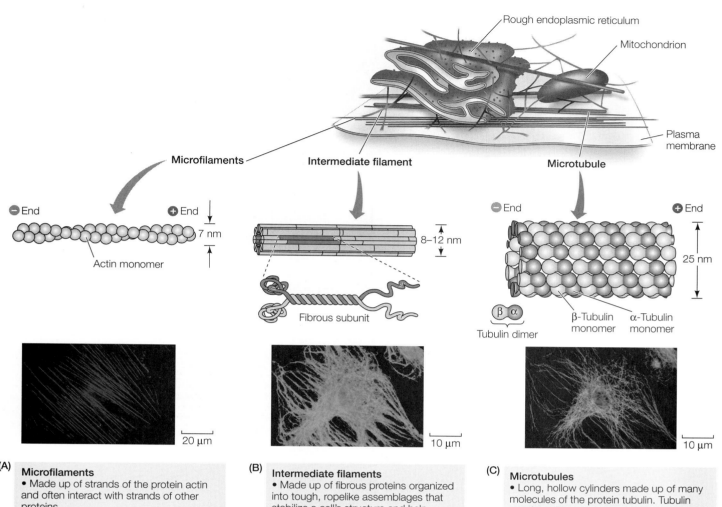

Rough endoplasmic reticulum

Mitochondrion

Plasma membrane

Microfilaments

⊖ End   ⊕ End

7 nm

Actin monomer

Intermediate filament

8–12 nm

Fibrous subunit

Microtubule

⊖ End   ⊕ End

25 nm

β α

Tubulin dimer

β-Tubulin monomer   α-Tubulin monomer

20 μm

10 μm

10 μm

(A) Microfilaments
- Made up of strands of the protein actin and often interact with strands of other proteins.
- They change cell shape and drive cellular motion, including contraction, cytoplasmic streaming, and the "pinched" shape changes that occur during cell division.
- Microfilaments and myosin strands together drive muscle action.

(B) Intermediate filaments
- Made up of fibrous proteins organized into tough, ropelike assemblages that stabilize a cell's structure and help maintain its shape.
- Some intermediate filaments help to hold neighboring cells together. Others make up the nuclear lamina.

(C) Microtubules
- Long, hollow cylinders made up of many molecules of the protein tubulin. Tubulin consists of two subunits, α-tubulin and β-tubulin.
- Microtubules lengthen or shorten by adding or subtracting tubulin dimers.
- Microtubule shortening moves chromosomes.
- Interactions between microtubules drive the movement of cells.
- Microtubules serve as "tracks" for the movement of vesicles.

4.20 The Cytoskeleton *(Page 87)*

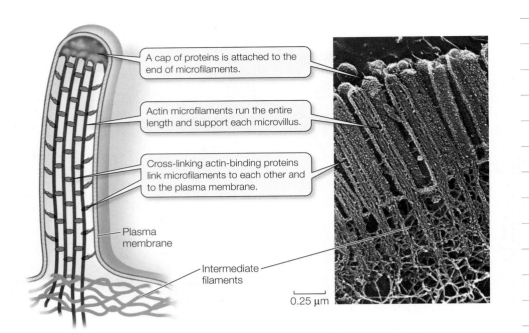

A cap of proteins is attached to the end of microfilaments.

Actin microfilaments run the entire length and support each microvillus.

Cross-linking actin-binding proteins link microfilaments to each other and to the plasma membrane.

Plasma membrane

Intermediate filaments

0.25 μm

4.21 Microfilaments for Support *(Page 88)*

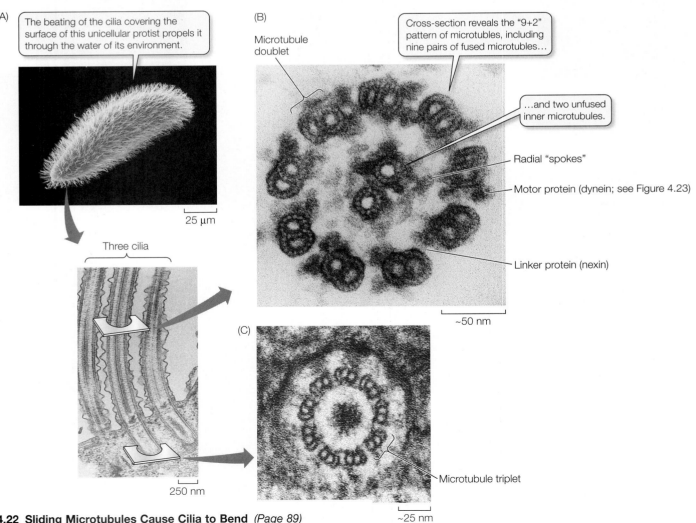

(A) The beating of the cilia covering the surface of this unicellular protist propels it through the water of its environment.

25 μm

Three cilia

250 nm

(B) Cross-section reveals the "9+2" pattern of microtubules, including nine pairs of fused microtubules…

Microtubule doublet

…and two unfused inner microtubules.

Radial "spokes"

Motor protein (dynein; see Figure 4.23)

Linker protein (nexin)

~50 nm

(C)

Microtubule triplet

~25 nm

4.22 Sliding Microtubules Cause Cilia to Bend *(Page 89)*

(A)

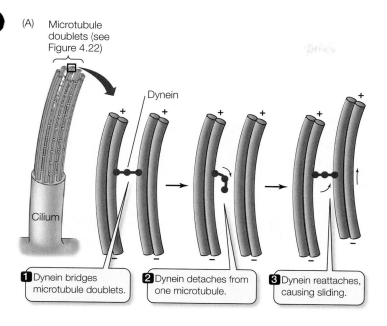

Microtubule doublets (see Figure 4.22)

Dynein

Cilium

1 Dynein bridges microtubule doublets.

2 Dynein detaches from one microtubule.

3 Dynein reattaches, causing sliding.

(B)

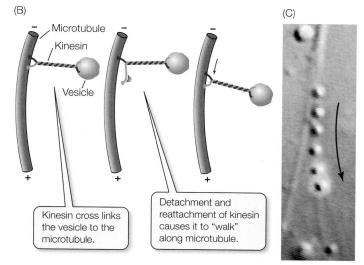

Microtubule

Kinesin

Vesicle

Kinesin cross links the vesicle to the microtubule.

Detachment and reattachment of kinesin causes it to "walk" along microtubule.

(C)

4.23 Motor Proteins Drive Vesicles along Microtubules
(Page 90)

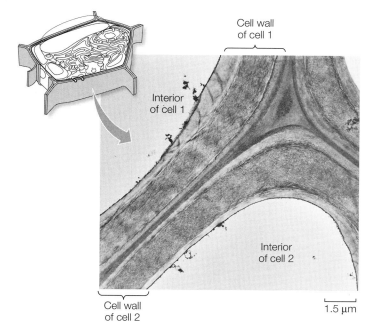

Cell wall of cell 1

Interior of cell 1

Interior of cell 2

Cell wall of cell 2

1.5 μm

4.24 The Plant Cell Wall *(Page 91)*

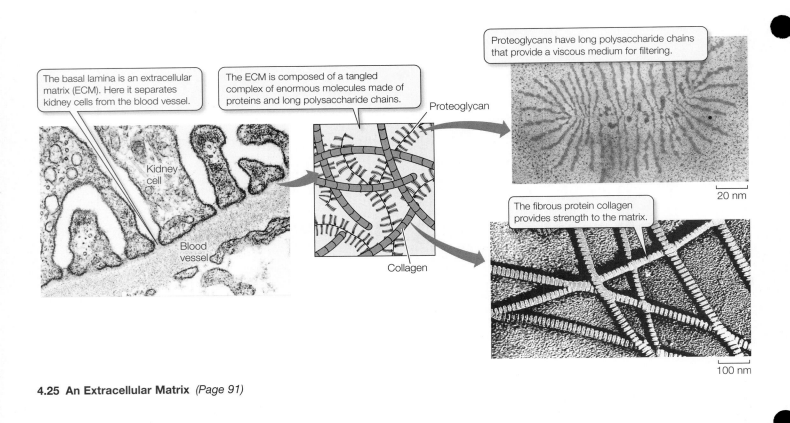

The basal lamina is an extracellular matrix (ECM). Here it separates kidney cells from the blood vessel.

The ECM is composed of a tangled complex of enormous molecules made of proteins and long polysaccharide chains.

Proteoglycans have long polysaccharide chains that provide a viscous medium for filtering.

The fibrous protein collagen provides strength to the matrix.

Kidney cell

Blood vessel

Proteoglycan

Collagen

20 nm

100 nm

4.25 An Extracellular Matrix *(Page 91)*

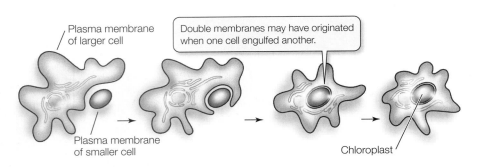

Plasma membrane of larger cell

Double membranes may have originated when one cell engulfed another.

Plasma membrane of smaller cell

Chloroplast

4.26 The Endosymbiosis Theory *(Page 92)*

CHAPTER 5 The Dynamic Cell Membrane

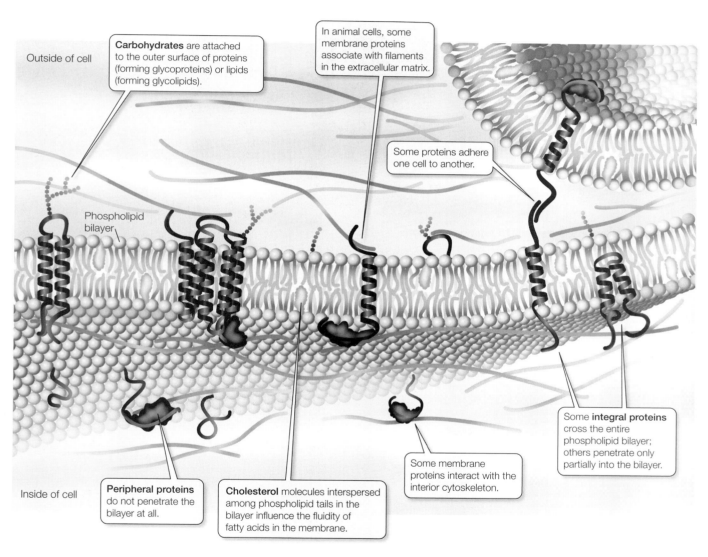

Outside of cell

Carbohydrates are attached to the outer surface of proteins (forming glycoproteins) or lipids (forming glycolipids).

In animal cells, some membrane proteins associate with filaments in the extracellular matrix.

Phospholipid bilayer

Some proteins adhere one cell to another.

Some **integral proteins** cross the entire phospholipid bilayer; others penetrate only partially into the bilayer.

Inside of cell

Peripheral proteins do not penetrate the bilayer at all.

Cholesterol molecules interspersed among phospholipid tails in the bilayer influence the fluidity of fatty acids in the membrane.

Some membrane proteins interact with the interior cytoskeleton.

5.1 The Fluid Mosaic Model *(Page 98)*

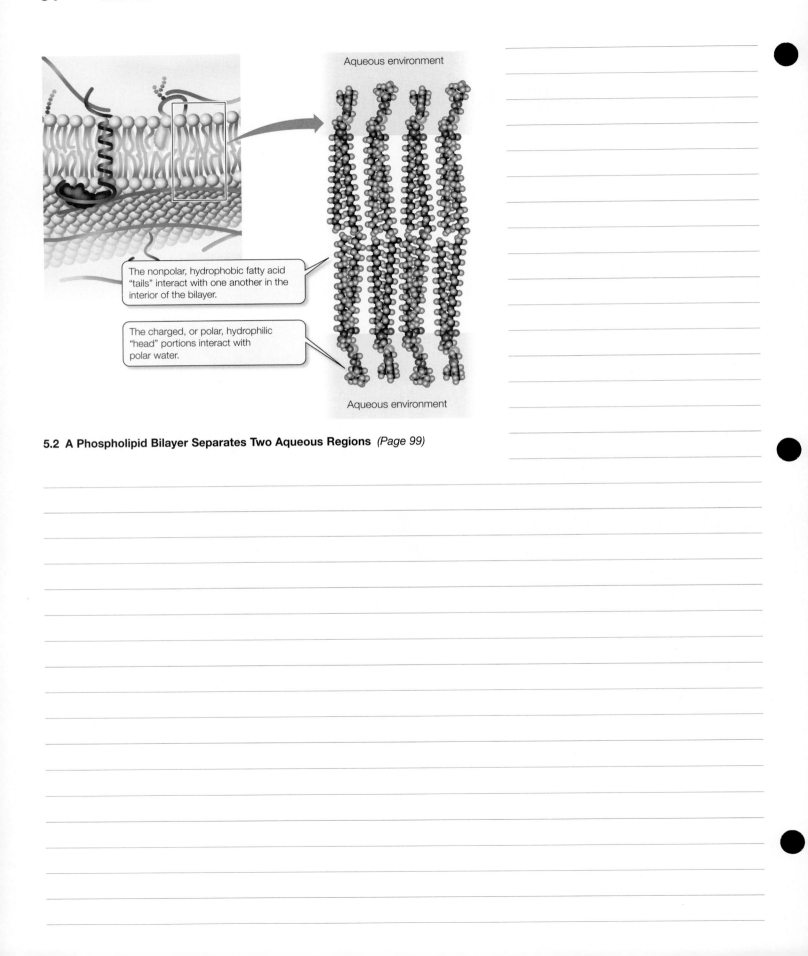

Aqueous environment

The nonpolar, hydrophobic fatty acid "tails" interact with one another in the interior of the bilayer.

The charged, or polar, hydrophilic "head" portions interact with polar water.

Aqueous environment

5.2 A Phospholipid Bilayer Separates Two Aqueous Regions *(Page 99)*

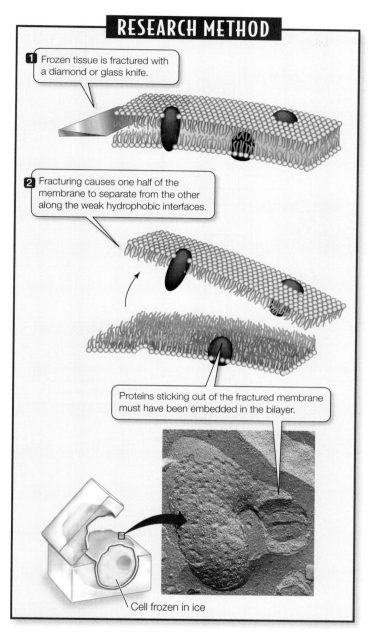

1 Frozen tissue is fractured with a diamond or glass knife.

2 Fracturing causes one half of the membrane to separate from the other along the weak hydrophobic interfaces.

Proteins sticking out of the fractured membrane must have been embedded in the bilayer.

Cell frozen in ice

5.3 Membrane Proteins Revealed by the Freeze-Fracture Technique *(Page 100)*

Hydrophilic R groups in exposed parts of the protein interact with aqueous environments.

Outside of cell (aqueous)

Hydrophobic interior of bilayer

Hydrophobic R groups interact with the hydrophobic core of the membrane, away from water.

Inside of cell (aqueous)

5.4 Interactions of Integral Membrane Proteins *(Page 100)*

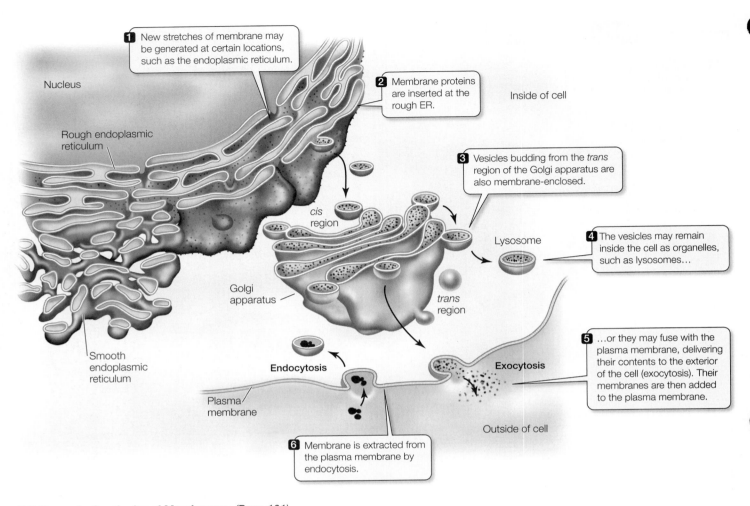

1 New stretches of membrane may be generated at certain locations, such as the endoplasmic reticulum.

Nucleus

Rough endoplasmic reticulum

2 Membrane proteins are inserted at the rough ER.

Inside of cell

3 Vesicles budding from the *trans* region of the Golgi apparatus are also membrane-enclosed.

cis region

Lysosome

4 The vesicles may remain inside the cell as organelles, such as lysosomes…

Golgi apparatus

trans region

Smooth endoplasmic reticulum

Endocytosis

Exocytosis

5 …or they may fuse with the plasma membrane, delivering their contents to the exterior of the cell (exocytosis). Their membranes are then added to the plasma membrane.

Plasma membrane

Outside of cell

6 Membrane is extracted from the plasma membrane by endocytosis.

5.5 Dynamic Continuity of Membranes *(Page 101)*

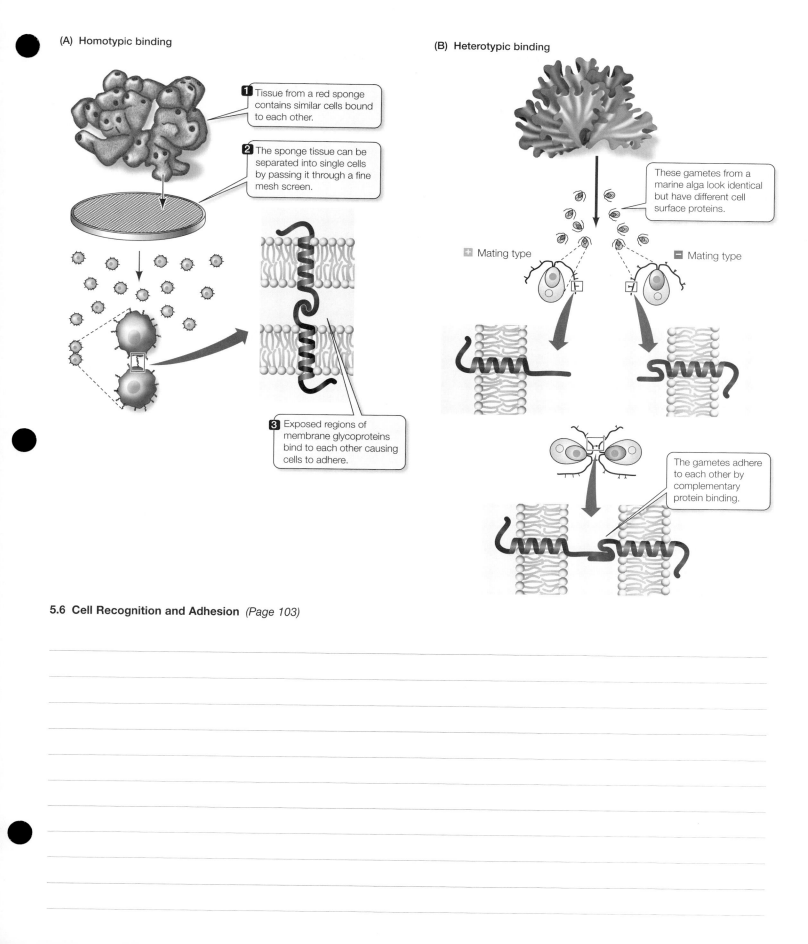

(A) Homotypic binding

1 Tissue from a red sponge contains similar cells bound to each other.

2 The sponge tissue can be separated into single cells by passing it through a fine mesh screen.

3 Exposed regions of membrane glycoproteins bind to each other causing cells to adhere.

(B) Heterotypic binding

These gametes from a marine alga look identical but have different cell surface proteins.

⊞ Mating type   ⊟ Mating type

The gametes adhere to each other by complementary protein binding.

5.6 Cell Recognition and Adhesion *(Page 103)*

(A)

Plasma membranes

Intercellular space

Junctional proteins (interlocking)

The proteins of **tight junctions** form a "quilted" seal, barring the movement of dissolved materials through the space between epithelial cells.

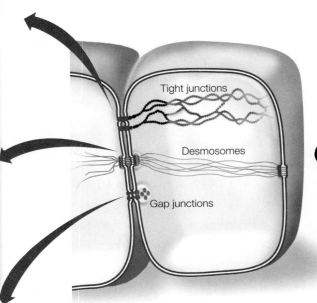

Tight junctions

Desmosomes

Gap junctions

(B)

Plasma membranes

Intercellular space

Cytoplasmic plaque

Adhesion proteins

Keratin fiber (cytoskeleton filaments)

Desmosomes link adjacent cells tightly but permit materials to move around them in the intercellular space.

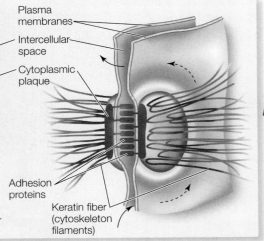

(C)

Plasma membranes

Intercellular space

Hydrophilic channel

Molecules pass between cells

Connexons (channel proteins)

Gap junctions let adjacent cells communicate.

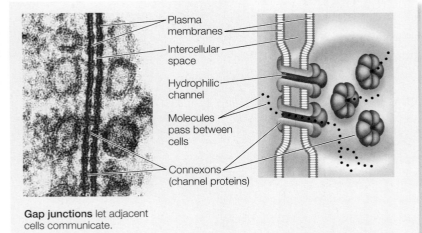

5.7 Junctions Link Animal Cells Together *(Page 104)*

EXPERIMENT

HYPOTHESIS: Diffusion leads to a uniform distribution of solutes.

METHOD   **RESULTS**

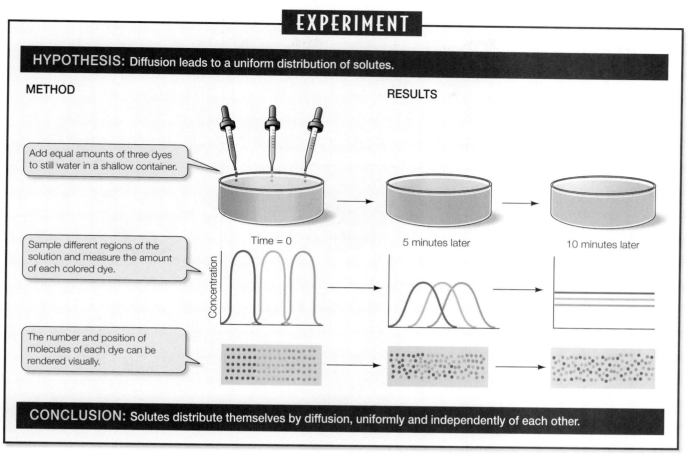

Add equal amounts of three dyes to still water in a shallow container.

Sample different regions of the solution and measure the amount of each colored dye.

The number and position of molecules of each dye can be rendered visually.

Time = 0   5 minutes later   10 minutes later

Concentration

CONCLUSION: Solutes distribute themselves by diffusion, uniformly and independently of each other.

5.8 Diffusion Leads to Uniform Distribution of Solutes *(Page 106)*

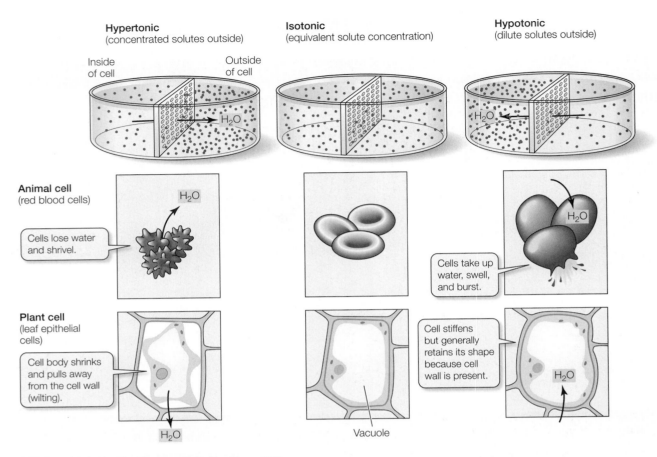

5.9 Osmosis Can Modify the Shapes of Cells *(Page 107)*

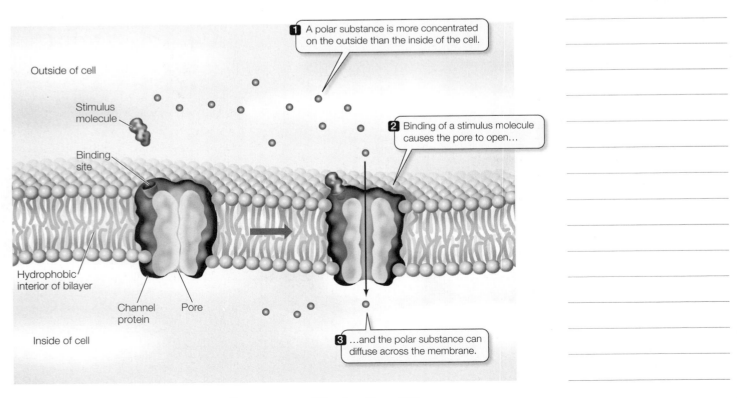

5.10 A Gated Channel Protein Opens in Response to a Stimulus *(Page 108)*

(A) Side view

Outside of cell

Potassium ions fit uniquely inside the funnel.

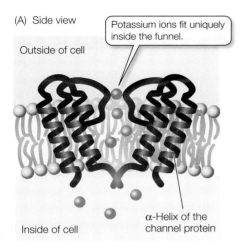

Inside of cell

α-Helix of the channel protein

(B) "Top down" view

K⁺

5.11 The Potassium Channel *(Page 109)*

(A)

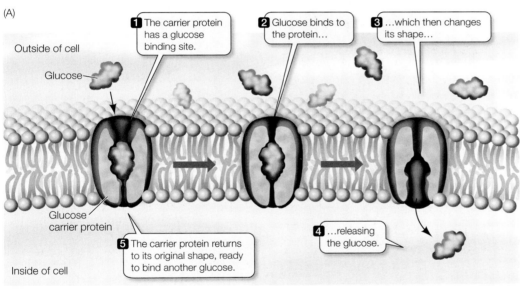

Outside of cell

Glucose

Glucose carrier protein

1 The carrier protein has a glucose binding site.

2 Glucose binds to the protein…

3 …which then changes its shape…

4 …releasing the glucose.

5 The carrier protein returns to its original shape, ready to bind another glucose.

Inside of cell

(B)

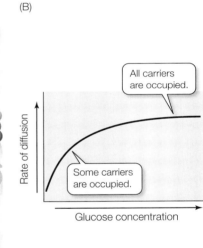

All carriers are occupied.

Some carriers are occupied.

Rate of diffusion

Glucose concentration

5.12 A Carrier Protein Facilitates Diffusion *(Page 110)*

TABLE 5.1

Membrane Transport Mechanisms

TRANSPORT MECHANISM	EXTERNAL ENERGY REQUIRED?	DRIVING FORCE	MEMBRANE PROTEIN REQUIRED?	SPECIFICITY
Simple diffusion	No	With concentration gradient	No	Not specific
Facilitated diffusion	No	With concentration gradient	Yes	Specific
Active transport	Yes	ATP hydrolysis (against concentration gradient)	Yes	Specific

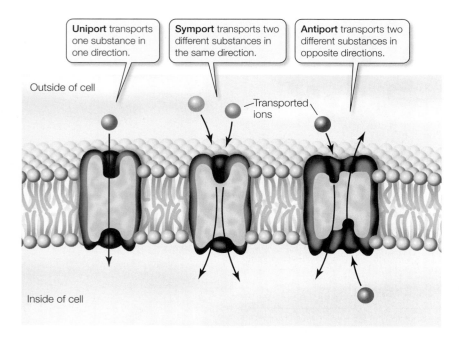

Uniport transports one substance in one direction.

Symport transports two different substances in the same direction.

Antiport transports two different substances in opposite directions.

Outside of cell

Transported ions

Inside of cell

5.13 Three Types of Proteins for Active Transport *(Page 111)*

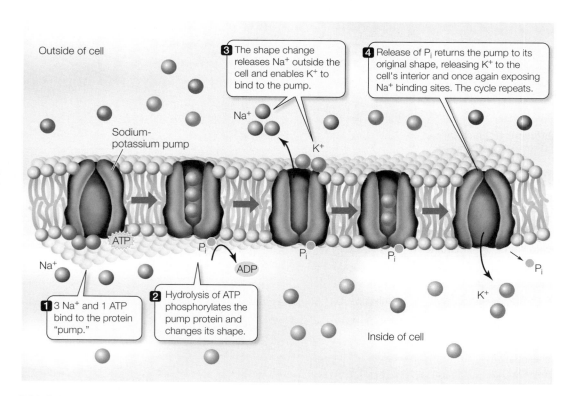

Outside of cell

3 The shape change releases Na$^+$ outside the cell and enables K$^+$ to bind to the pump.

4 Release of P$_i$ returns the pump to its original shape, releasing K$^+$ to the cell's interior and once again exposing Na$^+$ binding sites. The cycle repeats.

Na$^+$

K$^+$

Sodium-potassium pump

ATP

P$_i$

ADP

P$_i$

P$_i$

P$_i$

Na$^+$

K$^+$

1 3 Na$^+$ and 1 ATP bind to the protein "pump."

2 Hydrolysis of ATP phosphorylates the pump protein and changes its shape.

Inside of cell

5.14 Primary Active Transport: The Sodium–Potassium Pump *(Page 112)*

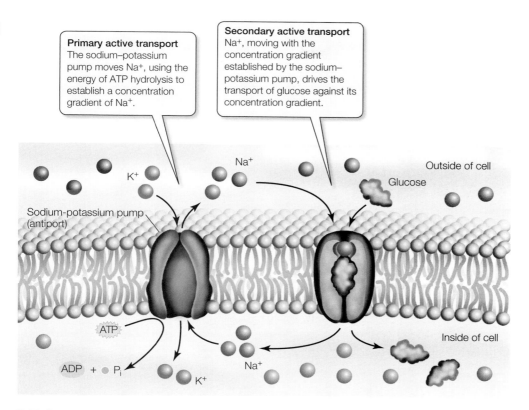

Primary active transport
The sodium–potassium pump moves Na$^+$, using the energy of ATP hydrolysis to establish a concentration gradient of Na$^+$.

Secondary active transport
Na$^+$, moving with the concentration gradient established by the sodium–potassium pump, drives the transport of glucose against its concentration gradient.

Na$^+$

Outside of cell

Glucose

K$^+$

Sodium-potassium pump (antiport)

ATP

ADP + P$_i$

K$^+$

Na$^+$

Inside of cell

5.15 Secondary Active Transport *(Page 112)*

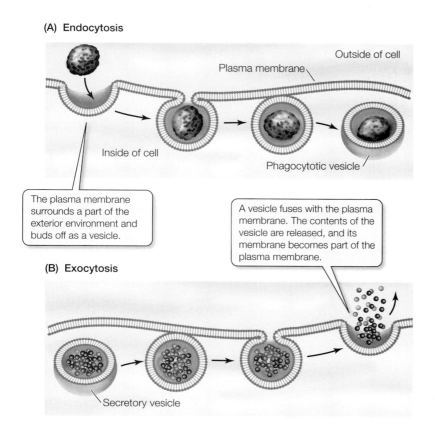

(A) Endocytosis

Outside of cell

Plasma membrane

Inside of cell

Phagocytotic vesicle

The plasma membrane surrounds a part of the exterior environment and buds off as a vesicle.

A vesicle fuses with the plasma membrane. The contents of the vesicle are released, and its membrane becomes part of the plasma membrane.

(B) Exocytosis

Secretory vesicle

5.16 Endocytosis and Exocytosis *(Page 113)*

(A) Energy transformation

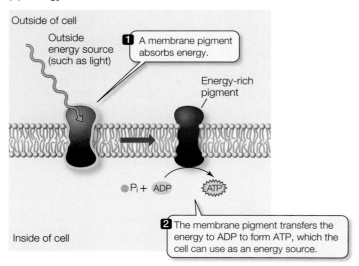

(B) Organizing chemical reactions

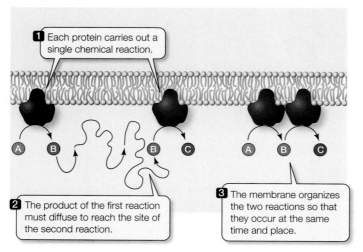

(C) Information processing

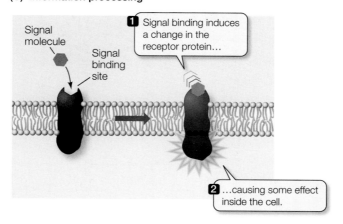

5.18 More Membrane Functions *(Page 115)*

CHAPTER 6 Energy, Enzymes, and Metabolism

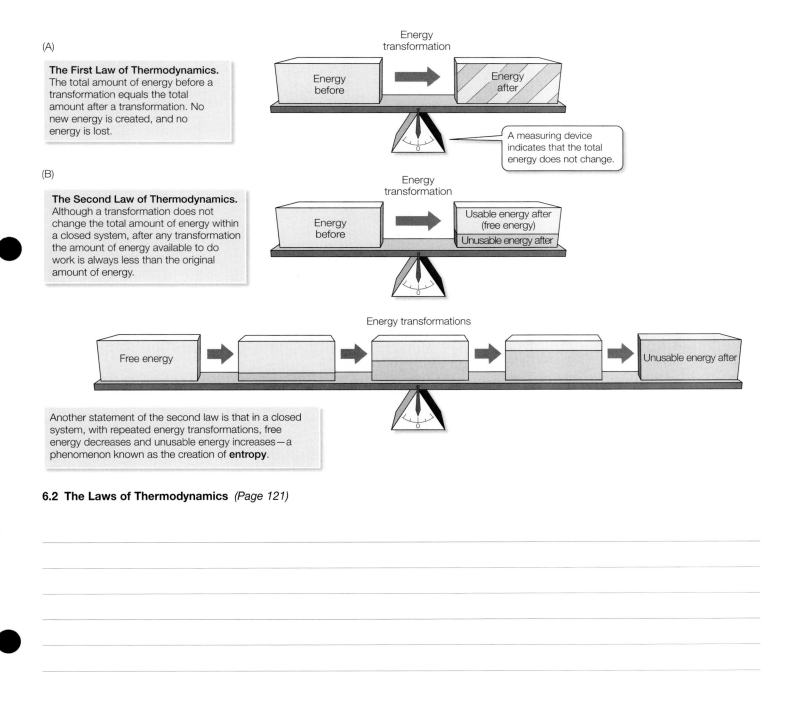

(A)

The First Law of Thermodynamics.
The total amount of energy before a transformation equals the total amount after a transformation. No new energy is created, and no energy is lost.

Energy transformation

Energy before → Energy after

A measuring device indicates that the total energy does not change.

(B)

The Second Law of Thermodynamics.
Although a transformation does not change the total amount of energy within a closed system, after any transformation the amount of energy available to do work is always less than the original amount of energy.

Energy transformation

Energy before → Usable energy after (free energy) / Unusable energy after

Energy transformations

Free energy → → → → Unusable energy after

Another statement of the second law is that in a closed system, with repeated energy transformations, free energy decreases and unusable energy increases—a phenomenon known as the creation of **entropy**.

6.2 The Laws of Thermodynamics *(Page 121)*

(A) Exergonic reaction

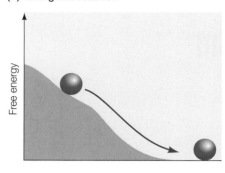

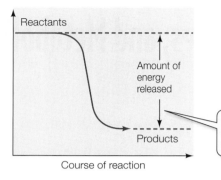

In an exergonic reaction, *energy is released* as the reactants form lower-energy products. ΔG is negative.

(B) Endergonic reaction

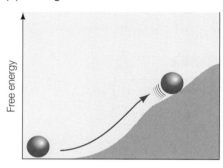

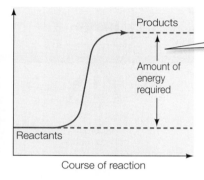

Energy must be added for an endergonic reaction, in which reactants are converted to products with a higher energy level. ΔG is positive.

6.3 Exergonic and Endergonic Reactions *(Page 122)*

_____

_____

_____

_____

_____

_____

_____

_____

_____

_____

_____

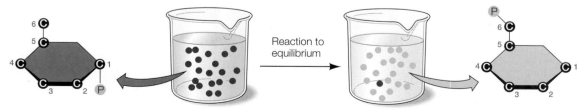

100% Glucose 1-phosphate
(0.02 *M* concentration)

95% Glucose 6-phosphate (0.019 *M* concentration)
5% Glucose 1-phosphate (0.001 *M* concentration)

6.4 Chemical Reactions Run to Equilibrium *(Page 123)*

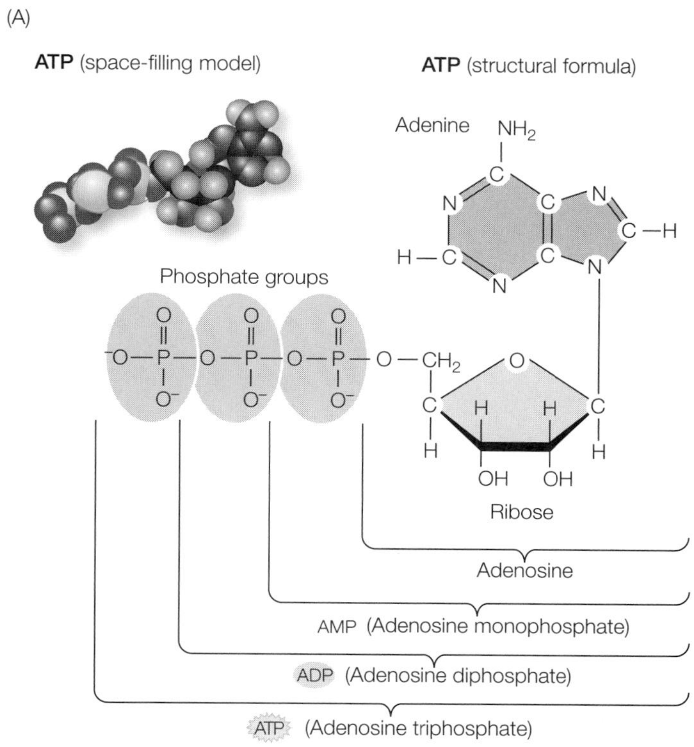

(A)

ATP (space-filling model)

ATP (structural formula)

Adenine NH$_2$

Phosphate groups

Ribose

Adenosine

AMP (Adenosine monophosphate)

ADP (Adenosine diphosphate)

ATP (Adenosine triphosphate)

(B)

6.5 ATP *(Page 124)*

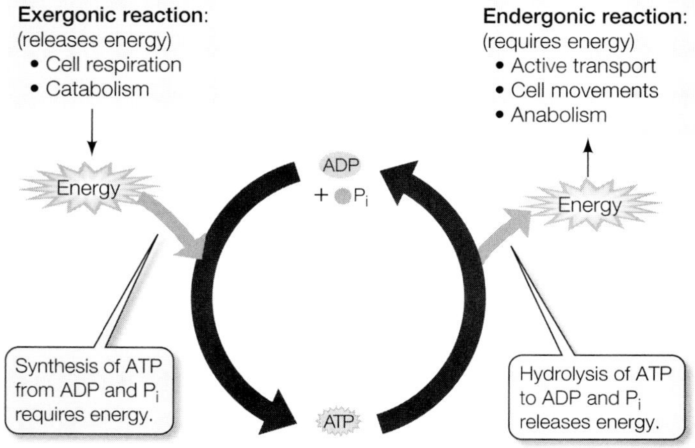

Exergonic reaction:
(releases energy)
  • Cell respiration
  • Catabolism

Energy

ADP
+ ● P$_i$

Endergonic reaction:
(requires energy)
  • Active transport
  • Cell movements
  • Anabolism

Energy

Synthesis of ATP
from ADP and P$_i$
requires energy.

ATP

Hydrolysis of ATP
to ADP and P$_i$
releases energy.

6.6 Coupling of Reactions *(Page 125)*

Exergonic reaction
(releases energy)

ATP hydrolysis

ATP + H_2O → ADP + •P_i

> The negative ΔG indicates an exergonic reaction.

$\Delta G = -7.3$ kcal/mol

Energy

Endergonic reaction
(requires energy)

> The positive ΔG indicates an endergonic reaction.

$\Delta G = +3.4$ kcal/mol

R—C + NH_4^+ → R—C

Glutamate   Glutamine

Net $\Delta G = -3.9$ kcal/mol

> The coupled reaction has an overall negative ΔG, indicating an exergonic reaction and that proceeds toward completion.

6.7 Coupling of ATP Hydrolysis to an Endergonic Reaction
(Page 125)

(A)

Free energy

Energy barrier   Transition state

Reactants (stable) E_a

ΔG

> ΔG for the reaction is not affected by E_a.

Products

Course of reaction

> E_a is the activation energy required for a reaction to begin.

(B)

Free energy

> The ball needs a push (E_a) to get it out of the depression.

Stable state

Free energy

Less stable state (transition state)

> A ball that has received an input of activation energy can roll downhill spontaneously, releasing free energy.

6.8 Activation Energy Initiates Reactions *(Page 126)*

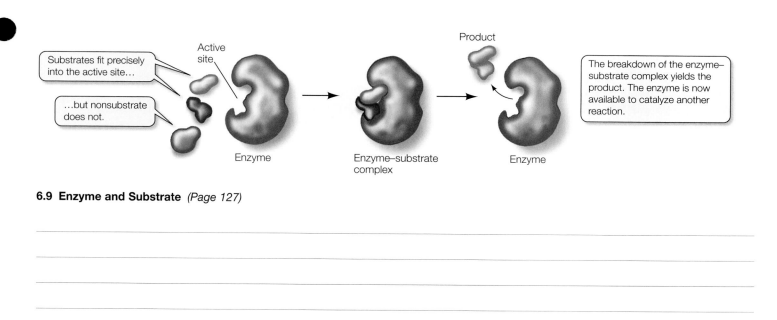

6.9 **Enzyme and Substrate** *(Page 127)*

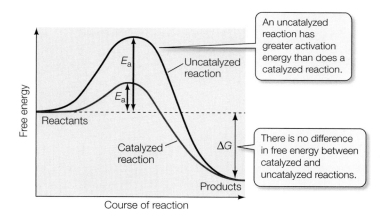

6.10 **Enzymes Lower the Energy Barrier** *(Page 127)*

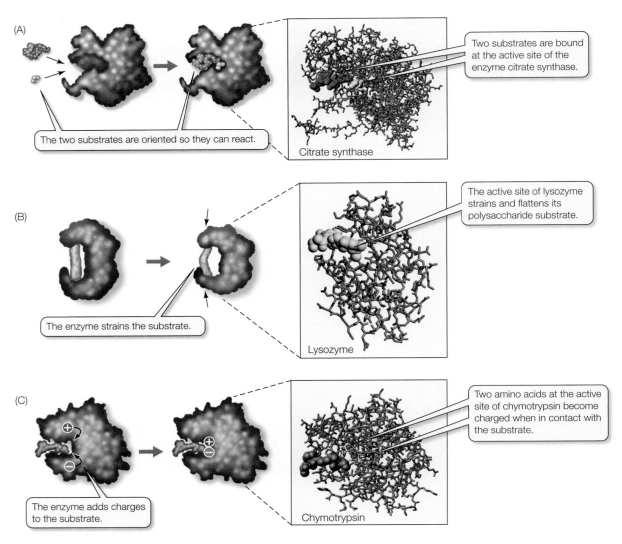

(A)

The two substrates are oriented so they can react.

Two substrates are bound at the active site of the enzyme citrate synthase.

Citrate synthase

(B)

The enzyme strains the substrate.

The active site of lysozyme strains and flattens its polysaccharide substrate.

Lysozyme

(C)

The enzyme adds charges to the substrate.

Two amino acids at the active site of chymotrypsin become charged when in contact with the substrate.

Chymotrypsin

6.11 Life at the Active Site *(Page 128)*

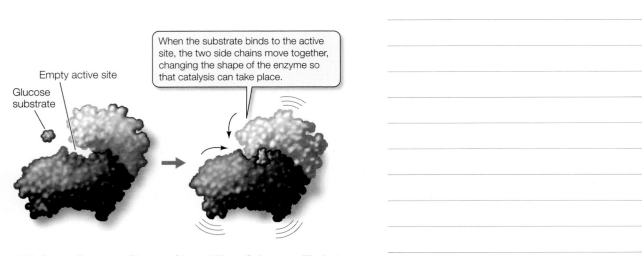

Empty active site

Glucose substrate

When the substrate binds to the active site, the two side chains move together, changing the shape of the enzyme so that catalysis can take place.

6.12 Some Enzymes Change Shape When Substrate Binds to Them *(Page 129)*

TABLE 6.1

Some Examples of Nonprotein "Partners" of Enzymes

TYPE OF MOLECULE	ROLE IN CATALYZED REACTIONS
COFACTORS	
Iron (Fe^{2+} or Fe^{3+})	Oxidation/reduction
Copper (Cu^+ or Cu^{2+})	Oxidation/reduction
Zinc (Zn^{2+})	Helps bind NAD
COENZYMES	
Biotin	Carries $-COO^-$
Coenzyme A	Carries $-CH_2-CH_3$
NAD	Carries electrons
FAD	Carries electrons
ATP	Provides/extracts energy
PROSTHETIC GROUPS	
Heme	Binds ions, O_2, and electrons; contains iron cofactor
Flavin	Binds electrons
Retinal	Converts light energy

(Page 129)

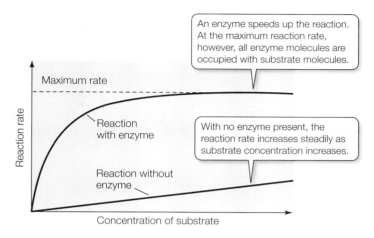

6.14 **Catalyzed Reactions Reach a Maximum Rate** *(Page 130)*

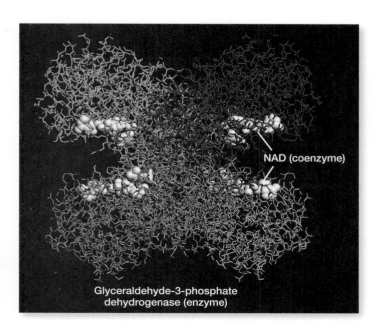

6.13 **An Enzyme with a Coenzyme** *(Page 130)*

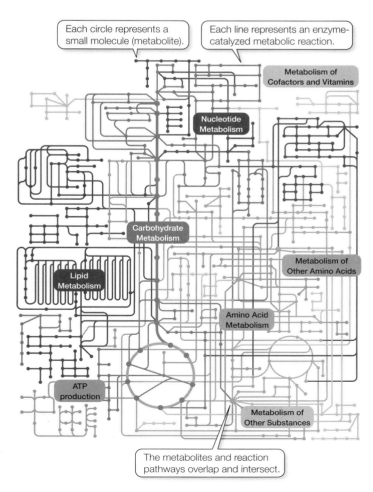

6.15 **Metabolic Pathways** *(Page 131)*

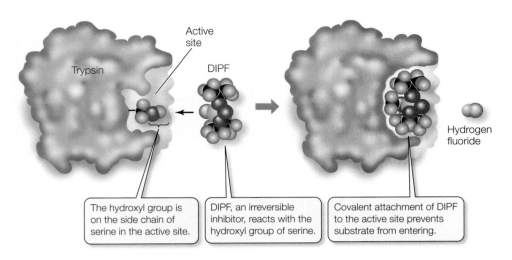

Trypsin

Active site

DIPF

Hydrogen fluoride

The hydroxyl group is on the side chain of serine in the active site.

DIPF, an irreversible inhibitor, reacts with the hydroxyl group of serine.

Covalent attachment of DIPF to the active site prevents substrate from entering.

6.16 Irreversible Inhibition *(Page 132)*

(A) Competitive inhibition

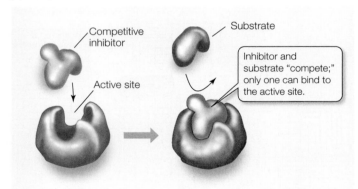

Competitive inhibitor

Substrate

Active site

Inhibitor and substrate "compete;" only one can bind to the active site.

(B) Noncompetitive inhibition

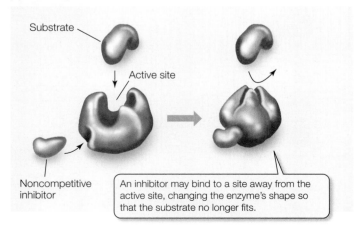

Substrate

Active site

Noncompetitive inhibitor

An inhibitor may bind to a site away from the active site, changing the enzyme's shape so that the substrate no longer fits.

6.17 Reversible Inhibition *(Page 132)*

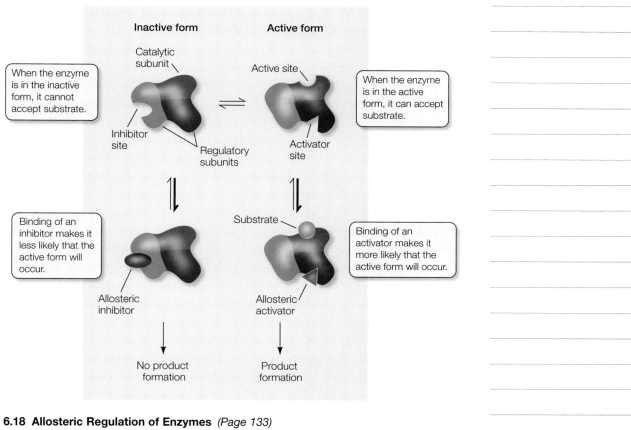

6.18 Allosteric Regulation of Enzymes *(Page 133)*

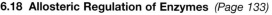

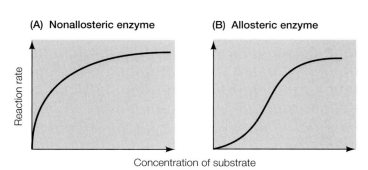

6.19 Allostery and Reaction Rate *(Page 133)*

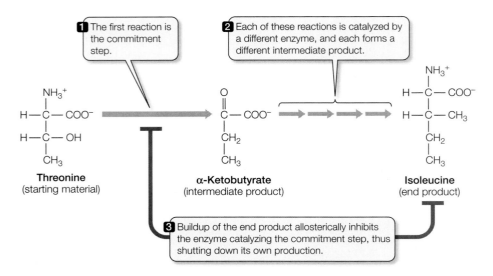

1 The first reaction is the commitment step.

2 Each of these reactions is catalyzed by a different enzyme, and each forms a different intermediate product.

NH$_3^+$
|
H—C—COO$^-$
|
H—C—OH
|
CH$_3$

Threonine
(starting material)

O
||
C—COO$^-$
|
CH$_2$
|
CH$_3$

α-Ketobutyrate
(intermediate product)

NH$_3^+$
|
H—C—COO$^-$
|
H—C—CH$_3$
|
CH$_2$
|
CH$_3$

Isoleucine
(end product)

3 Buildup of the end product allosterically inhibits the enzyme catalyzing the commitment step, thus shutting down its own production.

6.20 Feedback Inhibition of Metabolic Pathways *(Page 134)*

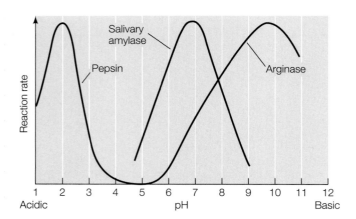

Salivary amylase

Pepsin

Arginase

Reaction rate

1 2 3 4 5 6 7 8 9 10 11 12
Acidic   pH   Basic

6.21 pH Affects Enzyme Activity *(Page 134)*

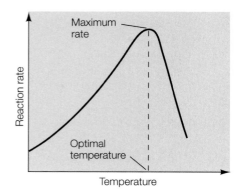

Maximum rate

Reaction rate

Optimal temperature

Temperature

6.22 Temperature Affects Enzyme Activity *(Page 134)*

CHAPTER 7 Pathways That Harvest Chemical Energy

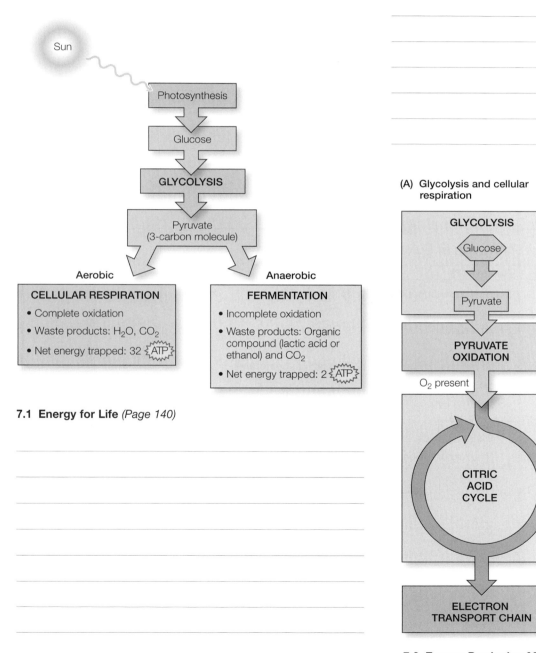

7.1 **Energy for Life** *(Page 140)*

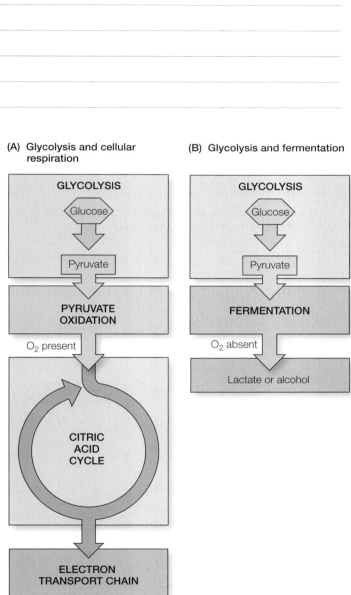

7.2 **Energy-Producing Metabolic Pathways** *(Page 141)*

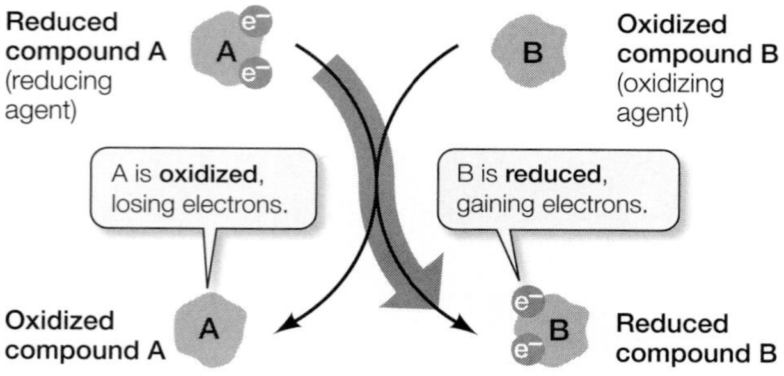

Reduced
compound A
(reducing
agent)

A is **oxidized**,
losing electrons.

Oxidized
compound B
(oxidizing
agent)

B is **reduced**,
gaining electrons.

Oxidized
compound A

Reduced
compound B

7.3 Oxidation and Reduction Are Coupled *(Page 141)*

TABLE 7.1

Cellular Locations for Energy Pathways in Eukaryotes and Prokaryotes

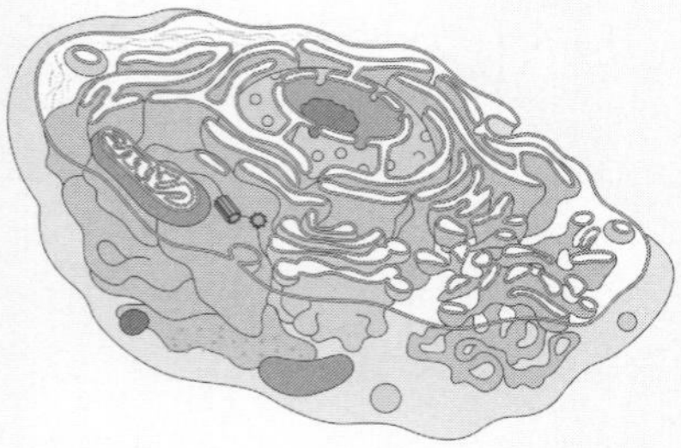

EUKARYOTES	PROKARYOTES
External to mitochondrion	**In cytoplasm**
Glycolysis	Glycolysis
Fermentation	Fermentation
	Citric acid cycle
Inside mitochondrion	**On plasma membrane**
Inner membrane	Pyruvate oxidation
Electron transport chain	Electron transport chain
Matrix	
Citric acid cycle	
Pyruvate oxidation	

(Page 141)

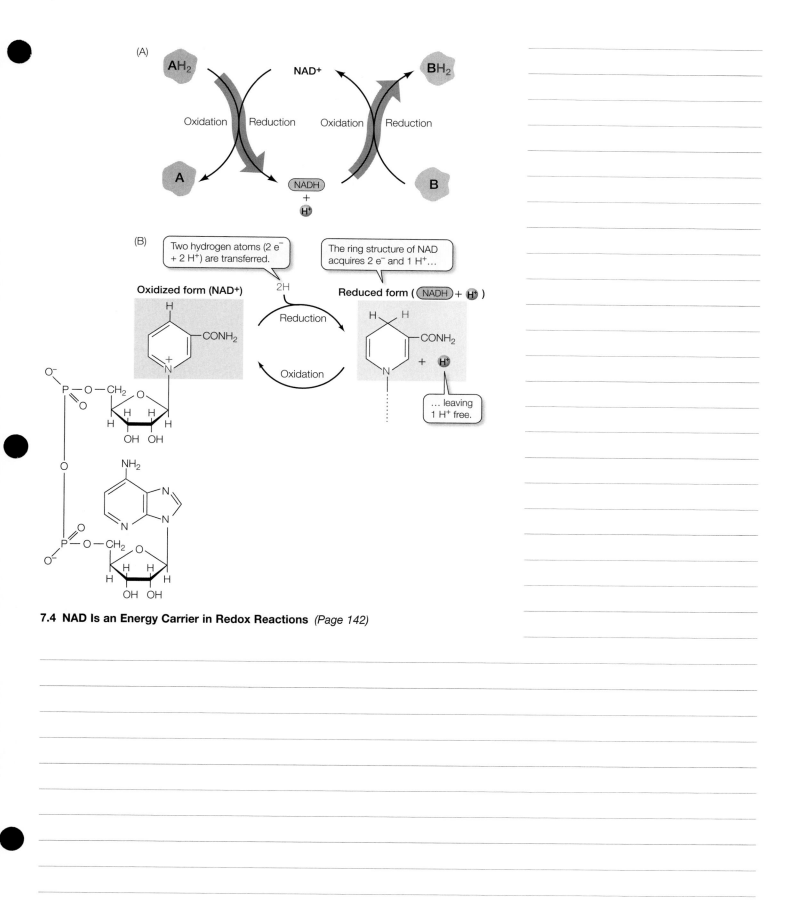

7.4 NAD Is an Energy Carrier in Redox Reactions *(Page 142)*

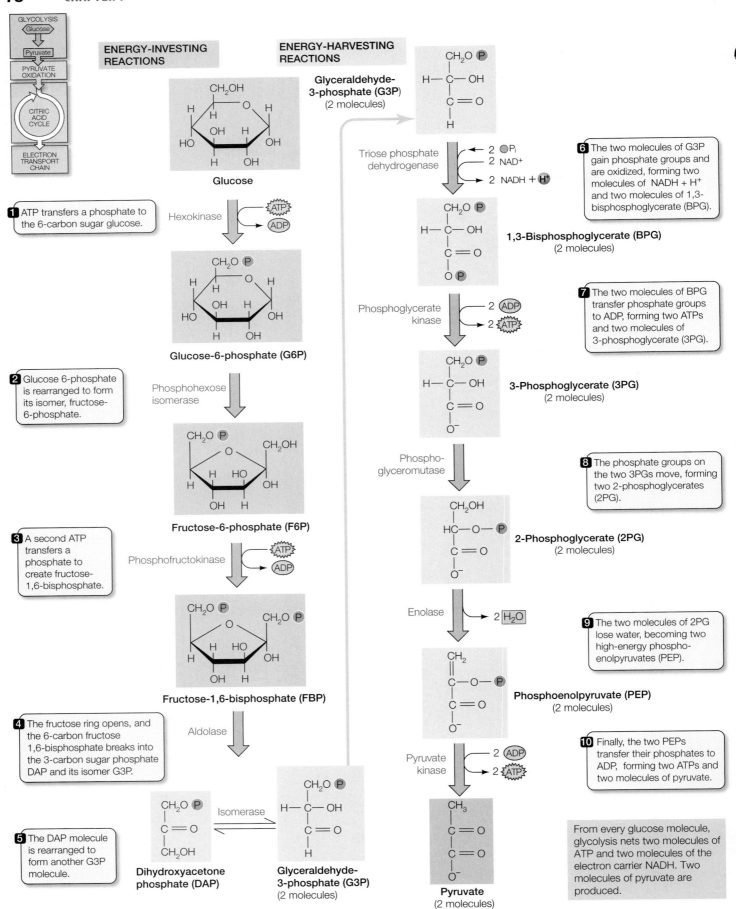

ENERGY-INVESTING REACTIONS

ENERGY-HARVESTING REACTIONS

Glyceraldehyde-3-phosphate (G3P) (2 molecules)

Glucose

1 ATP transfers a phosphate to the 6-carbon sugar glucose.

Hexokinase

Glucose-6-phosphate (G6P)

2 Glucose 6-phosphate is rearranged to form its isomer, fructose-6-phosphate.

Phosphohexose isomerase

Fructose-6-phosphate (F6P)

3 A second ATP transfers a phosphate to create fructose-1,6-bisphosphate.

Phosphofructokinase

Fructose-1,6-bisphosphate (FBP)

4 The fructose ring opens, and the 6-carbon fructose 1,6-bisphosphate breaks into the 3-carbon sugar phosphate DAP and its isomer G3P.

Aldolase

5 The DAP molecule is rearranged to form another G3P molecule.

Isomerase

Dihydroxyacetone phosphate (DAP)

Glyceraldehyde-3-phosphate (G3P) (2 molecules)

Triose phosphate dehydrogenase

2 P_i
2 NAD^+
2 $NADH + H^+$

6 The two molecules of G3P gain phosphate groups and are oxidized, forming two molecules of $NADH + H^+$ and two molecules of 1,3-bisphosphoglycerate (BPG).

1,3-Bisphosphoglycerate (BPG) (2 molecules)

Phosphoglycerate kinase

2 ADP
2 ATP

7 The two molecules of BPG transfer phosphate groups to ADP, forming two ATPs and two molecules of 3-phosphoglycerate (3PG).

3-Phosphoglycerate (3PG) (2 molecules)

Phospho-glyceromutase

8 The phosphate groups on the two 3PGs move, forming two 2-phosphoglycerates (2PG).

2-Phosphoglycerate (2PG) (2 molecules)

Enolase

2 H_2O

9 The two molecules of 2PG lose water, becoming two high-energy phospho-enolpyruvates (PEP).

Phosphoenolpyruvate (PEP) (2 molecules)

Pyruvate kinase

2 ADP
2 ATP

10 Finally, the two PEPs transfer their phosphates to ADP, forming two ATPs and two molecules of pyruvate.

From every glucose molecule, glycolysis nets two molecules of ATP and two molecules of the electron carrier NADH. Two molecules of pyruvate are produced.

Pyruvate (2 molecules)

7.5 Glycolysis Converts Glucose into Pyruvate *(Page 143)*

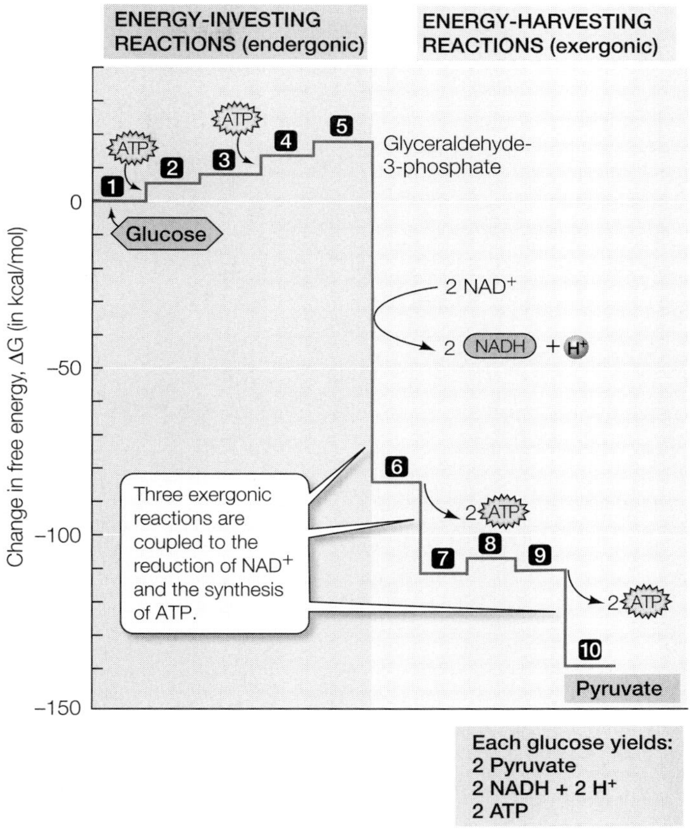
ENERGY-INVESTING REACTIONS (endergonic)
ENERGY-HARVESTING REACTIONS (exergonic)
Change in free energy, ΔG (in kcal/mol)
0
−50
−100
−150
ATP
ATP
1
2
3
4
5
Glyceraldehyde-3-phosphate
Glucose
2 NAD+
2 NADH + H+
6
2 ATP
7
8
9
2 ATP
10
Pyruvate
Three exergonic reactions are coupled to the reduction of NAD+ and the synthesis of ATP.
Each glucose yields:
2 Pyruvate
2 NADH + 2 H+
2 ATP

7.6 Changes in Free Energy During Glycolysis *(Page 144)*

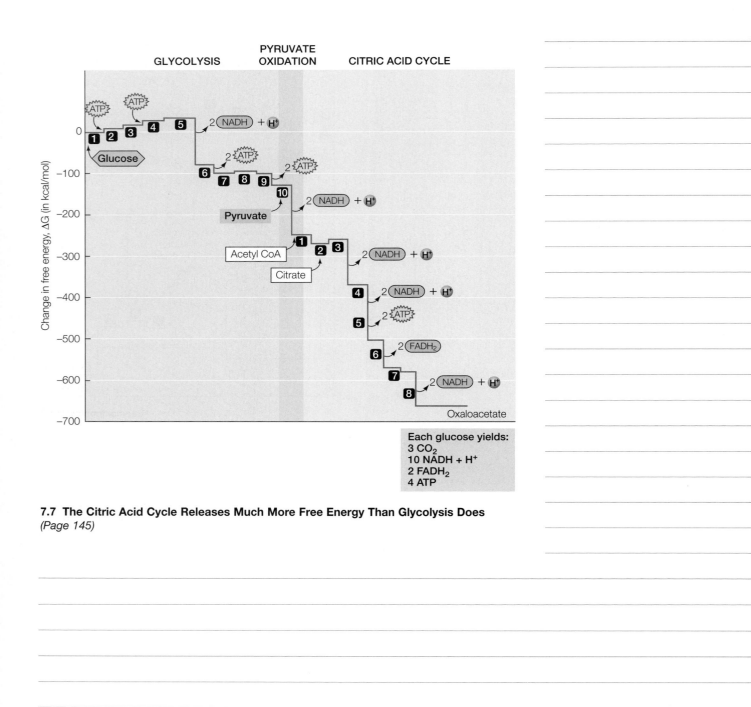

7.7 The Citric Acid Cycle Releases Much More Free Energy Than Glycolysis Does
(Page 145)

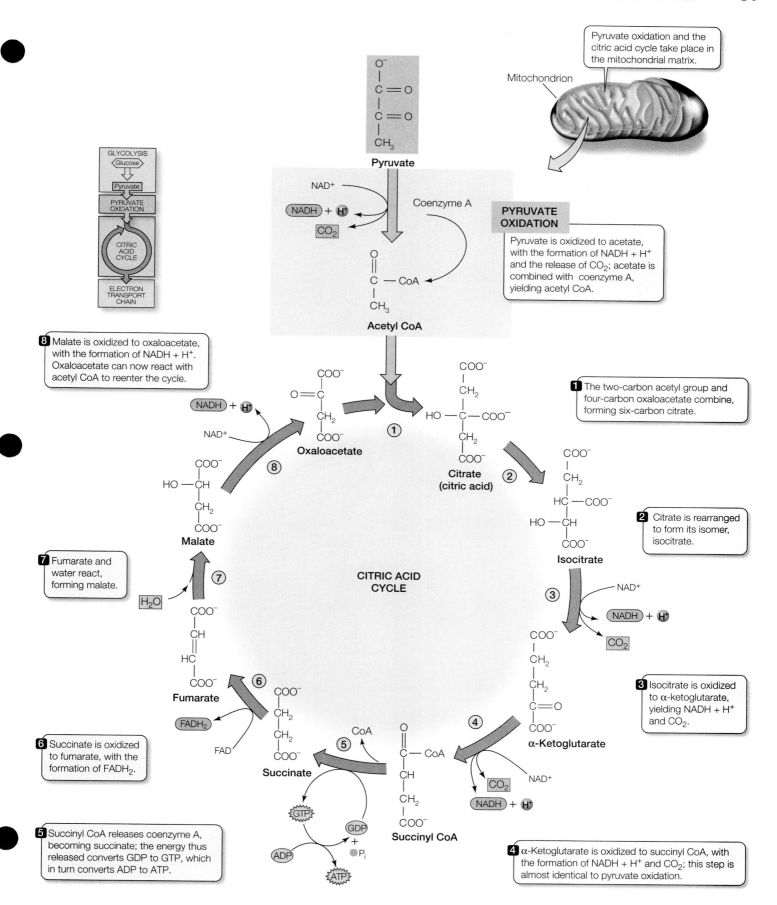

Pyruvate oxidation and the citric acid cycle take place in the mitochondrial matrix.

Mitochondrion

Pyruvate

NAD⁺

NADH + H⁺

CO₂

Coenzyme A

PYRUVATE OXIDATION

Pyruvate is oxidized to acetate, with the formation of NADH + H⁺ and the release of CO_2; acetate is combined with coenzyme A, yielding acetyl CoA.

GLYCOLYSIS
Glucose
Pyruvate
PYRUVATE OXIDATION
CITRIC ACID CYCLE
ELECTRON TRANSPORT CHAIN

Acetyl CoA

8 Malate is oxidized to oxaloacetate, with the formation of NADH + H⁺. Oxaloacetate can now react with acetyl CoA to reenter the cycle.

NADH + H⁺

NAD⁺

Oxaloacetate

Citrate (citric acid)

1 The two-carbon acetyl group and four-carbon oxaloacetate combine, forming six-carbon citrate.

Malate

Isocitrate

2 Citrate is rearranged to form its isomer, isocitrate.

7 Fumarate and water react, forming malate.

H_2O

CITRIC ACID CYCLE

NAD⁺

NADH + H⁺

CO₂

Fumarate

3 Isocitrate is oxidized to α-ketoglutarate, yielding NADH + H⁺ and CO_2.

6 Succinate is oxidized to fumarate, with the formation of $FADH_2$.

FADH₂

FAD

Succinate

CoA

α-Ketoglutarate

CO₂

NAD⁺

NADH + H⁺

Succinyl CoA

GTP

GDP + Pᵢ

ADP

ATP

5 Succinyl CoA releases coenzyme A, becoming succinate; the energy thus released converts GDP to GTP, which in turn converts ADP to ATP.

4 α-Ketoglutarate is oxidized to succinyl CoA, with the formation of NADH + H⁺ and CO_2; this step is almost identical to pyruvate oxidation.

7.8 Pyruvate Oxidation and the Citric Acid Cycle *(Page 146)*

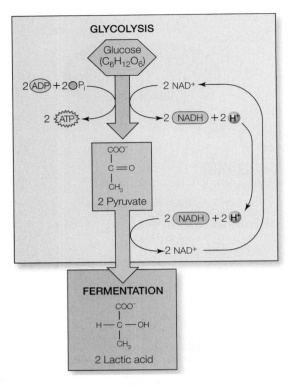

7.9 Lactic Acid Fermentation *(Page 147)*

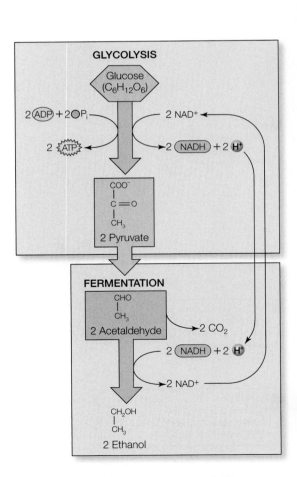

7.10 Alcoholic Fermentation *(Page 148)*

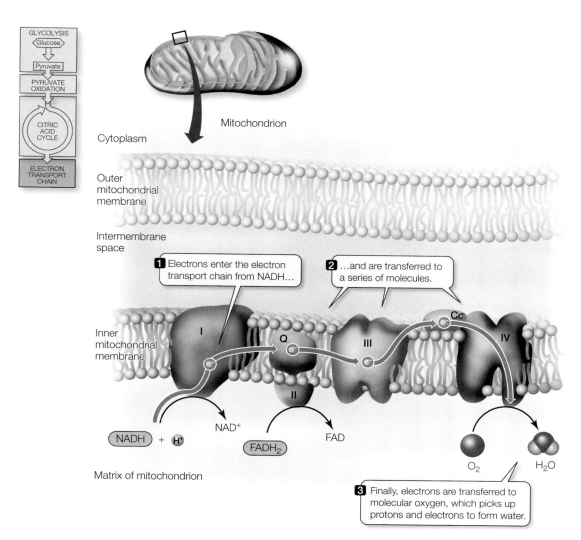

1 Electrons enter the electron transport chain from NADH...

2 ...and are transferred to a series of molecules.

3 Finally, electrons are transferred to molecular oxygen, which picks up protons and electrons to form water.

7.11 The Oxidation of NADH + H+ *(Page 149)*

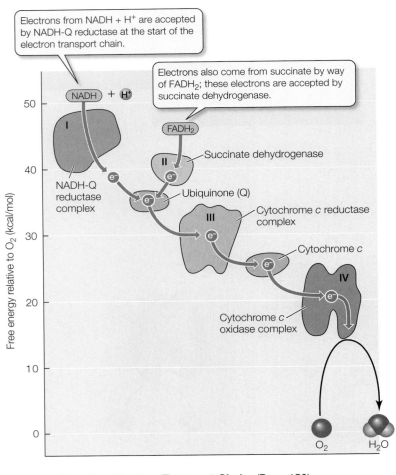

Electrons from NADH + H⁺ are accepted by NADH-Q reductase at the start of the electron transport chain.

Electrons also come from succinate by way of FADH₂; these electrons are accepted by succinate dehydrogenase.

7.12 The Complete Electron Transport Chain *(Page 150)*

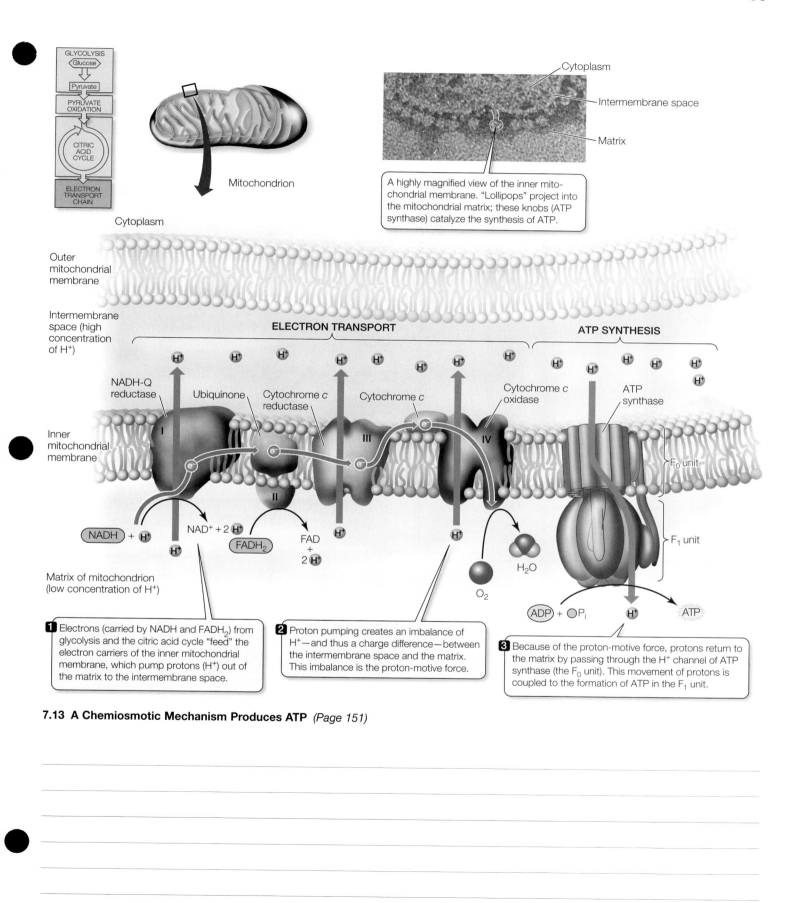

GLYCOLYSIS
Glucose
Pyruvate
PYRUVATE OXIDATION
CITRIC ACID CYCLE
ELECTRON TRANSPORT CHAIN

Cytoplasm

Mitochondrion

Cytoplasm
Intermembrane space
Matrix

A highly magnified view of the inner mitochondrial membrane. "Lollipops" project into the mitochondrial matrix; these knobs (ATP synthase) catalyze the synthesis of ATP.

Cytoplasm

Outer mitochondrial membrane

Intermembrane space (high concentration of H^+)

ELECTRON TRANSPORT

ATP SYNTHESIS

NADH-Q reductase
Ubiquinone
Cytochrome c reductase
Cytochrome c
Cytochrome c oxidase
ATP synthase

Inner mitochondrial membrane

I
II
III
IV
F_0 unit
F_1 unit

NADH + H^+
$NAD^+ + 2 H^+$
FADH$_2$
FAD + 2 H^+
H^+
H^+
H_2O
O_2
ADP + P_i
H^+
ATP

Matrix of mitochondrion (low concentration of H^+)

1 Electrons (carried by NADH and FADH$_2$) from glycolysis and the citric acid cycle "feed" the electron carriers of the inner mitochondrial membrane, which pump protons (H^+) out of the matrix to the intermembrane space.

2 Proton pumping creates an imbalance of H^+—and thus a charge difference—between the intermembrane space and the matrix. This imbalance is the proton-motive force.

3 Because of the proton-motive force, protons return to the matrix by passing through the H^+ channel of ATP synthase (the F_0 unit). This movement of protons is coupled to the formation of ATP in the F_1 unit.

7.13 A Chemiosmotic Mechanism Produces ATP *(Page 151)*

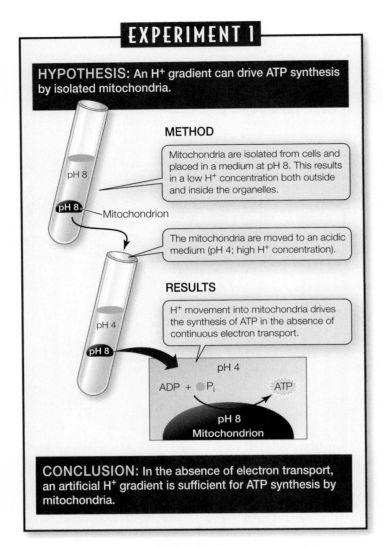

EXPERIMENT 1

HYPOTHESIS: An H+ gradient can drive ATP synthesis by isolated mitochondria.

METHOD

pH 8

pH 8 — Mitochondrion

Mitochondria are isolated from cells and placed in a medium at pH 8. This results in a low H+ concentration both outside and inside the organelles.

The mitochondria are moved to an acidic medium (pH 4; high H+ concentration).

RESULTS

pH 4

pH 8

H+ movement into mitochondria drives the synthesis of ATP in the absence of continuous electron transport.

pH 4

ADP + P_i → ATP

pH 8
Mitochondrion

CONCLUSION: In the absence of electron transport, an artificial H+ gradient is sufficient for ATP synthesis by mitochondria.

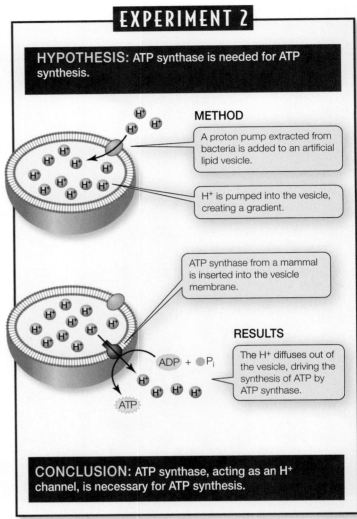

EXPERIMENT 2

HYPOTHESIS: ATP synthase is needed for ATP synthesis.

METHOD

H+

A proton pump extracted from bacteria is added to an artificial lipid vesicle.

H+ is pumped into the vesicle, creating a gradient.

ATP synthase from a mammal is inserted into the vesicle membrane.

RESULTS

ADP + P_i

The H+ diffuses out of the vesicle, driving the synthesis of ATP by ATP synthase.

ATP

CONCLUSION: ATP synthase, acting as an H+ channel, is necessary for ATP synthesis.

7.14 Two Experiments Demonstrate the Chemiosmotic Mechanism *(Page 152)*

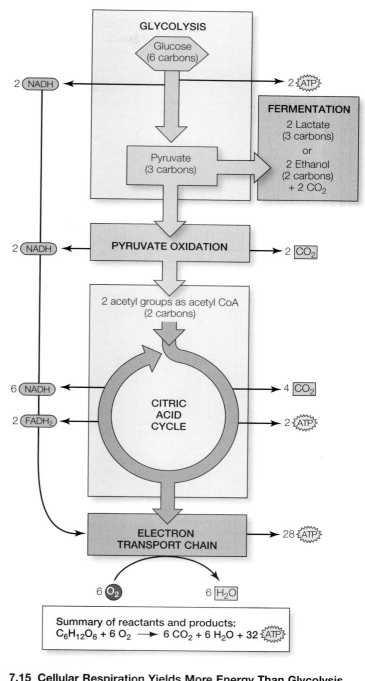

7.15 Cellular Respiration Yields More Energy Than Glycolysis Does *(Page 153)*

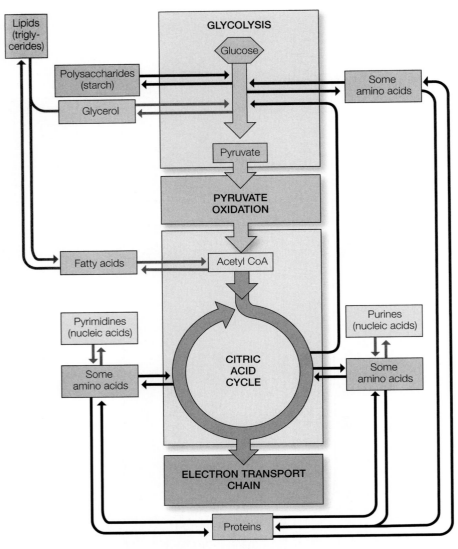

7.16 Relationships among the Major Metabolic Pathways of the Cell *(Page 154)*

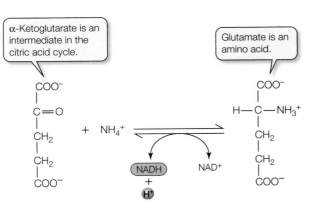

α-Ketoglutarate is an intermediate in the citric acid cycle.

Glutamate is an amino acid.

7.17 Coupling Metabolic Pathways *(Page 155)*

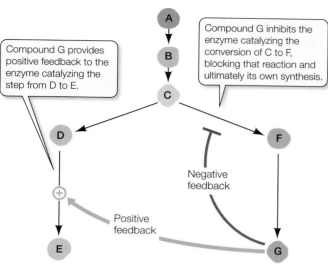

7.18 Regulation by Negative and Positive Feedback
(Page 156)

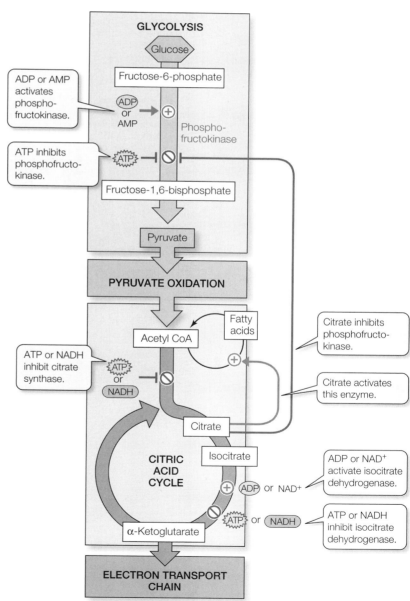

7.19 Allosteric Regulation of Glycolysis and the Citric Acid Cycle
(Page 156)

8 Photosynthesis: Energy from Sunlight

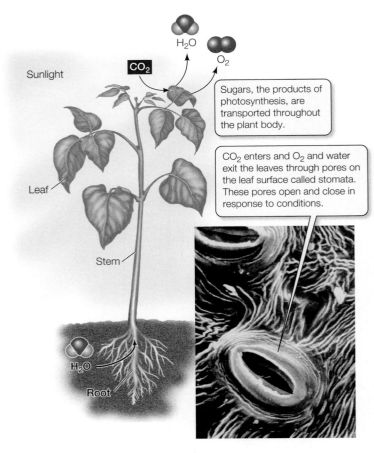

8.1 The Ingredients for Photosynthesis *(Page 162)*

Sunlight

H_2O

CO_2

O_2

Leaf

Stem

H_2O

Root

Sugars, the products of photosynthesis, are transported throughout the plant body.

CO_2 enters and O_2 and water exit the leaves through pores on the leaf surface called stomata. These pores open and close in response to conditions.

EXPERIMENT

HYPOTHESIS: The oxygen released by photosynthesis comes from water rather than from CO_2.

Experiment 1

$H_2{}^{18}O$, CO_2

Experiment 2

H_2O, $C{}^{18}O_2$

METHOD

Give plants isotope-labeled water, and unlabeled CO_2.

Give plants isotope-labeled carbon dioxide, and unlabeled water.

${}^{18}O_2$

O_2

RESULTS

The oxygen released is labeled.

The oxygen released is unlabeled.

CONCLUSION: Water is the source of the O_2 produced by photosynthesis.

8.2 Water Is the Source of the Oxygen Produced by Photosynthesis *(Page 162)*

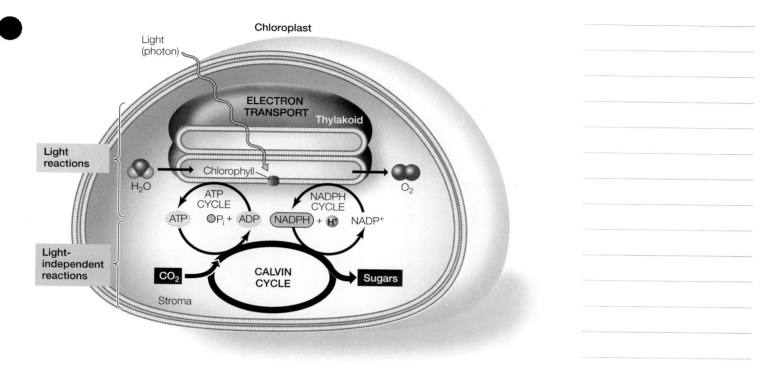

Chloroplast

8.3 An Overview of Photosynthesis *(Page 163)*

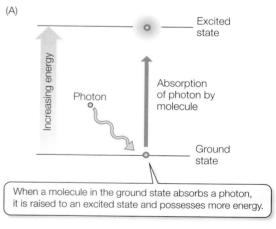

(A)

Increasing energy

Photon

Absorption
of photon by
molecule

Excited
state

Ground
state

When a molecule in the ground state absorbs a photon,
it is raised to an excited state and possesses more energy.

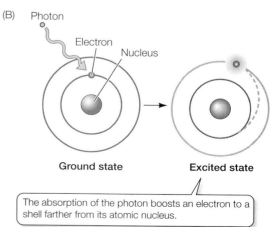

(B)   Photon

Electron

Nucleus

Ground state

Excited state

The absorption of the photon boosts an electron to a
shell farther from its atomic nucleus.

8.4 Exciting a Molecule *(Page 164)*

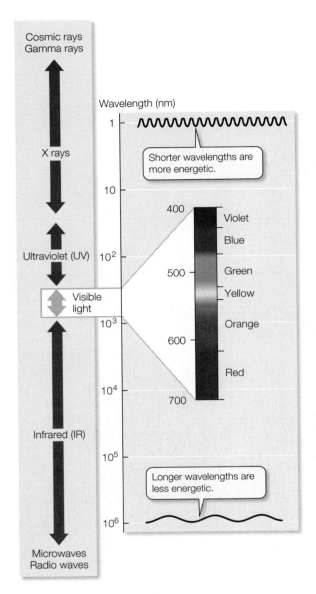

8.5 The Electromagnetic Spectrum *(Page 164)*

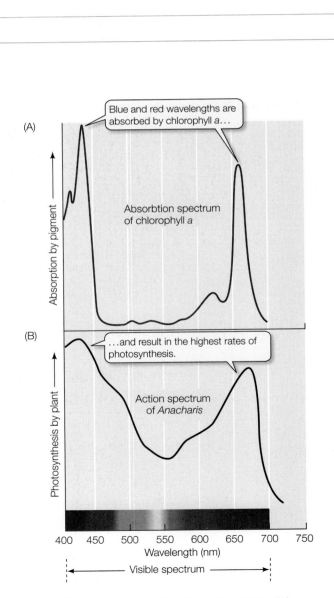

8.6 Absorption and Action Spectra *(Page 165)*

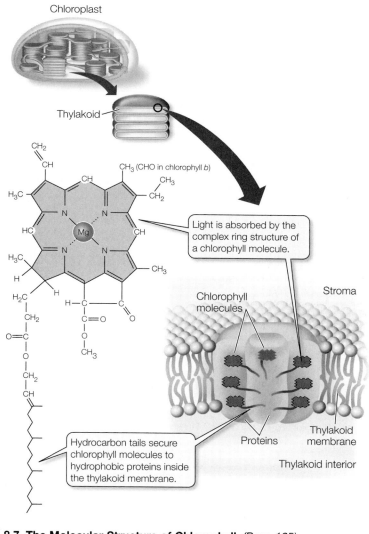

Chloroplast

Thylakoid

CH₂

CH

CH₃ (CHO in chlorophyll *b*)

CH₃

H₃C

CH₂

N   N

Mg

HC   CH

N   N

H₃C   CH₃

H

H₂C

H — C — C

CH₂   C=O

O=C   O

O   CH₃

CH₂

CH

Light is absorbed by the complex ring structure of a chlorophyll molecule.

Chlorophyll molecules

Stroma

Thylakoid membrane

Thylakoid interior

Proteins

Hydrocarbon tails secure chlorophyll molecules to hydrophobic proteins inside the thylakoid membrane.

8.7 The Molecular Structure of Chlorophyll *(Page 165)*

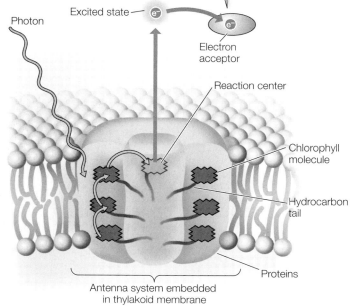

The energized electron from the chlorophyll molecules can be passed on to an electron acceptor to reduce it.

Excited state — e⁻

Photon

Electron acceptor

Reaction center

Chlorophyll molecule

Hydrocarbon tail

Proteins

Antenna system embedded in thylakoid membrane

8.8 Energy Transfer and Electron Transport *(Page 166)*

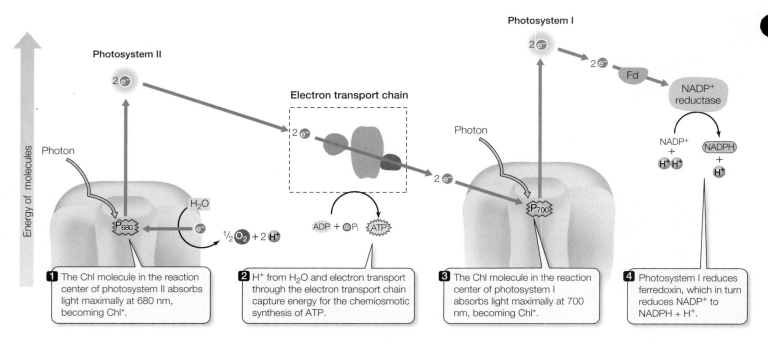

8.9 Noncyclic Electron Transport Uses Two Photosystems *(Page 167)*

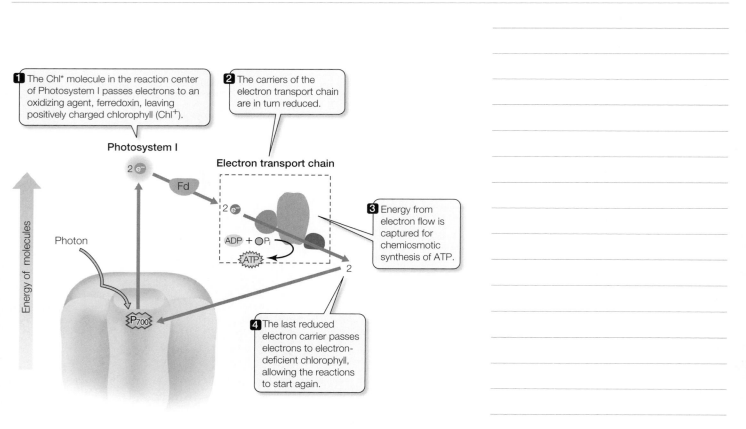

8.10 Cyclic Electron Transport Traps Light Energy as ATP *(Page 168)*

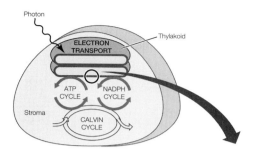

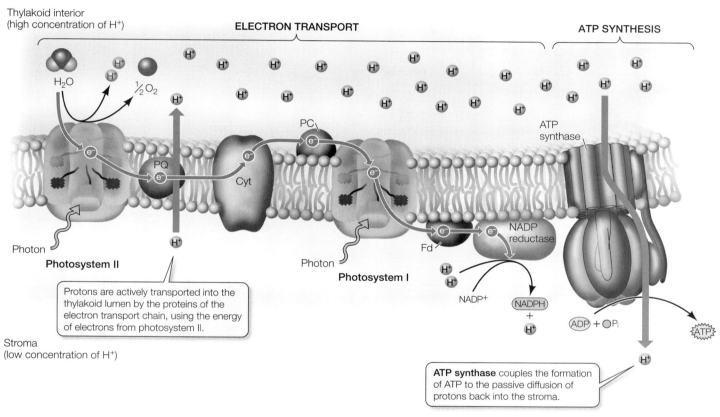

Thylakoid interior
(high concentration of H⁺)

ELECTRON TRANSPORT

ATP SYNTHESIS

H_2O

$\frac{1}{2}O_2$

PC

Cyt

PQ

ATP
synthase

Photon

Photosystem II

Photon

Photosystem I

Fd

NADP
reductase

Protons are actively transported into the
thylakoid lumen by the proteins of the
electron transport chain, using the energy
of electrons from photosystem II.

Stroma
(low concentration of H⁺)

$NADP^+$

NADPH
+
H⁺

ADP + P_i

ATP

ATP synthase couples the formation
of ATP to the passive diffusion of
protons back into the stroma.

8.11 Chloroplasts Form ATP Chemiosmotically *(Page 169)*

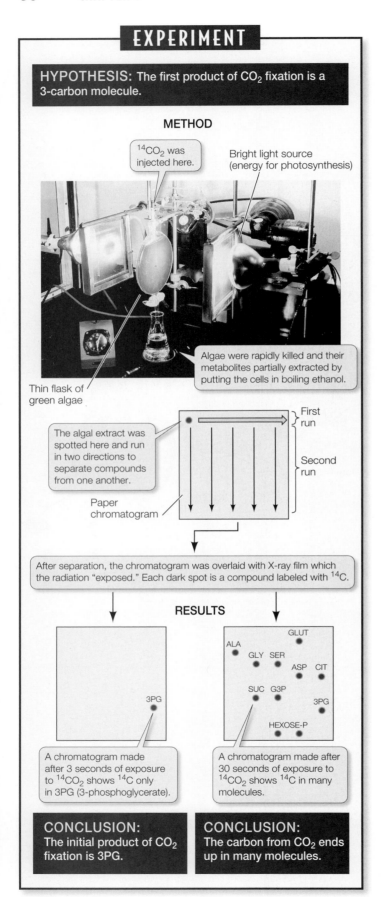

EXPERIMENT

HYPOTHESIS: The first product of CO_2 fixation is a 3-carbon molecule.

METHOD

$^{14}CO_2$ was injected here.

Bright light source (energy for photosynthesis)

Thin flask of green algae

Algae were rapidly killed and their metabolites partially extracted by putting the cells in boiling ethanol.

The algal extract was spotted here and run in two directions to separate compounds from one another.

First run

Second run

Paper chromatogram

After separation, the chromatogram was overlaid with X-ray film which the radiation "exposed." Each dark spot is a compound labeled with ^{14}C.

RESULTS

3PG

GLUT
ALA
GLY   SER
ASP   CIT
SUC   G3P
3PG
HEXOSE-P

A chromatogram made after 3 seconds of exposure to $^{14}CO_2$ shows ^{14}C only in 3PG (3-phosphoglycerate).

A chromatogram made after 30 seconds of exposure to $^{14}CO_2$ shows ^{14}C in many molecules.

CONCLUSION: The initial product of CO_2 fixation is 3PG.

CONCLUSION: The carbon from CO_2 ends up in many molecules.

COO⁻ Carboxyl group

H—C—OH

H—C—O—P

H

3-Phosphoglycerate (3PG)

Page 170 In-Text Art

8.12 Tracing the Pathway of CO$_2$ *(Page 170)*

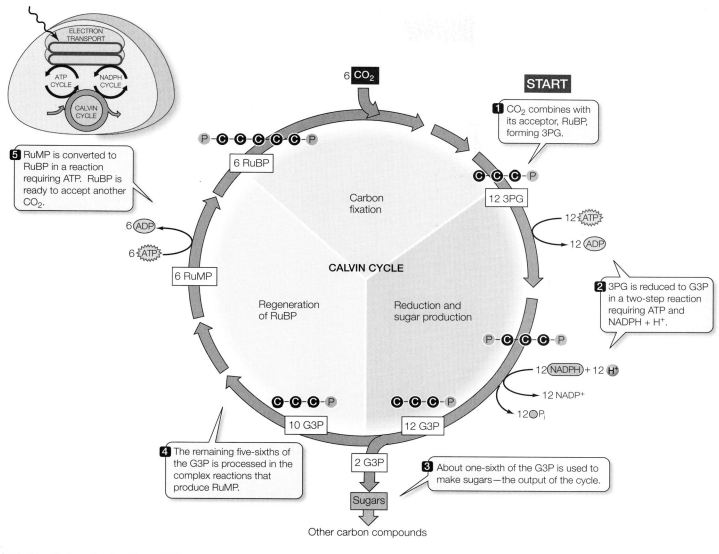

5 RuMP is converted to RuBP in a reaction requiring ATP. RuBP is ready to accept another CO_2.

START

1 CO_2 combines with its acceptor, RuBP, forming 3PG.

6 RuBP

Carbon fixation

CALVIN CYCLE

12 3PG

12 ATP

12 ADP

2 3PG is reduced to G3P in a two-step reaction requiring ATP and NADPH + H^+.

6 ADP

6 ATP

6 RuMP

Regeneration of RuBP

Reduction and sugar production

12 NADPH + 12 H^+

12 NADP+

12 P_i

4 The remaining five-sixths of the G3P is processed in the complex reactions that produce RuMP.

10 G3P

12 G3P

2 G3P

3 About one-sixth of the G3P is used to make sugars—the output of the cycle.

Sugars

Other carbon compounds

8.13 The Calvin Cycle *(Page 171)*

Glyceraldehyde 3-phosphate (G3P)

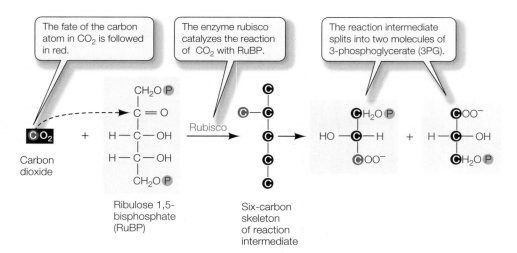

The fate of the carbon atom in CO_2 is followed in red.

The enzyme rubisco catalyzes the reaction of CO_2 with RuBP.

The reaction intermediate splits into two molecules of 3-phosphoglycerate (3PG).

CO_2
Carbon dioxide

Ribulose 1,5-bisphosphate (RuBP)

Rubisco

Six-carbon skeleton of reaction intermediate

8.14 RuBP Is the Carbon Dioxide Acceptor *(Page 171)*

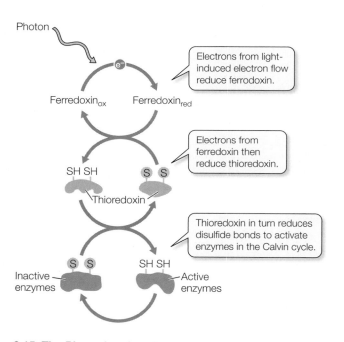

Photon

Electrons from light-induced electron flow reduce ferrodoxin.

Ferredoxin$_{ox}$ Ferredoxin$_{red}$

Electrons from ferredoxin then reduce thioredoxin.

SH SH   S   S

Thioredoxin

Thioredoxin in turn reduces disulfide bonds to activate enzymes in the Calvin cycle.

S   S   SH SH

Inactive enzymes   Active enzymes

8.15 The Photochemical Reactions Stimulate the Calvin Cycle
(Page 172)

(A) Arrangement of cells in a C_3 leaf

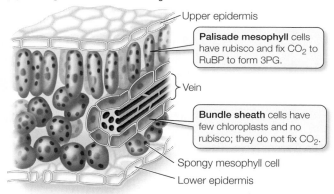

Upper epidermis

Palisade mesophyll cells have rubisco and fix CO_2 to RuBP to form 3PG.

Vein

Bundle sheath cells have few chloroplasts and no rubisco; they do not fix CO_2.

Spongy mesophyll cell

Lower epidermis

(B) Arrangement of cells in a C_4 leaf

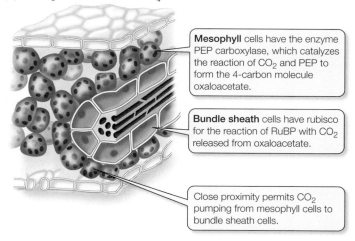

Mesophyll cells have the enzyme PEP carboxylase, which catalyzes the reaction of CO_2 and PEP to form the 4-carbon molecule oxaloacetate.

Bundle sheath cells have rubisco for the reaction of RuBP with CO_2 released from oxaloacetate.

Close proximity permits CO_2 pumping from mesophyll cells to bundle sheath cells.

8.17 Leaf Anatomy of C_3 and C_4 Plants *(Page 173)*

(A)

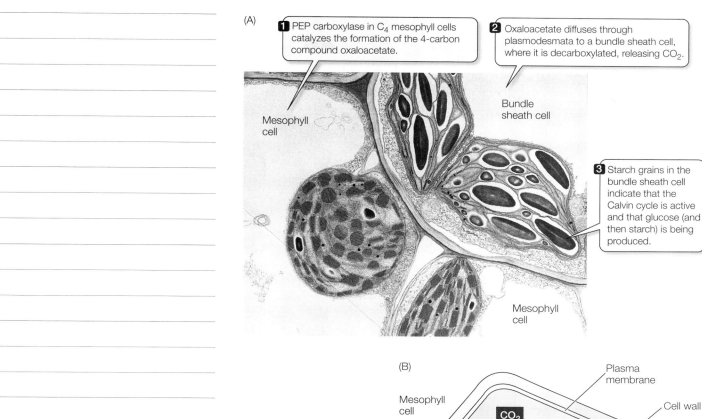

1 PEP carboxylase in C_4 mesophyll cells catalyzes the formation of the 4-carbon compound oxaloacetate.

2 Oxaloacetate diffuses through plasmodesmata to a bundle sheath cell, where it is decarboxylated, releasing CO_2.

Mesophyll cell

Bundle sheath cell

3 Starch grains in the bundle sheath cell indicate that the Calvin cycle is active and that glucose (and then starch) is being produced.

Mesophyll cell

(B)

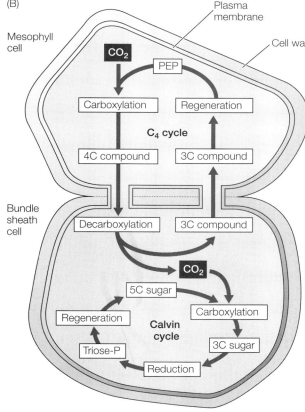

Plasma membrane

Mesophyll cell

Cell wall

CO_2

PEP

Carboxylation

Regeneration

C_4 cycle

4C compound

3C compound

Bundle sheath cell

Decarboxylation

3C compound

CO_2

5C sugar

Regeneration

Carboxylation

Calvin cycle

Triose-P

3C sugar

Reduction

8.18 The Anatomy and Biochemistry of C_4 Carbon Fixation *(Page 174)*

TABLE 8.1

Comparison of Photosynthesis in C_3 and C_4 Plants

VARIABLE	C_3 PLANTS	C_4 PLANTS
Photorespiration	Extensive	Minimal
Perform Calvin cycle?	Yes	Yes
Primary CO_2 acceptor	RuBP	PEP
CO_2-fixing enzyme	Rubisco (RuBP carboxylase/ oxygenase)	PEP carboxylase and rubisco
First product of CO_2 fixation	3PG (3-carbon compound)	Oxaloacetate (4-carbon compound)
Affinity of carboxylase for CO_2	Moderate	High
Photosynthetic cells of leaf	Mesophyll	Mesophyll + bundle sheath
Classes of chloroplasts	One	Two

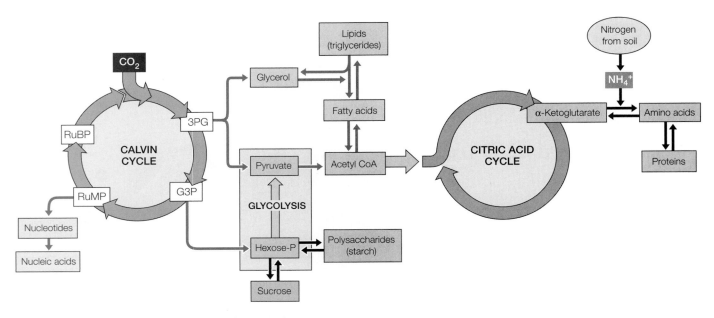

8.19 Metabolic Interactions in a Plant Cell *(Page 176)*

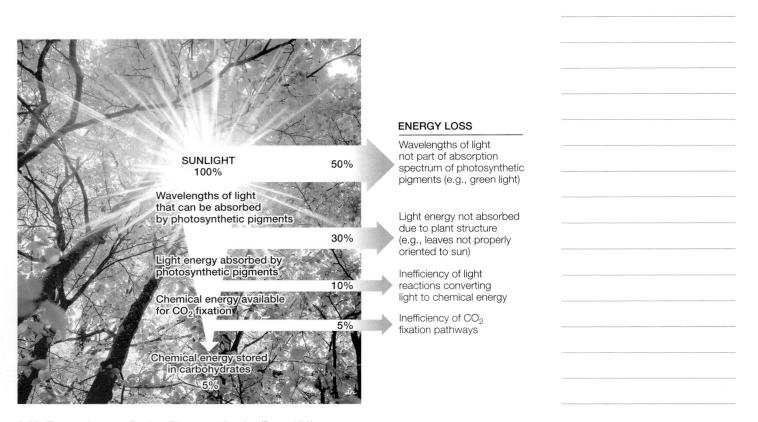

8.20 Energy Losses During Photosynthesis *(Page 176)*

9 Chromosomes, the Cell Cycle, and Cell Division

CHAPTER

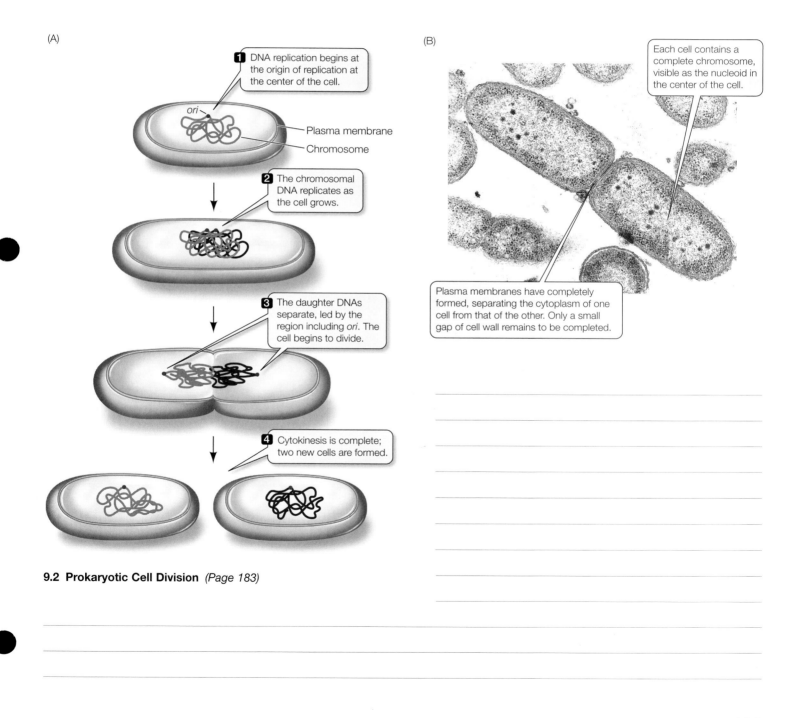

(A)

1 DNA replication begins at the origin of replication at the center of the cell.

ori

Plasma membrane

Chromosome

2 The chromosomal DNA replicates as the cell grows.

3 The daughter DNAs separate, led by the region including *ori*. The cell begins to divide.

4 Cytokinesis is complete; two new cells are formed.

(B)

Each cell contains a complete chromosome, visible as the nucleoid in the center of the cell.

Plasma membranes have completely formed, separating the cytoplasm of one cell from that of the other. Only a small gap of cell wall remains to be completed.

9.2 Prokaryotic Cell Division *(Page 183)*

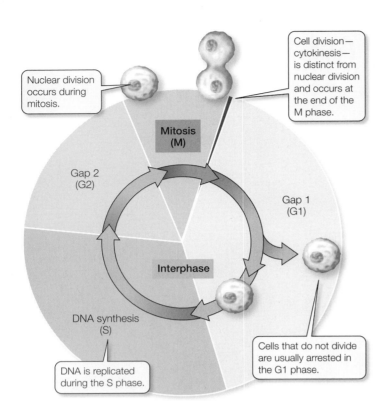

9.3 The Eukaryotic Cell Cycle *(Page 184)*

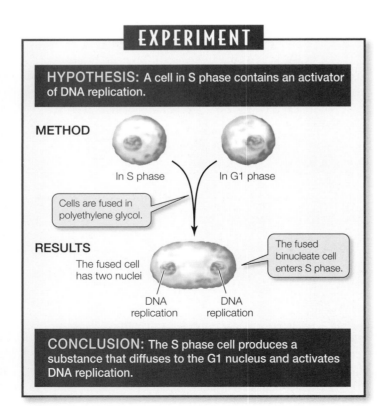

9.4 Regulation of the Cell Cycle *(Page 185)*

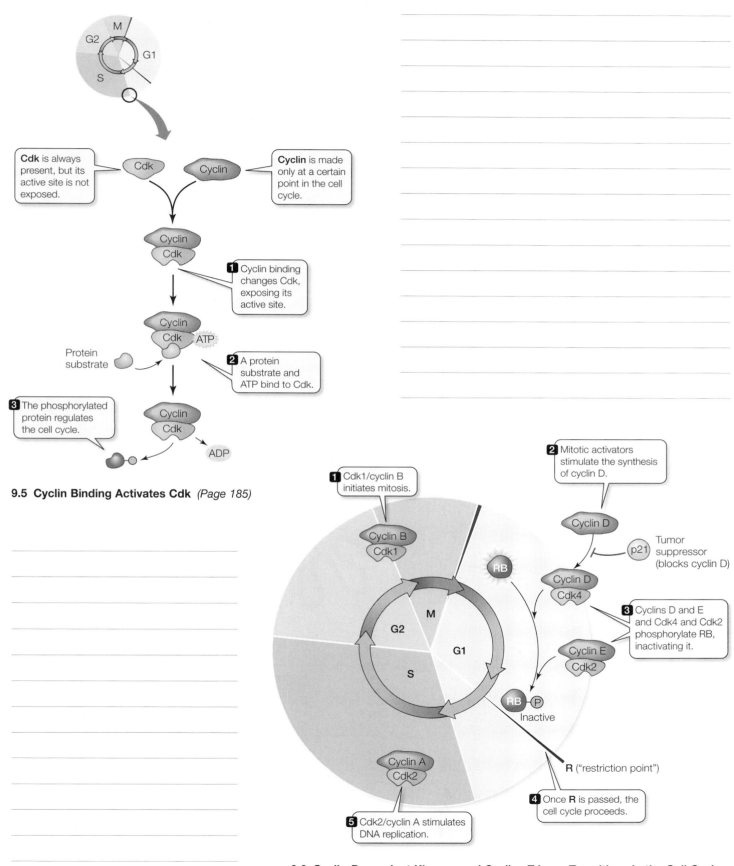

Cdk is always present, but its active site is not exposed.

Cyclin is made only at a certain point in the cell cycle.

1 Cyclin binding changes Cdk, exposing its active site.

2 A protein substrate and ATP bind to Cdk.

Protein substrate

ATP

3 The phosphorylated protein regulates the cell cycle.

ADP

9.5 Cyclin Binding Activates Cdk *(Page 185)*

1 Cdk1/cyclin B initiates mitosis.

2 Mitotic activators stimulate the synthesis of cyclin D.

p21 Tumor suppressor (blocks cyclin D)

3 Cyclins D and E and Cdk4 and Cdk2 phosphorylate RB, inactivating it.

RB

Cyclin B
Cdk1

Cyclin D

Cyclin D
Cdk4

Cyclin E
Cdk2

M

G2

G1

S

RB ⓟ
Inactive

R ("restriction point")

4 Once **R** is passed, the cell cycle proceeds.

Cyclin A
Cdk2

5 Cdk2/cyclin A stimulates DNA replication.

9.6 Cyclin-Dependent Kinases and Cyclins Trigger Transitions in the Cell Cycle *(Page 186)*

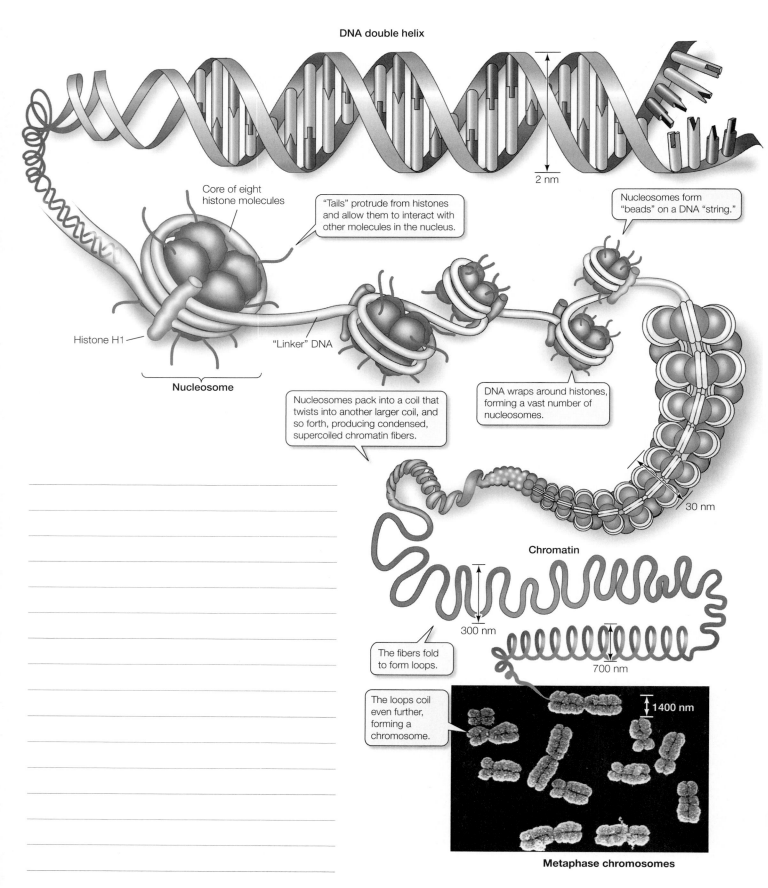

DNA double helix

2 nm

Core of eight histone molecules

"Tails" protrude from histones and allow them to interact with other molecules in the nucleus.

Nucleosomes form "beads" on a DNA "string."

Histone H1

"Linker" DNA

Nucleosome

Nucleosomes pack into a coil that twists into another larger coil, and so forth, producing condensed, supercoiled chromatin fibers.

DNA wraps around histones, forming a vast number of nucleosomes.

30 nm

Chromatin

The fibers fold to form loops.

300 nm

700 nm

The loops coil even further, forming a chromosome.

1400 nm

Metaphase chromosomes

9.8 DNA is Packed into a Mitotic Chromosome *(Page 188)*

(A)

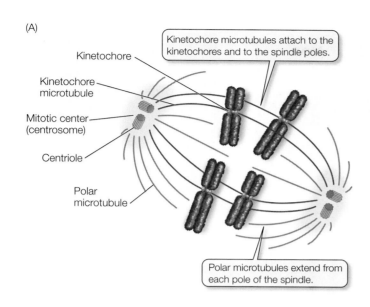

Kinetochore

Kinetochore microtubule

Mitotic center (centrosome)

Centriole

Polar microtubule

Kinetochore microtubules attach to the kinetochores and to the spindle poles.

Polar microtubules extend from each pole of the spindle.

(B)

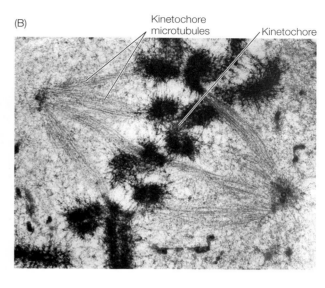

Kinetochore microtubules

Kinetochore

9.9 The Mitotic Spindle Consists of Microtubules *(Page 189)*

Interphase

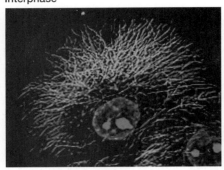

Prophase

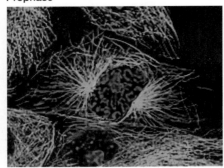

Prometaphase

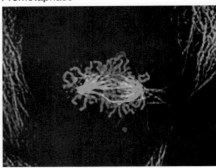

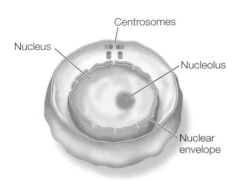

Centrosomes

Nucleus

Nucleolus

Nuclear
envelope

1 During the S phase of interphase, the nucleus replicates its DNA and centrosomes.

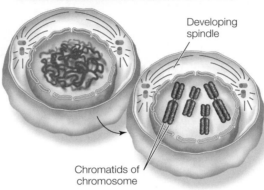

Developing
spindle

Chromatids of
chromosome

2 The chromatin coils and supercoils, become more and more compact, condensing into visible chromosomes. The chromosomes consist of identical, paired sister chromatids.

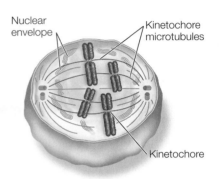

Nuclear
envelope

Kinetochore
microtubules

Kinetochore

3 The nuclear envelope breaks down. Kinetochore microtubules appear and connect the kinetochores to the poles.

9.10 Mitosis *(Page 190)*

Metaphase

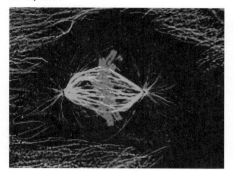

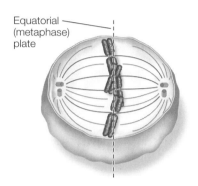

Equatorial (metaphase) plate

4 The centromeres become aligned in a plane at the cell's equator.

Anaphase

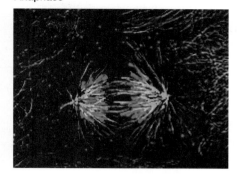

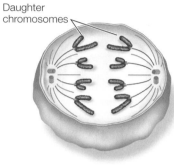

Daughter chromosomes

5 The paired sister chromatids separate, and the new daughter chromosomes begin to move toward the poles.

Telophase

6 Daughter chromosomes reach the poles. As telophase concludes, the nuclear envelopes and nucleoli re-form, chromatin becomes diffuse, and the cell again enters interphase.

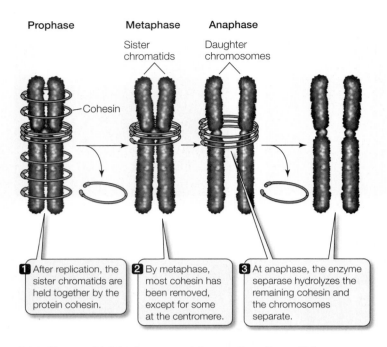

Prophase   Metaphase   Anaphase

Sister chromatids

Daughter chromosomes

Cohesin

1 After replication, the sister chromatids are held together by the protein cohesin.

2 By metaphase, most cohesin has been removed, except for some at the centromere.

3 At anaphase, the enzyme separase hydrolyzes the remaining cohesin and the chromosomes separate.

9.11 Chromatid Attachment and Separation *(Page 191)*

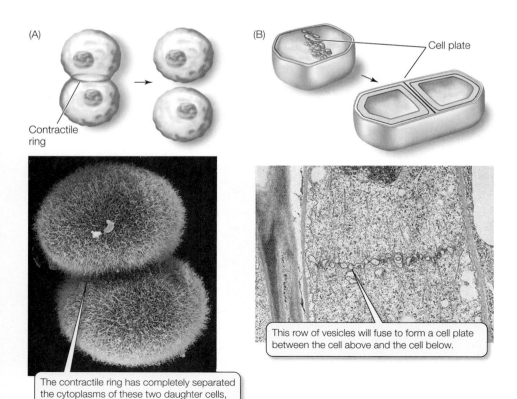

(A)

Contractile ring

(B)

Cell plate

This row of vesicles will fuse to form a cell plate between the cell above and the cell below.

The contractile ring has completely separated the cytoplasms of these two daughter cells, although their surfaces remain in contact.

9.12 Cytokinesis Differs in Animal and Plant Cells *(Page 192)*

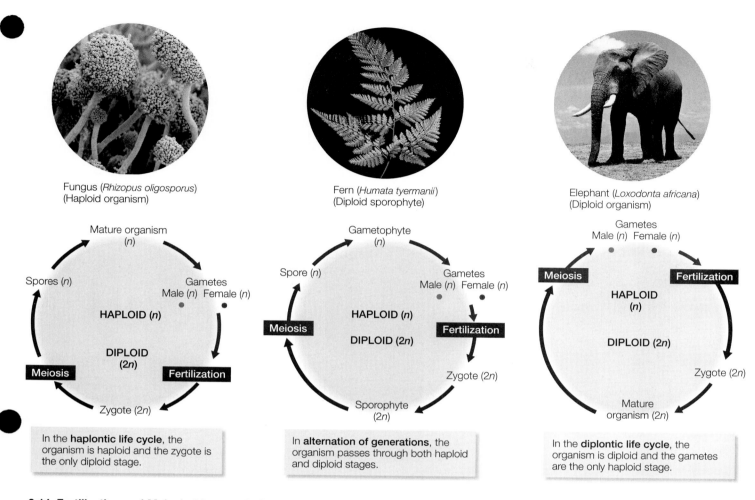

Fungus (*Rhizopus oligosporus*)
(Haploid organism)

Fern (*Humata tyermanii*)
(Diploid sporophyte)

Elephant (*Loxodonta africana*)
(Diploid organism)

In the **haplontic life cycle**, the organism is haploid and the zygote is the only diploid stage.

In **alternation of generations**, the organism passes through both haploid and diploid stages.

In the **diplontic life cycle**, the organism is diploid and the gametes are the only haploid stage.

9.14 Fertilization and Meiosis Alternate in Sexual Reproduction
(Page 194)

TABLE 9.1

Numbers of Pairs of Chromosomes in Some Plant and Animal Species

COMMON NAME	SPECIES	NUMBER OF CHROMOSOME PAIRS
Mosquito	*Culex pipiens*	3
Housefly	*Musca domestica*	6
Toad	*Bufo americanus*	11
Rice	*Oryza sativa*	12
Frog	*Rana pipiens*	13
Alligator	*Alligator mississippiensis*	16
Rhesus monkey	*Macaca mulatta*	21
Wheat	*Triticum aestivum*	21
Human	*Homo sapiens*	23
Potato	*Solanum tuberosum*	24
Donkey	*Equus asinus*	31
Horse	*Equus caballus*	32
Dog	*Canis familiaris*	39
Carp	*Cyprinus carpio*	52

(Page 194)

MEIOSIS I

Early prophase I

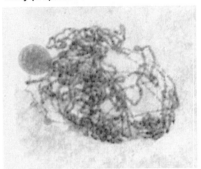

Mid-prophase I

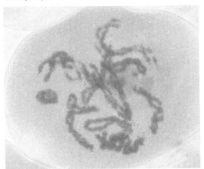

Late prophase I–prometaphase

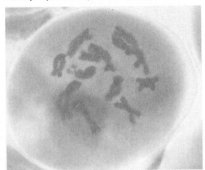

Centrosomes

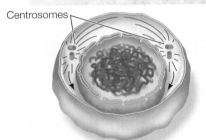

Pairs of homologs

Tetrad

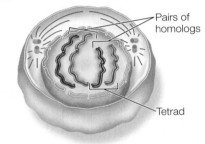

1 The chromatin begins to condense following interphase.

2 Synapsis aligns homologs, and chromosomes condense further.

3 The chromosomes continue to coil and shorten. Crossing over results in an exchange of genetic material. In prometaphase the nuclear envelope breaks down.

MEIOSIS II

Prophase II

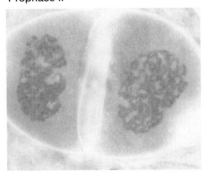

Metaphase II

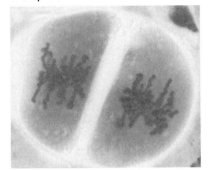

Anaphase II

Equatorial plate

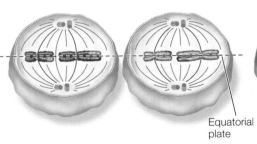

7 The chromosomes condense again, following a brief interphase (interkinesis) in which DNA does not replicate.

8 The centrosomes of the paired chromatids line up at the equatorial plates of each cell.

9 The chromatids finally separate, becoming chromosomes in their own right, and are pulled to opposite poles. Because of crossing over in prophase I, each new cell will have a different genetic makeup.

9.16 Meiosis *(Page 196)*

Metaphase I

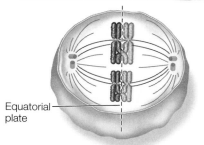

Equatorial plate

4 The homologous pairs line up on the equatorial (metaphase) plate.

Anaphase I

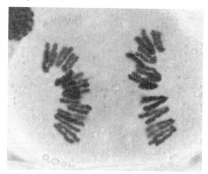

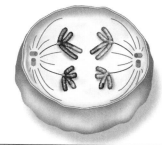

5 The homologous chromosomes (each with two chromatids) move to opposite poles of the cell.

Telophase I

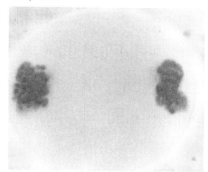

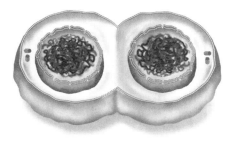

6 The chromosomes gather into nuclei, and the original cell divides.

Telophase II

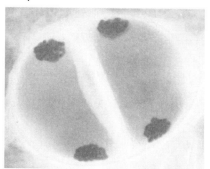

10 The chromosomes gather into nuclei, and the cells divide.

Products

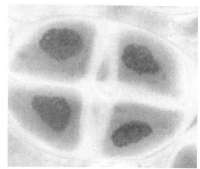

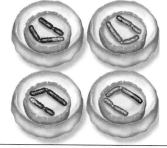

11 Each of the four cells has a nucleus with a haploid number of chromosomes.

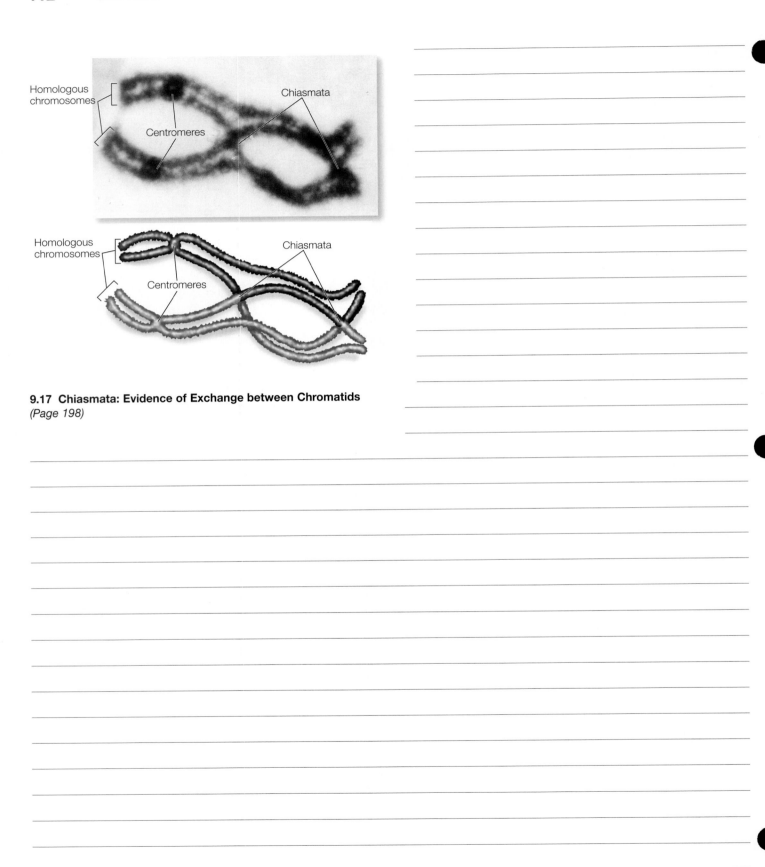

9.17 Chiasmata: Evidence of Exchange between Chromatids
(Page 198)

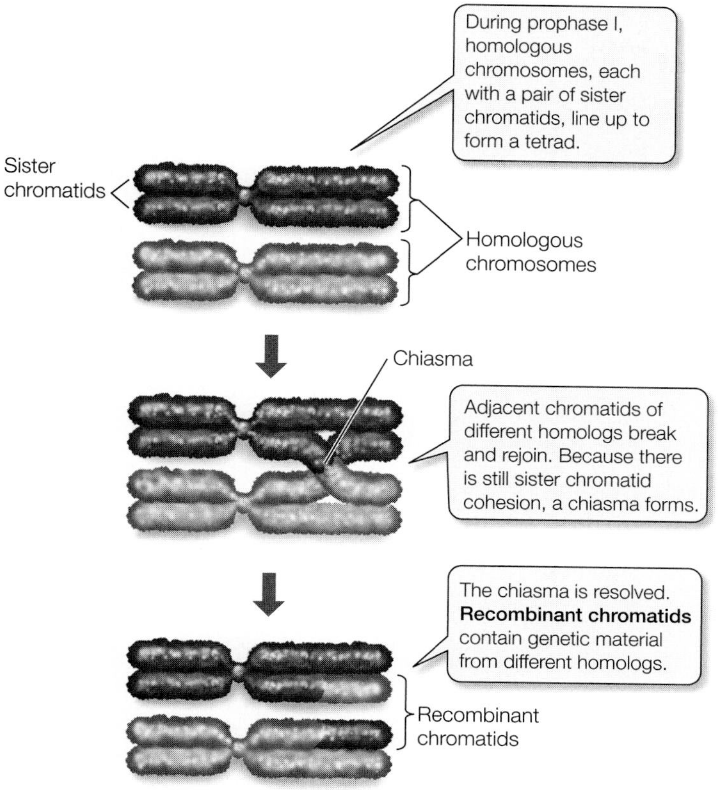

During prophase I, homologous chromosomes, each with a pair of sister chromatids, line up to form a tetrad.

Sister chromatids

Homologous chromosomes

Chiasma

Adjacent chromatids of different homologs break and rejoin. Because there is still sister chromatid cohesion, a chiasma forms.

The chiasma is resolved. **Recombinant chromatids** contain genetic material from different homologs.

Recombinant chromatids

9.18 Crossing Over Forms Genetically Diverse Chromosomes
(Page 198)

MITOSIS

Parent cell (2n) — Prophase — Metaphase — Anaphase

1 No synapsis of homologous chromosomes.

2 Individual chromosomes align at the equatorial plate.

3 Centromeres separate. Sister chromatids separate during anaphase, becoming daughter chromosomes.

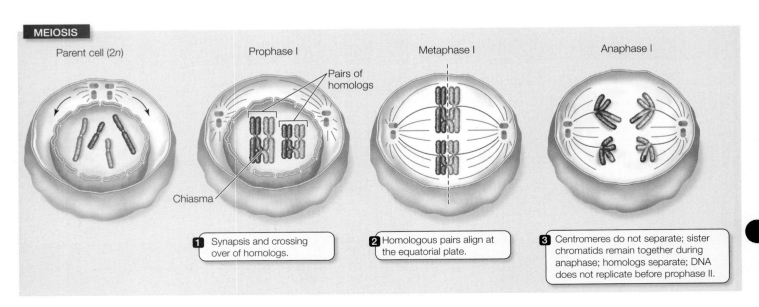

MEIOSIS

Parent cell (2n) — Prophase I — Metaphase I — Anaphase I

Pairs of homologs

Chiasma

1 Synapsis and crossing over of homologs.

2 Homologous pairs align at the equatorial plate.

3 Centromeres do not separate; sister chromatids remain together during anaphase; homologs separate; DNA does not replicate before prophase II.

9.19 Mitosis and Meiosis: A Comparison *(Page 200)*

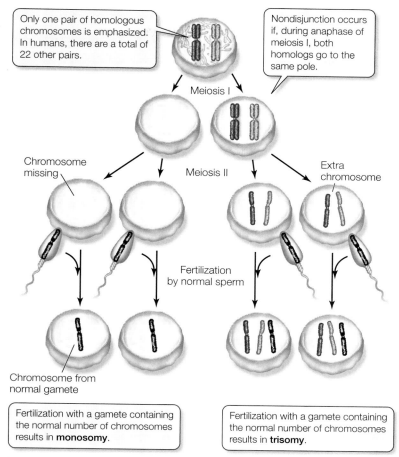

Only one pair of homologous chromosomes is emphasized. In humans, there are a total of 22 other pairs.

Nondisjunction occurs if, during anaphase of meiosis I, both homologs go to the same pole.

Meiosis I

Chromosome missing

Meiosis II

Extra chromosome

Fertilization by normal sperm

Chromosome from normal gamete

Fertilization with a gamete containing the normal number of chromosomes results in **monosomy**.

Fertilization with a gamete containing the normal number of chromosomes results in **trisomy**.

9.20 Nondisjunction Leads to Aneuploidy *(Page 201)*

Two daughter cells (each 2n)

2n   2n

Mitosis is a mechanism for constancy: The parent nucleus produces two identical daughter nuclei.

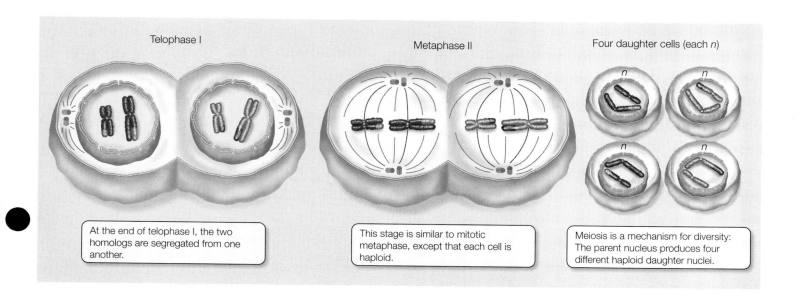

Telophase I

Metaphase II

Four daughter cells (each n)

n   n

n   n

At the end of telophase I, the two homologs are segregated from one another.

This stage is similar to mitotic metaphase, except that each cell is haploid.

Meiosis is a mechanism for diversity: The parent nucleus produces four different haploid daughter nuclei.

TABLE 9.2		
Two Different Ways for Cells to Die		
	NECROSIS	**APOPTOSIS**
Stimuli	Low O_2, toxins, ATP depletion, damage	Specific, genetically programmed physiological signals
ATP required	No	Yes
Cellular pattern	Swelling, organelle disruption, tissue death	Chromatin condensation, membrane blebbing, single-cell death
DNA breakdown	Random fragments	Nucleosome-sized fragments
Plasma membrane	Bursts	Blebbed (see Figure 9.21A)
Fate of dead cells	Ingested by white blood cells	Ingested by neighboring cells
Reaction in tissue	Inflammation	No inflammation

(Page 202)

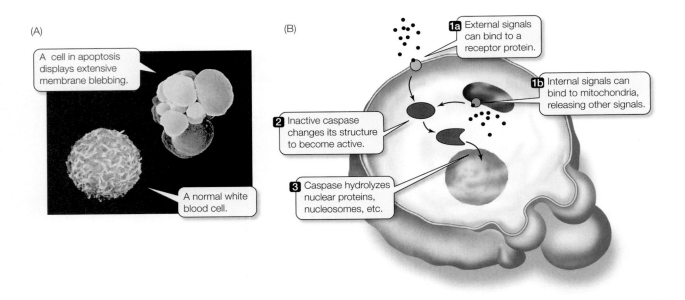

(A)

A cell in apoptosis displays extensive membrane blebbing.

A normal white blood cell.

(B)

1a External signals can bind to a receptor protein.

1b Internal signals can bind to mitochondria, releasing other signals.

2 Inactive caspase changes its structure to become active.

3 Caspase hydrolyzes nuclear proteins, nucleosomes, etc.

9.21 Apoptosis: Programmed Cell Death *(Page 203)*

CHAPTER 10 Genetics: Mendel and Beyond

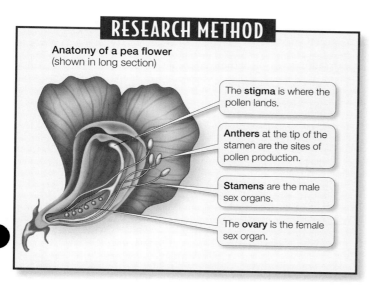

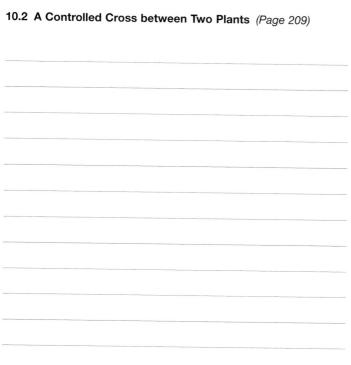

RESEARCH METHOD

Anatomy of a pea flower
(shown in long section)

The **stigma** is where the pollen lands.

Anthers at the tip of the stamen are the sites of pollen production.

Stamens are the male sex organs.

The **ovary** is the female sex organ.

10.2 A Controlled Cross between Two Plants *(Page 209)*

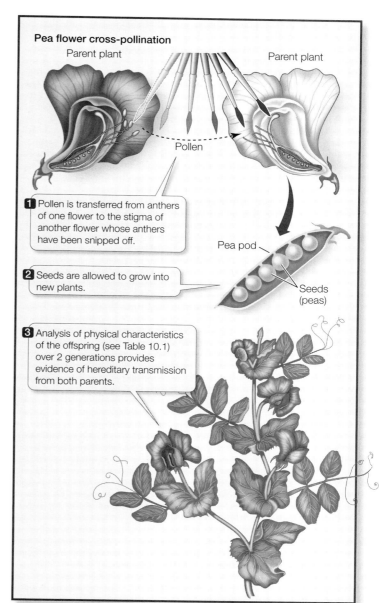

Pea flower cross-pollination

Parent plant

Parent plant

Pollen

1 Pollen is transferred from anthers of one flower to the stigma of another flower whose anthers have been snipped off.

Pea pod

2 Seeds are allowed to grow into new plants.

Seeds (peas)

3 Analysis of physical characteristics of the offspring (see Table 10.1) over 2 generations provides evidence of hereditary transmission from both parents.

TABLE 10.1

Mendel's Results from Monohybrid Crosses

PARENTAL GENERATION PHENOTYPES			F₂ GENERATION PHENOTYPES			
DOMINANT	RECESSIVE		DOMINANT	RECESSIVE	TOTAL	RATIO
Spherical seeds × Wrinkled seeds			5,474	1,850	7,324	2.96:1
Yellow seeds × Green seeds			6,022	2,001	8,023	3.01:1
 Purple flowers × White flowers			 705	224	929	3.15:1
 Inflated pods × Constricted pods			 882	299	1,181	2.95:1
 Green pods × Yellow pods			 428	152	580	2.82:1
 Axial flowers × Terminal flowers			 651	207	858	3.14:1
Tall stems × Dwarf stems (1 m) (0.3 m)			 787	277	1,064	2.84:1

(Page 210)

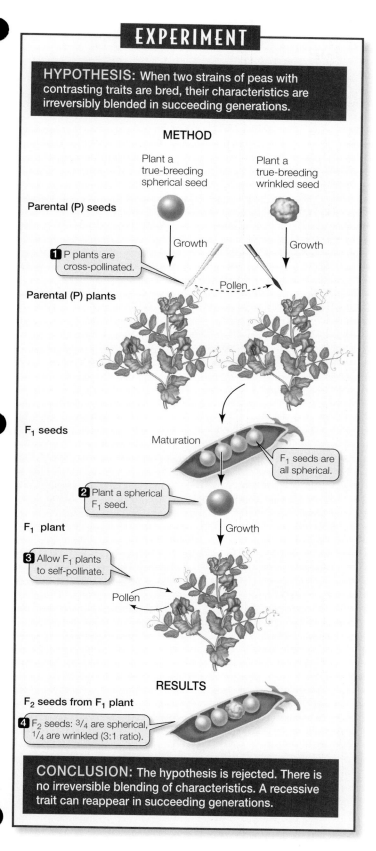

EXPERIMENT

HYPOTHESIS: When two strains of peas with contrasting traits are bred, their characteristics are irreversibly blended in succeeding generations.

METHOD

Plant a true-breeding spherical seed

Plant a true-breeding wrinkled seed

Parental (P) seeds

1 P plants are cross-pollinated.

Parental (P) plants

Growth   Growth

Pollen

F₁ seeds

Maturation

F₁ seeds are all spherical.

2 Plant a spherical F₁ seed.

F₁ plant

3 Allow F₁ plants to self-pollinate.

Growth

Pollen

RESULTS

F₂ seeds from F₁ plant

4 F₂ seeds: 3/4 are spherical, 1/4 are wrinkled (3:1 ratio).

CONCLUSION: The hypothesis is rejected. There is no irreversible blending of characteristics. A recessive trait can reappear in succeeding generations.

10.3 Mendel's Monohybrid Experiments *(Page 211)*

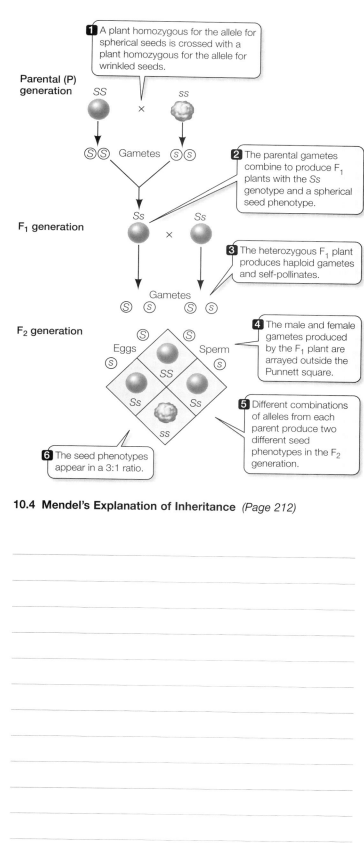

1 A plant homozygous for the allele for spherical seeds is crossed with a plant homozygous for the allele for wrinkled seeds.

Parental (P) generation

SS × ss

Gametes

2 The parental gametes combine to produce F₁ plants with the Ss genotype and a spherical seed phenotype.

F₁ generation

Ss × Ss

3 The heterozygous F₁ plant produces haploid gametes and self-pollinates.

Gametes

F₂ generation

4 The male and female gametes produced by the F₁ plant are arrayed outside the Punnett square.

Eggs   Sperm

SS

Ss Ss

ss

5 Different combinations of alleles from each parent produce two different seed phenotypes in the F₂ generation.

6 The seed phenotypes appear in a 3:1 ratio.

10.4 Mendel's Explanation of Inheritance *(Page 212)*

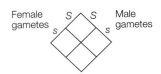

Page 212 In-Text Art

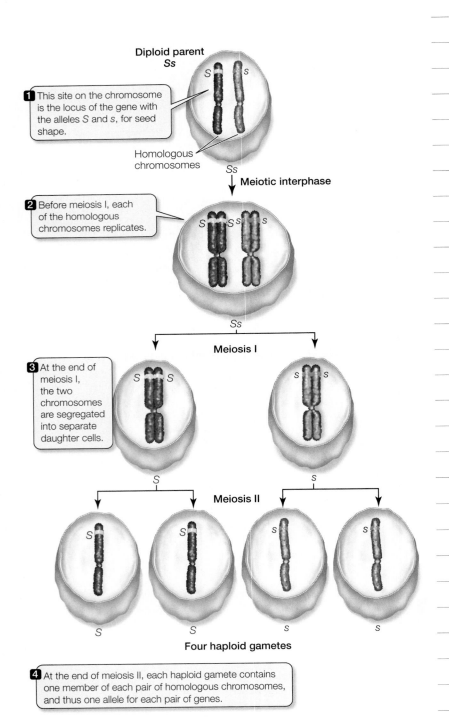

10.5 Meiosis Accounts for the Segregation of Alleles *(Page 212)*

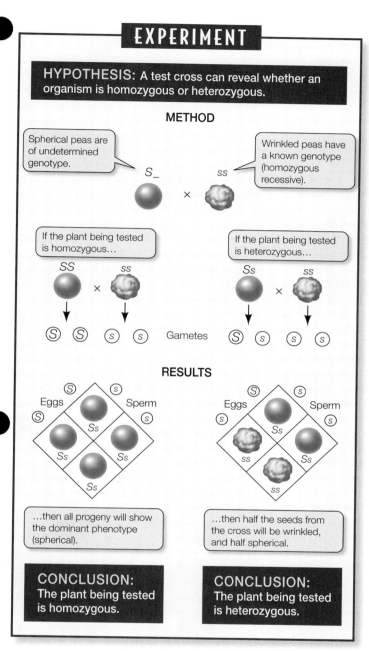

EXPERIMENT

HYPOTHESIS: A test cross can reveal whether an organism is homozygous or heterozygous.

METHOD

Spherical peas are of undetermined genotype.

*S_*

Wrinkled peas have a known genotype (homozygous recessive).

ss

×

If the plant being tested is homozygous…

SS × *ss*

If the plant being tested is heterozygous…

Ss × *ss*

Ⓢ Ⓢ Ⓢ Ⓢ   Gametes   Ⓢ Ⓢ Ⓢ Ⓢ

RESULTS

Ⓢ Ⓢ
Eggs   Sperm
Ⓢ
Ss
Ss   *Ss*
Ss

Ⓢ Ⓢ
Eggs   Sperm
Ⓢ
Ss
ss   *Ss*
ss

…then all progeny will show the dominant phenotype (spherical).

…then half the seeds from the cross will be wrinkled, and half spherical.

CONCLUSION: The plant being tested is homozygous.

CONCLUSION: The plant being tested is heterozygous.

10.6 Homozygous or Heterozygous? *(Page 213)*

Parental (P) generation

SSYY × *ssyy*

SsYy

F₁ generation

Ⓢ*Y* Ⓢ*y* Ⓢ*Y* Ⓢ*y*
Gametes

When F₁ plants self-pollinate, the gametes combine randomly to produce an F₂ generation with four phenotypes in a 9:3:3:1 ratio.

F₂ generation

Ⓢ*Y* Ⓢ*Y*
Eggs Ⓢ*y*   Ⓢ*y* Sperm
Ⓢ*Y*   Ⓢ*Y*
Ⓢ*y*   Ⓢ*y*

SSYY
SSYy   *SSYy*
SsYY   *SSyy*   *SsYY*
SsYy   *SsYy*   *SsYy*   *SsYy*
Ssyy   *ssYY*   *Ssyy*
ssYy   *ssYy*
ssyy

10.7 Independent Assortment *(Page 214)*

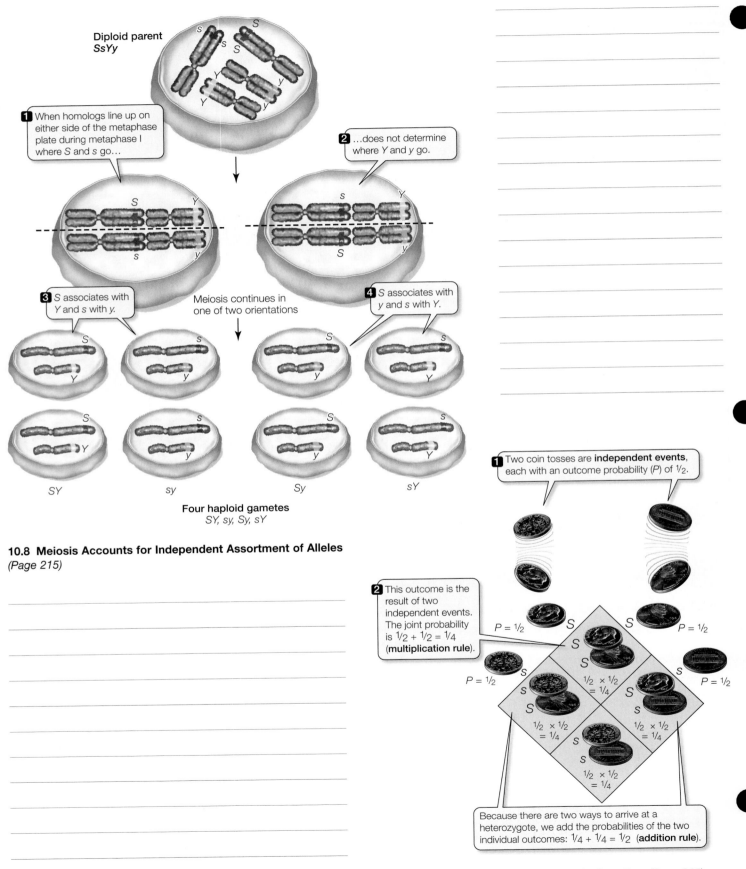

Diploid parent
SsYy

1 When homologs line up on either side of the metaphase plate during metaphase I where *S* and *s* go...

2 ...does not determine where *Y* and *y* go.

Meiosis continues in one of two orientations

3 *S* associates with *Y* and *s* with *y.*

4 *S* associates with *y* and *s* with *Y.*

SY   sy   Sy   sY

Four haploid gametes
SY, sy, Sy, sY

10.8 Meiosis Accounts for Independent Assortment of Alleles
(Page 215)

1 Two coin tosses are **independent events**, each with an outcome probability (*P*) of ¹/₂.

2 This outcome is the result of two independent events. The joint probability is ¹/₂ + ¹/₂ = ¹/₄ (**multiplication rule**).

$P = $ ¹/₂ $P = $ ¹/₂

$P = $ ¹/₂ $P = $ ¹/₂

¹/₂ × ¹/₂ = ¹/₄

¹/₂ × ¹/₂ = ¹/₄   ¹/₂ × ¹/₂ = ¹/₄

¹/₂ × ¹/₂ = ¹/₄

Because there are two ways to arrive at a heterozygote, we add the probabilities of the two individual outcomes: ¹/₄ + ¹/₄ = ¹/₂ (**addition rule**).

10.9 Using Probability Calculations in Genetics *(Page 215)*

(A) Dominant inheritance

Generation I (parents)

Generation II

Generation III

> Every affected individual has an affected parent.

> About ½ of the offspring (of both sexes) of an affected parent are affected.

Oldest   Youngest

Siblings

(B) Recessive inheritance

1 One parent is heterozygous…

2 …and the recessive allele is passed on to ½ of the phenotypically normal offspring.

Generation I (parents)

Generation II

Generation III

3 These cousins are both heterozygous.

Generation IV

4 Mating of heterozygous recessive parents may produce homozygous recessive (affected) offspring.

	Unaffected	Affected	Heterozygote (unaffected phenotype)
Female			
Male			
Mating			
Mating between relatives			

10.10 Pedigree Analysis and Inheritance *(Page 216)*

Possible genotypes	CC, Cc^{ch}, Cc^h, Cc	$c^{ch}c^{ch}$	$c^{ch}c^h, c^{ch}c$	c^hc^h, c^hc	cc
Phenotype	Dark gray	Chinchilla	Light gray	Point restricted	Albino

10.11 Inheritance of Coat Color in Rabbits *(Page 218)*

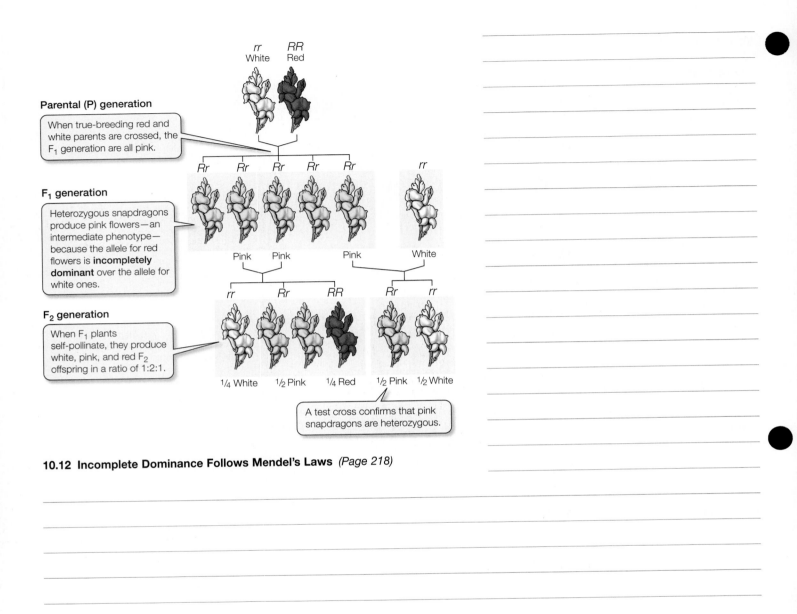

Parental (P) generation

When true-breeding red and white parents are crossed, the F₁ generation are all pink.

F₁ generation

Heterozygous snapdragons produce pink flowers—an intermediate phenotype—because the allele for red flowers is **incompletely dominant** over the allele for white ones.

F₂ generation

When F₁ plants self-pollinate, they produce white, pink, and red F₂ offspring in a ratio of 1:2:1.

A test cross confirms that pink snapdragons are heterozygous.

10.12 Incomplete Dominance Follows Mendel's Laws *(Page 218)*

Blood type of cells	Genotype	Antibodies made by body	Reaction to added antibodies	
			Anti-A	Anti-B
A	$I^A I^A$ or $I^A i^o$	Anti-B		
B	$I^B I^B$ or $I^B i^o$	Anti-A		
AB	$I^A I^B$	Neither anti-A nor anti-B		
O	$i^o i^o$	Both anti-A and anti-B		

Red blood cells that do not react with antibody remain evenly dispersed.

Red blood cells that react with antibody clump together (speckled appearance).

10.13 ABO Blood Reactions Are Important in Transfusions *(Page 219)*

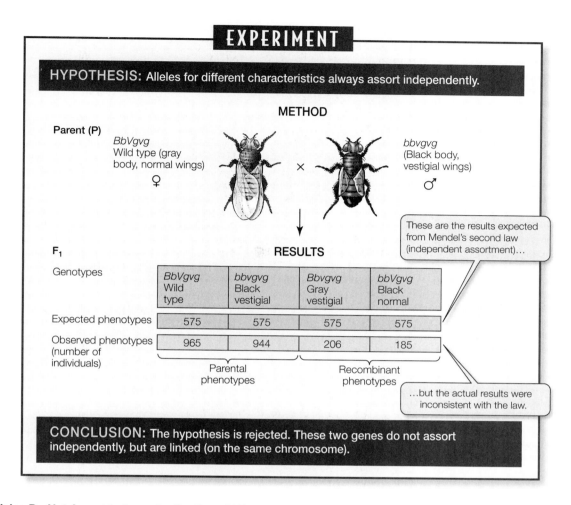

10.18 Some Alleles Do Not Assort Independently *(Page 222)*

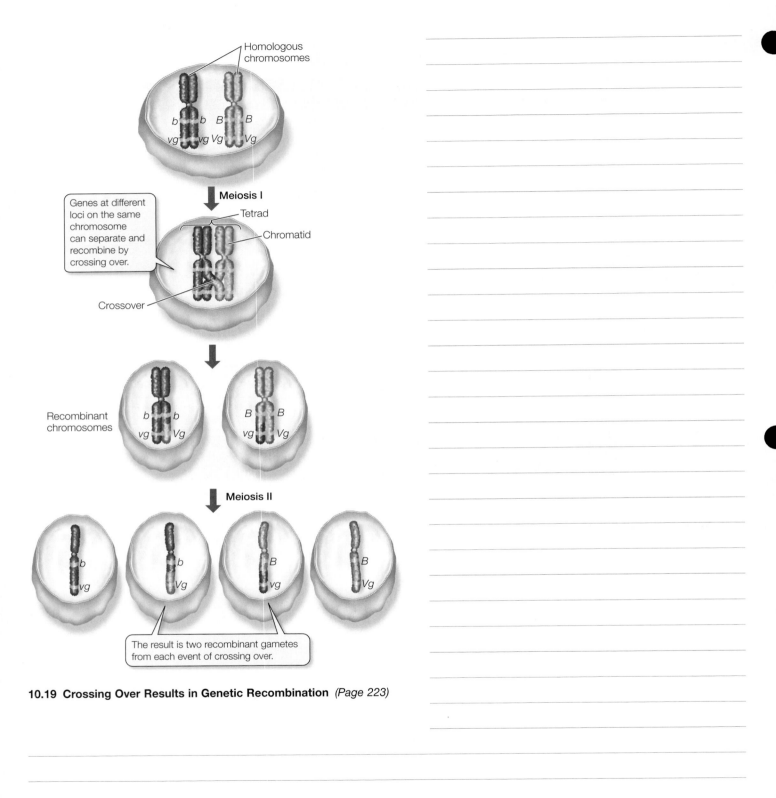

10.19 Crossing Over Results in Genetic Recombination *(Page 223)*

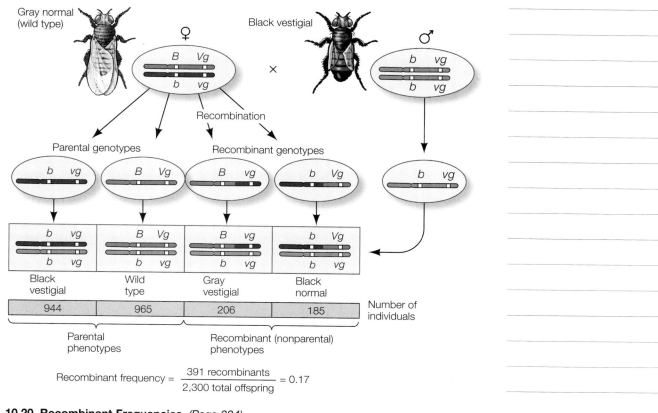

Recombinant frequency = $\dfrac{391\ \text{recombinants}}{2{,}300\ \text{total offspring}} = 0.17$

10.20 Recombinant Frequencies *(Page 224)*

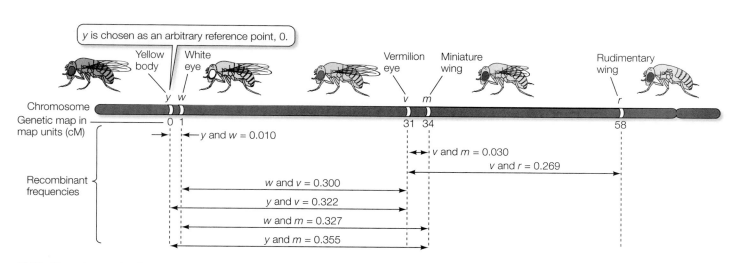

10.21 Steps toward a Genetic Map *(Page 224)*

1 At the outset, we have no idea of the individual distances between the genes, and there are several possible sequences (*a-b-c*, *a-c-b*, *b-a-c*).

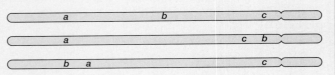

We make a cross *AABB* × *aabb*, and obtain an F_1 generation with a genotype *AaBb*. We test cross these *AaBb* individuals with *aabb*. Here are the genotypes of the first 1,000 progeny:

450 *AaBb*, 450 *aabb*, 50 *Aabb*, and 50 *aaBb*.
(parental types)　　(recombinant types)

2 How far apart are the *a* and *b* genes?

What is the recombinant frequency? Which are the recombinant types, and which are the parental types?

Recombinant frequency (*a* to *b*) = (50 + 50)/1,000 = 0.1
So the map distance is

Map distance = 100 × recombinant frequency =
100 × 0.1 = 10 cM

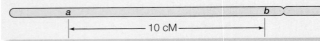

3 How far apart are the *a* and *c* genes?

Now we make a cross *AACC* × *aacc*, obtain an F_1 generation, and test cross it, obtaining

460 *AaCc*, 460 *aacc*, 40 *Aacc*, and 40 *aaCc*

Recombinant frequency (*a* to *c*) = (40 + 40)/1,000 = 0.08

Map distance = 100 × recombinant frequency =
100 × 0.08 = 8 cM

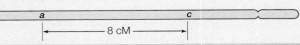

4 How far apart are the *b* and *c* genes?

We make a cross *BBCC* × *bbcc*, obtain an F_1 generation, and test cross it, obtaining

490 *BbCc*, 490 *bbcc*, 10 *Bbcc*, and 10 *bbCc*

Recombinant frequency (*b* to *c*) = (10 + 10)/1,000 = 0.02

Map distance = 100 × recombinant frequency =
100 × 0.02 = 2 cM

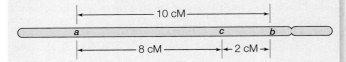

5 Which of the three genes is between the other two?
Because *a* and *b* are the farthest apart, *c* must be between them.

These numbers add up perfectly. In most real cases, they will not add up perfectly because of multiple crossovers.

10.22 Map These Genes *(Page 225)*

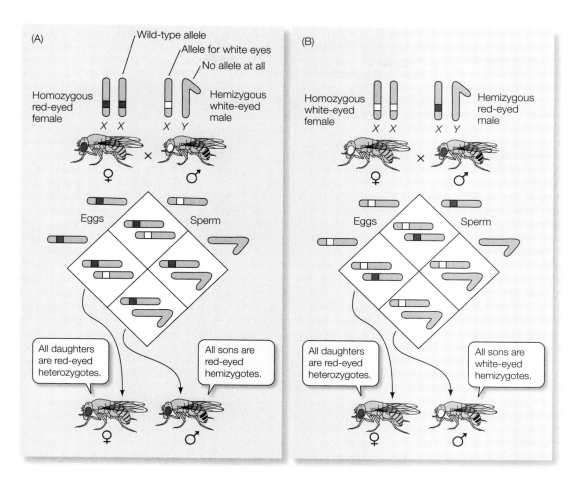

10.23 Eye Color Is a Sex-Linked Trait in _Drosophila_ _(Page 227)_

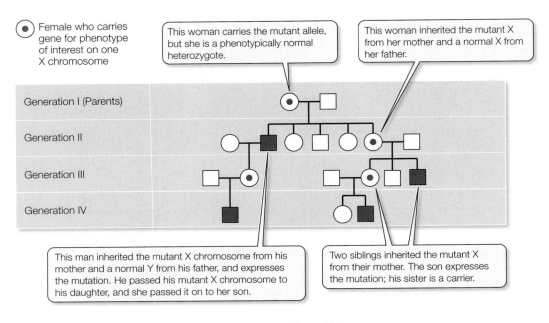

10.24 Red-Green Color Blindness Is a Sex-Linked Trait in Humans *(Page 228)*

11 DNA and Its Role in Heredity

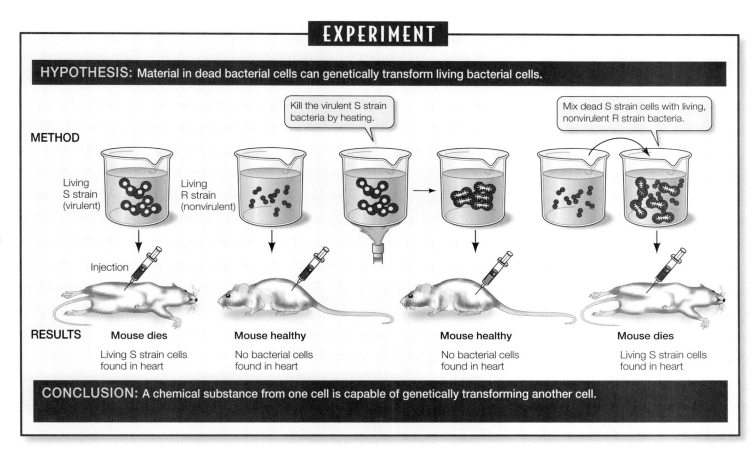

EXPERIMENT

HYPOTHESIS: Material in dead bacterial cells can genetically transform living bacterial cells.

Kill the virulent S strain bacteria by heating.

Mix dead S strain cells with living, nonvirulent R strain bacteria.

METHOD

Living S strain (virulent)   Living R strain (nonvirulent)

Injection

RESULTS   Mouse dies   Mouse healthy   Mouse healthy   Mouse dies

Living S strain cells found in heart   No bacterial cells found in heart   No bacterial cells found in heart   Living S strain cells found in heart

CONCLUSION: A chemical substance from one cell is capable of genetically transforming another cell.

11.1 Genetic Transformation of Nonvirulent Pneumococci *(Page 234)*

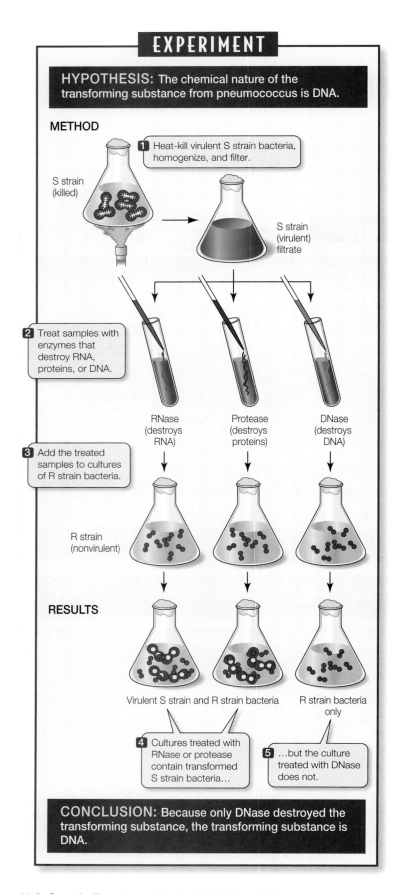

EXPERIMENT

HYPOTHESIS: The chemical nature of the transforming substance from pneumococcus is DNA.

METHOD

1 Heat-kill virulent S strain bacteria, homogenize, and filter.

S strain (killed)

S strain (virulent) filtrate

2 Treat samples with enzymes that destroy RNA, proteins, or DNA.

RNase (destroys RNA)

Protease (destroys proteins)

DNase (destroys DNA)

3 Add the treated samples to cultures of R strain bacteria.

R strain (nonvirulent)

RESULTS

Virulent S strain and R strain bacteria

R strain bacteria only

4 Cultures treated with RNase or protease contain transformed S strain bacteria...

5 ...but the culture treated with DNase does not.

CONCLUSION: Because only DNase destroyed the transforming substance, the transforming substance is DNA.

11.2 Genetic Transformation by DNA *(Page 235)*

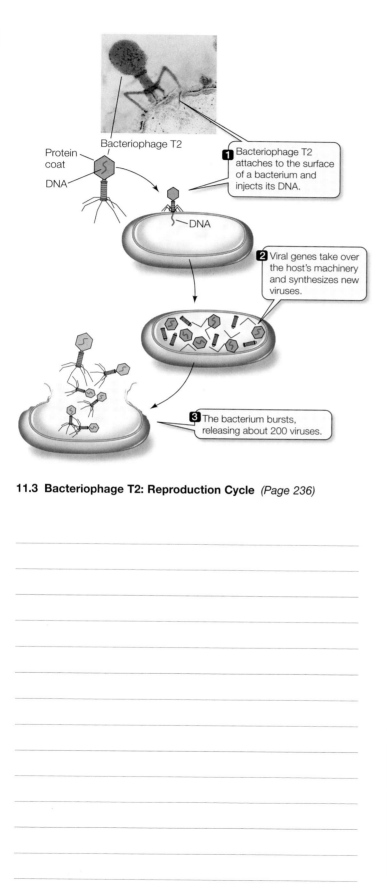

11.3 Bacteriophage T2: Reproduction Cycle *(Page 236)*

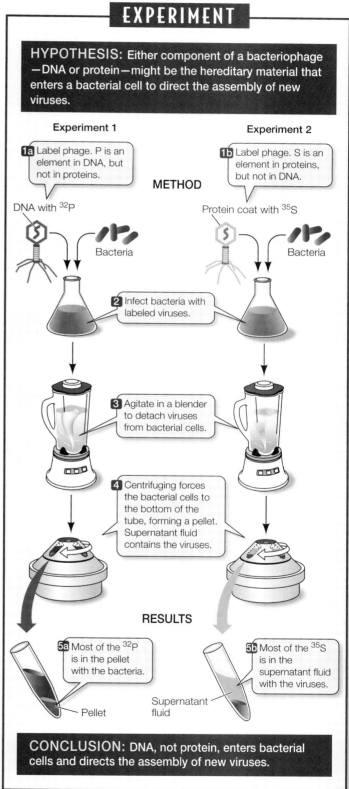

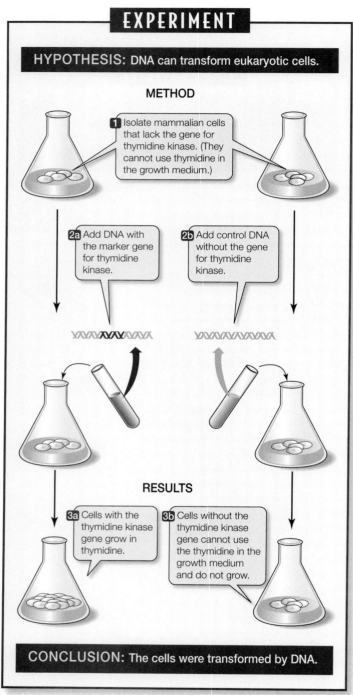

11.5 Transfection in Eukaryotic Cells *(Page 237)*

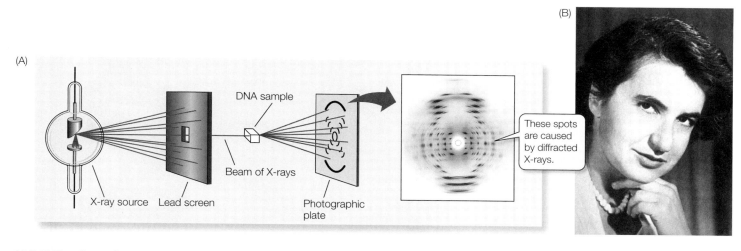

11.6 X-Ray Crystallography Helped Reveal the Structure of DNA *(Page 238)*

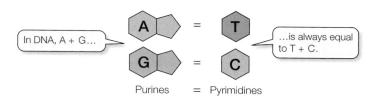

11.7 Chargaff's Rule *(Page 239)*

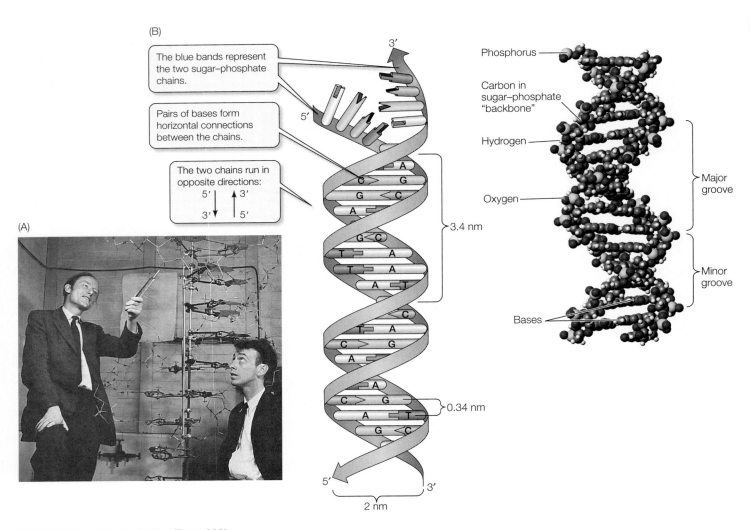

(B)

The blue bands represent the two sugar–phosphate chains.

Pairs of bases form horizontal connections between the chains.

The two chains run in opposite directions:

5′ → 3′
3′ ← 5′

3.4 nm

0.34 nm

2 nm

(A)

Phosphorus

Carbon in sugar–phosphate "backbone"

Hydrogen

Oxygen

Bases

Major groove

Minor groove

11.8 DNA Is a Double Helix *(Page 239)*

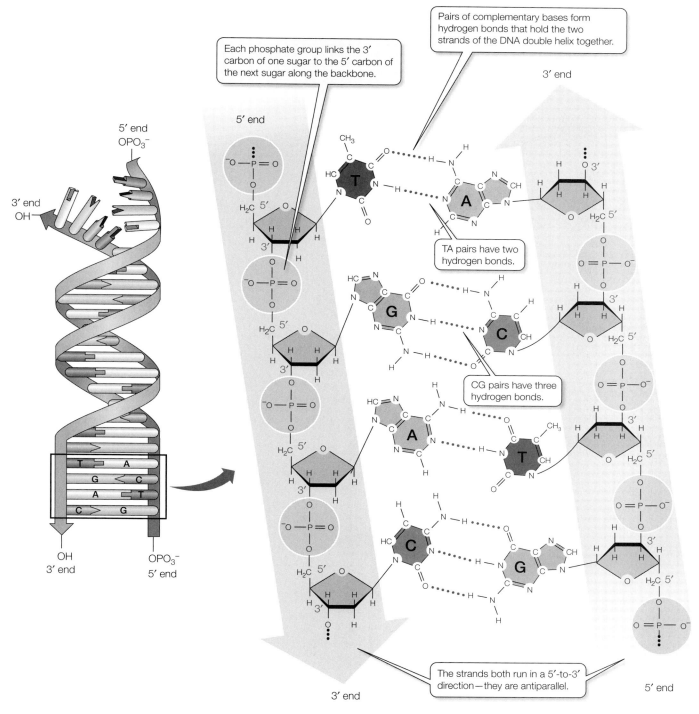

11.9 Base Pairing in DNA Is Complementary *(Page 240)*

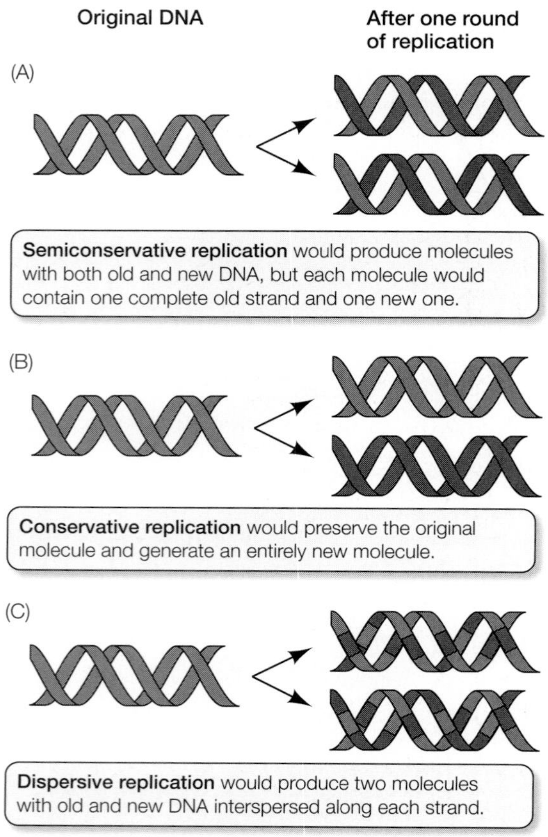

Original DNA   After one round
                                of replication

(A)

Semiconservative replication would produce molecules
with both old and new DNA, but each molecule would
contain one complete old strand and one new one.

(B)

Conservative replication would preserve the original
molecule and generate an entirely new molecule.

(C)

Dispersive replication would produce two molecules
with old and new DNA interspersed along each strand.

11.10 Three Models for DNA Replication *(Page 242)*

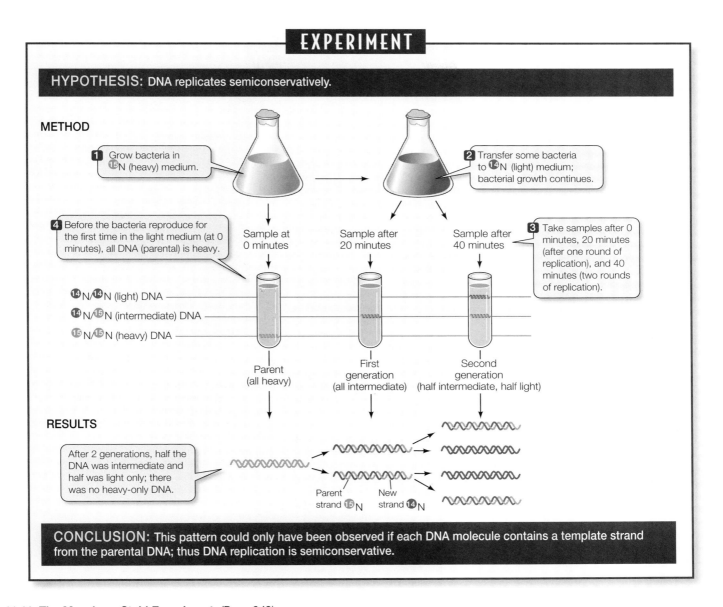

11.11 The Meselson–Stahl Experiment *(Page 243)*

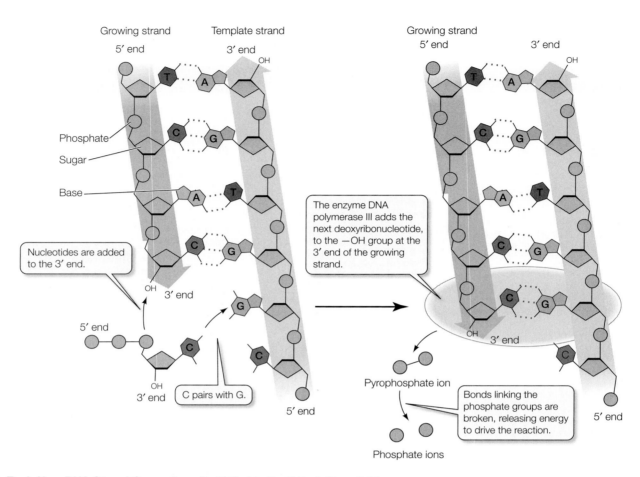

11.12 Each New DNA Strand Grows from Its 5′ End to Its 3′ End *(Page 244)*

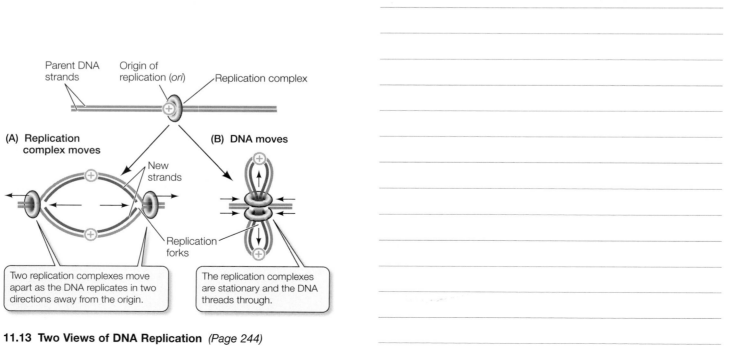

11.13 Two Views of DNA Replication *(Page 244)*

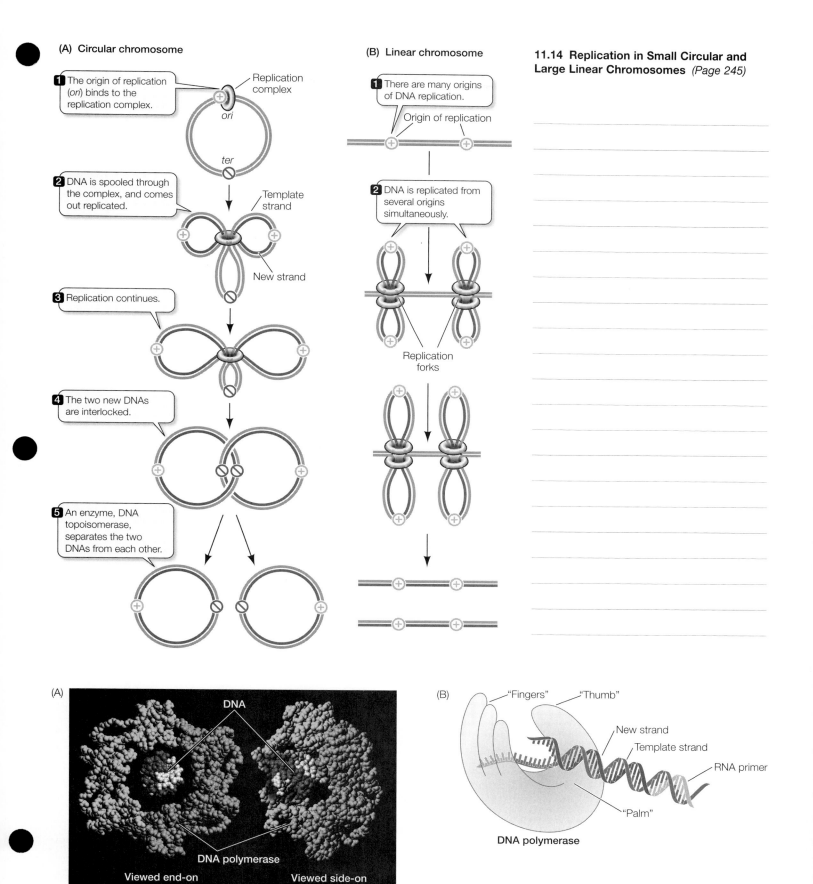

(A) Circular chromosome

1 The origin of replication (*ori*) binds to the replication complex.

Replication complex

ori

ter

2 DNA is spooled through the complex, and comes out replicated.

Template strand

New strand

3 Replication continues.

4 The two new DNAs are interlocked.

5 An enzyme, DNA topoisomerase, separates the two DNAs from each other.

(B) Linear chromosome

1 There are many origins of DNA replication.

Origin of replication

2 DNA is replicated from several origins simultaneously.

Replication forks

11.14 Replication in Small Circular and Large Linear Chromosomes (Page 245)

(A)

DNA

DNA polymerase

Viewed end-on   Viewed side-on

(B)

"Fingers"   "Thumb"

New strand

Template strand

RNA primer

"Palm"

DNA polymerase

11.15 DNA Polymerase Binds to the Template Strand (Page 245)

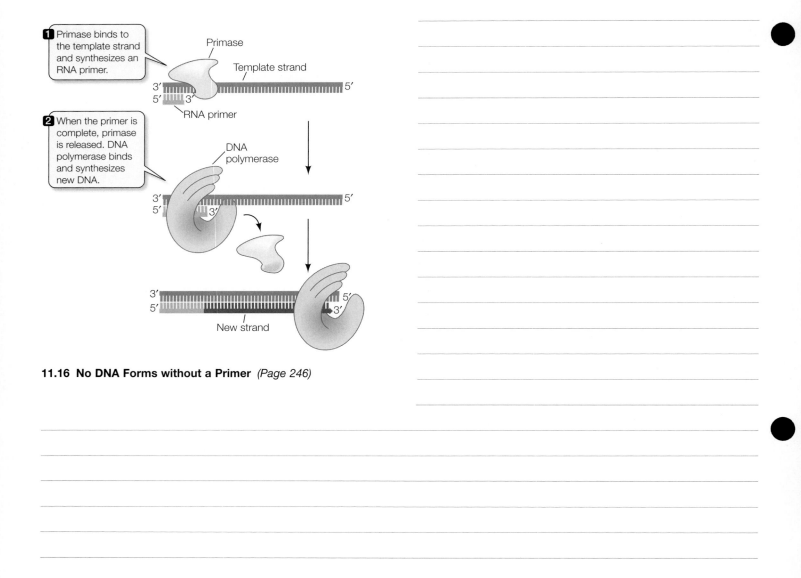

1 Primase binds to the template strand and synthesizes an RNA primer.

Primase

Template strand

3′
5′   3′

RNA primer

2 When the primer is complete, primase is released. DNA polymerase binds and synthesizes new DNA.

DNA polymerase

3′
5′   3′

3′
5′

New strand

11.16 No DNA Forms without a Primer *(Page 246)*

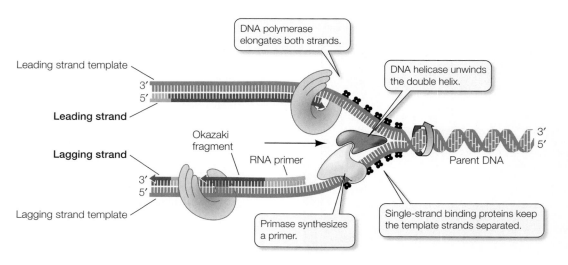

DNA polymerase elongates both strands.

DNA helicase unwinds the double helix.

Leading strand template

3′
5′

Leading strand

Lagging strand

Okazaki fragment

RNA primer

3′
5′

Parent DNA

Lagging strand

3′
5′

Lagging strand template

Primase synthesizes a primer.

Single-strand binding proteins keep the template strands separated.

11.17 Many Proteins Collaborate in the Replication Complex *(Page 246)*

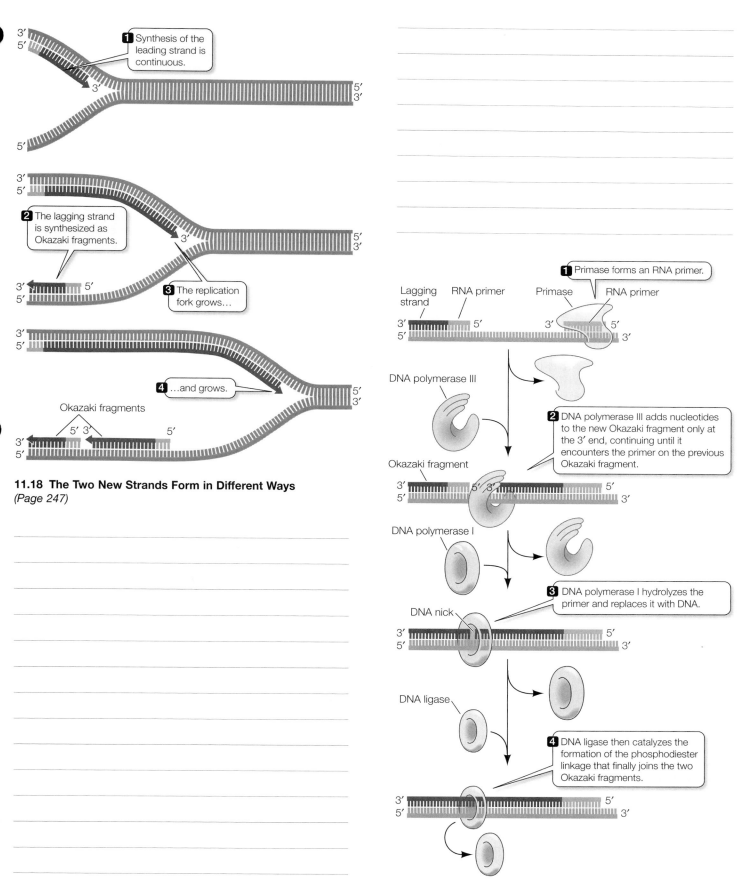

11.18 The Two New Strands Form in Different Ways
(Page 247)

11.19 The Lagging Strand Story *(Page 247)*

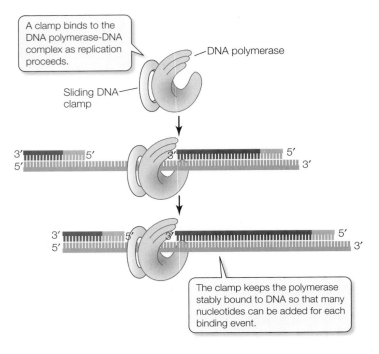

11.20 A Sliding DNA Clamp Increases the Efficiency of DNA Polymerization *(Page 248)*

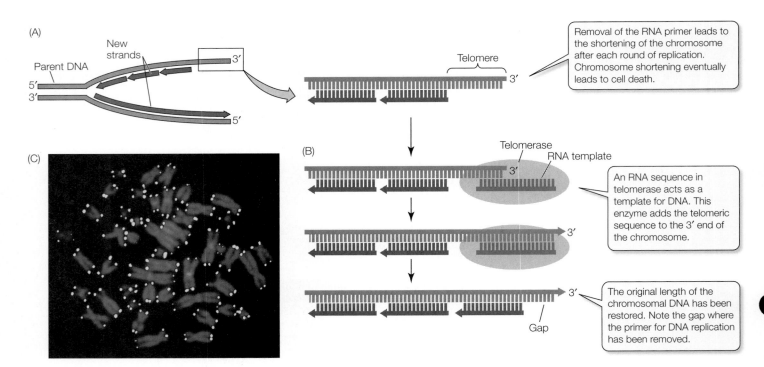

11.21 Telomeres and Telomerase *(Page 248)*

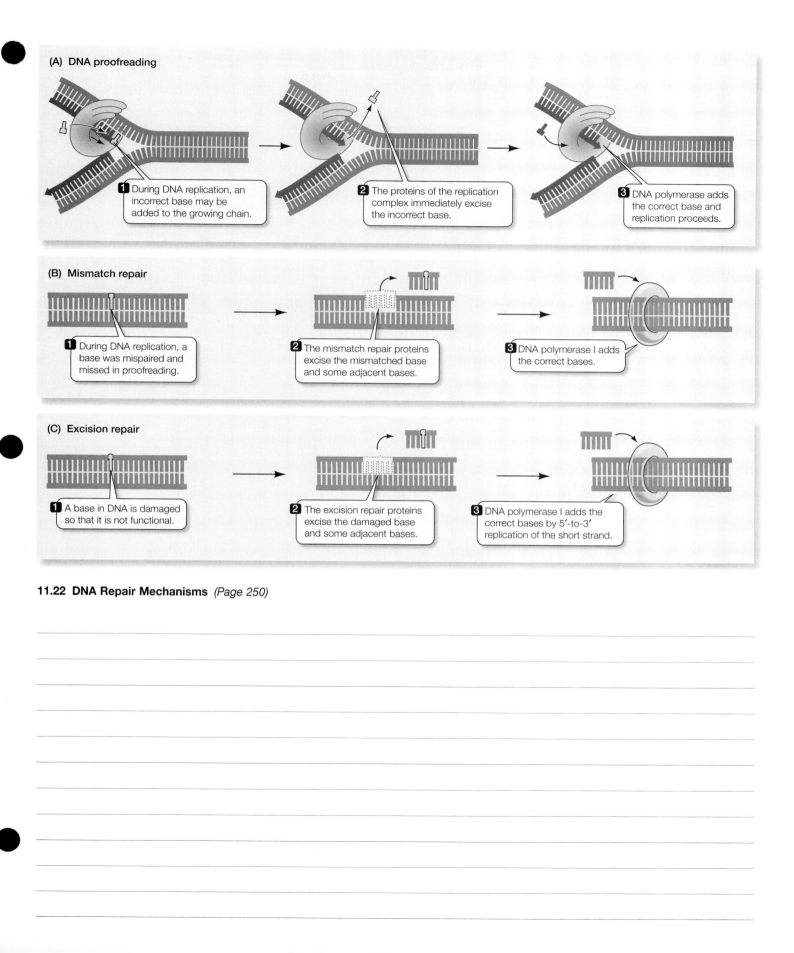

11.22 DNA Repair Mechanisms *(Page 250)*

RESEARCH METHOD

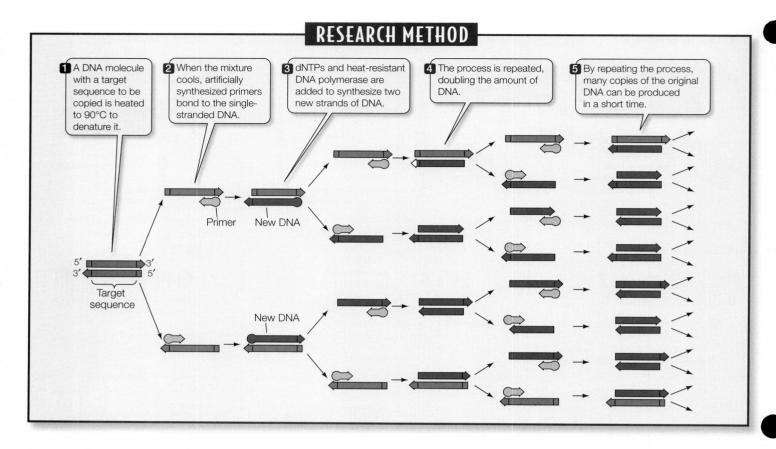

1 A DNA molecule with a target sequence to be copied is heated to 90°C to denature it.

2 When the mixture cools, artificially synthesized primers bond to the single-stranded DNA.

3 dNTPs and heat-resistant DNA polymerase are added to synthesize two new strands of DNA.

4 The process is repeated, doubling the amount of DNA.

5 By repeating the process, many copies of the original DNA can be produced in a short time.

Primer   New DNA

5′   3′
3′   5′

Target sequence

New DNA

11.23 The Polymerase Chain Reaction *(Page 251)*

RESEARCH METHOD

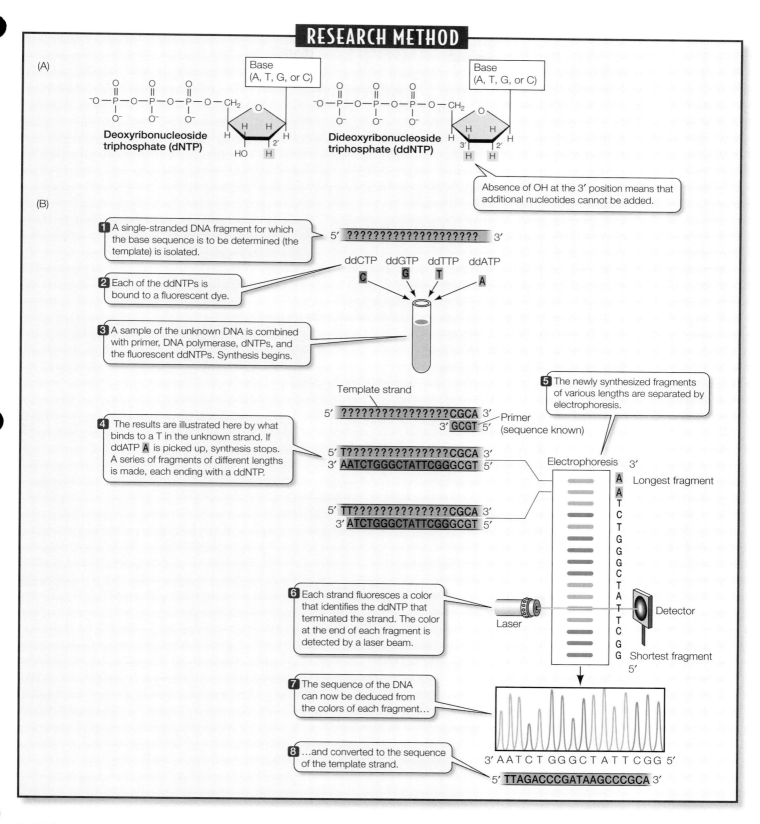

(A)

Deoxyribonucleoside triphosphate (dNTP)

Base (A, T, G, or C)

Dideoxyribonucleoside triphosphate (ddNTP)

Base (A, T, G, or C)

Absence of OH at the 3′ position means that additional nucleotides cannot be added.

(B)

1 A single-stranded DNA fragment for which the base sequence is to be determined (the template) is isolated.

5′ ???????????????????? 3′

ddCTP   ddGTP   ddTTP   ddATP

C   G   T   A

2 Each of the ddNTPs is bound to a fluorescent dye.

3 A sample of the unknown DNA is combined with primer, DNA polymerase, dNTPs, and the fluorescent ddNTPs. Synthesis begins.

Template strand

5′ ????????????????CGCA 3′
3′ GCGT 5′   Primer (sequence known)

4 The results are illustrated here by what binds to a T in the unknown strand. If ddATP A is picked up, synthesis stops. A series of fragments of different lengths is made, each ending with a ddNTP.

5′ T???????????????CGCA 3′
3′ AATCTGGGCTATTCGGGCGT 5′

5′ TT??????????????CGCA 3′
3′ ATCTGGGCTATTCGGGCGT 5′

5 The newly synthesized fragments of various lengths are separated by electrophoresis.

Electrophoresis   3′

A   Longest fragment
A
T
C
T
G
G
G
C
T
A
T
T
C
G
G   Shortest fragment
5′

Laser

Detector

6 Each strand fluoresces a color that identifies the ddNTP that terminated the strand. The color at the end of each fragment is detected by a laser beam.

7 The sequence of the DNA can now be deduced from the colors of each fragment…

8 …and converted to the sequence of the template strand.

3′ A A T C T G G G C T A T T C G G 5′
5′ TTAGACCCGATAAGCCCGCA 3′

11.24 Sequencing DNA *(Page 252)*

CHAPTER 12 From DNA to Protein: Genotype to Phenotype

EXPERIMENT

HYPOTHESIS: Genes determine enzymes in a biochemical pathway.

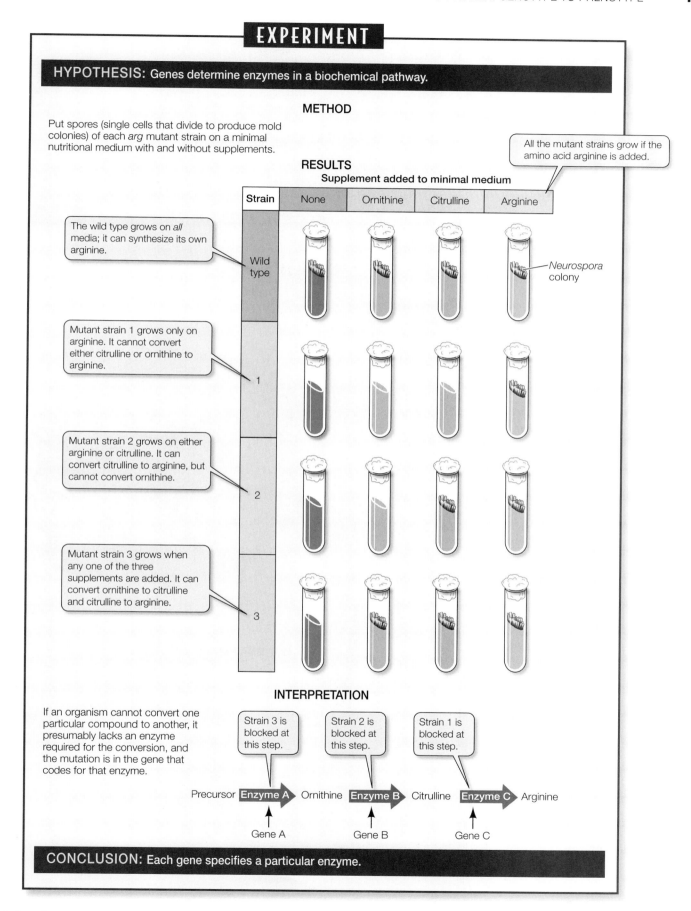

METHOD

Put spores (single cells that divide to produce mold colonies) of each *arg* mutant strain on a minimal nutritional medium with and without supplements.

All the mutant strains grow if the amino acid arginine is added.

RESULTS

Supplement added to minimal medium

The wild type grows on *all* media; it can synthesize its own arginine.

Mutant strain 1 grows only on arginine. It cannot convert either citrulline or ornithine to arginine.

Mutant strain 2 grows on either arginine or citrulline. It can convert citrulline to arginine, but cannot convert ornithine.

Mutant strain 3 grows when any one of the three supplements are added. It can convert ornithine to citrulline and citrulline to arginine.

Neurospora colony

INTERPRETATION

If an organism cannot convert one particular compound to another, it presumably lacks an enzyme required for the conversion, and the mutation is in the gene that codes for that enzyme.

Strain 3 is blocked at this step.

Strain 2 is blocked at this step.

Strain 1 is blocked at this step.

Precursor → Enzyme A → Ornithine → Enzyme B → Citrulline → Enzyme C → Arginine

Gene A   Gene B   Gene C

CONCLUSION: Each gene specifies a particular enzyme.

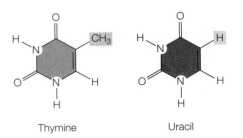

Thymine   Uracil

Page 260 In-Text Art

Page 261 In-Text Art (1)

Page 261 In-Text Art (2)

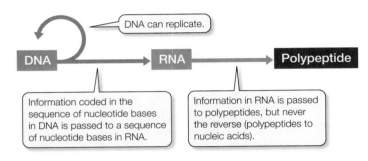

12.2 The Central Dogma *(Page 260)*

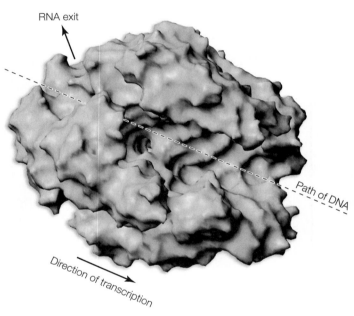

12.4 RNA Polymerase *(Page 262)*

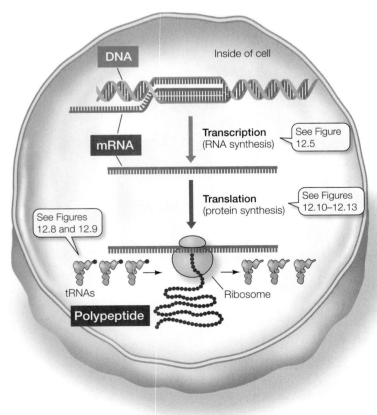

12.3 From Gene to Protein *(Page 261)*

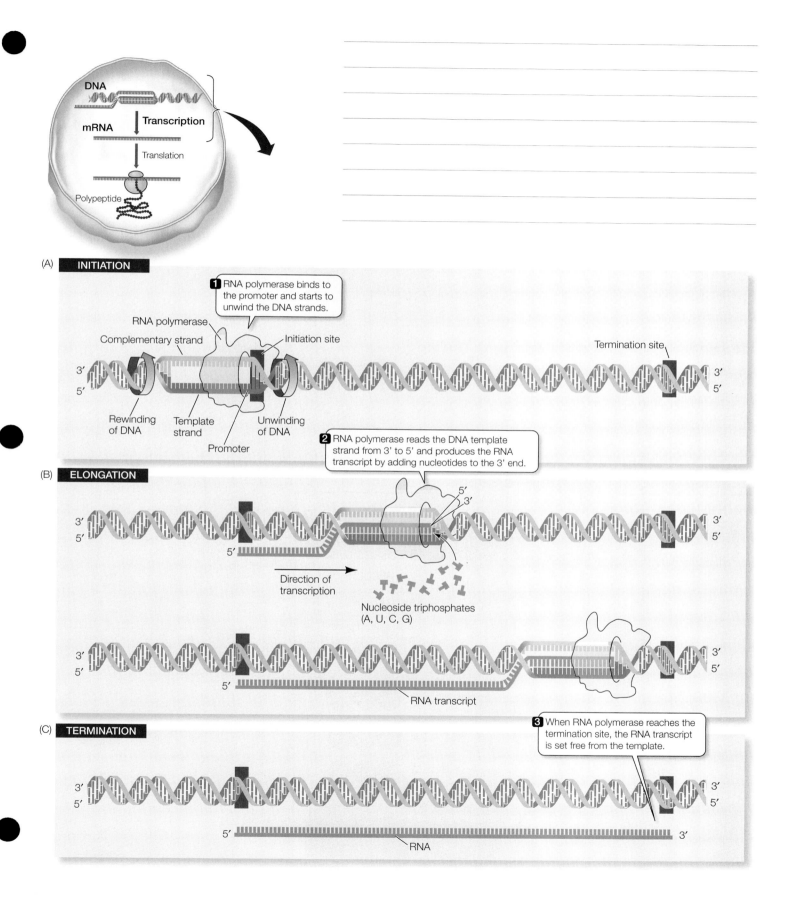

(A) **INITIATION**

1 RNA polymerase binds to the promoter and starts to unwind the DNA strands.

RNA polymerase

Complementary strand

Initiation site

Termination site

3′
5′

3′
5′

Rewinding of DNA

Template strand

Unwinding of DNA

Promoter

2 RNA polymerase reads the DNA template strand from 3′ to 5′ and produces the RNA transcript by adding nucleotides to the 3′ end.

(B) **ELONGATION**

3′
5′

5′
3′

3′
5′

5′

Direction of transcription

Nucleoside triphosphates (A, U, C, G)

3′
5′

3′
5′

5′

RNA transcript

3 When RNA polymerase reaches the termination site, the RNA transcript is set free from the template.

(C) **TERMINATION**

3′
5′

3′
5′

5′

3′

RNA

12.5 DNA Is Transcribed to Form RNA *(Page 263)*

Second letter

		U		C		A		G		
First letter	**U**	UUU UUC	Phenyl-alanine	UCU UCC UCA UCG	Serine	UAU UAC	Tyrosine	UGU UGC	Cysteine	U C A G
		UUA UUG	Leucine			UAA UAG	Stop codon Stop codon	UGA UGG	Stop codon Tryptophan	
	C	CUU CUC CUA CUG	Leucine	CCU CCC CCA CCG	Proline	CAU CAC	Histidine	CGU CGC CGA CGG	Arginine	U C A G
						CAA CAG	Glutamine			
	A	AUU AUC AUA	Isoleucine	ACU ACC ACA ACG	Threonine	AAU AAC	Asparagine	AGU AGC	Serine	U C A G
		AUG	Methionine; start codon			AAA AAG	Lysine	AGA AGG	Arginine	
	G	GUU GUC GUA GUG	Valine	GCU GCC GCA GCG	Alanine	GAU GAC	Aspartic acid	GGU GGC GGA GGG	Glycine	U C A G
						GAA GAG	Glutamic acid			

(Third letter)

12.6 The Genetic Code *(Page 264)*

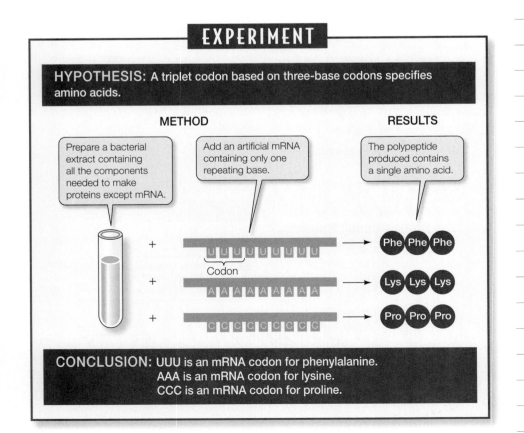

EXPERIMENT

HYPOTHESIS: A triplet codon based on three-base codons specifies amino acids.

METHOD

Prepare a bacterial extract containing all the components needed to make proteins except mRNA.

Add an artificial mRNA containing only one repeating base.

RESULTS

The polypeptide produced contains a single amino acid.

U U U U U U U U U U → Phe Phe Phe

Codon

A A A A A A A A A A → Lys Lys Lys

C C C C C C C C C C → Pro Pro Pro

CONCLUSION: UUU is an mRNA codon for phenylalanine.
AAA is an mRNA codon for lysine.
CCC is an mRNA codon for proline.

12.7 Deciphering the Genetic Code *(Page 265)*

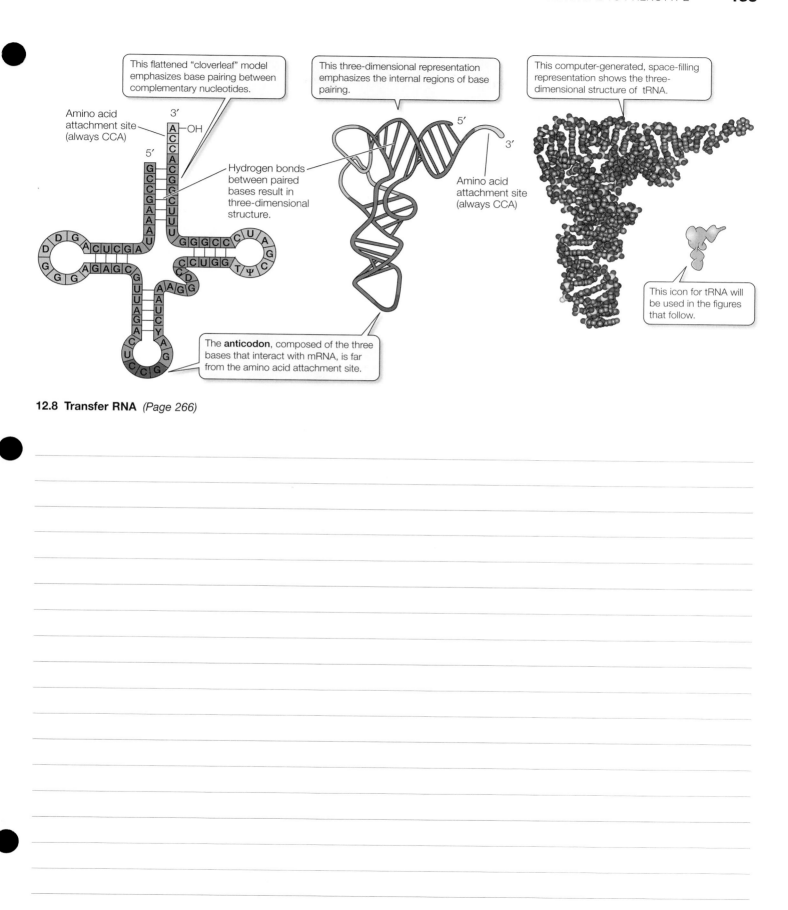

This flattened "cloverleaf" model emphasizes base pairing between complementary nucleotides.

This three-dimensional representation emphasizes the internal regions of base pairing.

This computer-generated, space-filling representation shows the three-dimensional structure of tRNA.

Amino acid attachment site (always CCA)

3′

5′

Hydrogen bonds between paired bases result in three-dimensional structure.

5′

3′

Amino acid attachment site (always CCA)

This icon for tRNA will be used in the figures that follow.

The **anticodon**, composed of the three bases that interact with mRNA, is far from the amino acid attachment site.

12.8 Transfer RNA *(Page 266)*

(A)

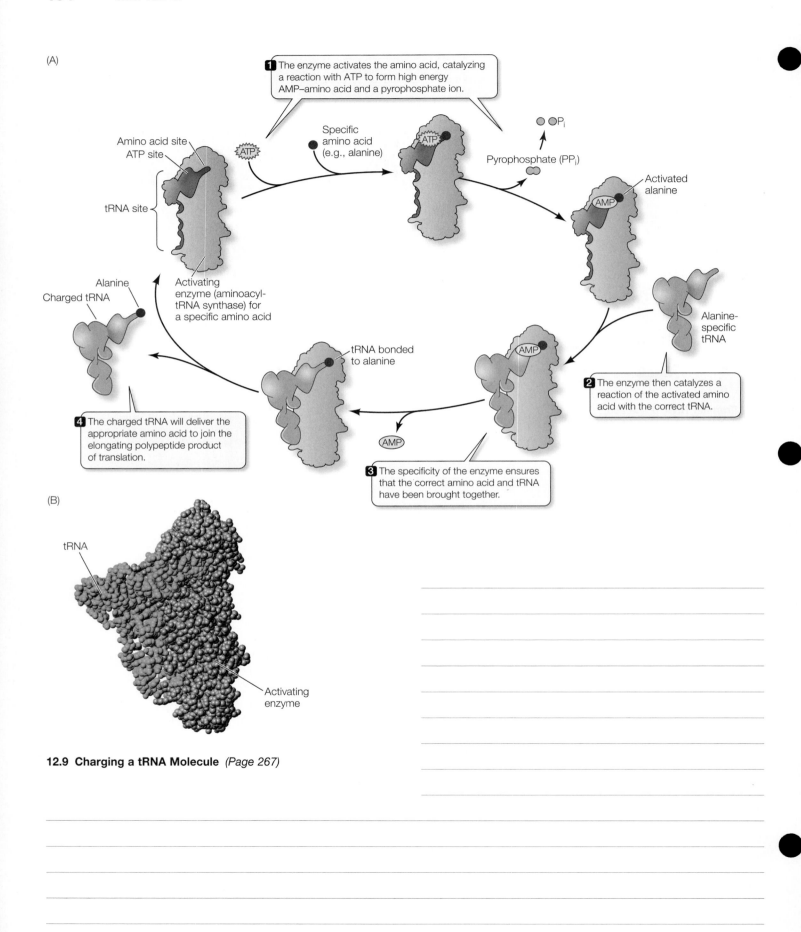

1 The enzyme activates the amino acid, catalyzing a reaction with ATP to form high energy AMP–amino acid and a pyrophosphate ion.

Amino acid site
ATP site

tRNA site

Specific amino acid (e.g., alanine)

Pyrophosphate (PP$_i$)

P$_i$

Activated alanine

Alanine

Charged tRNA

Activating enzyme (aminoacyl-tRNA synthase) for a specific amino acid

Alanine-specific tRNA

2 The enzyme then catalyzes a reaction of the activated amino acid with the correct tRNA.

tRNA bonded to alanine

AMP

4 The charged tRNA will deliver the appropriate amino acid to join the elongating polypeptide product of translation.

AMP

3 The specificity of the enzyme ensures that the correct amino acid and tRNA have been brought together.

(B)

tRNA

Activating enzyme

12.9 Charging a tRNA Molecule *(Page 267)*

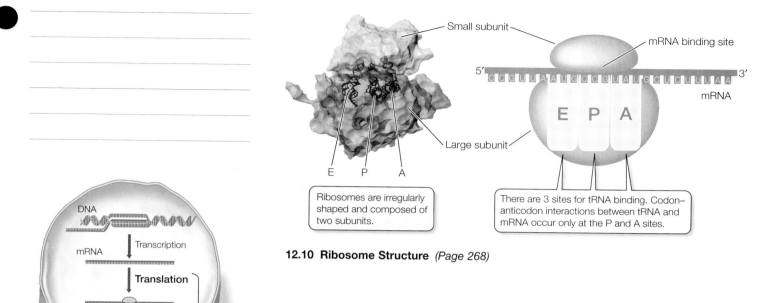

Small subunit

mRNA binding site

5′   3′

mRNA

E   P   A

E   P   A

Large subunit

Ribosomes are irregularly shaped and composed of two subunits.

There are 3 sites for tRNA binding. Codon–anticodon interactions between tRNA and mRNA occur only at the P and A sites.

12.10 Ribosome Structure *(Page 268)*

DNA

mRNA   ↓ Transcription

↓ **Translation**

Polypeptide

INITIATION

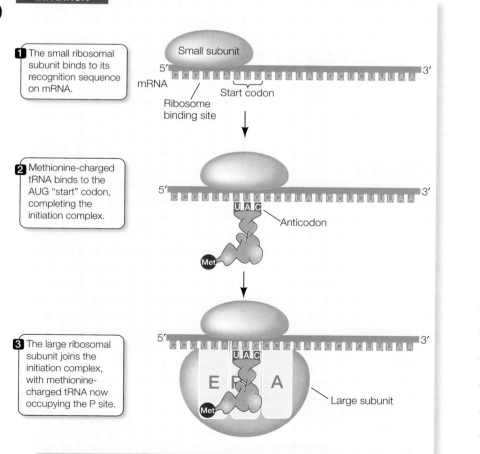

1 The small ribosomal subunit binds to its recognition sequence on mRNA.

Small subunit

5′   3′

mRNA

Start codon

Ribosome binding site

2 Methionine-charged tRNA binds to the AUG "start" codon, completing the initiation complex.

5′   3′

U A C

Anticodon

Met

3 The large ribosomal subunit joins the initiation complex, with methionine-charged tRNA now occupying the P site.

5′   3′

U A C

E   P   A

Met

Large subunit

12.11 The Initiation of Translation *(Page 269)*

ELONGATION

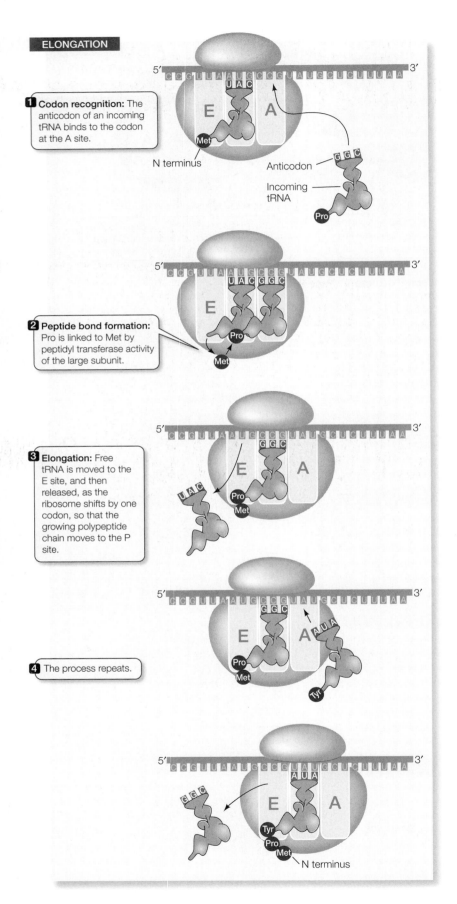

1 **Codon recognition:** The anticodon of an incoming tRNA binds to the codon at the A site.

N terminus

Anticodon

Incoming tRNA

2 **Peptide bond formation:** Pro is linked to Met by peptidyl transferase activity of the large subunit.

3 **Elongation:** Free tRNA is moved to the E site, and then released, as the ribosome shifts by one codon, so that the growing polypeptide chain moves to the P site.

4 The process repeats.

N terminus

12.12 The Elongation of Translation *(Page 270)*

TABLE 12.1		
Signals that Start and Stop Transcription and Translation		
	TRANSCRIPTION	**TRANSLATION**
Initiation	Promoter sequence in DNA	AUG start codon in mRNA
Termination	Terminator sequence in DNA	UAA, UAG, or UGA stop codon in mRNA

(Page 270)

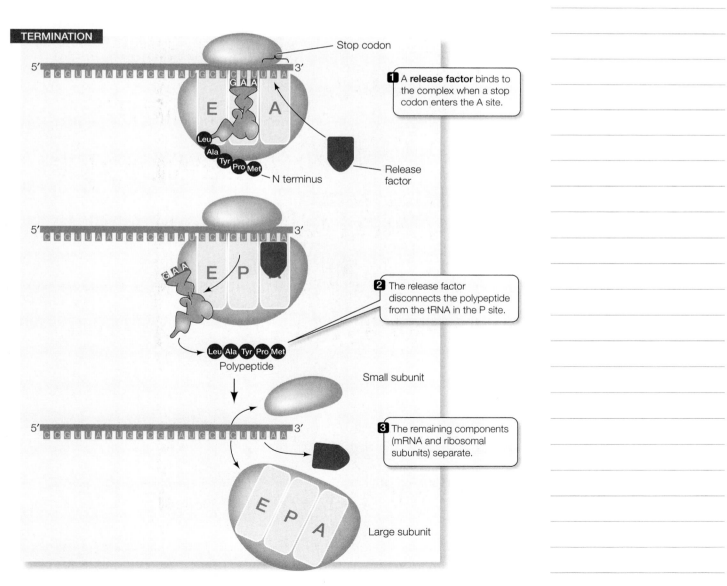

TERMINATION

Stop codon

1 A **release factor** binds to the complex when a stop codon enters the A site.

Release factor

N terminus

2 The release factor disconnects the polypeptide from the tRNA in the P site.

Polypeptide

Small subunit

3 The remaining components (mRNA and ribosomal subunits) separate.

Large subunit

12.13 The Termination of Translation *(Page 271)*

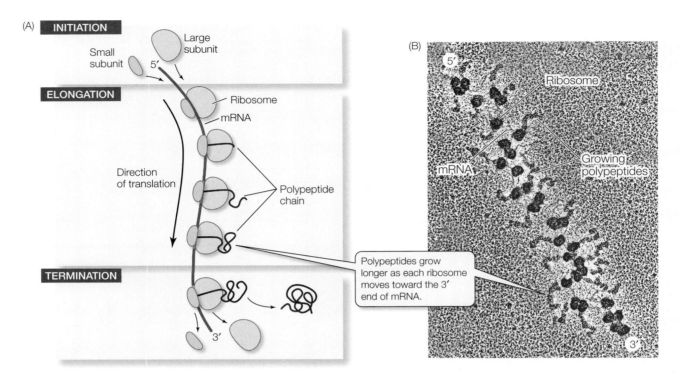

12.14 A Polysome *(Page 271)*

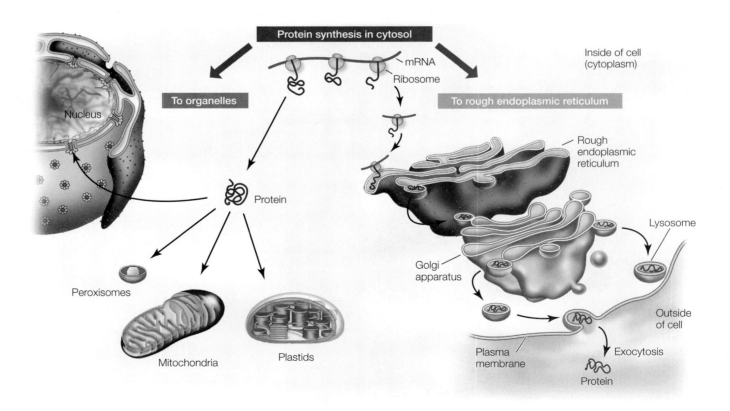

12.15 Destinations for Newly Translated Polypeptides in a Eukaryotic Cell *(Page 272)*

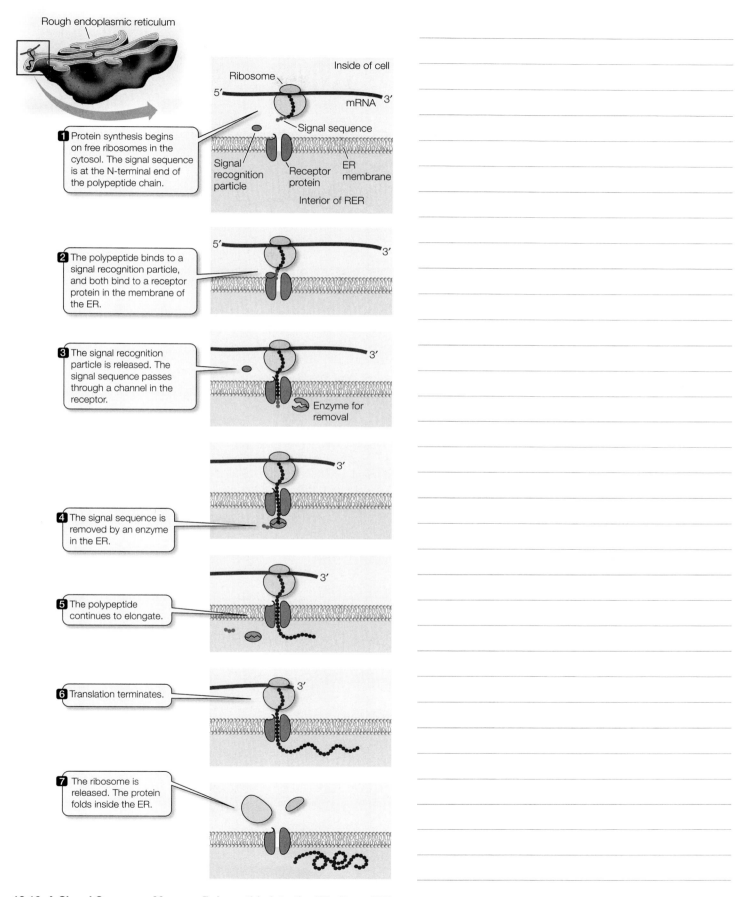

Rough endoplasmic reticulum

1 Protein synthesis begins on free ribosomes in the cytosol. The signal sequence is at the N-terminal end of the polypeptide chain.

Inside of cell

Ribosome

5′

mRNA

3′

Signal sequence

Signal recognition particle

Receptor protein

ER membrane

Interior of RER

2 The polypeptide binds to a signal recognition particle, and both bind to a receptor protein in the membrane of the ER.

5′

3′

3 The signal recognition particle is released. The signal sequence passes through a channel in the receptor.

3′

Enzyme for removal

4 The signal sequence is removed by an enzyme in the ER.

3′

5 The polypeptide continues to elongate.

3′

6 Translation terminates.

3′

7 The ribosome is released. The protein folds inside the ER.

12.16 A Signal Sequence Moves a Polypeptide into the ER *(Page 273)*

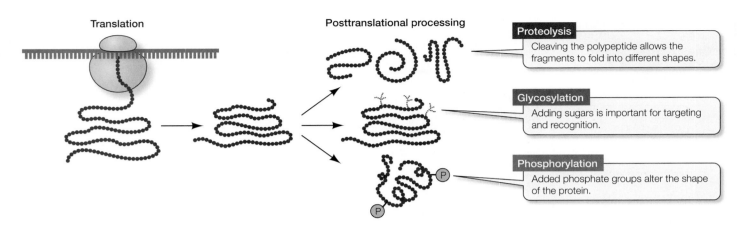

Translation   Posttranslational processing

Proteolysis
Cleaving the polypeptide allows the fragments to fold into different shapes.

Glycosylation
Adding sugars is important for targeting and recognition.

Phosphorylation
Added phosphate groups alter the shape of the protein.

12.17 Posttranslational Modifications of Proteins *(Page 274)*

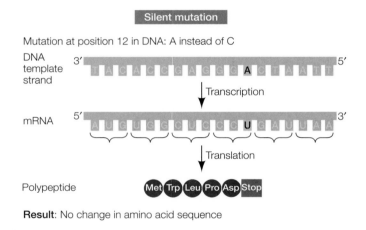

Silent mutation

Mutation at position 12 in DNA: A instead of C

DNA template strand   3′ TAGACCGAGGG**A**GTAATT 5′

↓ Transcription

mRNA   5′ AUGUGGCUCGGG**U**GAUUAA 3′

↓ Translation

Polypeptide   Met Trp Leu Pro Asp Stop

Result: No change in amino acid sequence

Page 275 In-Text Art (1)

Missense mutation

Mutation at position 14 in DNA: A instead of T

DNA template strand   3′ TAGACCGAGGGC**A**AATT 5′

↓ Transcription

mRNA   5′ AUGUGGCUCCCG**U**UUAA 3′

↓ Translation

Polypeptide   Met Trp Leu Pro Val Stop

Result: Amino acid change at position 5: Val instead of Asp

Page 275 In-Text Art (2)

Nonsense mutation

Mutation at position 5 in DNA: T instead of C

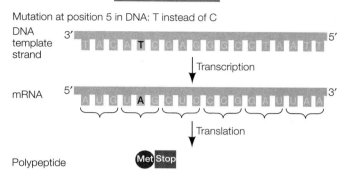

Result: Only one amino acid translated; no protein made

Page 276 In-Text Art (1)

Frame-shift mutation

Mutation by insertion of T between bases 6 and 7 in DNA

DNA template strand 3′ T A C A C C C G A G G G C C T A A T 5′

DNA template strand 3′ T A C A C C **T** G A G G G C C T A A T 5′

↓ Transcription

mRNA 5′ A U G U G G **A C U C C C G G A U U A A** 3′

↓ Translation

Polypeptide Met Trp Thr Pro Gly Leu

Result: All amino acids changed beyond the insertion

Page 276 In-Text Art (2)

(A) **Deletion** is the loss of a chromosome segment.

A B C D E F G → A B E F G

C D (lost)

(B) **Duplication and deletion** result when homologous chromosomes break at different points... ...and swap segments.

A B C D E F G → A B E F G

A B C D E F G → A B C D C D E F G

(C) **Inversion** results when a broken segment is inserted in reverse order.

A B C D E F G → A B E D C F G

(D) **Reciprocal translocation** results when nonhomologous chromosomes exchange segments.

A B C D E F G → A B L M N O

H I J K L M N O → H I J K C D E F G

12.19 Chromosomal Mutations *(Page 277)*

(A) A spontaneous mutation

Cytosine
(common tautomer)

Cytosine
(rare tautomer)

This C cannot hydrogen-bond with G but instead pairs with A.

(B) An induced mutation

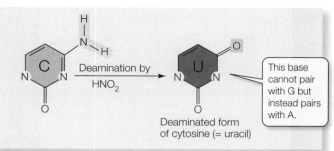

Deamination by
HNO₂

Deaminated form
of cytosine (= uracil)

This base cannot pair with G but instead pairs with A.

(C) The consequences of either mutation

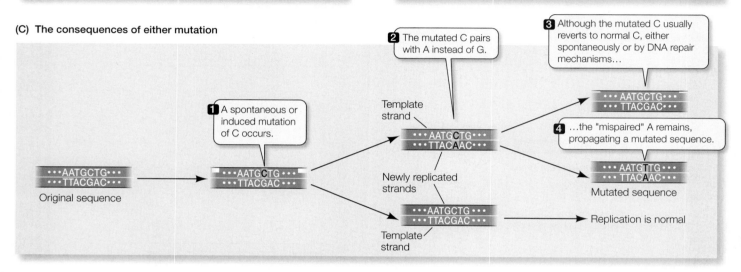

1 A spontaneous or induced mutation of C occurs.

2 The mutated C pairs with A instead of G.

3 Although the mutated C usually reverts to normal C, either spontaneously or by DNA repair mechanisms…

4 …the "mispaired" A remains, propagating a mutated sequence.

Original sequence

Template strand

Newly replicated strands

Template strand

···AATGCTG···
···TTACGAC···

···AATGCTG···
···TTACGAC···

···AATGCTG···
···TTACAAC···

···AATGCTG···
···TTACGAC···

···AATGCTG···
···TTACGAC···

···AATGTTG···
···TTACAAC···

Mutated sequence

Replication is normal

12.20 Spontaneous and Induced Mutations *(Page 278)*

13 The Genetics of Viruses and Prokaryotes

TABLE 13.1

Relative Sizes of Microorganisms

MICROORGANISM	TYPE	TYPICAL SIZE RANGE (μm^3)
Protists	Eukaryote	5,000–50,000
Photosynthetic bacteria	Prokaryote	5–50
Spirochetes	Prokaryote	0.1–2.0
Mycoplasmas	Prokaryote	0.01–0.1
Poxviruses	Virus	0.01
Influenza virus	Virus	0.0005
Poliovirus	Virus	0.00001

(Page 284)

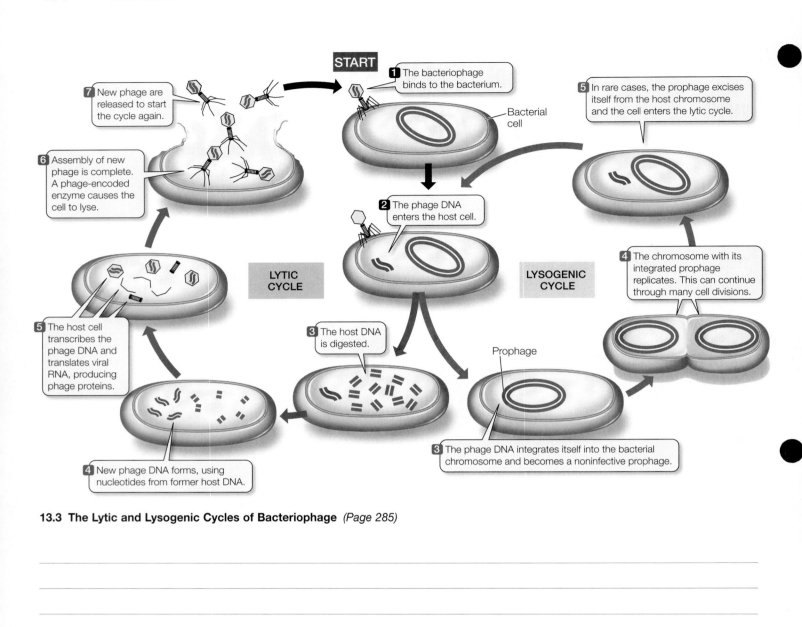

START

1 The bacteriophage binds to the bacterium.

Bacterial cell

2 The phage DNA enters the host cell.

LYTIC CYCLE

LYSOGENIC CYCLE

3 The host DNA is digested.

5 In rare cases, the prophage excises itself from the host chromosome and the cell enters the lytic cycle.

4 The chromosome with its integrated prophage replicates. This can continue through many cell divisions.

7 New phage are released to start the cycle again.

6 Assembly of new phage is complete. A phage-encoded enzyme causes the cell to lyse.

5 The host cell transcribes the phage DNA and translates viral RNA, producing phage proteins.

4 New phage DNA forms, using nucleotides from former host DNA.

Prophage

3 The phage DNA integrates itself into the bacterial chromosome and becomes a noninfective prophage.

13.3 The Lytic and Lysogenic Cycles of Bacteriophage *(Page 285)*

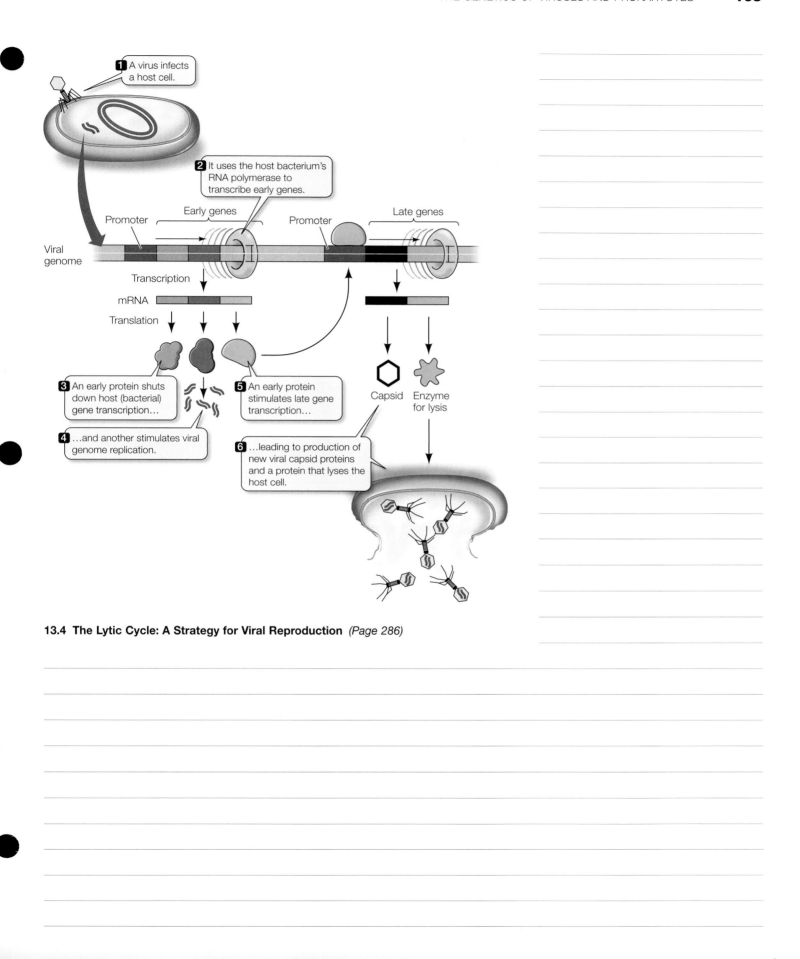

13.4 The Lytic Cycle: A Strategy for Viral Reproduction *(Page 286)*

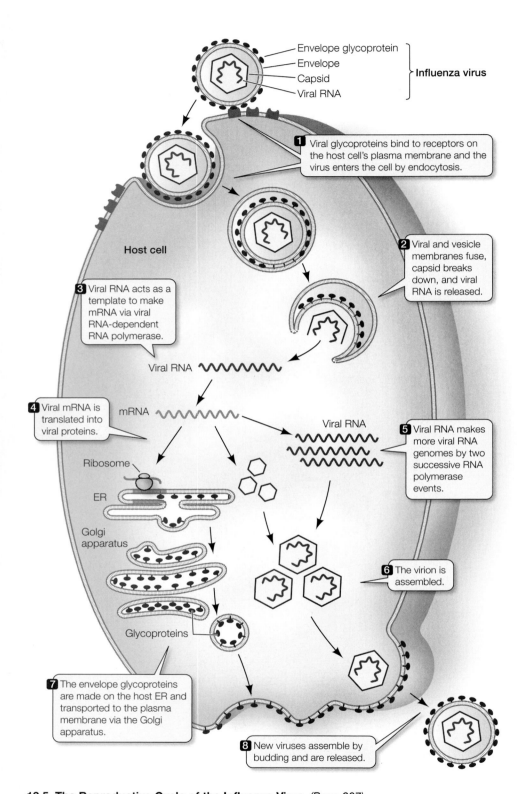

13.5 The Reproductive Cycle of the Influenza Virus *(Page 287)*

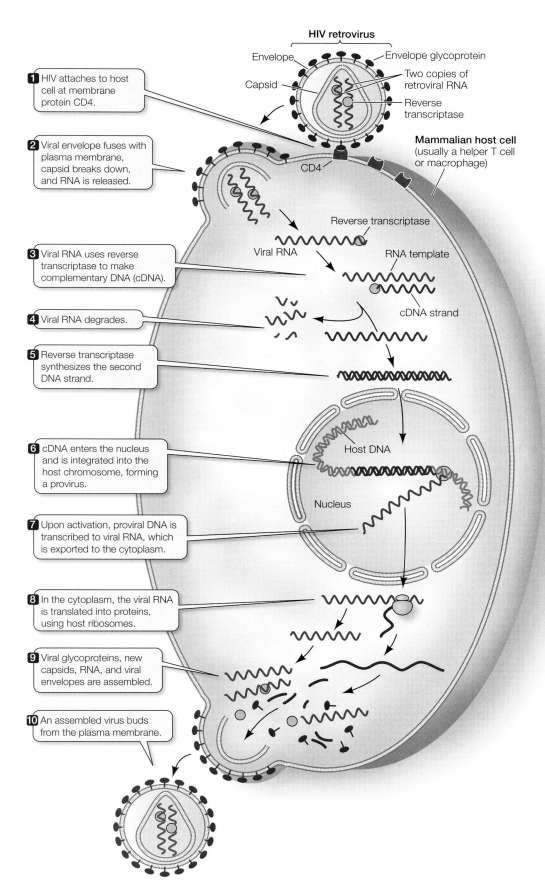

HIV retrovirus

Envelope

Capsid

Envelope glycoprotein

Two copies of retroviral RNA

Reverse transcriptase

1 HIV attaches to host cell at membrane protein CD4.

2 Viral envelope fuses with plasma membrane, capsid breaks down, and RNA is released.

CD4

Mammalian host cell (usually a helper T cell or macrophage)

Reverse transcriptase

Viral RNA

RNA template

3 Viral RNA uses reverse transcriptase to make complementary DNA (cDNA).

cDNA strand

4 Viral RNA degrades.

5 Reverse transcriptase synthesizes the second DNA strand.

Host DNA

6 cDNA enters the nucleus and is integrated into the host chromosome, forming a provirus.

Nucleus

7 Upon activation, proviral DNA is transcribed to viral RNA, which is exported to the cytoplasm.

8 In the cytoplasm, the viral RNA is translated into proteins, using host ribosomes.

9 Viral glycoproteins, new capsids, RNA, and viral envelopes are assembled.

10 An assembled virus buds from the plasma membrane.

13.6 The Reproductive Cycle of HIV *(Page 288)*

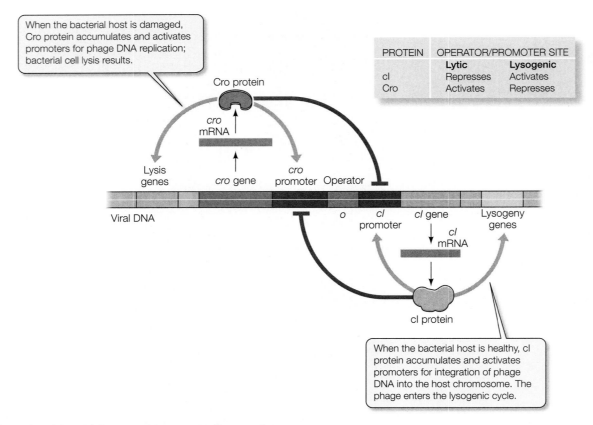

When the bacterial host is damaged, Cro protein accumulates and activates promoters for phage DNA replication; bacterial cell lysis results.

Cro protein

cro mRNA

PROTEIN	OPERATOR/PROMOTER SITE	
	Lytic	**Lysogenic**
cI	Represses	Activates
Cro	Activates	Represses

Lysis genes

cro gene

cro promoter

Operator

Viral DNA

o

cI promoter

cI gene

cI mRNA

Lysogeny genes

cI protein

When the bacterial host is healthy, cI protein accumulates and activates promoters for integration of phage DNA into the host chromosome. The phage enters the lysogenic cycle.

13.8 Control of Phage λ Lysis and Lysogeny *(Page 290)*

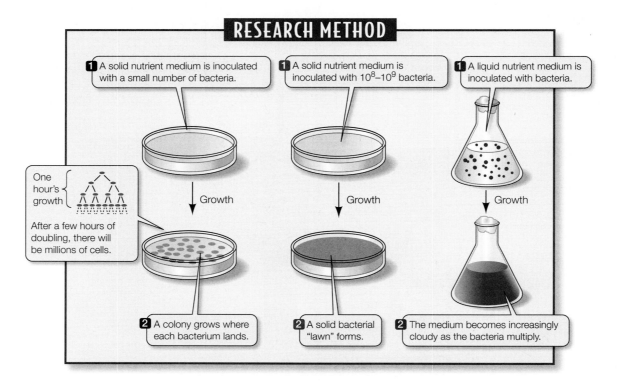

RESEARCH METHOD

1 A solid nutrient medium is inoculated with a small number of bacteria.

1 A solid nutrient medium is inoculated with 10^8–10^9 bacteria.

1 A liquid nutrient medium is inoculated with bacteria.

One hour's growth

After a few hours of doubling, there will be millions of cells.

Growth

Growth

Growth

2 A colony grows where each bacterium lands.

2 A solid bacterial "lawn" forms.

2 The medium becomes increasingly cloudy as the bacteria multiply.

13.9 Growing Bacteria in the Laboratory *(Page 291)*

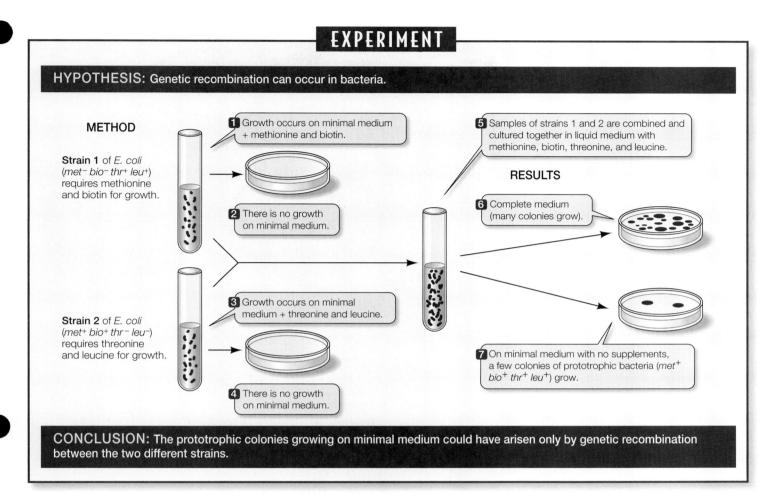

EXPERIMENT

HYPOTHESIS: Genetic recombination can occur in bacteria.

METHOD

Strain 1 of *E. coli* (*met⁻ bio⁻ thr⁺ leu⁺*) requires methionine and biotin for growth.

1 Growth occurs on minimal medium + methionine and biotin.

2 There is no growth on minimal medium.

Strain 2 of *E. coli* (*met⁺ bio⁺ thr⁻ leu⁻*) requires threonine and leucine for growth.

3 Growth occurs on minimal medium + threonine and leucine.

4 There is no growth on minimal medium.

5 Samples of strains 1 and 2 are combined and cultured together in liquid medium with methionine, biotin, threonine, and leucine.

RESULTS

6 Complete medium (many colonies grow).

7 On minimal medium with no supplements, a few colonies of prototrophic bacteria (*met⁺ bio⁺ thr⁺ leu⁺*) grow.

CONCLUSION: The prototrophic colonies growing on minimal medium could have arisen only by genetic recombination between the two different strains.

13.10 Lederberg and Tatum's Experiment *(Page 292)*

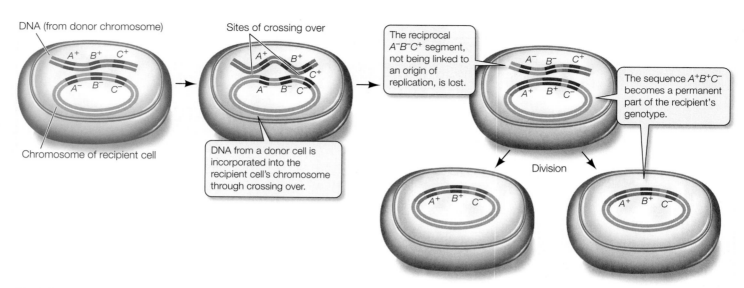

13.12 Recombination Following Conjugation *(Page 293)*

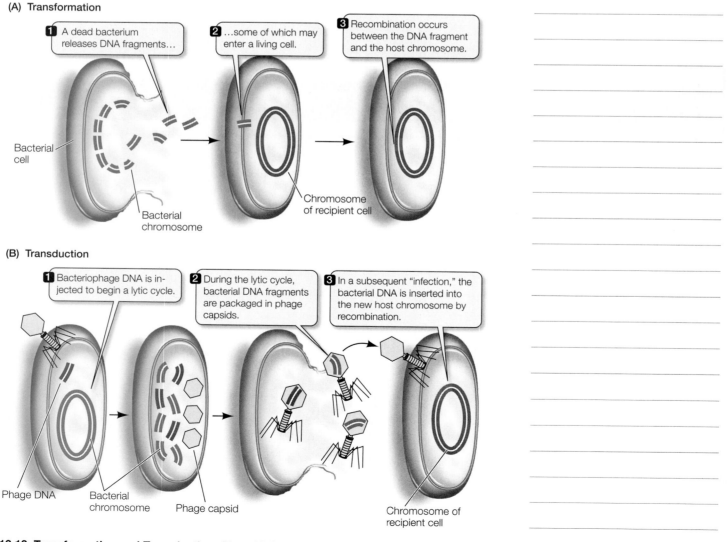

13.13 Transformation and Transduction *(Page 293)*

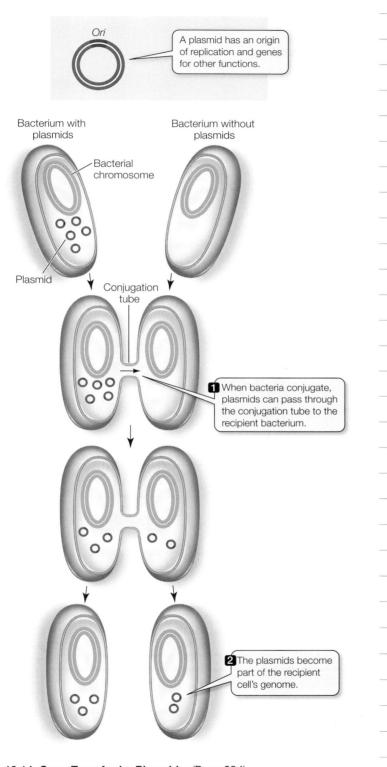

Ori

A plasmid has an origin of replication and genes for other functions.

Bacterium with plasmids

Bacterium without plasmids

Bacterial chromosome

Plasmid

Conjugation tube

1 When bacteria conjugate, plasmids can pass through the conjugation tube to the recipient bacterium.

2 The plasmids become part of the recipient cell's genome.

13.14 Gene Transfer by Plasmids *(Page 294)*

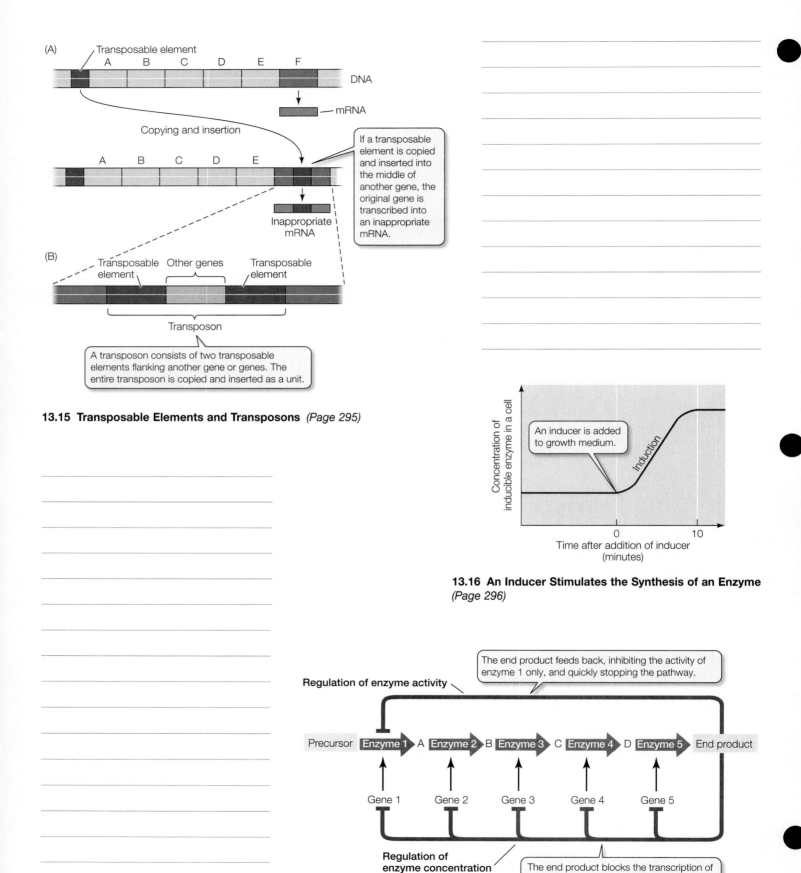

(A)

Transposable element

A B C D E F

DNA

↓

mRNA

Copying and insertion

A B C D E

If a transposable element is copied and inserted into the middle of another gene, the original gene is transcribed into an inappropriate mRNA.

↓

Inappropriate mRNA

(B)

Transposable element   Other genes   Transposable element

Transposon

A transposon consists of two transposable elements flanking another gene or genes. The entire transposon is copied and inserted as a unit.

13.15 Transposable Elements and Transposons *(Page 295)*

An inducer is added to growth medium.

Induction

Concentration of inducible enzyme in a cell

0   10

Time after addition of inducer (minutes)

13.16 An Inducer Stimulates the Synthesis of an Enzyme *(Page 296)*

The end product feeds back, inhibiting the activity of enzyme 1 only, and quickly stopping the pathway.

Regulation of enzyme activity

Precursor → Enzyme 1 → A → Enzyme 2 → B → Enzyme 3 → C → Enzyme 4 → D → Enzyme 5 → End product

Gene 1   Gene 2   Gene 3   Gene 4   Gene 5

Regulation of enzyme concentration

The end product blocks the transcription of all five genes. No enzymes are produced.

13.17 Two Ways to Regulate a Metabolic Pathway *(Page 297)*

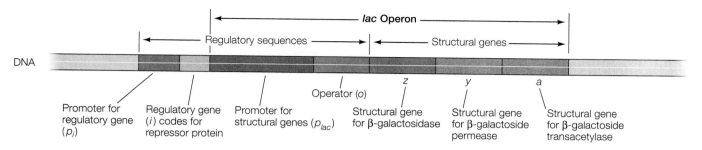

13.18 The *lac* Operon of *E. coli* (Page 297)

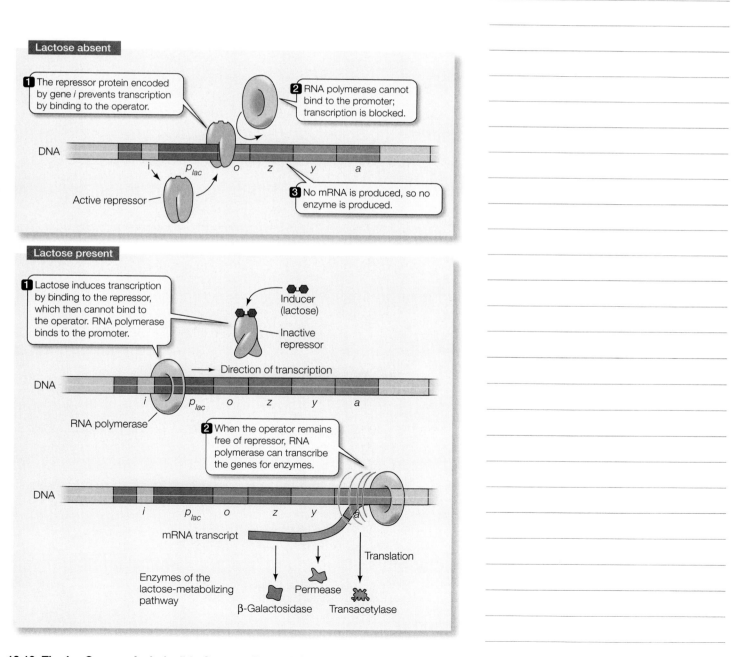

13.19 The *lac* Operon: An Inducible System (Page 298)

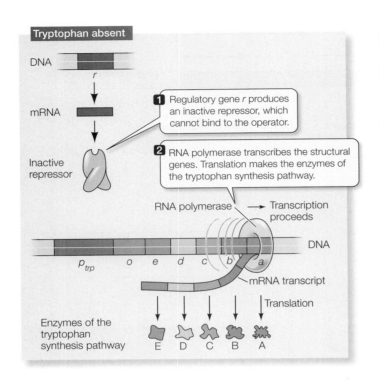

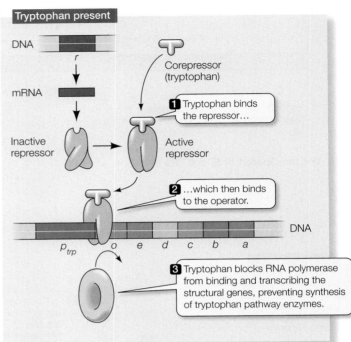

13.20 The *trp* Operon: A Repressible System *(Page 299)*

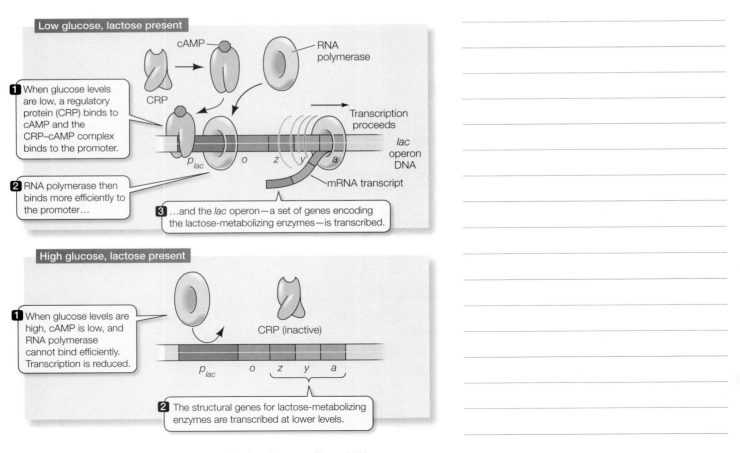

13.21 Catabolite Repression Regulates the *lac* Operon *(Page 299)*

TABLE 13.2

Positive and Negative Regulation in the *lac Operon*[a]

GLUCOSE	cAMP LEVELS	RNA POLYMERASE BINDING TO PROMOTER	LACTOSE	*lac* REPRESSOR	TRANSCRIPTION OF *lac* GENES?	LACTOSE USED BY CELLS?
Present	Low	Absent	Absent	Active and bound to operator	No	No
Present	Low	Present, not efficient	Present	Inactive and not bound to operator	Low level	No
Absent	High	Present, very efficient	Present	Inactive and not bound to operator	High level	Yes
Absent	High	Absent	Absent	Active and bound to operator	No	No

[a]Negative regulators are in red type.

(Page 300)

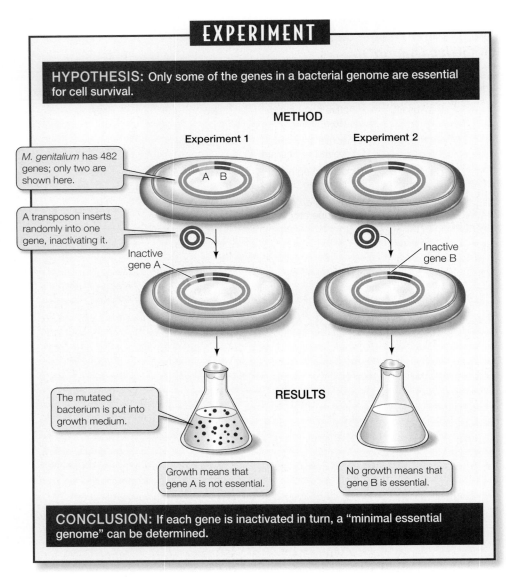

13.23 Using Transposon Mutagenesis to Determine the Minimal Genome *(Page 302)*

CHAPTER 14 The Eukaryotic Genome and Its Expression

TABLE 14.1

A Comparison of Prokaryotic and Eukaryotic Genes and Genomes

CHARACTERISTIC	PROKARYOTES	EUKARYOTES
Genome size (base pairs)	10^4–10^7	10^8–10^{11}
Repeated sequences	Few	Many
Noncoding DNA within coding sequences	Rare	Common
Transcription and translation separated in cell	No	Yes
DNA segregated within a nucleus	No	Yes
DNA bound to proteins	Some	Extensive
Promoters	Yes	Yes
Enhancers/silencers	Rare	Common
Capping and tailing of mRNA	No	Yes
RNA splicing required (spliceosomes)	Rare	Common
Number of chromosomes in genome	One	Many

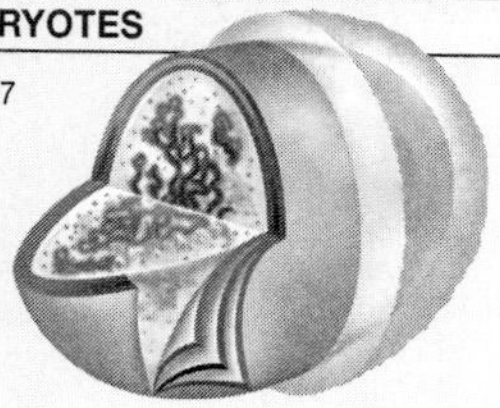

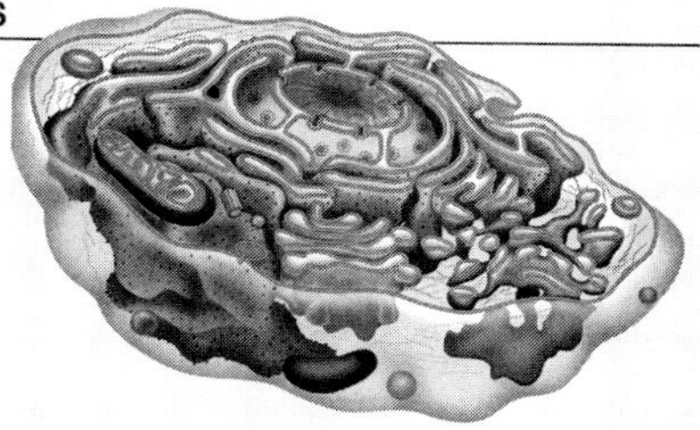

(Page 308)

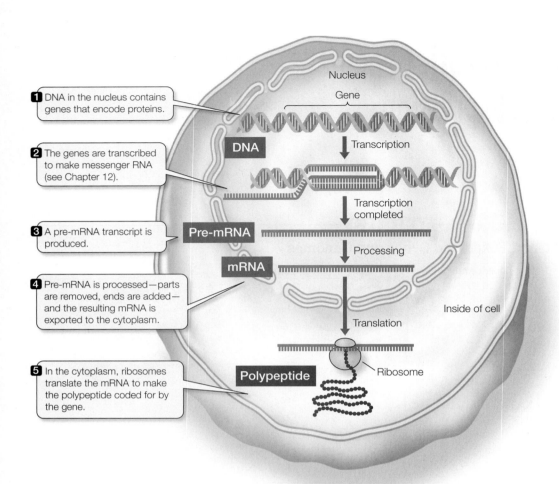

1 DNA in the nucleus contains genes that encode proteins.

DNA

2 The genes are transcribed to make messenger RNA (see Chapter 12).

3 A pre-mRNA transcript is produced.

Pre-mRNA

mRNA

4 Pre-mRNA is processed—parts are removed, ends are added—and the resulting mRNA is exported to the cytoplasm.

5 In the cytoplasm, ribosomes translate the mRNA to make the polypeptide coded for by the gene.

Polypeptide

Nucleus

Gene

Transcription

Transcription completed

Processing

Translation

Inside of cell

Ribosome

14.1 Eukaryotic mRNA is Transcribed in the Nucleus but Translated in the Cytoplasm
(Page 308)

TABLE 14.2

Comparison of the Genomes of *E. coli* and Yeast

	E. COLI	YEAST
Genome length (base pairs)	4,640,000	12,068,000
Number of protein-coding genes	4,300	5,800
Proteins with roles in:		
Metabolism	650	650
Energy production/storage	240	175
Membrane transport	280	250
DNA replication/repair/ recombination	120	175
Transcription	230	400
Translation	180	350
Protein targeting/secretion	35	430
Cell structure	180	250

TABLE 14.3

C. elegans Genes Essential to Multicellularity

FUNCTION	PROTEIN/DOMAIN	NUMBER OF GENES
Transcription control	Zinc finger; homeobox	540
RNA processing	RNA binding domains	100
Nerve impulse transmission	Gated ion channels	80
Tissue formation	Collagens	170
Cell interactions	Extracellular domains; glycotransferases	330
Cell–cell signaling	G protein-linked receptors; protein kinases; protein phosphatases	1,290

(Page 310)

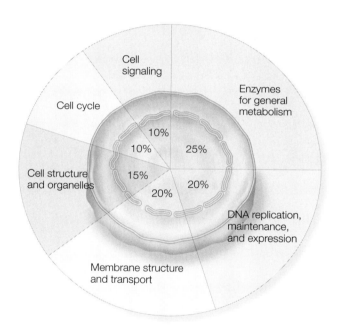

14.2 Functions of the Eukaryotic Genome *(Page 310)*

TABLE 14.4

Arabidopsis Genes Unique to Plants

FUNCTION	NUMBER OF GENES
Cell wall and growth	420
Water channels	300
Photosynthesis	139
Defense and metabolism	94

(Page 310)

TABLE 14.5		
Comparison of the Rice and *Arabidopsis* Genomes		
	PERCENTAGE OF GENOME	
FUNCTION	**RICE**	***ARABIDOPSIS***
Cell structure	9	10
Enzymes	21	20
Ligand binding	10	10
DNA binding	10	10
Signal transduction	3	3
Membrane transport	5	5
Cell growth and maintenance	24	22
Other functions	18	20

(Page 311)

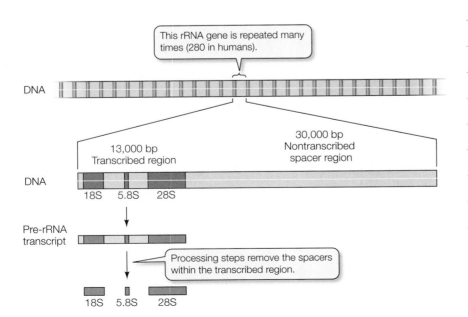

14.3 A Moderately Repetitive Sequence Codes for rRNA *(Page 312)*

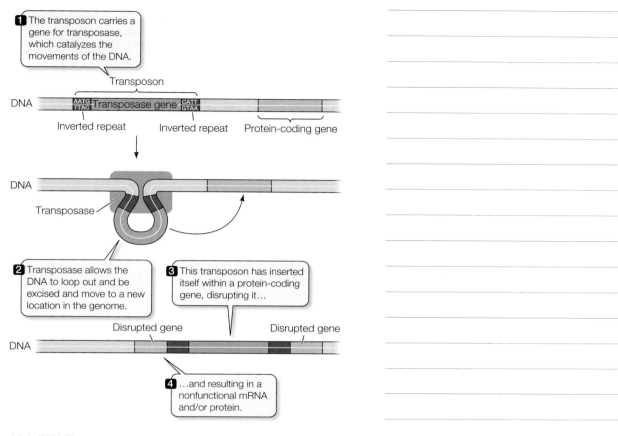

14.4 DNA Transposons and Transposition *(Page 312)*

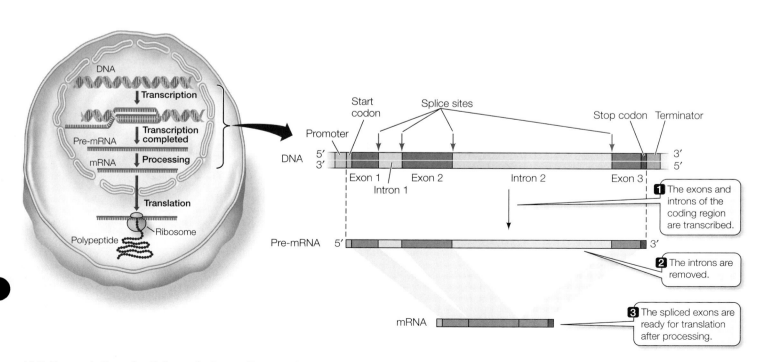

14.5 Transcription of a Eukaryotic Gene *(Page 313)*

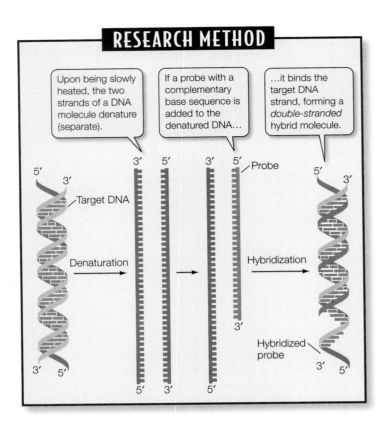

14.6 Nucleic Acid Hybridization *(Page 314)*

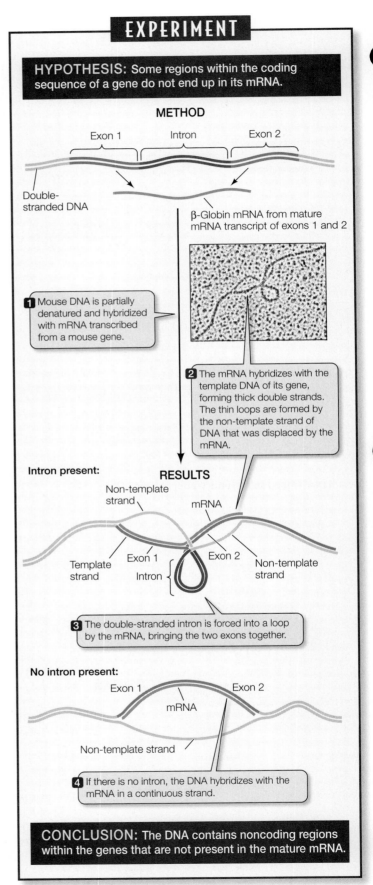

14.7 Nucleic Acid Hybridization Revealed the Existence of Introns *(Page 314)*

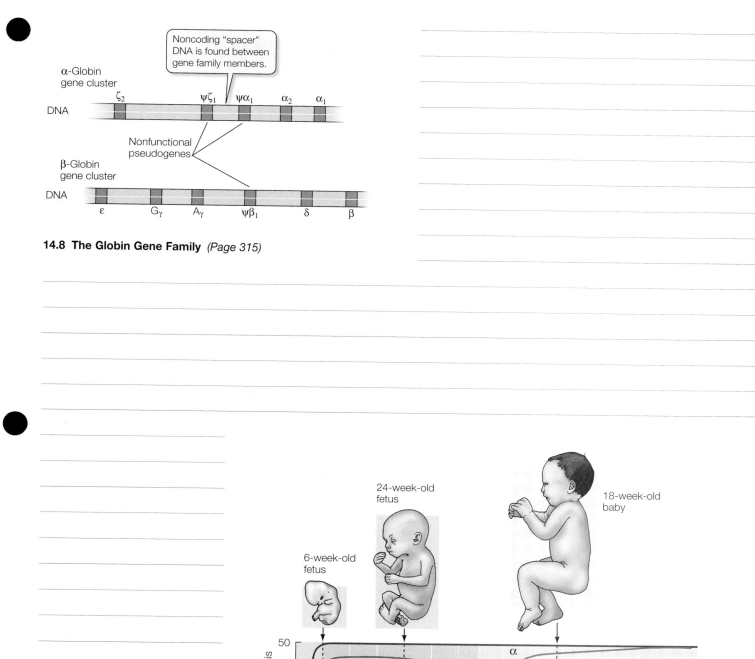

14.8 The Globin Gene Family *(Page 315)*

14.9 Differential Expression in the Globin Gene Family *(Page 316)*

14.10 Processing the Ends of Eukaryotic Pre-mRNA *(Page 317)*

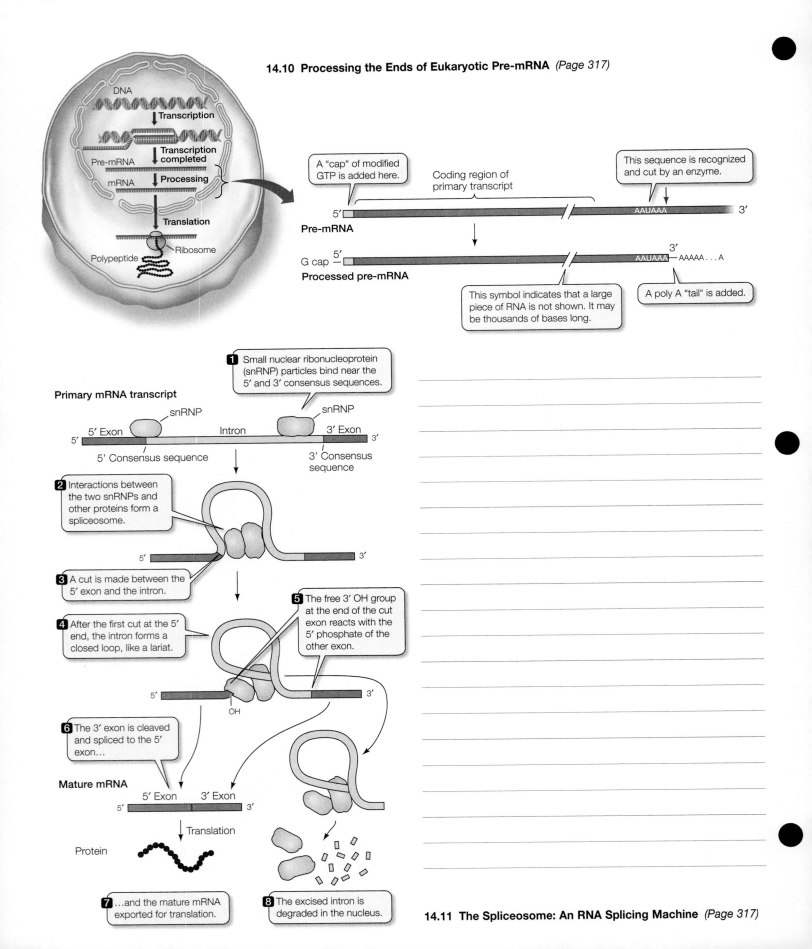

14.11 The Spliceosome: An RNA Splicing Machine *(Page 317)*

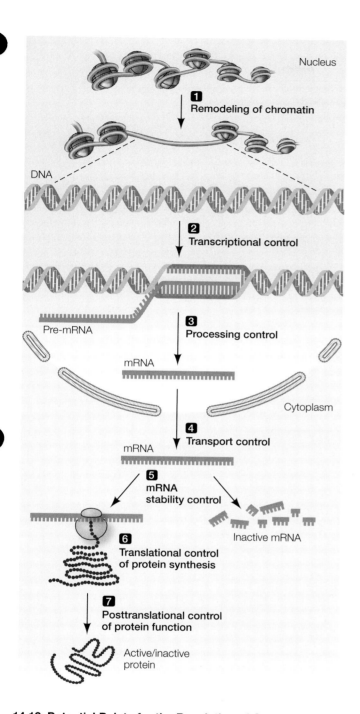

14.12 Potential Points for the Regulation of Gene Expression
(Page 318)

Nucleus

1 Remodeling of chromatin

DNA

2 Transcriptional control

Pre-mRNA

3 Processing control

mRNA

Cytoplasm

4 Transport control

mRNA

5 mRNA stability control

Inactive mRNA

6 Translational control of protein synthesis

7 Posttranslational control of protein function

Active/inactive protein

Promoter

Initiation site for transcription

DNA

TATAT ATATA

TATA box

TFIID

TFIID

1 The first transcription factor, TFIID, binds to the promoter at the TATA box…

B

TFIID

B

2 …and another transcription factor joins it.

F

RNA polymerase II

TFIID

F

B

3 RNA polymerase II binds only after several transcription factors are already bound to DNA.

E

H

4 More transcription factors are added…

H

TFIID

E

F

RNA polymerase II

B

5 …and the RNA polymerase is ready to transcribe RNA.

14.13 The Initiation of Transcription in Eukaryotes *(Page 319)*

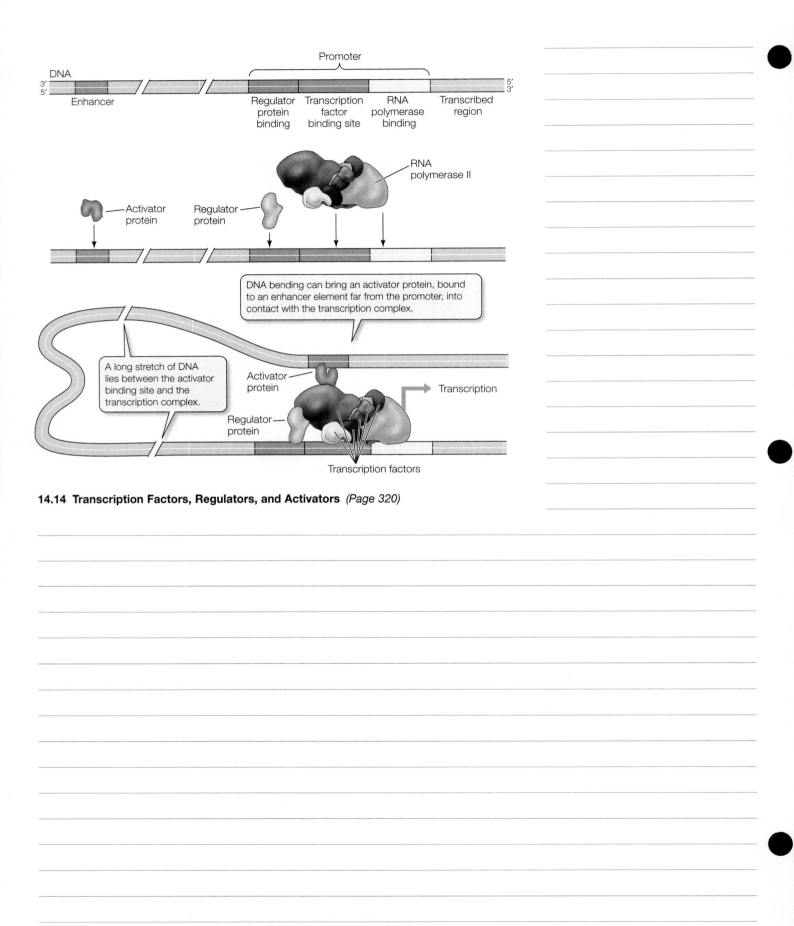

14.14 Transcription Factors, Regulators, and Activators *(Page 320)*

Helix-turn-helix motif

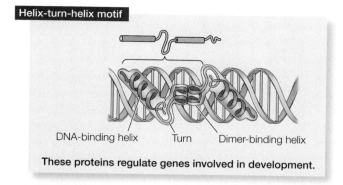

DNA-binding helix — Turn — Dimer-binding helix

These proteins regulate genes involved in development.

Leucine zipper motif

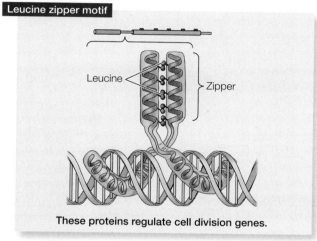

Leucine — Zipper

These proteins regulate cell division genes.

Zinc finger motif

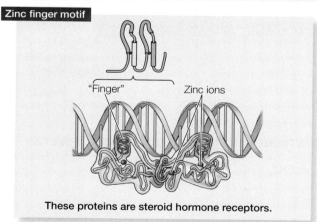

"Finger" — Zinc ions

These proteins are steroid hormone receptors.

Helix-loop-helix motif

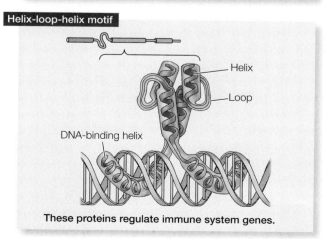

Helix — Loop — DNA-binding helix

These proteins regulate immune system genes.

14.15 Protein–DNA Interactions *(Page 321)*

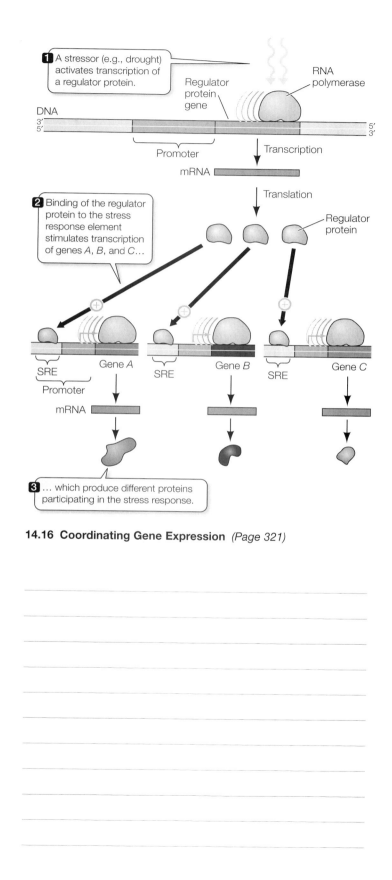

1 A stressor (e.g., drought) activates transcription of a regulator protein.

Regulator protein gene

RNA polymerase

DNA
3'
5'
5'
3'

Promoter

Transcription

mRNA

Translation

2 Binding of the regulator protein to the stress response element stimulates transcription of genes *A*, *B*, and *C*...

Regulator protein

SRE — Gene *A*
Promoter

SRE — Gene *B*

SRE — Gene *C*

mRNA

3 ... which produce different proteins participating in the stress response.

14.16 Coordinating Gene Expression *(Page 321)*

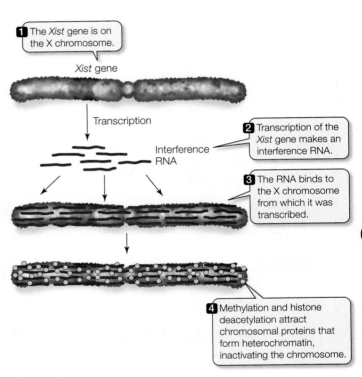

Lysine in histone   Acetyl-CoA   Acetyl-lysine

Page 322 In-Text Art

Initiation

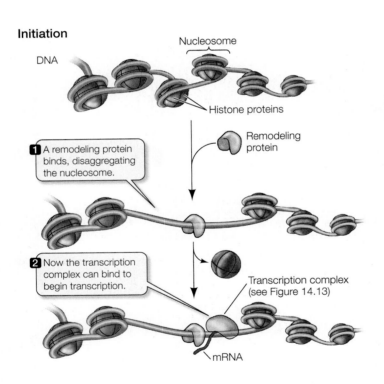

1 A remodeling protein binds, disaggregating the nucleosome.

Remodeling protein

2 Now the transcription complex can bind to begin transcription.

Transcription complex (see Figure 14.13)

mRNA

Elongation

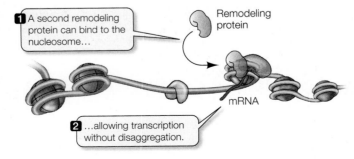

1 A second remodeling protein can bind to the nucleosome...

Remodeling protein

mRNA

2 ...allowing transcription without disaggregation.

14.17 Local Remodeling of Chromatin for Transcription
(Page 322)

1 The *Xist* gene is on the X chromosome.

Xist gene

Transcription

2 Transcription of the *Xist* gene makes an interference RNA.

Interference RNA

3 The RNA binds to the X chromosome from which it was transcribed.

4 Methylation and histone deacetylation attract chromosomal proteins that form heterochromatin, inactivating the chromosome.

14.19 A Model for X Chromosome Inactivation *(Page 323)*

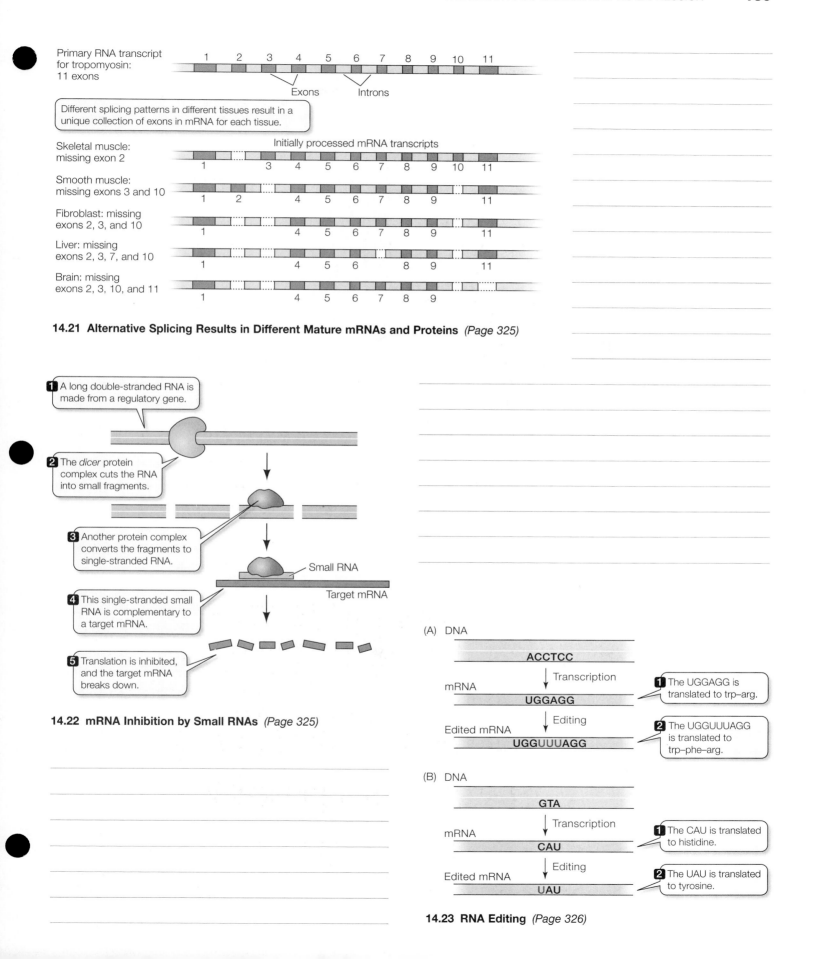

Primary RNA transcript for tropomyosin: 11 exons

1 2 3 4 5 6 7 8 9 10 11

Exons   Introns

Different splicing patterns in different tissues result in a unique collection of exons in mRNA for each tissue.

Skeletal muscle: missing exon 2

Initially processed mRNA transcripts

1   3 4 5 6 7 8 9 10 11

Smooth muscle: missing exons 3 and 10

1 2   4 5 6 7 8 9   11

Fibroblast: missing exons 2, 3, and 10

1   4 5 6 7 8 9   11

Liver: missing exons 2, 3, 7, and 10

1   4 5 6   8 9   11

Brain: missing exons 2, 3, 10, and 11

1   4 5 6 7 8 9

14.21 Alternative Splicing Results in Different Mature mRNAs and Proteins *(Page 325)*

1 A long double-stranded RNA is made from a regulatory gene.

2 The *dicer* protein complex cuts the RNA into small fragments.

3 Another protein complex converts the fragments to single-stranded RNA.

Small RNA

Target mRNA

4 This single-stranded small RNA is complementary to a target mRNA.

5 Translation is inhibited, and the target mRNA breaks down.

14.22 mRNA Inhibition by Small RNAs *(Page 325)*

(A) DNA

ACCTCC

mRNA   ↓ Transcription

UGGAGG

1 The UGGAGG is translated to trp–arg.

↓ Editing

Edited mRNA

UGGUUUAGG

2 The UGGUUUAGG is translated to trp–phe–arg.

(B) DNA

GTA

mRNA   ↓ Transcription

CAU

1 The CAU is translated to histidine.

↓ Editing

Edited mRNA

UAU

2 The UAU is translated to tyrosine.

14.23 RNA Editing *(Page 326)*

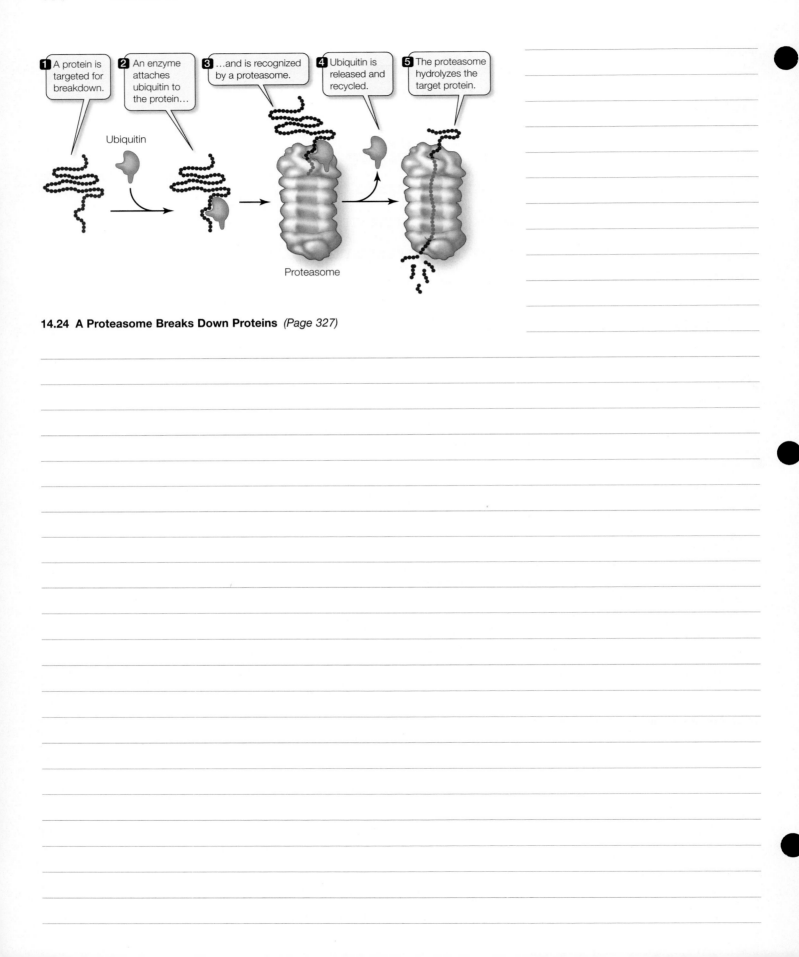

14.24 A Proteasome Breaks Down Proteins *(Page 327)*

15 Cell Signaling and Communication

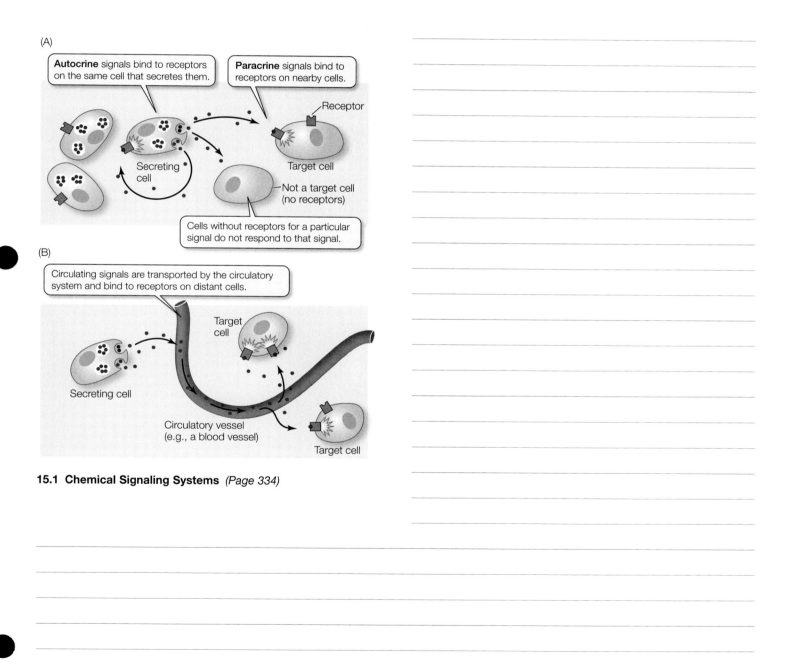

(A)

Autocrine signals bind to receptors on the same cell that secretes them.

Paracrine signals bind to receptors on nearby cells.

Receptor

Secreting cell

Target cell

Not a target cell (no receptors)

Cells without receptors for a particular signal do not respond to that signal.

(B)

Circulating signals are transported by the circulatory system and bind to receptors on distant cells.

Target cell

Secreting cell

Circulatory vessel (e.g., a blood vessel)

Target cell

15.1 Chemical Signaling Systems *(Page 334)*

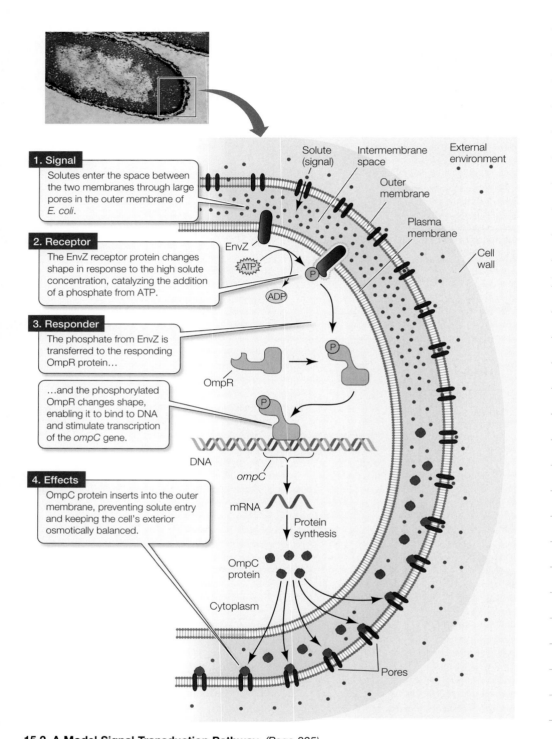

1. Signal

Solutes enter the space between the two membranes through large pores in the outer membrane of *E. coli*.

2. Receptor

The EnvZ receptor protein changes shape in response to the high solute concentration, catalyzing the addition of a phosphate from ATP.

3. Responder

The phosphate from EnvZ is transferred to the responding OmpR protein…

…and the phosphorylated OmpR changes shape, enabling it to bind to DNA and stimulate transcription of the *ompC* gene.

4. Effects

OmpC protein inserts into the outer membrane, preventing solute entry and keeping the cell's exterior osmotically balanced.

Solute (signal)

Intermembrane space

External environment

Outer membrane

Plasma membrane

Cell wall

EnvZ

ATP

ADP

P

OmpR

P

P

DNA

ompC

mRNA

Protein synthesis

OmpC protein

Cytoplasm

Pores

15.2 A Model Signal Transduction Pathway *(Page 335)*

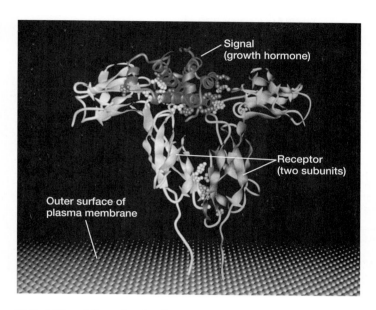

15.3 A Signal Bound to Its Receptor *(Page 336)*

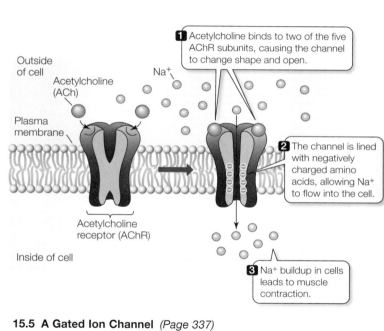

1 Acetylcholine binds to two of the five AChR subunits, causing the channel to change shape and open.

Outside of cell

Acetylcholine (ACh)

Na⁺

Plasma membrane

2 The channel is lined with negatively charged amino acids, allowing Na⁺ to flow into the cell.

Acetylcholine receptor (AChR)

Inside of cell

3 Na⁺ buildup in cells leads to muscle contraction.

15.5 A Gated Ion Channel *(Page 337)*

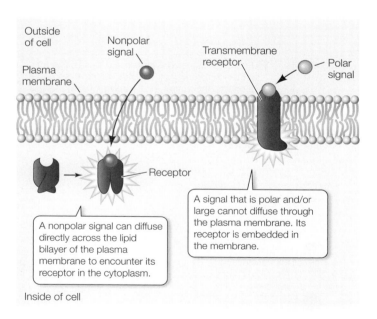

Outside of cell

Nonpolar signal

Transmembrane receptor

Polar signal

Plasma membrane

Receptor

A signal that is polar and/or large cannot diffuse through the plasma membrane. Its receptor is embedded in the membrane.

A nonpolar signal can diffuse directly across the lipid bilayer of the plasma membrane to encounter its receptor in the cytoplasm.

Inside of cell

15.4 Two Locations for Receptors *(Page 337)*

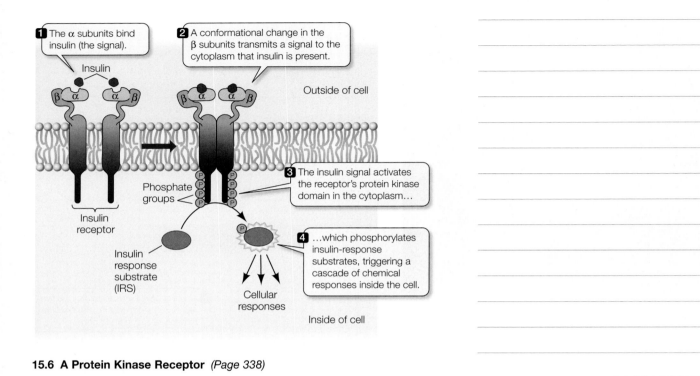

1 The α subunits bind insulin (the signal).

2 A conformational change in the β subunits transmits a signal to the cytoplasm that insulin is present.

Insulin

Outside of cell

Phosphate groups

Insulin receptor

3 The insulin signal activates the receptor's protein kinase domain in the cytoplasm...

Insulin response substrate (IRS)

4 ...which phosphorylates insulin-response substrates, triggering a cascade of chemical responses inside the cell.

Cellular responses

Inside of cell

15.6 A Protein Kinase Receptor *(Page 338)*

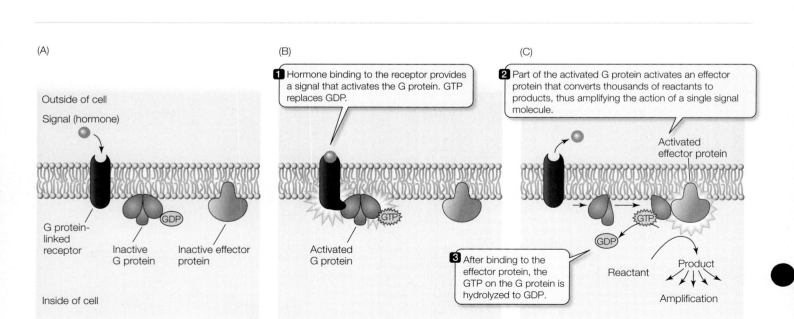

(A)

Outside of cell

Signal (hormone)

G protein-linked receptor

Inactive G protein

Inactive effector protein

Inside of cell

(B)

1 Hormone binding to the receptor provides a signal that activates the G protein. GTP replaces GDP.

Activated G protein

(C)

2 Part of the activated G protein activates an effector protein that converts thousands of reactants to products, thus amplifying the action of a single signal molecule.

Activated effector protein

3 After binding to the effector protein, the GTP on the G protein is hydrolyzed to GDP.

Reactant

Product

Amplification

15.7 A G Protein-Linked Receptor *(Page 338)*

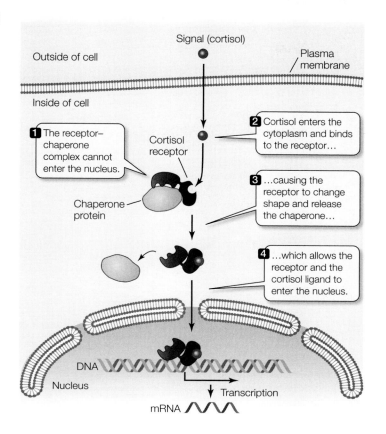

15.8 A Cytoplasmic Receptor *(Page 339)*

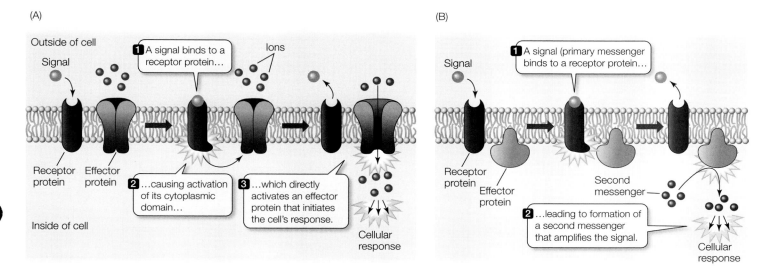

15.9 Direct and Indirect Signal Transduction *(Page 340)*

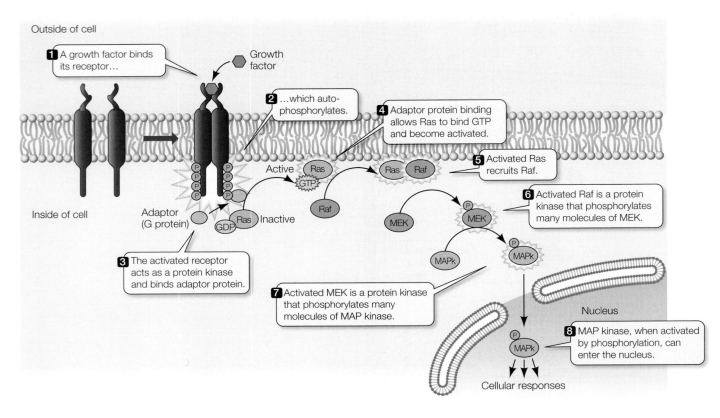

Outside of cell

1 A growth factor binds its receptor...

Growth factor

2 ...which auto-phosphorylates.

4 Adaptor protein binding allows Ras to bind GTP and become activated.

Active

5 Activated Ras recruits Raf.

6 Activated Raf is a protein kinase that phosphorylates many molecules of MEK.

Inside of cell

Adaptor (G protein)

GDP Ras Inactive

Ras GTP

Ras Raf

Raf

MEK

P
MEK

MAPk

P
MAPk

3 The activated receptor acts as a protein kinase and binds adaptor protein.

7 Activated MEK is a protein kinase that phosphorylates many molecules of MAP kinase.

Nucleus

P
MAPk

8 MAP kinase, when activated by phosphorylation, can enter the nucleus.

Cellular responses

15.10 A Protein Kinase Cascade *(Page 341)*

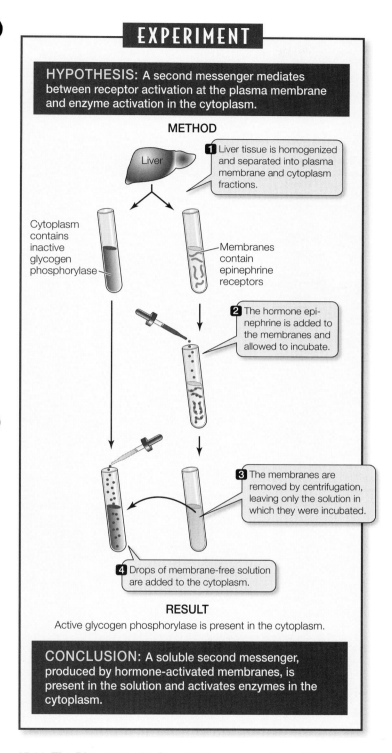

HYPOTHESIS: A second messenger mediates between receptor activation at the plasma membrane and enzyme activation in the cytoplasm.

METHOD

Liver

1 Liver tissue is homogenized and separated into plasma membrane and cytoplasm fractions.

Cytoplasm contains inactive glycogen phosphorylase

Membranes contain epinephrine receptors

2 The hormone epinephrine is added to the membranes and allowed to incubate.

3 The membranes are removed by centrifugation, leaving only the solution in which they were incubated.

4 Drops of membrane-free solution are added to the cytoplasm.

RESULT

Active glycogen phosphorylase is present in the cytoplasm.

CONCLUSION: A soluble second messenger, produced by hormone-activated membranes, is present in the solution and activates enzymes in the cytoplasm.

15.11 The Discovery of a Second Messenger *(Page 342)*

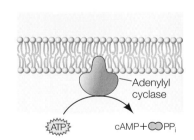

Adenylyl cyclase

ATP

$cAMP +$ PP$_i$

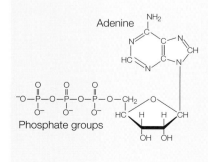

Adenine

NH$_2$

Phosphate groups

ATP

NH$_2$

Cyclic AMP (cAMP)

15.12 The Formation of Cyclic AMP *(Page 342)*

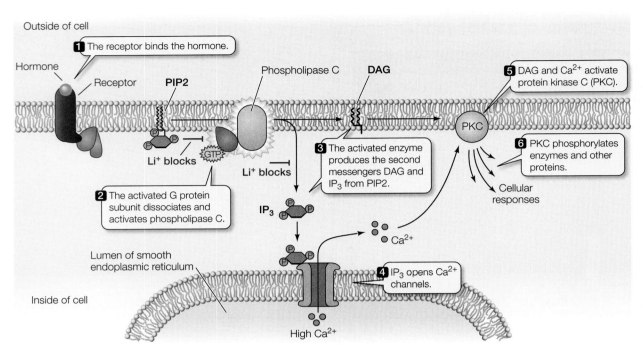

Outside of cell

1 The receptor binds the hormone.

Hormone

Receptor

PIP2

Phospholipase C

DAG

5 DAG and Ca^{2+} activate protein kinase C (PKC).

Li⁺ blocks

GTP

Li⁺ blocks

PKC

6 PKC phosphorylates enzymes and other proteins.

2 The activated G protein subunit dissociates and activates phospholipase C.

3 The activated enzyme produces the second messengers DAG and IP_3 from PIP2.

Cellular responses

IP_3

Ca^{2+}

Lumen of smooth endoplasmic reticulum

Inside of cell

4 IP_3 opens Ca^{2+} channels.

High Ca^{2+}

15.13 The IP_3/DAG Second-Messenger System *(Page 343)*

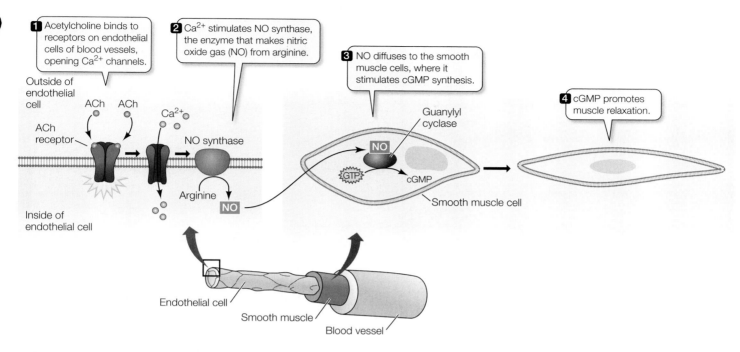

1 Acetylcholine binds to receptors on endothelial cells of blood vessels, opening Ca^{2+} channels.

2 Ca^{2+} stimulates NO synthase, the enzyme that makes nitric oxide gas (NO) from arginine.

3 NO diffuses to the smooth muscle cells, where it stimulates cGMP synthesis.

4 cGMP promotes muscle relaxation.

Outside of endothelial cell

ACh   ACh

ACh receptor

Ca^{2+}

NO synthase

Arginine

NO

Inside of endothelial cell

Guanylyl cyclase

NO

GTP

cGMP

Smooth muscle cell

Endothelial cell

Smooth muscle

Blood vessel

15.15 Nitric Oxide as a Second Messenger *(Page 344)*

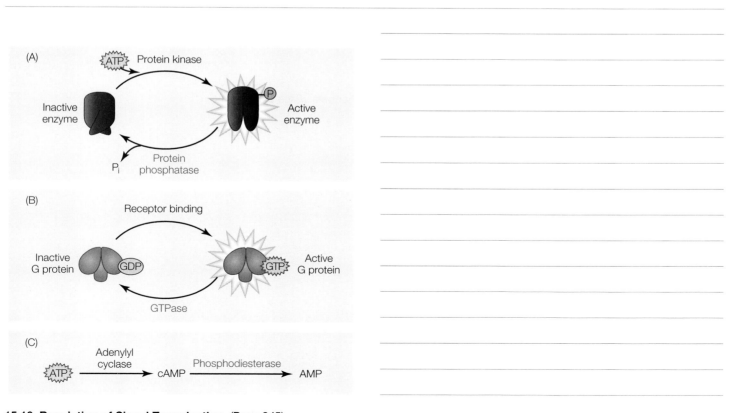

(A)

ATP   Protein kinase

Inactive enzyme

P

Active enzyme

P_i Protein phosphatase

(B)

Receptor binding

Inactive G protein   GDP

GTP   Active G protein

GTPase

(C)

ATP   Adenylyl cyclase → cAMP   Phosphodiesterase → AMP

15.16 Regulation of Signal Transduction *(Page 345)*

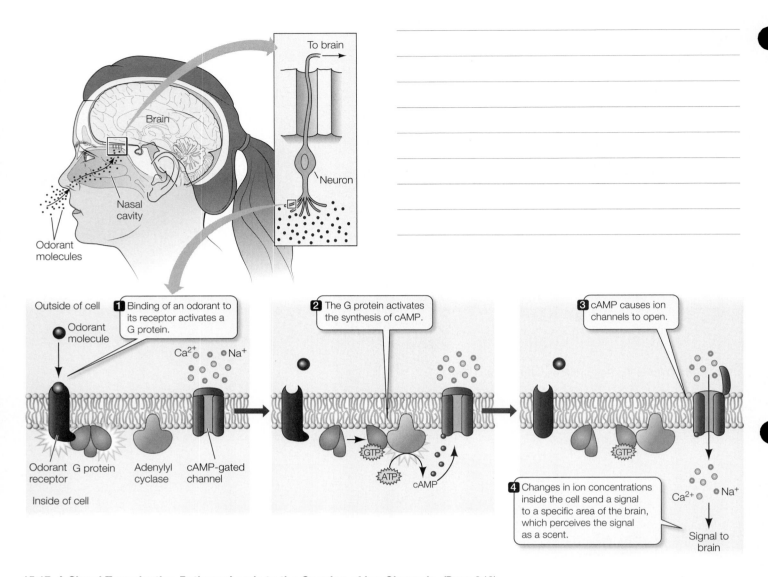

15.17 A Signal Transduction Pathway Leads to the Opening of Ion Channels *(Page 346)*

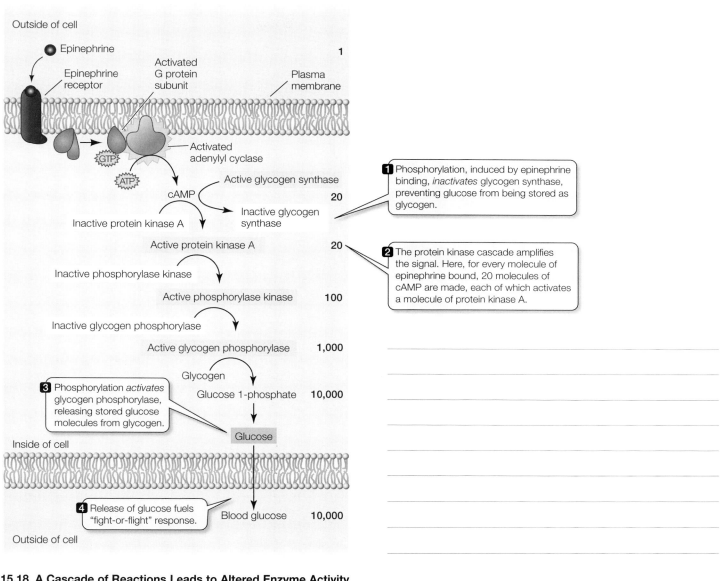

Outside of cell

Epinephrine

Epinephrine receptor

Activated G protein subunit

Plasma membrane

1

Activated adenylyl cyclase

GTP

ATP

cAMP

Active glycogen synthase

20

Inactive protein kinase A

Inactive glycogen synthase

Active protein kinase A

20

Inactive phosphorylase kinase

Active phosphorylase kinase

100

Inactive glycogen phosphorylase

Active glycogen phosphorylase

1,000

Glycogen

Glucose 1-phosphate

10,000

Glucose

Inside of cell

Blood glucose

10,000

Outside of cell

1 Phosphorylation, induced by epinephrine binding, *inactivates* glycogen synthase, preventing glucose from being stored as glycogen.

2 The protein kinase cascade amplifies the signal. Here, for every molecule of epinephrine bound, 20 molecules of cAMP are made, each of which activates a molecule of protein kinase A.

3 Phosphorylation *activates* glycogen phosphorylase, releasing stored glucose molecules from glycogen.

4 Release of glucose fuels "fight-or-flight" response.

15.18 A Cascade of Reactions Leads to Altered Enzyme Activity *(Page 347)*

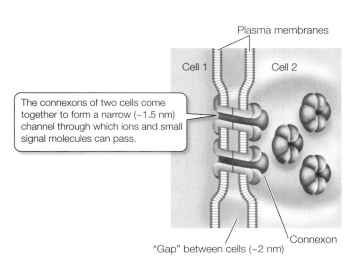

Plasma membranes

Cell 1

Cell 2

The connexons of two cells come together to form a narrow (~1.5 nm) channel through which ions and small signal molecules can pass.

"Gap" between cells (~2 nm)

Connexon

15.19 Gap Junctions Connect Animal Cells *(Page 348)*

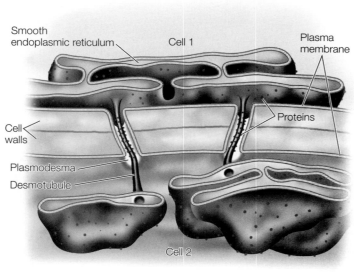

Smooth
endoplasmic reticulum

Cell 1

Plasma
membrane

Proteins

Cell
walls

Plasmodesma

Desmotubule

Cell 2

15.20 Plasmodesmata Connect Plant Cells *(Page 349)*

CHAPTER 16 Recombinant DNA and Biotechnology

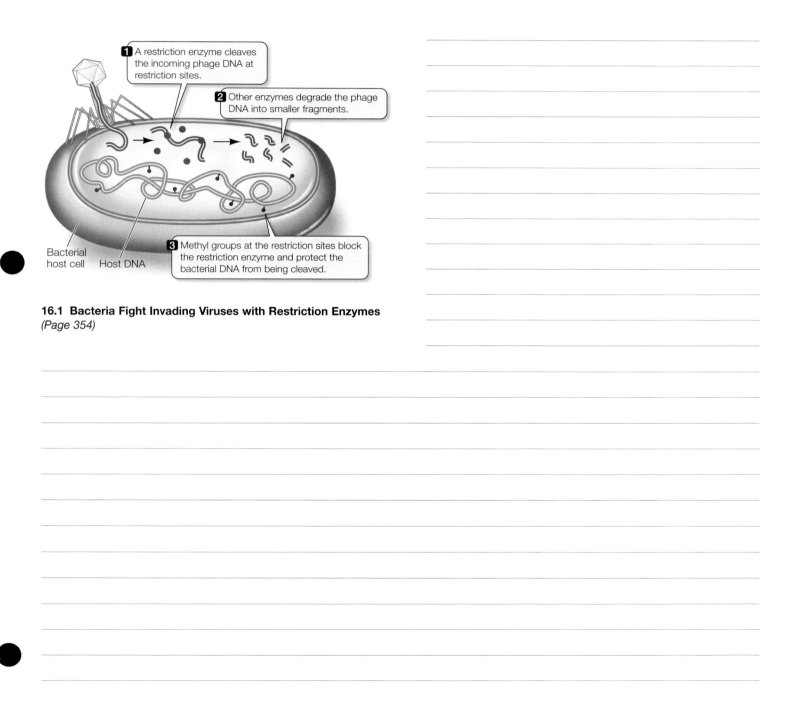

1 A restriction enzyme cleaves the incoming phage DNA at restriction sites.

2 Other enzymes degrade the phage DNA into smaller fragments.

3 Methyl groups at the restriction sites block the restriction enzyme and protect the bacterial DNA from being cleaved.

Bacterial host cell   Host DNA

16.1 Bacteria Fight Invading Viruses with Restriction Enzymes
(Page 354)

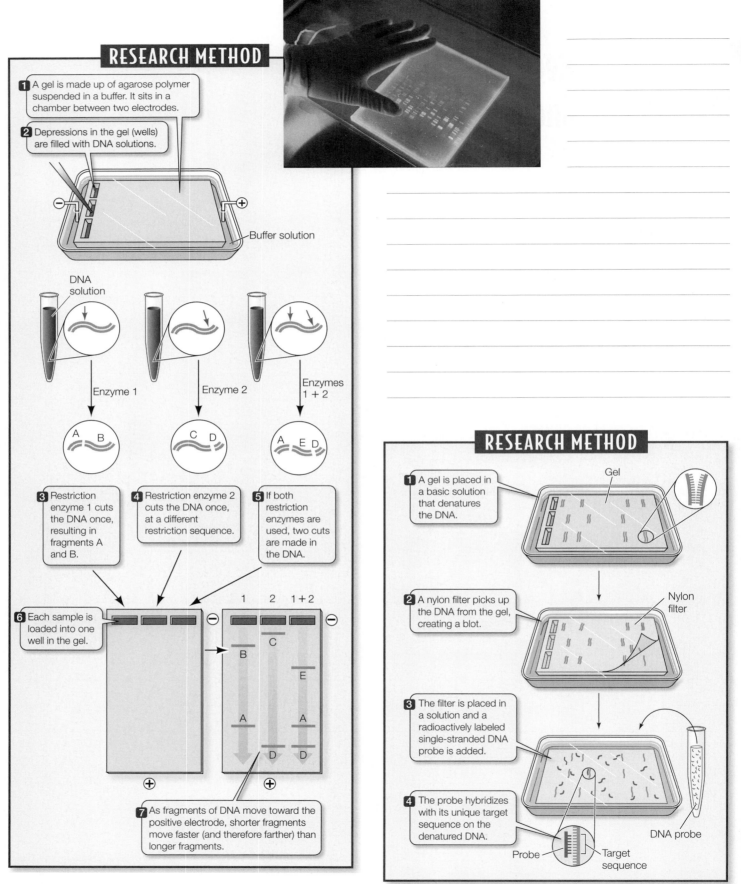

RESEARCH METHOD

1. A gel is made up of agarose polymer suspended in a buffer. It sits in a chamber between two electrodes.

2. Depressions in the gel (wells) are filled with DNA solutions.

Buffer solution

DNA solution

Enzyme 1

Enzyme 2

Enzymes 1 + 2

A   B

C   D

A   E   D

3. Restriction enzyme 1 cuts the DNA once, resulting in fragments A and B.

4. Restriction enzyme 2 cuts the DNA once, at a different restriction sequence.

5. If both restriction enzymes are used, two cuts are made in the DNA.

6. Each sample is loaded into one well in the gel.

1   2   1 + 2

B

C

E

A   A

D   D

7. As fragments of DNA move toward the positive electrode, shorter fragments move faster (and therefore farther) than longer fragments.

16.2 Separating Fragments of DNA by Gel Electrophoresis
(Page 355)

RESEARCH METHOD

1. A gel is placed in a basic solution that denatures the DNA.

Gel

2. A nylon filter picks up the DNA from the gel, creating a blot.

Nylon filter

3. The filter is placed in a solution and a radioactively labeled single-stranded DNA probe is added.

4. The probe hybridizes with its unique target sequence on the denatured DNA.

DNA probe

Probe

Target sequence

16.3 Analyzing DNA Fragments by Southern Blotting *(Page 355)*

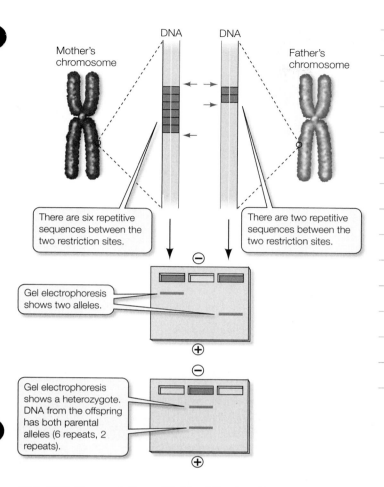

Mother's chromosome

DNA   DNA

Father's chromosome

There are six repetitive sequences between the two restriction sites.

There are two repetitive sequences between the two restriction sites.

Gel electrophoresis shows two alleles.

Gel electrophoresis shows a heterozygote. DNA from the offspring has both parental alleles (6 repeats, 2 repeats).

16.4 DNA Fingerprinting with Short Tandem Repeats
(Page 356)

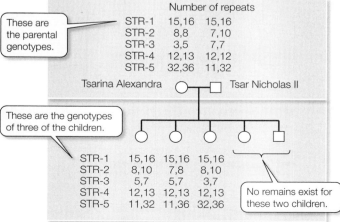

	Number of repeats	
STR-1	15,16	15,16
STR-2	8,8	7,10
STR-3	3,5	7,7
STR-4	12,13	12,12
STR-5	32,36	11,32

These are the parental genotypes.

Tsarina Alexandra   Tsar Nicholas II

These are the genotypes of three of the children.

STR-1	15,16	15,16	15,16
STR-2	8,10	7,8	8,10
STR-3	5,7	5,7	3,7
STR-4	12,13	12,13	12,13
STR-5	11,32	11,36	32,36

No remains exist for these two children.

16.5 DNA Fingerprinting the Russian Royal Family *(Page 357)*

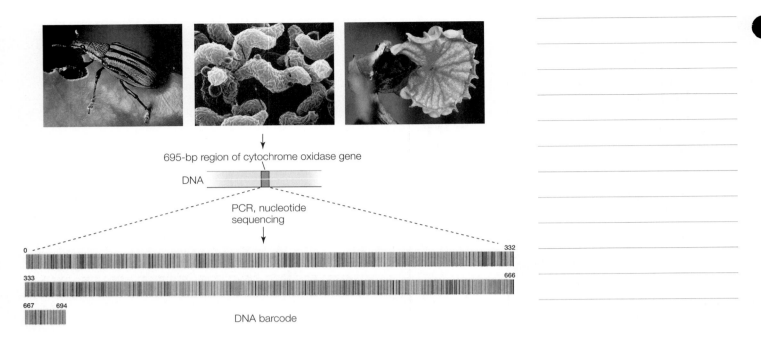

695-bp region of cytochrome oxidase gene

DNA

PCR, nucleotide
sequencing

0   332

333   666

667   694

DNA barcode

16.6 A DNA Barcode *(Page 357)*

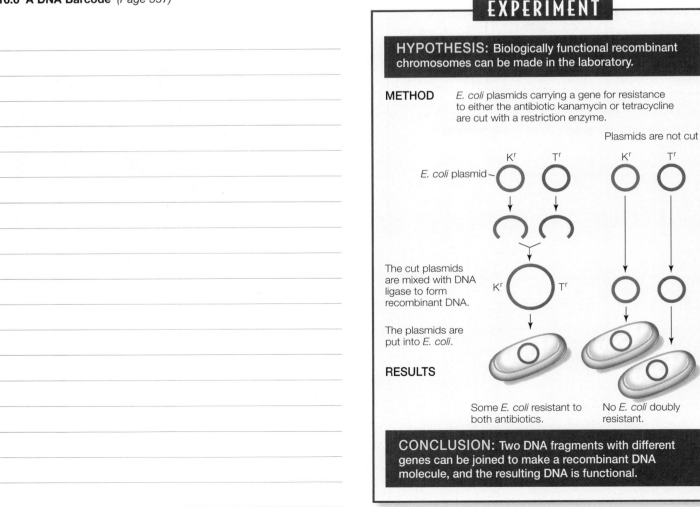

EXPERIMENT

HYPOTHESIS: Biologically functional recombinant chromosomes can be made in the laboratory.

METHOD *E. coli* plasmids carrying a gene for resistance to either the antibiotic kanamycin or tetracycline are cut with a restriction enzyme.

Plasmids are not cut

K^r T^r K^r T^r

E. coli plasmid

The cut plasmids are mixed with DNA ligase to form recombinant DNA. K^r T^r

The plasmids are put into *E. coli*.

RESULTS

Some *E. coli* resistant to both antibiotics.   No *E. coli* doubly resistant.

CONCLUSION: Two DNA fragments with different genes can be joined to make a recombinant DNA molecule, and the resulting DNA is functional.

16.7 Making Recombinant DNA *(Page 358)*

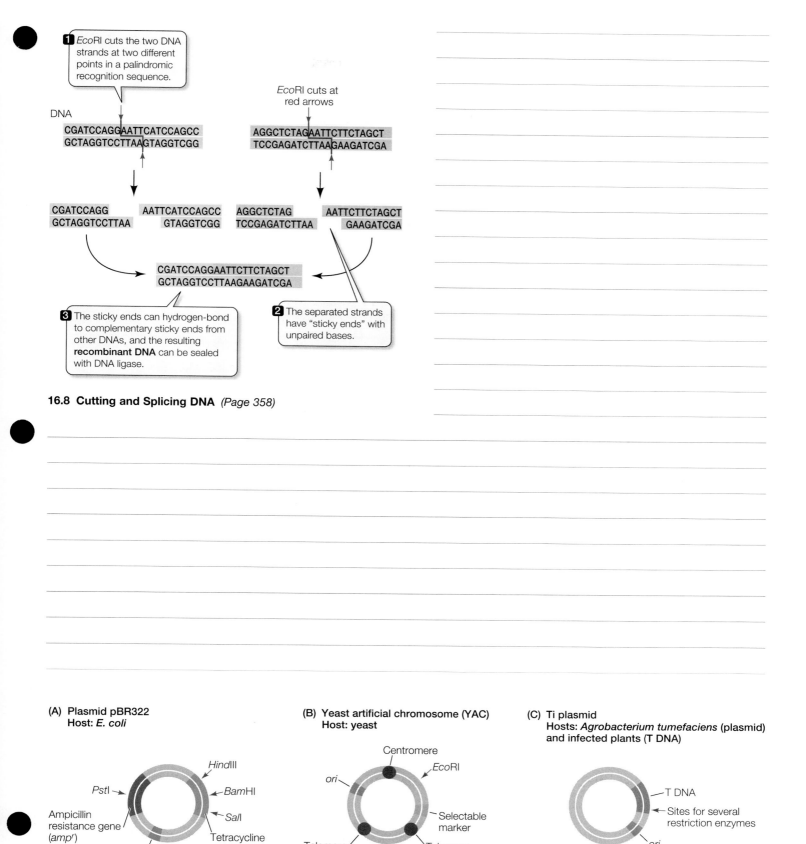

1 *Eco*RI cuts the two DNA strands at two different points in a palindromic recognition sequence.

DNA

CGATCCAGGAATTCATCCAGCC
GCTAGGTCCTTAAGTAGGTCGG

*Eco*RI cuts at red arrows

AGGCTCTAGAATTCTTCTAGCT
TCCGAGATCTTAAGAAGATCGA

CGATCCAGG
GCTAGGTCCTTAA

AATTCATCCAGCC
GTAGGTCGG

AGGCTCTAG
TCCGAGATCTTAA

AATTCTTCTAGCT
GAAGATCGA

CGATCCAGGAATTCTTCTAGCT
GCTAGGTCCTTAAGAAGATCGA

3 The sticky ends can hydrogen-bond to complementary sticky ends from other DNAs, and the resulting **recombinant DNA** can be sealed with DNA ligase.

2 The separated strands have "sticky ends" with unpaired bases.

16.8 Cutting and Splicing DNA *(Page 358)*

(A) Plasmid pBR322
Host: *E. coli*

*Hind*III
*Bam*HI
*Sal*I
*Pst*I
Ampicillin
resistance gene
(*amp*ʳ)
Origin of
replication (*ori*)
Tetracycline
resistance gene
(*tet*ʳ)

(B) Yeast artificial chromosome (YAC)
Host: yeast

Centromere
*Eco*RI
ori
Selectable
marker
Telomere
Telomere
*Bam*HI

(C) Ti plasmid
Hosts: *Agrobacterium tumefaciens* (plasmid)
and infected plants (T DNA)

T DNA
Sites for several
restriction enzymes
ori

↓ Recognition sites for restriction enzymes

16.9 Vectors for Carrying DNA into Cells *(Page 360)*

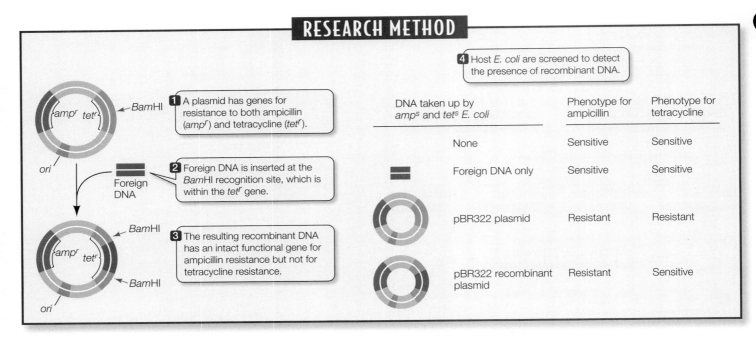

16.10 Marking Recombinant DNA by Inactivating a Gene *(Page 361)*

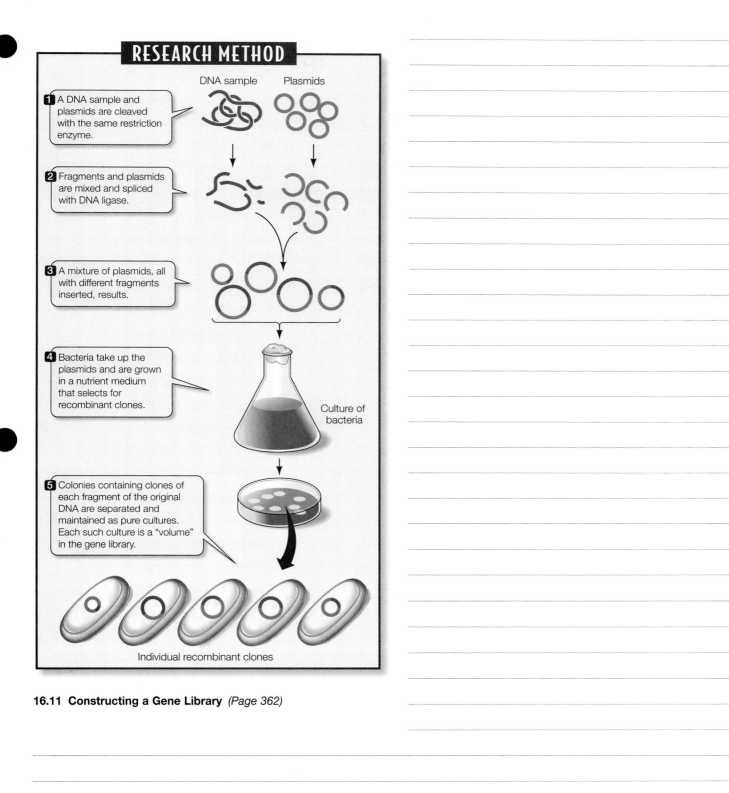

RESEARCH METHOD

DNA sample   Plasmids

1 A DNA sample and plasmids are cleaved with the same restriction enzyme.

2 Fragments and plasmids are mixed and spliced with DNA ligase.

3 A mixture of plasmids, all with different fragments inserted, results.

4 Bacteria take up the plasmids and are grown in a nutrient medium that selects for recombinant clones.

Culture of bacteria

5 Colonies containing clones of each fragment of the original DNA are separated and maintained as pure cultures. Each such culture is a "volume" in the gene library.

Individual recombinant clones

16.11 Constructing a Gene Library *(Page 362)*

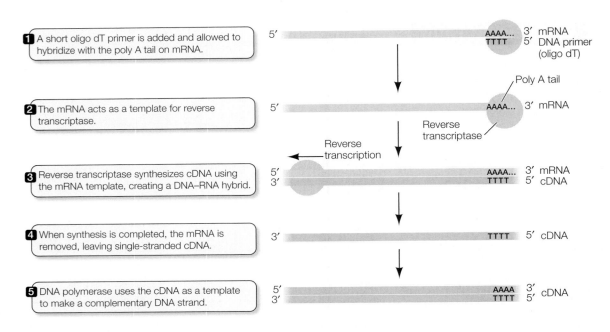

1 A short oligo dT primer is added and allowed to hybridize with the poly A tail on mRNA.

5′ ━━━━━━━━━━━━━━━━ AAAA... 3′ mRNA
                                   TTTT 5′ DNA primer (oligo dT)

2 The mRNA acts as a template for reverse transcriptase.

Poly A tail

5′ ━━━━━━━━━━━━━━━━ AAAA... 3′ mRNA

Reverse transcriptase

3 Reverse transcriptase synthesizes cDNA using the mRNA template, creating a DNA–RNA hybrid.

Reverse transcription

5′ ━━━━━━━━━━━━━━━━ AAAA... 3′ mRNA
3′ ━━━━━━━━━━━━━━━━ TTTT 5′ cDNA

4 When synthesis is completed, the mRNA is removed, leaving single-stranded cDNA.

3′ ━━━━━━━━━━━━━━━━ TTTT 5′ cDNA

5 DNA polymerase uses the cDNA as a template to make a complementary DNA strand.

5′ ━━━━━━━━━━━━━━━━ AAAA 3′ cDNA
3′ ━━━━━━━━━━━━━━━━ TTTT 5′

16.12 Synthesizing Complementary DNA *(Page 363)*

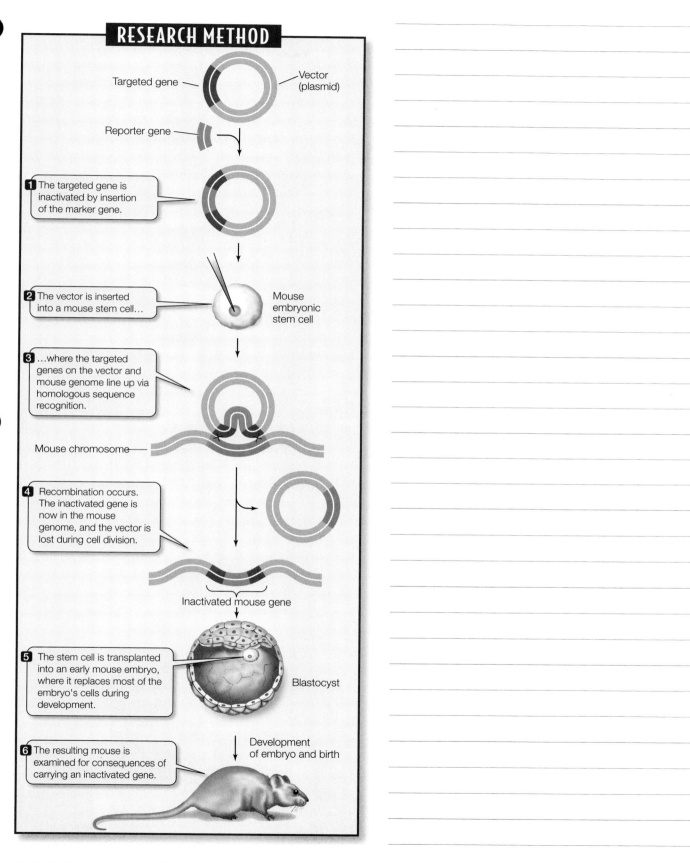

RESEARCH METHOD

Targeted gene ——— ◯ ——— Vector (plasmid)

Reporter gene ———

1 The targeted gene is inactivated by insertion of the marker gene.

2 The vector is inserted into a mouse stem cell...

Mouse embryonic stem cell

3 ...where the targeted genes on the vector and mouse genome line up via homologous sequence recognition.

Mouse chromosome ———

4 Recombination occurs. The inactivated gene is now in the mouse genome, and the vector is lost during cell division.

Inactivated mouse gene

5 The stem cell is transplanted into an early mouse embryo, where it replaces most of the embryo's cells during development.

Blastocyst

Development of embryo and birth

6 The resulting mouse is examined for consequences of carrying an inactivated gene.

16.13 Making a Knockout Mouse *(Page 364)*

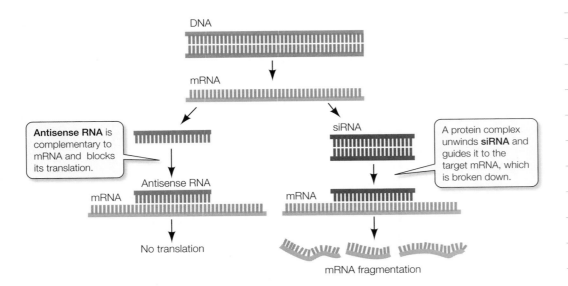

16.14 Using Antisense RNA and RNAi to Block Translation of mRNA *(Page 365)*

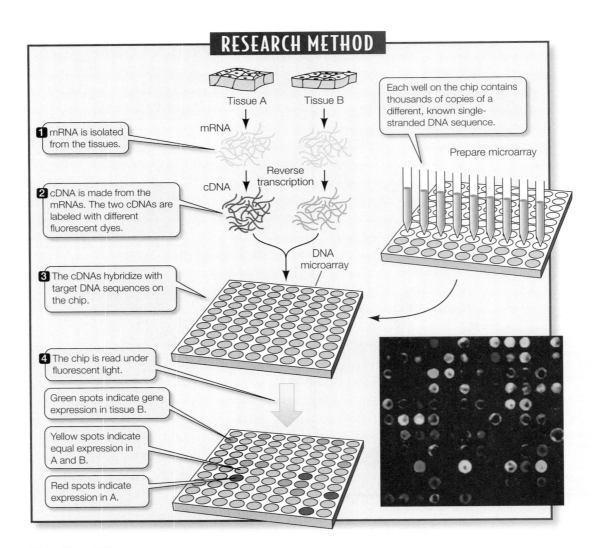

16.15 DNA on a Chip *(Page 366)*

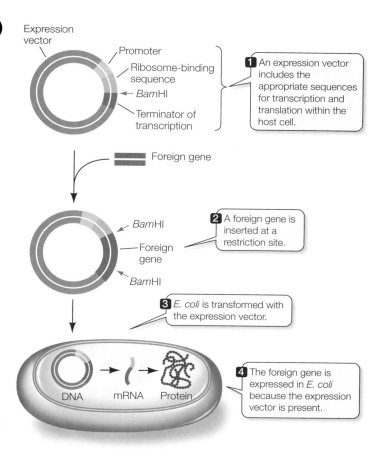

16.16 An Expression Vector Allows a Transgene to Be Expressed in a Host Cell *(Page 367)*

TABLE 16.1

Some Medically Useful Products of Biotechnology

PRODUCT	USE
Colony-stimulating factor	Stimulates production of white blood cells in patients with cancer and AIDS
Erythropoietin	Prevents anemia in patients undergoing kidney dialysis and cancer therapy
Factor VIII	Replaces clotting factor missing in patients with hemophilia A
Growth hormone	Replaces missing hormone in people of short stature
Insulin	Stimulates glucose uptake from blood in people with insulin-dependent (Type I) diabetes
Platelet-derived growth factor	Stimulates wound healing
Tissue plasminogen activator	Dissolves blood clots after heart attacks and strokes
Vaccine proteins: Hepatitis B, herpes, influenza, Lyme disease, meningitis, pertussis, etc.	Prevent and treat infectious diseases

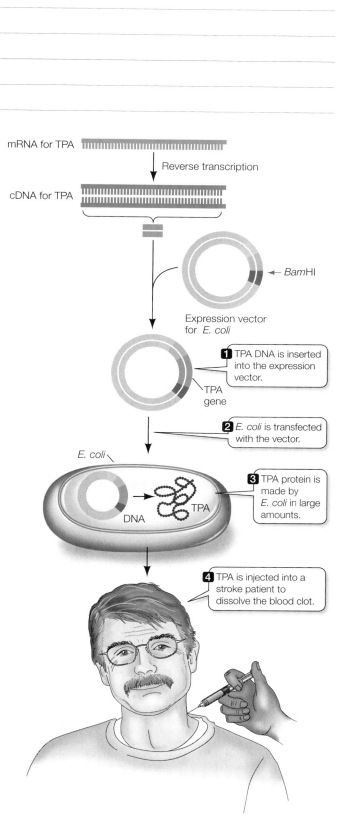

16.17 Tissue Plasminogen Activator: From Protein to Gene to Drug *(Page 368)*

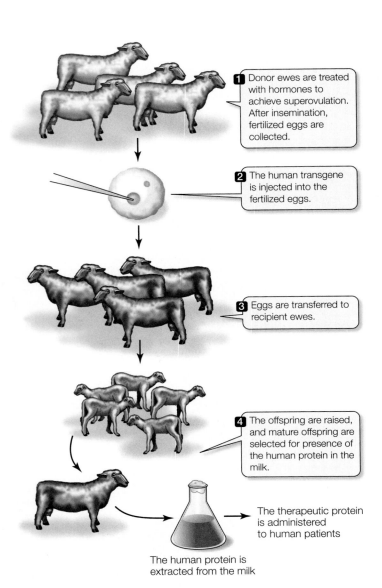

1 Donor ewes are treated with hormones to achieve superovulation. After insemination, fertilized eggs are collected.

2 The human transgene is injected into the fertilized eggs.

3 Eggs are transferred to recipient ewes.

4 The offspring are raised, and mature offspring are selected for presence of the human protein in the milk.

The human protein is extracted from the milk

The therapeutic protein is administered to human patients

16.18 Pharming *(Page 369)*

TABLE 16.2

Agricultural Applications of Biotechnology under Development

PROBLEM	TECHNOLOGY/GENES
Improving the environmental adaptations of plants	Genes for drought tolerance, salt tolerance
Improving nutritional traits	High-lysine seeds
Improving crops after harvest	Delay of fruit ripening; sweeter vegetables
Using plants as bioreactors	Plastics, oils, and drugs produced in plants
Controlling crop pests	Herbicide tolerance; resistance to viruses, bacteria, fungi, insects

17 Genome Sequencing, Molecular Biology, and Medicine

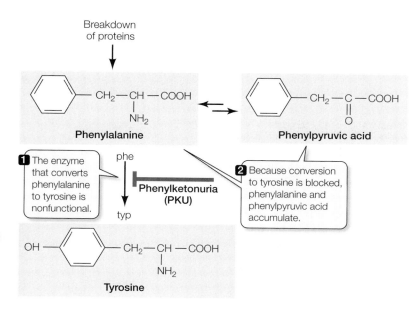

17.1 One Gene, One Enzyme *(Page 376)*

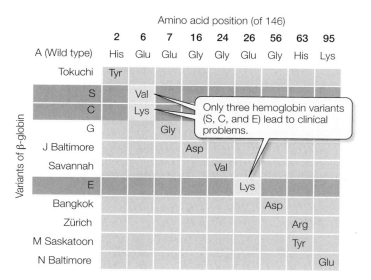

17.2 Hemoglobin Polymorphism *(Page 376)*

(A) Hypercholesterolemia

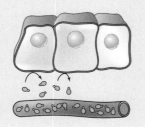

Normal liver cell: Cholesterol, as part of low-density lipoprotein (LDL), enters the cell after LDL binds to a receptor.

Familial hypercholesterolemia: Absence of a functional LDL receptor prevents cholesterol from entering the cells, and it accumulates in the blood.

Liver cells

LDL receptor

LDL

LDL in blood

(B) Cystic fibrosis

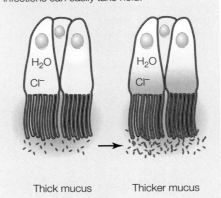

Normal cell lining the airway: Cl⁻ leaves the cell through an ion channel. Water follows by osmosis. The resulting moist thin mucus allows cilia to beat and sweep away foreign particles, including bacteria.

Cystic fibrosis: Lack of a Cl⁻ channel causes a thick, viscous mucus to form. Cilia cannot beat properly and remove bacteria; infections can easily take hold.

Lung cells

H_2O

Cl⁻ channel

Cilia

Cl⁻

H_2O

Cl⁻

H_2O

Bacteria

Cl⁻
H_2O

Thick mucus   Thin mucus

Thick mucus   Thicker mucus

17.3 Genetic Diseases of Membrane Proteins *(Page 377)*

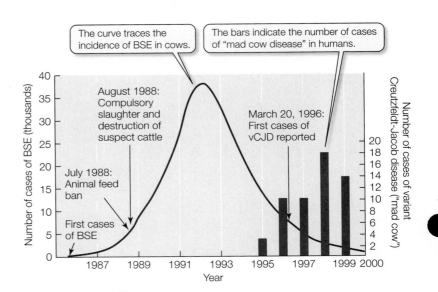

The curve traces the incidence of BSE in cows.

The bars indicate the number of cases of "mad cow disease" in humans.

August 1988: Compulsory slaughter and destruction of suspect cattle

March 20, 1996: First cases of vCJD reported

July 1988: Animal feed ban

First cases of BSE

Number of cases of BSE (thousands)

Number of cases of variant Creutzfeldt-Jacob disease ("mad cow")

Year

17.4 Mad Cow Disease in Britain *(Page 378)*

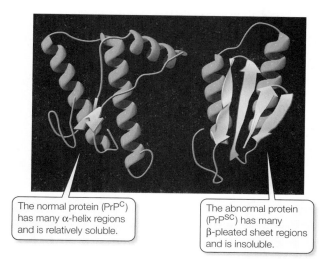

The normal protein (PrP^C) has many α-helix regions and is relatively soluble.

The abnormal protein (PrP^SC) has many β-pleated sheet regions and is insoluble.

17.5 Prion Diseases are Disorders of Protein Conformation
(Page 378)

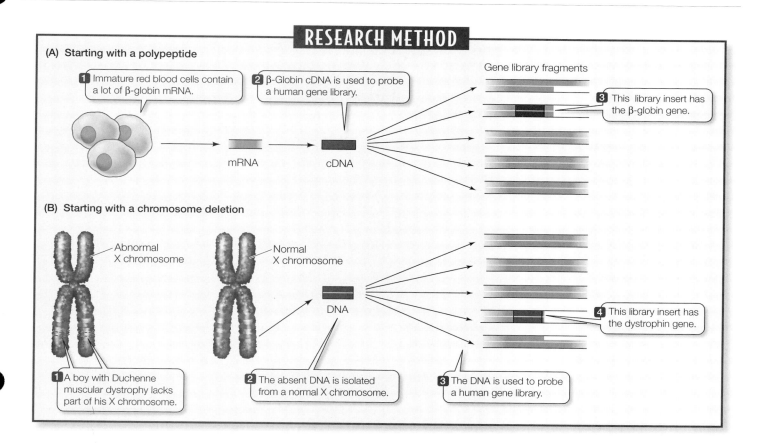

RESEARCH METHOD

(A) Starting with a polypeptide

1 Immature red blood cells contain a lot of β-globin mRNA.

2 β-Globin cDNA is used to probe a human gene library.

Gene library fragments

3 This library insert has the β-globin gene.

mRNA

cDNA

(B) Starting with a chromosome deletion

Abnormal X chromosome

Normal X chromosome

DNA

4 This library insert has the dystrophin gene.

1 A boy with Duchenne muscular dystrophy lacks part of his X chromosome.

2 The absent DNA is isolated from a normal X chromosome.

3 The DNA is used to probe a human gene library.

17.7 Two Strategies for Isolating Human Genes *(Page 380)*

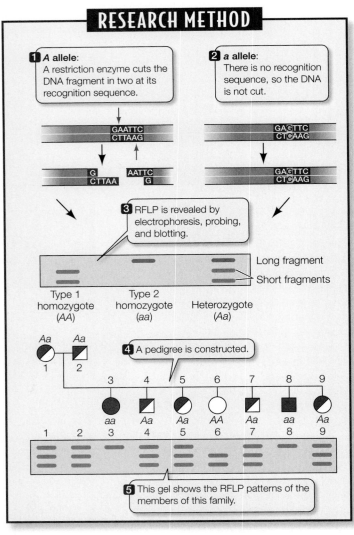

RESEARCH METHOD

1 *A* allele:
A restriction enzyme cuts the DNA fragment in two at its recognition sequence.

2 *a* allele:
There is no recognition sequence, so the DNA is not cut.

GAATTC
CTTAAG

GAGTTC
CTCAAG

G   AATTC
CTTAA   G

GAGTTC
CTCAAG

3 RFLP is revealed by electrophoresis, probing, and blotting.

Long fragment
Short fragments

Type 1 homozygote (*AA*)
Type 2 homozygote (*aa*)
Heterozygote (*Aa*)

Aa   *Aa*
1   2

4 A pedigree is constructed.

3   4   5   6   7   8   9

aa   *Aa*   *Aa*   *AA*   *Aa*   *aa*   *Aa*
3   4   5   6   7   8   9

1   2   3   4   5   6   7   8   9

5 This gel shows the RFLP patterns of the members of this family.

17.8 RFLP Mapping *(Page 381)*

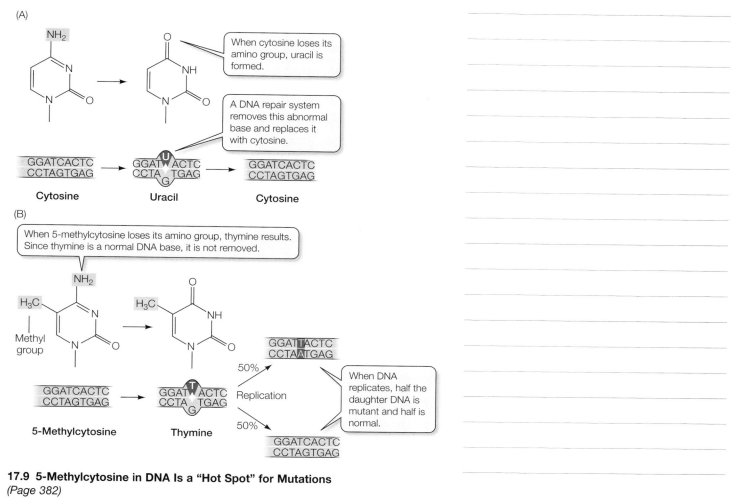

17.9 5-Methylcytosine in DNA Is a "Hot Spot" for Mutations
(Page 382)

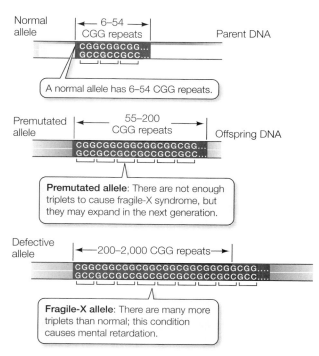

17.10 The CGG Repeats in the *FMR1* Gene Expand with Each Generation *(Page 383)*

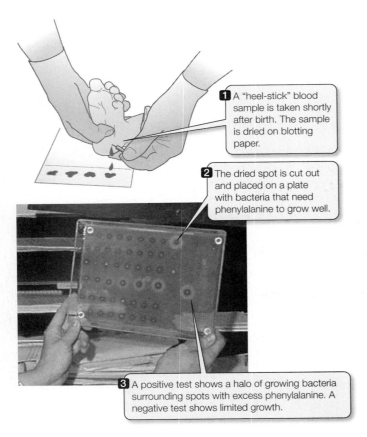

1 A "heel-stick" blood sample is taken shortly after birth. The sample is dried on blotting paper.

2 The dried spot is cut out and placed on a plate with bacteria that need phenylalanine to grow well.

3 A positive test shows a halo of growing bacteria surrounding spots with excess phenylalanine. A negative test shows limited growth.

17.11 Genetic Screening of Newborns for Phenylketonuria
(Page 384)

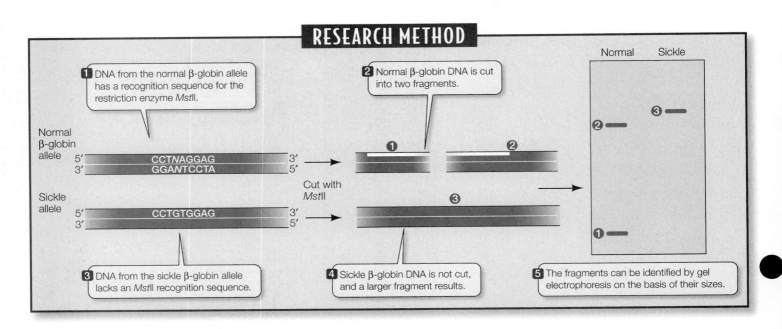

RESEARCH METHOD

1 DNA from the normal β-globin allele has a recognition sequence for the restriction enzyme *Mst*II.

2 Normal β-globin DNA is cut into two fragments.

Normal
β-globin
allele
5′ CCTNAGGAG 3′
3′ GGANTCCTA 5′

Cut with
*Mst*II

Sickle
allele
5′ CCTGTGGAG 3′
3′ 5′

Normal   Sickle

3 DNA from the sickle β-globin allele lacks an *Mst*II recognition sequence.

4 Sickle β-globin DNA is not cut, and a larger fragment results.

5 The fragments can be identified by gel electrophoresis on the basis of their sizes.

17.12 DNA Testing by Allele-Specific Cleavage *(Page 385)*

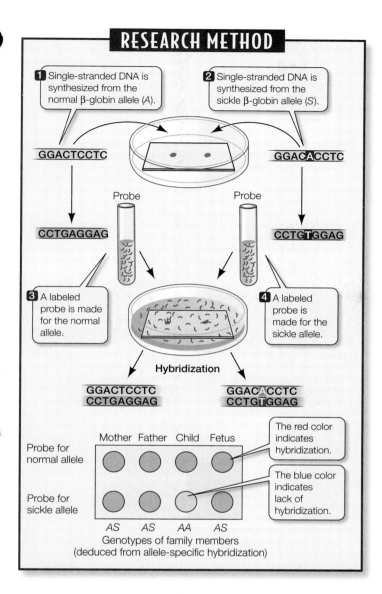

17.13 DNA Testing by Allele-Specific Oligonucleotide Hybridization
(Page 386)

TABLE 17.1

TABLE 17.1

Human Cancers Known To Be Caused by Viruses

CANCER	ASSOCIATED VIRUS
Liver cancer	Hepatitis B virus
Lymphoma, nasopharyngeal cancer	Epstein–Barr virus
T cell leukemia	Human T cell leukemia virus (HTLV-I)
Anogenital cancers	Papillomavirus
Kaposi's sarcoma	Kaposi's sarcoma herpesvirus

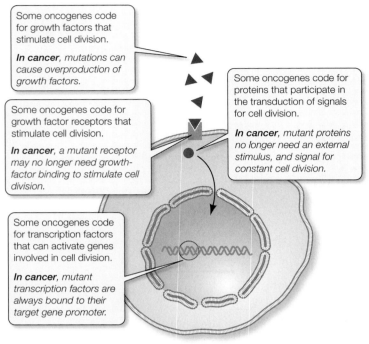

Some oncogenes code for growth factors that stimulate cell division.

In cancer, mutations can cause overproduction of growth factors.

Some oncogenes code for growth factor receptors that stimulate cell division.

In cancer, a mutant receptor may no longer need growth-factor binding to stimulate cell division.

Some oncogenes code for proteins that participate in the transduction of signals for cell division.

In cancer, mutant proteins no longer need an external stimulus, and signal for constant cell division.

Some oncogenes code for transcription factors that can activate genes involved in cell division.

In cancer, mutant transcription factors are always bound to their target gene promoter.

17.15 Oncogene Products Stimulate Cell Division *(Page 388)*

(A)

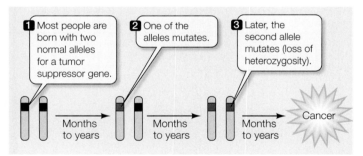

1 Most people are born with two normal alleles for a tumor suppressor gene.

2 One of the alleles mutates.

3 Later, the second allele mutates (loss of heterozygosity).

Months to years → Months to years → Months to years → Cancer

(B)

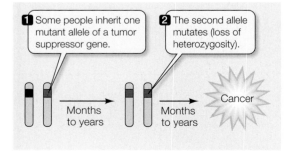

1 Some people inherit one mutant allele of a tumor suppressor gene.

2 The second allele mutates (loss of heterozygosity).

Months to years → Months to years → Cancer

17.16 The "Two-Hit" Hypothesis for Cancer *(Page 389)*

Some tumor suppressor genes code for cell adhesion/recognition proteins.

In cancer, mutations of these genes cause cells to lose adhesion to their neighbors and spread throughout the body.

Some tumor suppressor genes code for enzymes involved in DNA repair.

In cancer, mutant enzymes no longer repair DNA, and mutations accumulate.

Some tumor suppressor genes code for proteins that stop at the cell cycle in G1.

In cancer, mutant proteins no longer block cell division.

17.17 Tumor Suppressor Gene Products Inhibit Cell Division *(Page 389)*

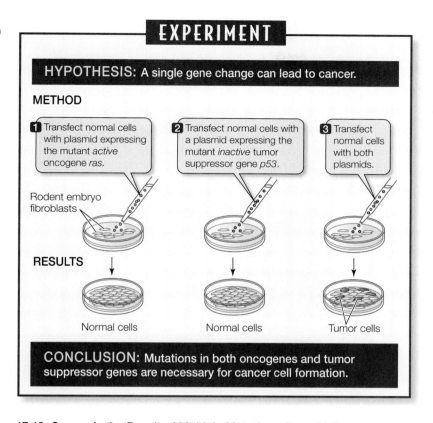

EXPERIMENT

HYPOTHESIS: A single gene change can lead to cancer.

METHOD

1 Transfect normal cells with plasmid expressing the mutant *active* oncogene *ras*.

2 Transfect normal cells with a plasmid expressing the mutant *inactive* tumor suppressor gene *p53*.

3 Transfect normal cells with both plasmids.

Rodent embryo fibroblasts

RESULTS

Normal cells

Normal cells

Tumor cells

CONCLUSION: Mutations in both oncogenes and tumor suppressor genes are necessary for cancer cell formation.

17.18 Cancer is the Result of Multiple Mutations *(Page 390)*

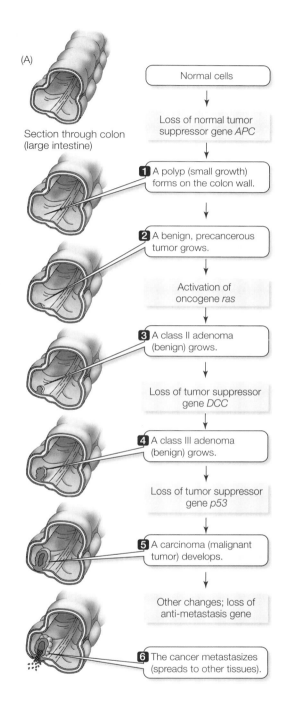

(A)

Section through colon (large intestine)

Normal cells

↓

Loss of normal tumor suppressor gene *APC*

↓

1 A polyp (small growth) forms on the colon wall.

↓

2 A benign, precancerous tumor grows.

↓

Activation of oncogene *ras*

↓

3 A class II adenoma (benign) grows.

↓

Loss of tumor suppressor gene *DCC*

↓

4 A class III adenoma (benign) grows.

↓

Loss of tumor suppressor gene *p53*

↓

5 A carcinoma (malignant tumor) develops.

↓

Other changes; loss of anti-metastasis gene

↓

6 The cancer metastasizes (spreads to other tissues).

(B)

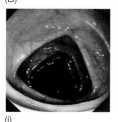

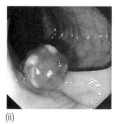

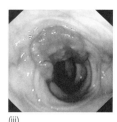

(i) (ii) (iii)

17.19 Multiple Mutations Transform a Normal Colon Epithelial Cell into a Cancer Cell *(Page 390)*

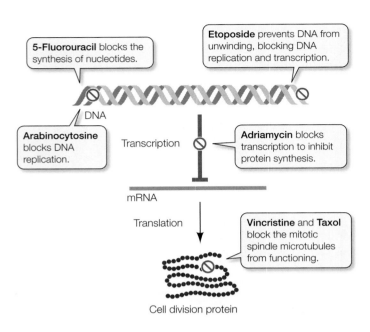

5-Fluorouracil blocks the synthesis of nucleotides.

Etoposide prevents DNA from unwinding, blocking DNA replication and transcription.

DNA

Arabinocytosine blocks DNA replication.

Transcription

Adriamycin blocks transcription to inhibit protein synthesis.

mRNA

Translation

Vincristine and **Taxol** block the mitotic spindle microtubules from functioning.

Cell division protein

17.20 Strategies for Killing Cancer Cells *(Page 391)*

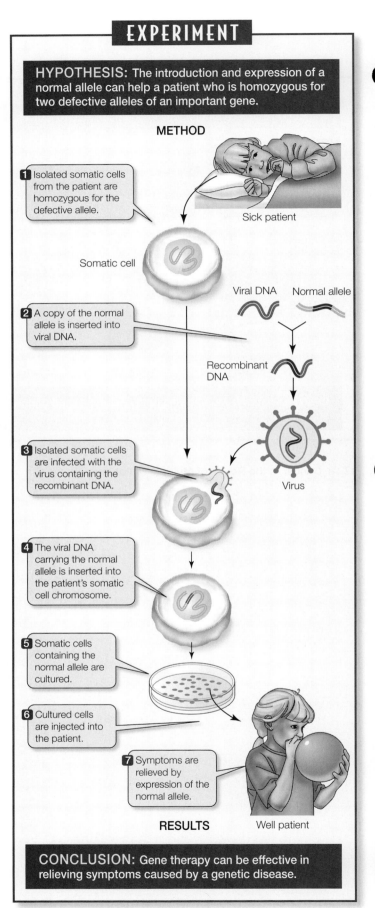

EXPERIMENT

HYPOTHESIS: The introduction and expression of a normal allele can help a patient who is homozygous for two defective alleles of an important gene.

METHOD

Sick patient

1 Isolated somatic cells from the patient are homozygous for the defective allele.

Somatic cell

Viral DNA   Normal allele

2 A copy of the normal allele is inserted into viral DNA.

Recombinant DNA

Virus

3 Isolated somatic cells are infected with the virus containing the recombinant DNA.

4 The viral DNA carrying the normal allele is inserted into the patient's somatic cell chromosome.

5 Somatic cells containing the normal allele are cultured.

6 Cultured cells are injected into the patient.

7 Symptoms are relieved by expression of the normal allele.

RESULTS   Well patient

CONCLUSION: Gene therapy can be effective in relieving symptoms caused by a genetic disease.

17.21 Gene Therapy: The *Ex Vivo* Approach *(Page 392)*

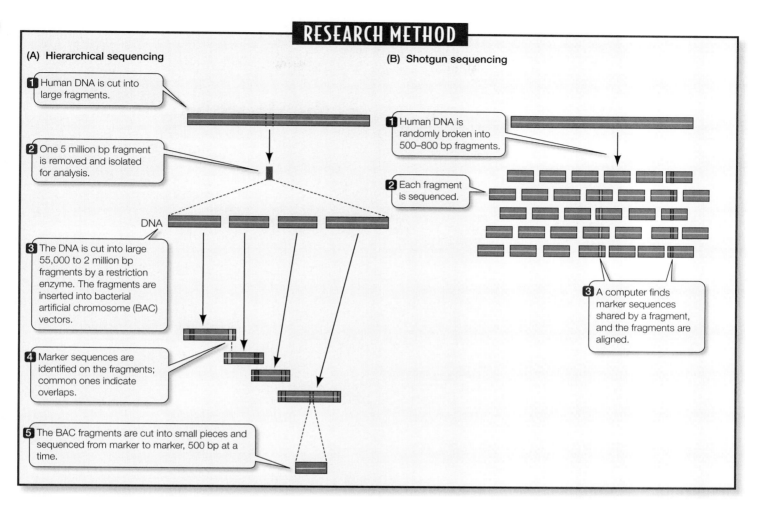

17.22 Two Approaches to Sequencing DNA *(Page 394)*

(A)

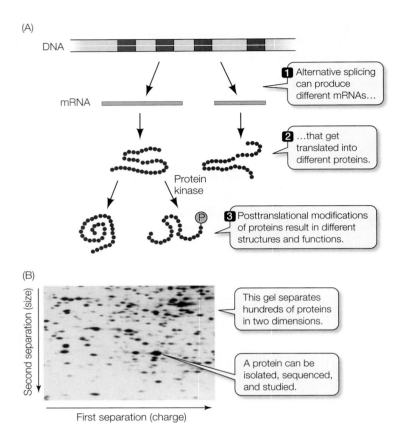

17.23 Proteomics *(Page 396)*

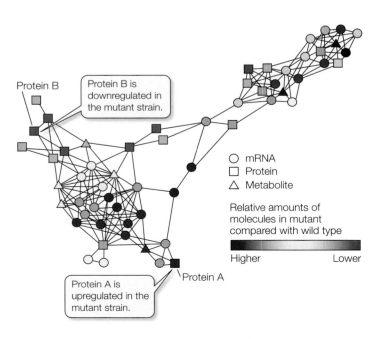

17.24 Applying Systems Biology *(Page 397)*

CHAPTER 18 Immunology: Gene Expression and Natural Defense Systems

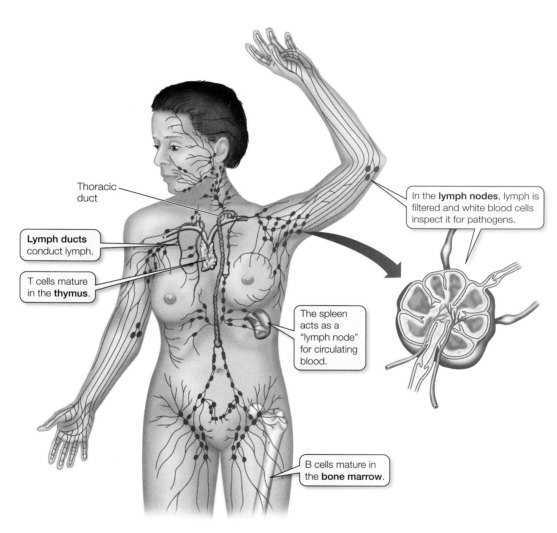

Thoracic duct

Lymph ducts conduct lymph.

T cells mature in the **thymus**.

The spleen acts as a "lymph node" for circulating blood.

In the **lymph nodes**, lymph is filtered and white blood cells inspect it for pathogens.

B cells mature in the **bone marrow**.

18.1 The Human Lymphatic System *(Page 402)*

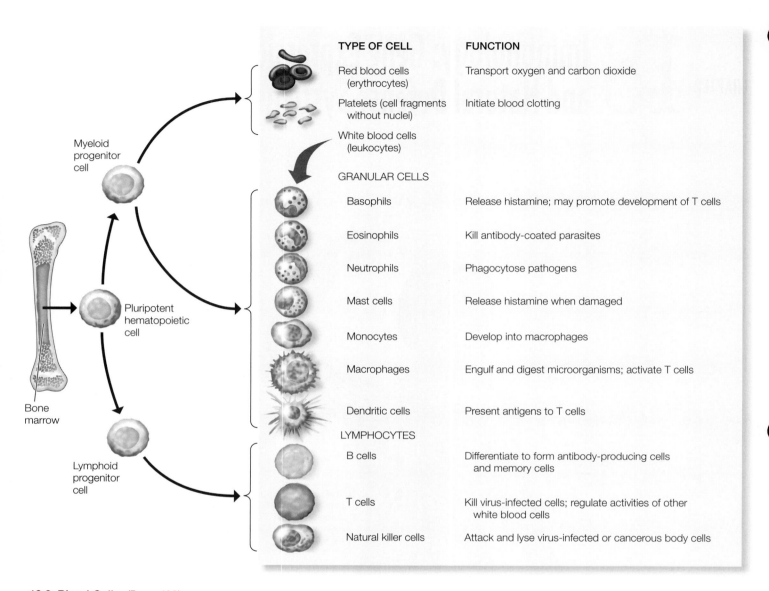

TYPE OF CELL	FUNCTION
Red blood cells (erythrocytes)	Transport oxygen and carbon dioxide
Platelets (cell fragments without nuclei)	Initiate blood clotting
White blood cells (leukocytes)	
GRANULAR CELLS	
Basophils	Release histamine; may promote development of T cells
Eosinophils	Kill antibody-coated parasites
Neutrophils	Phagocytose pathogens
Mast cells	Release histamine when damaged
Monocytes	Develop into macrophages
Macrophages	Engulf and digest microorganisms; activate T cells
Dendritic cells	Present antigens to T cells
LYMPHOCYTES	
B cells	Differentiate to form antibody-producing cells and memory cells
T cells	Kill virus-infected cells; regulate activities of other white blood cells
Natural killer cells	Attack and lyse virus-infected or cancerous body cells

Myeloid progenitor cell

Pluripotent hematopoietic cell

Bone marrow

Lymphoid progenitor cell

18.2 Blood Cells *(Page 403)*

TABLE 18.1

Human Nonspecific Defenses

DEFENSIVE MECHANISM	FUNCTION
Surface barriers	
Skin	Prevents entry of pathogens and foreign substances
Acid secretions	Inhibit bacterial growth on skin
Mucus	Prevents entry of pathogens; produces defensins that kill pathogens
Mucous secretions	Trap bacteria and other pathogens in digestive and respiratory tracts
Nasal hairs	Filter bacteria in nasal passages
Cilia	Move mucus and trapped materials away from respiratory passages
Gastric juice	Concentrated HCl and proteases destroy pathogens in stomach
Acid in vagina	Limits growth of fungi and bacteria in female reproductive tract
Tears, saliva	Lubricate and cleanse; contain lysozyme, which destroys bacteria
Nonspecific cellular, chemical, and coordinated defenses	
Normal flora	Compete with pathogens; may produce substances toxic to pathogens
Fever	Body-wide response inhibits microbial multiplication and speeds body repair processes
Coughing, sneezing	Expels pathogens from upper respiratory passages
Inflammatory response (involves leakage of blood plasma and phagocytes from capillaries)	Limits spread of pathogens to neighboring tissues; concentrates defenses; digests pathogens and dead tissue cells; released chemical mediators attract phagocytes and lymphocytes to site
Phagocytes (macrophages and neutrophils)	Engulf and destroy pathogens that enter body
Natural killer cells	Attack and lyse virus-infected or cancerous body cells
Antimicrobial proteins	
Interferons	Released by virus-infected cells to protect healthy tissue from viral infection; mobilize specific defenses
Complement proteins	Lyse microorganisms, enhance phagocytosis, and assist in inflammatory and antibody responses

(Page 404)

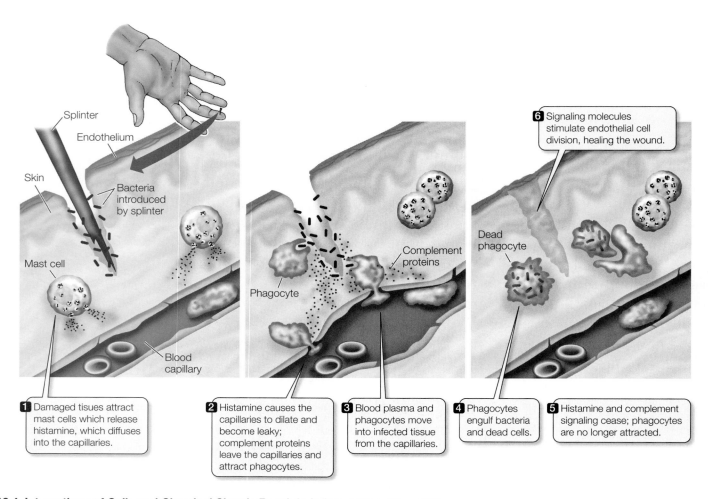

Splinter

Endothelium

Skin

Bacteria introduced by splinter

Mast cell

Phagocyte

Complement proteins

Blood capillary

Dead phagocyte

6 Signaling molecules stimulate endothelial cell division, healing the wound.

1 Damaged tisues attract mast cells which release histamine, which diffuses into the capillaries.

2 Histamine causes the capillaries to dilate and become leaky; complement proteins leave the capillaries and attract phagocytes.

3 Blood plasma and phagocytes move into infected tissue from the capillaries.

4 Phagocytes engulf bacteria and dead cells.

5 Histamine and complement signaling cease; phagocytes are no longer attracted.

18.4 Interactions of Cells and Chemical Signals Result in Inflammation *(Page 406)*

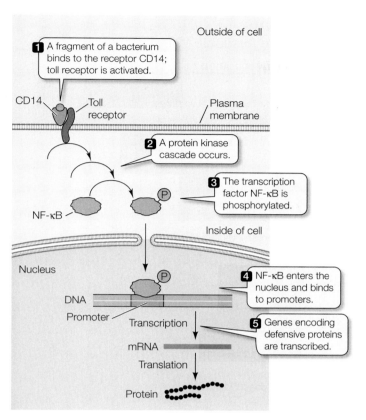

Outside of cell

1 A fragment of a bacterium binds to the receptor CD14; toll receptor is activated.

CD14 — Toll receptor

Plasma membrane

2 A protein kinase cascade occurs.

3 The transcription factor NF-κB is phosphorylated.

NF-κB

P

Inside of cell

Nucleus

P

DNA

Promoter

Transcription

4 NF-κB enters the nucleus and binds to promoters.

5 Genes encoding defensive proteins are transcribed.

mRNA

Translation

Protein

18.5 Cell Signaling and Defense *(Page 407)*

Antibodies react with antigenic determinants.

Antigenic determinants (epitopes) are small portions of antigens.

Antigen

Antigen

Page 407 In-Text Art

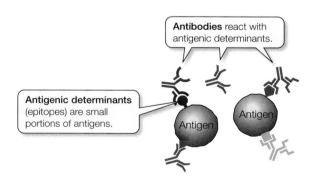

Each B cell makes a different, specific antibody and displays it on its cell surface.

1 This B cell makes an antibody that binds this specific antigenic determinant…

Antigenic determinant

Population of specific B cells

2 …which stimulates the cell to divide…

3 …resulting in a clone of cells.

Plasma cells

Memory cells

Antibodies

4 Some develop into plasma cells (effector B cells) that secrete the same antibody as the parent cell.

5 A few develop into non-secreting memory cells that divide at a low rate, perpetuating the clone.

18.6 Clonal Selection in B Cells *(Page 408)*

TABLE 18.2

Some Human Pathogens for Which Vaccines are Available

INFECTIOUS AGENT	DISEASE	VACCINATED POPULATION
Bacteria		
Bacillus anthracis	Anthrax	Those at risk in biological warfare
Bordetella pertussis	Whooping cough	Children and adults
Clostridium tetani	Tetanus	Children and adults
Corynebacterium diphtheriae	Diphtheria	Children
Haemophilus influenzae	Meningitis	Children
Mycobacterium tuberculosis	Tuberculosis	All people
Salmonella typhi	Typhoid fever	Areas exposed to agent
Streptococcus pneumoniae	Pneumonia	Elderly
Vibrio cholerae	Cholera	People in areas exposed to agent
Viruses		
Adenovirus	Respiratory disease	Military personnel
Hepatitis A	Liver disease	Areas exposed to agent
Hepatitis B	Liver disease, cancer	All people
Influenza virus	Flu	All people
Measles virus	Measles	Children and adolescents
Mumps virus	Mumps	Children and adolescents
Poliovirus	Polio	Children
Rabies virus	Rabies	Persons exposed to agent
Rubella virus	German measles	Children
Vaccinia virus	Smallpox	Laboratory workers, military personnel
Varicella-zoster virus	Chicken pox	Children

(Page 409)

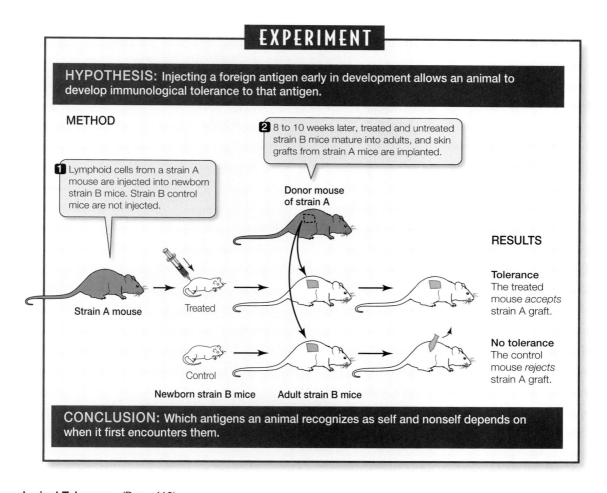

18.7 Immunological Tolerance *(Page 410)*

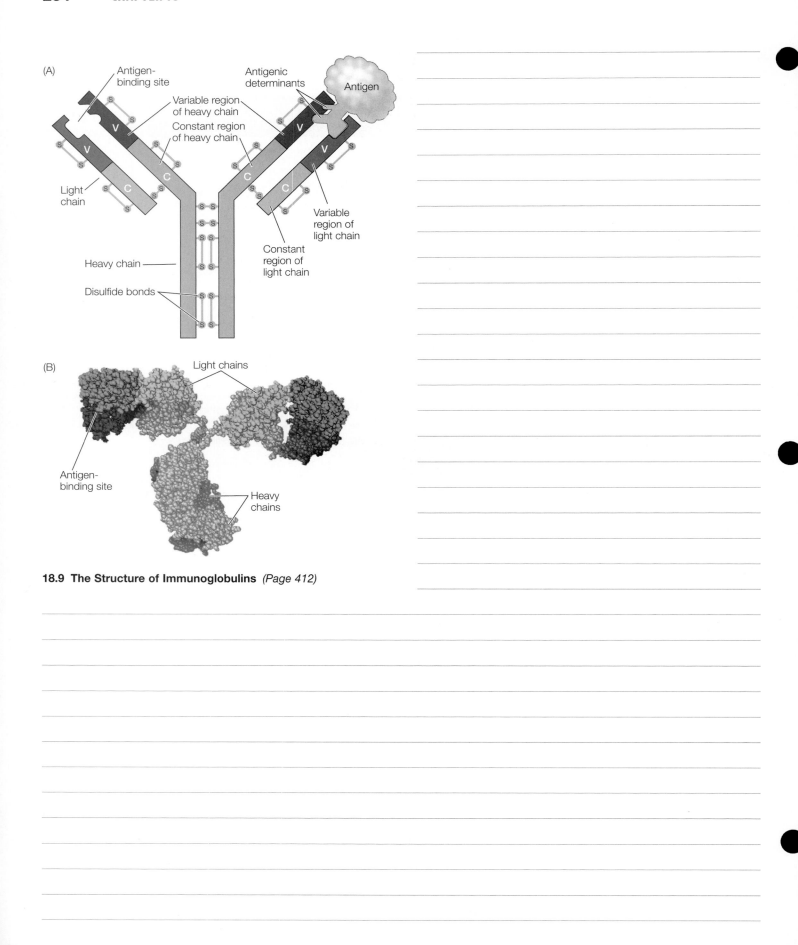

(A)

Antigen-binding site

Antigenic determinants

Antigen

Variable region of heavy chain

Constant region of heavy chain

Light chain

V

V

C

C

V

V

C

C

Variable region of light chain

Constant region of light chain

Heavy chain

Disulfide bonds

(B)

Light chains

Antigen-binding site

Heavy chains

18.9 The Structure of Immunoglobulins *(Page 412)*

TABLE 18.3

Antibody Classes

CLASS	GENERAL STRUCTURE		LOCATION	FUNCTION
IgG	Monomer		Free in blood plasma; about 80 percent of circulating antibodies	Most abundant antibody in primary and secondary immune responses; crosses placenta and provides passive immunization to fetus
IgM	Pentamer		Surface of B cell; free in blood plasma	Antigen receptor on B cell membrane; first class of antibodies released by B cells during primary response
IgD	Monomer		Surface of B cell	Cell surface receptor of mature B cell; important in B cell activation
IgA	Dimer		Saliva, tears, milk, and other body secretions	Protects mucosal surfaces; prevents attachment of pathogens to epithelial cells
IgE	Monomer		Secreted by plasma cells in skin and tissues lining gastrointestinal and respiratory tracts	When bound to antigens, binds to mast cells and basophils to trigger release of histamine that contributes to inflammation and some allergic responses

(Page 412)

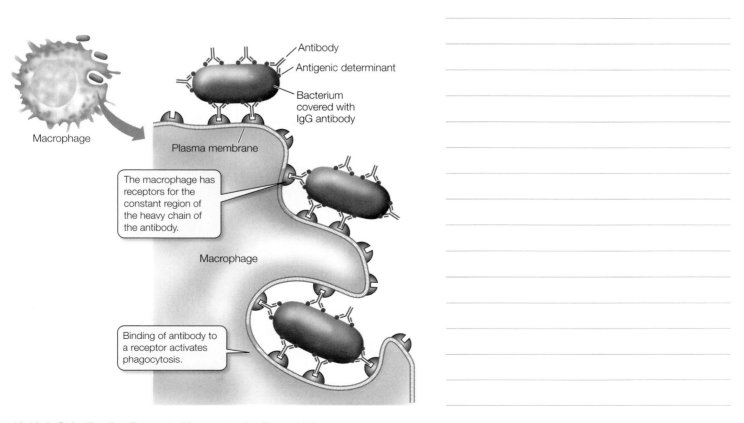

18.10 IgG Antibodies Promote Phagocytosis *(Page 413)*

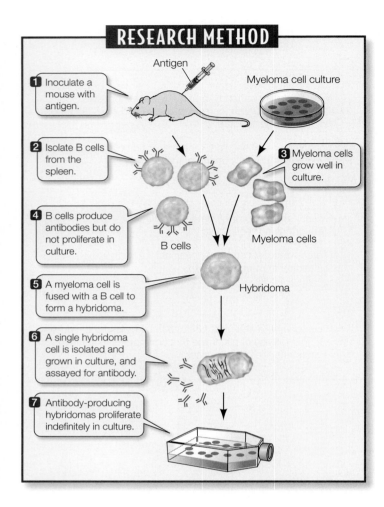

RESEARCH METHOD

Antigen

1 Inoculate a mouse with antigen.

Myeloma cell culture

2 Isolate B cells from the spleen.

3 Myeloma cells grow well in culture.

B cells

Myeloma cells

4 B cells produce antibodies but do not proliferate in culture.

5 A myeloma cell is fused with a B cell to form a hybridoma.

Hybridoma

6 A single hybridoma cell is isolated and grown in culture, and assayed for antibody.

7 Antibody-producing hybridomas proliferate indefinitely in culture.

18.11 Creating Hybridomas for the Production of Monoclonal Antibodies *(Page 413)*

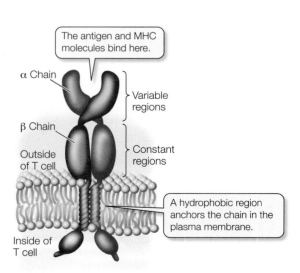

The antigen and MHC molecules bind here.

α Chain

Variable regions

β Chain

Constant regions

Outside of T cell

A hydrophobic region anchors the chain in the plasma membrane.

Inside of T cell

18.12 A T Cell Receptor *(Page 414)*

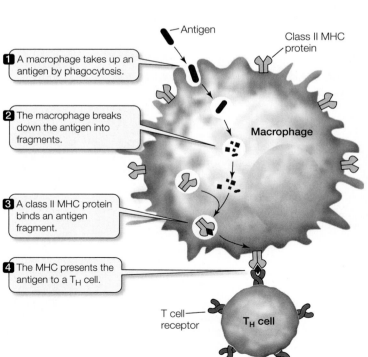

1 A macrophage takes up an antigen by phagocytosis.

Antigen

Class II MHC protein

2 The macrophage breaks down the antigen into fragments.

Macrophage

3 A class II MHC protein binds an antigen fragment.

4 The MHC presents the antigen to a T_H cell.

T cell receptor

T_H cell

18.13 Macrophages Are Antigen-Presenting Cells *(Page 415)*

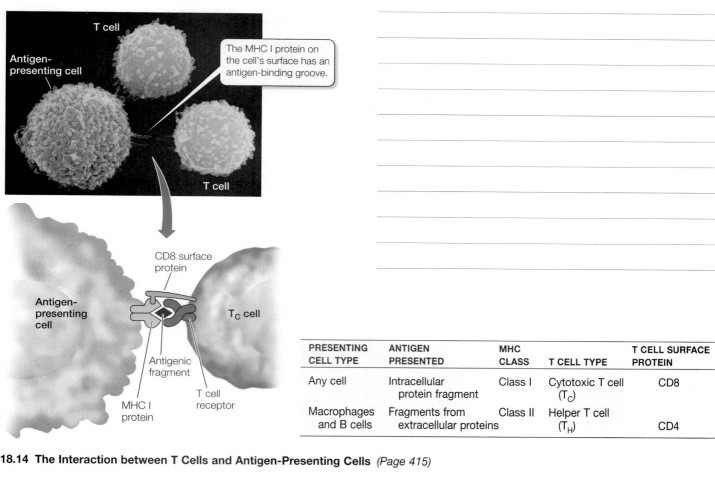

The MHC I protein on the cell's surface has an antigen-binding groove.

PRESENTING CELL TYPE	ANTIGEN PRESENTED	MHC CLASS	T CELL TYPE	T CELL SURFACE PROTEIN
Any cell	Intracellular protein fragment	Class I	Cytotoxic T cell (T_C)	CD8
Macrophages and B cells	Fragments from extracellular proteins	Class II	Helper T cell (T_H)	CD4

18.14 The Interaction between T Cells and Antigen-Presenting Cells *(Page 415)*

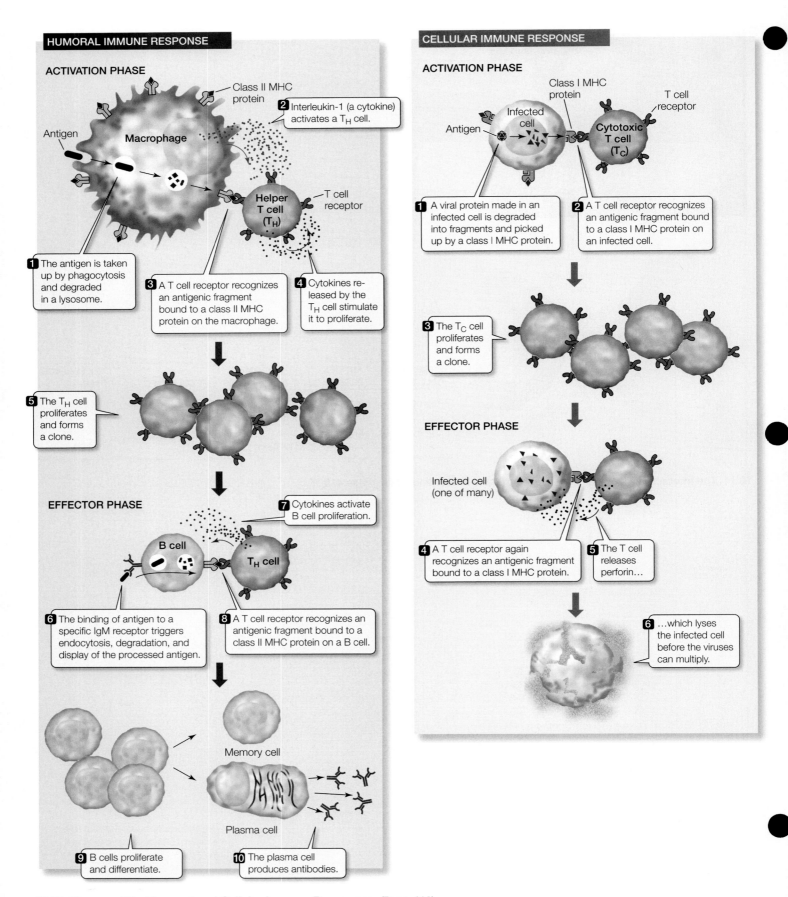

18.15 Phases of the Humoral and Cellular Immune Responses *(Page 416)*

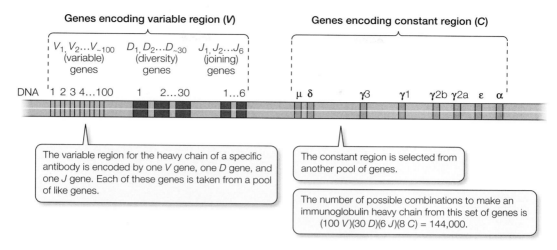

Genes encoding variable region (V)

Genes encoding constant region (C)

$V_1, V_2...V_{~100}$ (variable) genes

$D_1, D_2...D_{~30}$ (diversity) genes

$J_1, J_2...J_6$ (joining) genes

The variable region for the heavy chain of a specific antibody is encoded by one *V* gene, one *D* gene, and one *J* gene. Each of these genes is taken from a pool of like genes.

The constant region is selected from another pool of genes.

The number of possible combinations to make an immunoglobulin heavy chain from this set of genes is (100 *V*)(30 *D*)(6 *J*)(8 *C*) = 144,000.

18.16 Heavy-Chain Genes *(Page 418)*

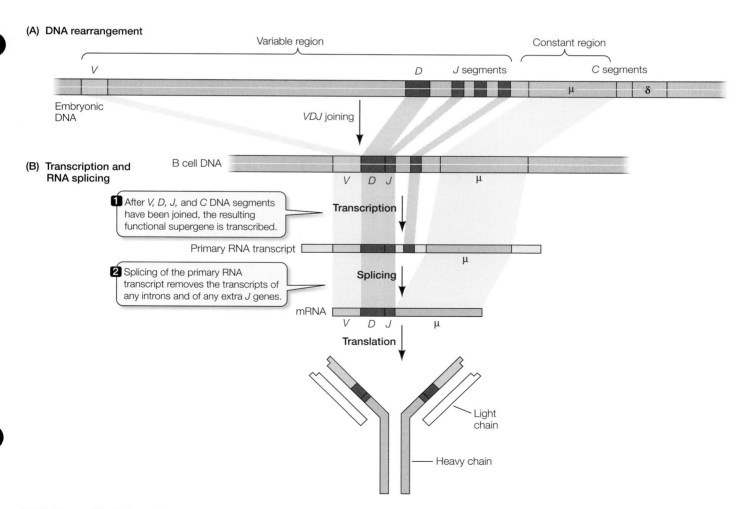

(A) DNA rearrangement

Variable region

Constant region

Embryonic DNA

VDJ joining

(B) Transcription and RNA splicing

B cell DNA

1 After *V, D, J,* and *C* DNA segments have been joined, the resulting functional supergene is transcribed.

Transcription

Primary RNA transcript

2 Splicing of the primary RNA transcript removes the transcripts of any introns and of any extra *J* genes.

Splicing

mRNA

Translation

Light chain

Heavy chain

18.17 Heavy-Chain Gene Rearrangement and Splicing *(Page 419)*

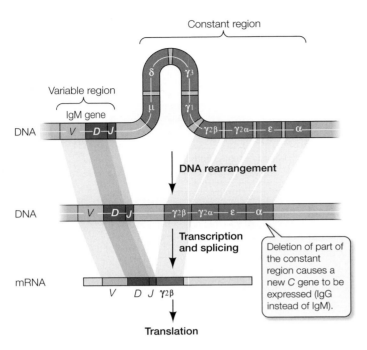

18.18 Class Switching *(Page 420)*

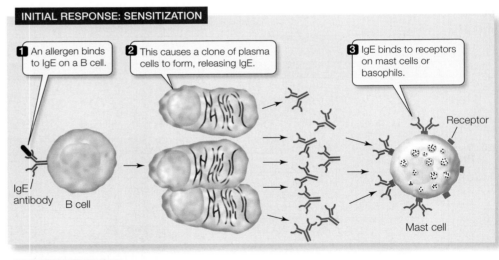

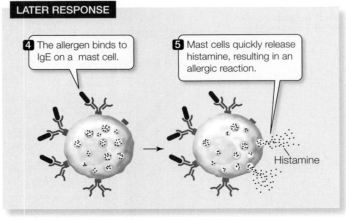

18.19 An Allergic Reaction *(Page 421)*

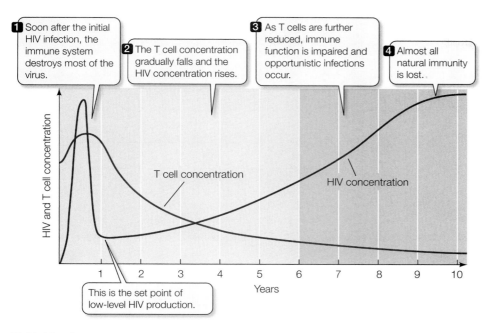

18.20 The Course of an HIV Infection *(Page 422)*

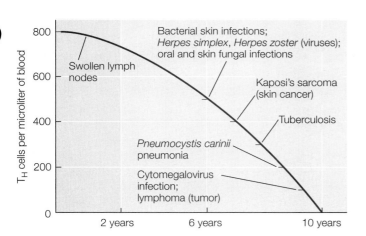

18.21 Relationship between T$_H$ Cell Count and Opportunistic Infections *(Page 422)*

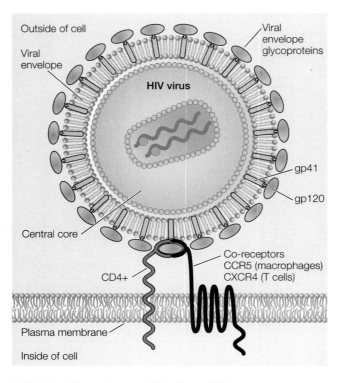

Outside of cell

Viral envelope glycoproteins

Viral envelope

HIV virus

gp41

gp120

Central core

Co-receptors
CCR5 (macrophages)
CXCR4 (T cells)

CD4+

Plasma membrane

Inside of cell

18.22 Two Receptors for HIV *(Page 423)*

CHAPTER 19 Differential Gene Expression in Development

ANIMAL DEVELOPMENT

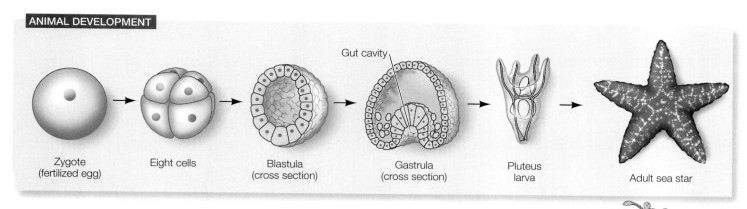

Gut cavity

Zygote (fertilized egg) → Eight cells → Blastula (cross section) → Gastrula (cross section) → Pluteus larva → Adult sea star

PLANT DEVELOPMENT

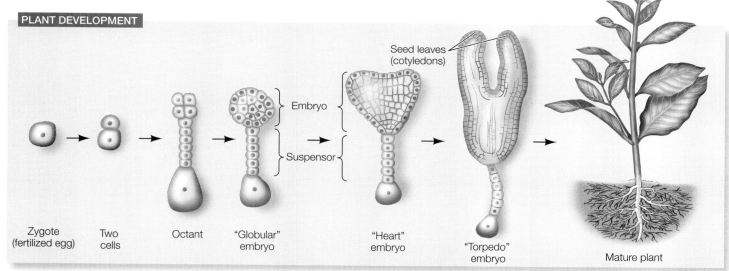

Seed leaves (cotyledons)

Embryo

Suspensor

Zygote (fertilized egg) → Two cells → Octant → "Globular" embryo → "Heart" embryo → "Torpedo" embryo → Mature plant

19.1 From Fertilized Egg to Adult *(Page 428)*

EXPERIMENT

HYPOTHESIS: The fate of the cells in an early amphibian embryo is irrevocably determined.

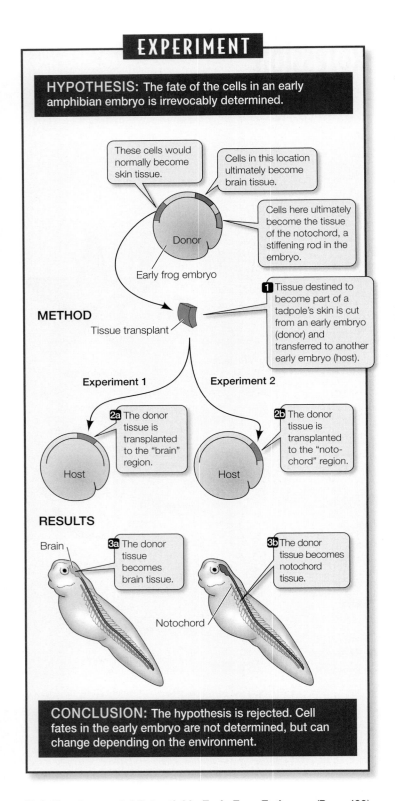

These cells would normally become skin tissue.

Cells in this location ultimately become brain tissue.

Cells here ultimately become the tissue of the notochord, a stiffening rod in the embryo.

Donor

Early frog embryo

METHOD

Tissue transplant

1 Tissue destined to become part of a tadpole's skin is cut from an early embryo (donor) and transferred to another early embryo (host).

Experiment 1

2a The donor tissue is transplanted to the "brain" region.

Host

Experiment 2

2b The donor tissue is transplanted to the "notochord" region.

Host

RESULTS

Brain

3a The donor tissue becomes brain tissue.

3b The donor tissue becomes notochord tissue.

Notochord

CONCLUSION: The hypothesis is rejected. Cell fates in the early embryo are not determined, but can change depending on the environment.

19.2 Developmental Potential in Early Frog Embryos *(Page 429)*

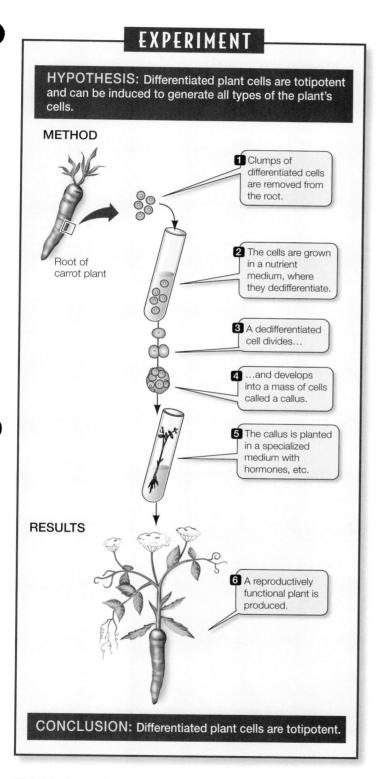

EXPERIMENT

HYPOTHESIS: Differentiated plant cells are totipotent and can be induced to generate all types of the plant's cells.

METHOD

Root of carrot plant

1 Clumps of differentiated cells are removed from the root.

2 The cells are grown in a nutrient medium, where they dedifferentiate.

3 A dedifferentiated cell divides...

4 ...and develops into a mass of cells called a callus.

5 The callus is planted in a specialized medium with hormones, etc.

RESULTS

6 A reproductively functional plant is produced.

CONCLUSION: Differentiated plant cells are totipotent.

19.3 Cloning a Plant *(Page 430)*

EXPERIMENT

HYPOTHESIS: Differentiated animal cells are totipotent.

METHOD

1 Cells are removed from the udder of a Dorset ewe.

Dorset sheep (#1)

2 An egg is removed from a Scottish blackface ewe.

Scottish blackface sheep (#2)

Nucleus

Micropipette

3 The nucleus is removed from the egg.

4 Udder cells are deprived of nutrients in culture to halt the cell cycle prior to DNA replication.

Donor nucleus (#1)

Enucleated egg (#2)

5 The udder cell (donor) and enucleated egg are fused.

6 Stimulating mitotic inducers causes the cell to divide.

7 An early embryo develops and is transplanted into a receptive ewe.

Scottish blackface sheep (#3)

RESULTS

8 The embryo develops and Dolly is born.

Dolly, a Dorset sheep, genetically identical to #1

CONCLUSION: Differentiated animal cells are totipotent in nuclear transplant experiments.

19.4 Cloning a Mammal *(Page 431)*

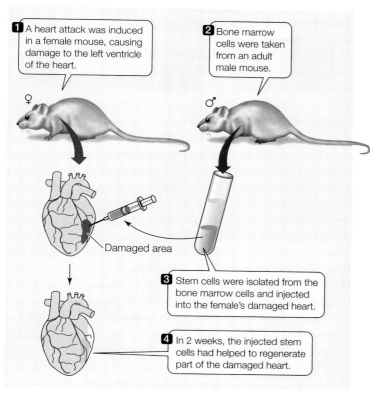

1 A heart attack was induced in a female mouse, causing damage to the left ventricle of the heart.

2 Bone marrow cells were taken from an adult male mouse.

Damaged area

3 Stem cells were isolated from the bone marrow cells and injected into the female's damaged heart.

4 In 2 weeks, the injected stem cells had helped to regenerate part of the damaged heart.

19.6 Repairing a Damaged Heart *(Page 433)*

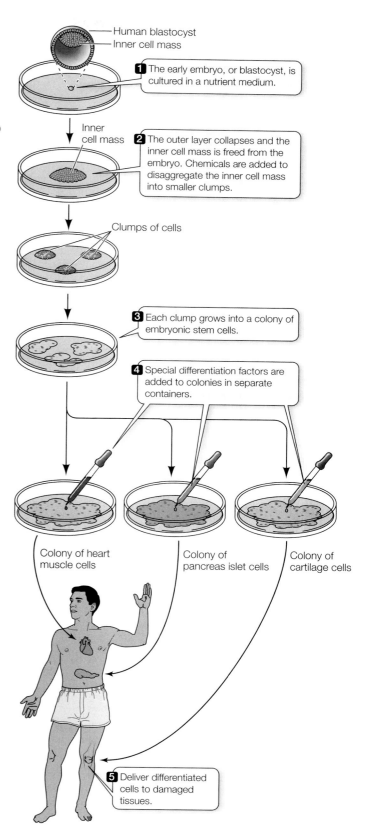

Human blastocyst
Inner cell mass

1 The early embryo, or blastocyst, is cultured in a nutrient medium.

Inner cell mass

2 The outer layer collapses and the inner cell mass is freed from the embryo. Chemicals are added to disaggregate the inner cell mass into smaller clumps.

Clumps of cells

3 Each clump grows into a colony of embryonic stem cells.

4 Special differentiation factors are added to colonies in separate containers.

Colony of heart muscle cells

Colony of pancreas islet cells

Colony of cartilage cells

5 Deliver differentiated cells to damaged tissues.

19.7 The Potential Use of Embryonic Stem Cells in Medicine *(Page 433)*

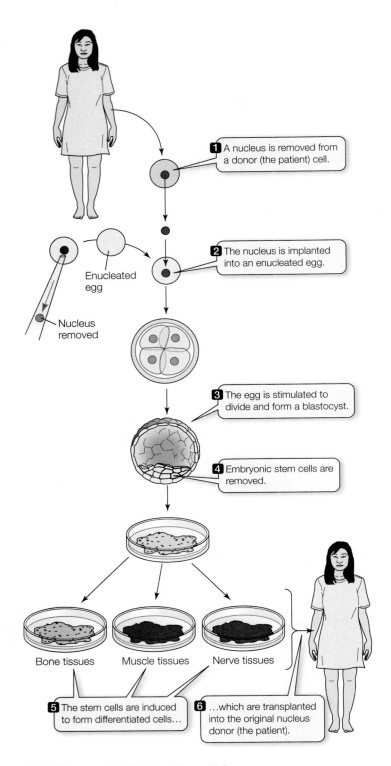

1 A nucleus is removed from a donor (the patient) cell.

Enucleated egg

Nucleus removed

2 The nucleus is implanted into an enucleated egg.

3 The egg is stimulated to divide and form a blastocyst.

4 Embryonic stem cells are removed.

Bone tissues   Muscle tissues   Nerve tissues

5 The stem cells are induced to form differentiated cells...

6 ...which are transplanted into the original nucleus donor (the patient).

19.8 Therapeutic Cloning *(Page 434)*

EXPERIMENT

HYPOTHESIS: Different regions in the fertilized egg and the embryo have different developmental fates.

METHOD

Top (animal pole)

Egg

Bottom (vegetal pole)

Fertilization and cell division

8-Cell stage

The embryo is bisected horizontally, separating upper cells from lower cells.

The embryo is bisected vertically, leaving each half with both upper and lower cells.

8-Cell stage

Glass needle

RESULTS

Top cells remain embryonic

Bottom cells produce abnormal larva

Two normal, but small, larvae are produced

CONCLUSION: The upper and lower halves of a sea urchin embryo differ in their developmental potential.

19.9 Asymmetry in the Early Sea Urchin Embryo *(Page 436)*

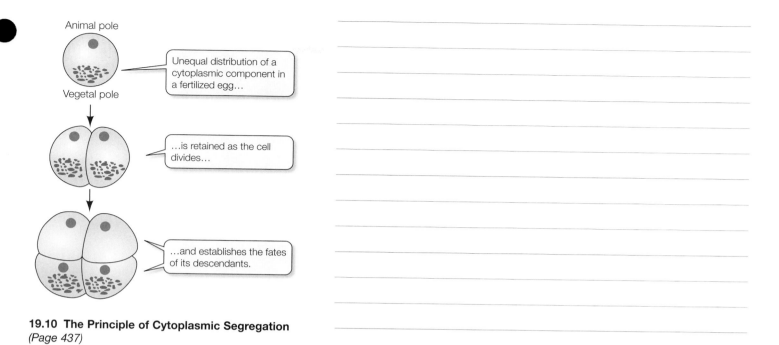

Animal pole

Unequal distribution of a cytoplasmic component in a fertilized egg…

Vegetal pole

…is retained as the cell divides…

…and establishes the fates of its descendants.

19.10 The Principle of Cytoplasmic Segregation
(Page 437)

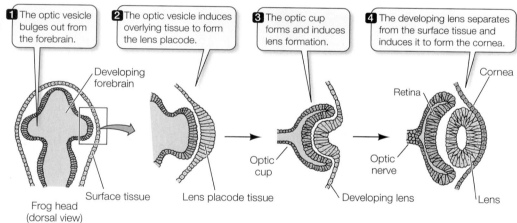

1 The optic vesicle bulges out from the forebrain.

2 The optic vesicle induces overlying tissue to form the lens placode.

3 The optic cup forms and induces lens formation.

4 The developing lens separates from the surface tissue and induces it to form the cornea.

Developing forebrain

Cornea

Retina

Optic cup

Optic nerve

Surface tissue

Frog head (dorsal view)

Lens placode tissue

Developing lens

Lens

19.11 Embryonic Inducers in the Vertebrate Eye *(Page 437)*

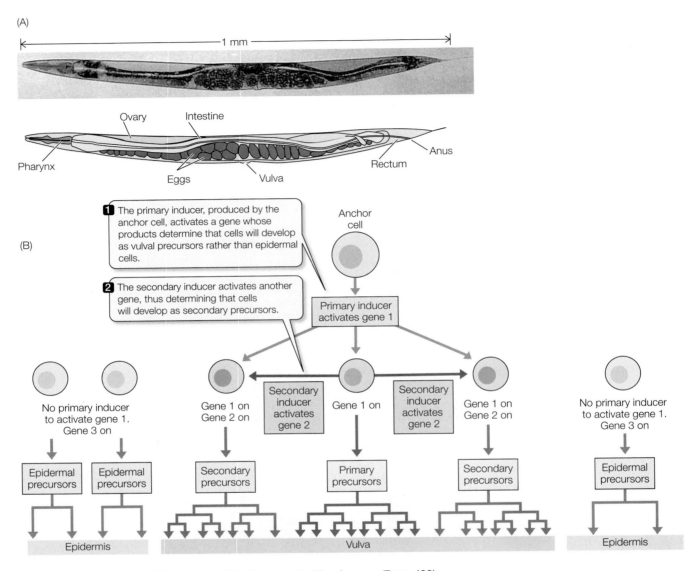

19.12 Induction during Vulval Development in *Caenorhabditis elegans* *(Page 438)*

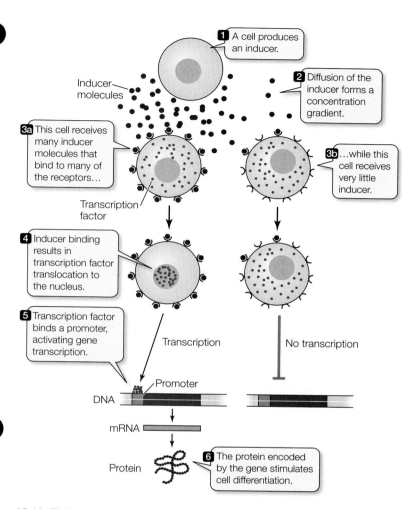

1 A cell produces an inducer.

Inducer molecules

2 Diffusion of the inducer forms a concentration gradient.

3a This cell receives many inducer molecules that bind to many of the receptors...

3b ...while this cell receives very little inducer.

Transcription factor

4 Inducer binding results in transcription factor translocation to the nucleus.

5 Transcription factor binds a promoter, activating gene transcription.

Transcription

No transcription

Promoter

DNA

mRNA

Protein

6 The protein encoded by the gene stimulates cell differentiation.

19.13 Embryonic Induction *(Page 439)*

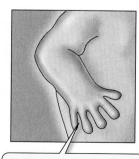

41 days after fertilization: Genes for apoptosis are expressed in the tissue between the digits.

56 days after fertilization: Apoptosis is complete. Cells of the digits have absorbed the remains of the dead cells.

19.14 Apoptosis Removes the Tissue between Human Fingers *(Page 440)*

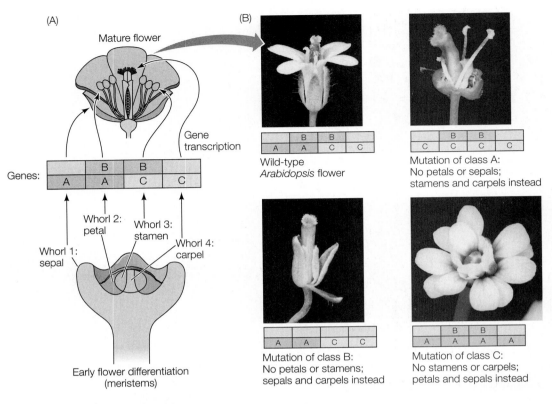

(A) Mature flower

Gene transcription

Genes:

	B	B	
A	A	C	C

Whorl 2: petal
Whorl 3: stamen
Whorl 4: carpel
Whorl 1: sepal

Early flower differentiation (meristems)

(B)

Wild-type *Arabidopsis* flower

	B	B	
A	A	C	C

Mutation of class A: No petals or sepals; stamens and carpels instead

	B	B	
C	C	C	C

Mutation of class B: No petals or stamens; sepals and carpels instead

A	A	C	C

Mutation of class C: No stamens or carpels; petals and sepals instead

	B	B	
A	A	A	A

19.15 Organ Identity Genes in *Arabidopsis* Flowers *(Page 440)*

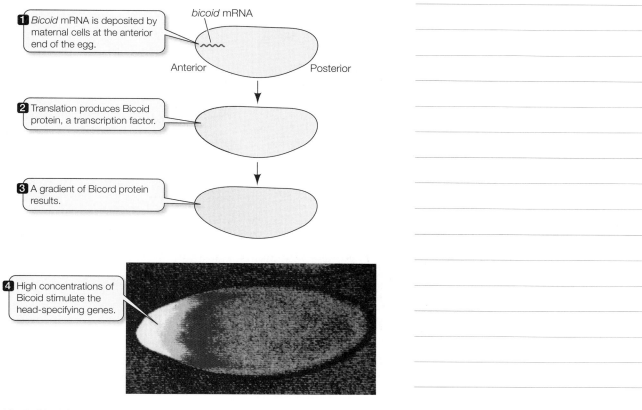

1 *Bicoid* mRNA is deposited by maternal cells at the anterior end of the egg.

bicoid mRNA

Anterior   Posterior

2 Translation produces Bicoid protein, a transcription factor.

3 A gradient of Bicord protein results.

4 High concentrations of Bicoid stimulate the head-specifying genes.

19.17 Bicoid Protein Provides Positional Information *(Page 443)*

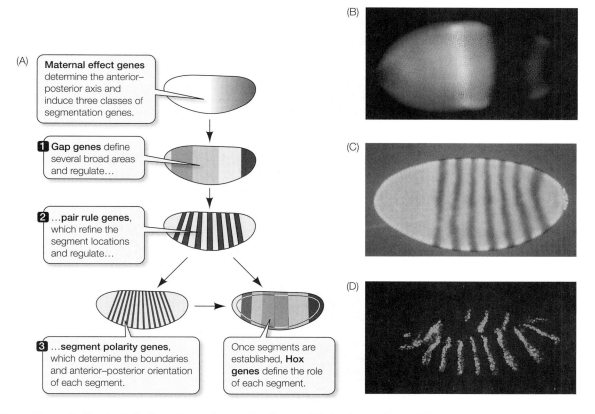

(A) **Maternal effect genes** determine the anterior–posterior axis and induce three classes of segmentation genes.

1 **Gap genes** define several broad areas and regulate…

2 …**pair rule genes**, which refine the segment locations and regulate…

3 …**segment polarity genes**, which determine the boundaries and anterior–posterior orientation of each segment.

Once segments are established, **Hox genes** define the role of each segment.

(B)

(C)

(D)

19.18 A Gene Cascade Controls Pattern Formation in the *Drosophila* Embryo *(Page 443)*

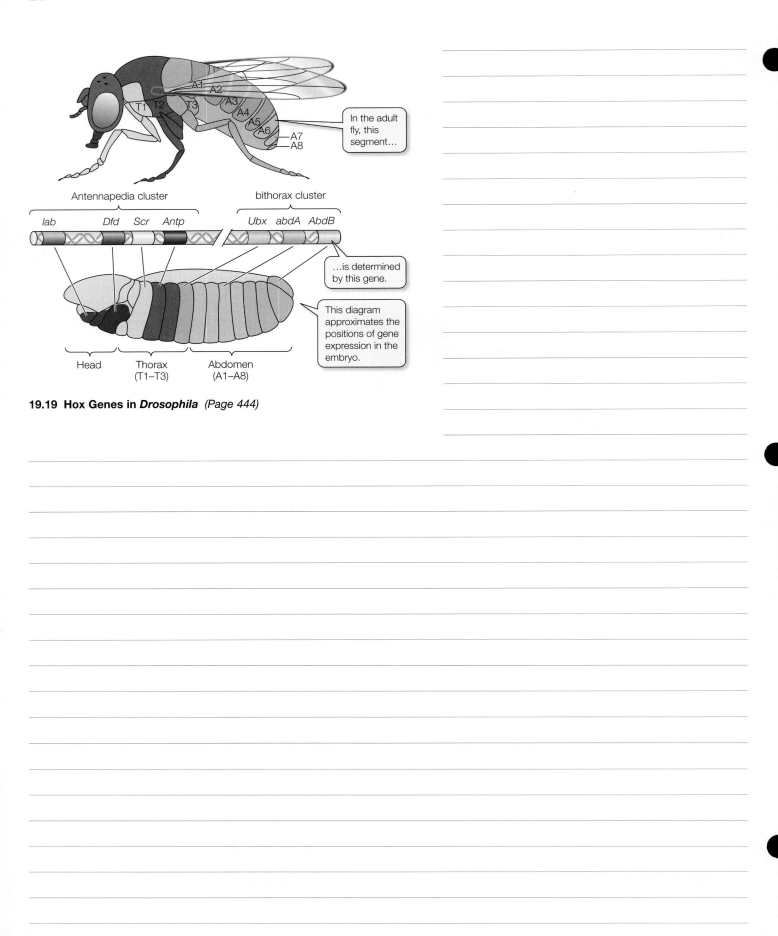

Antennapedia cluster   bithorax cluster

lab   *Dfd*   *Scr*   *Antp*   *Ubx*   *abdA*   *AbdB*

In the adult fly, this segment…

…is determined by this gene.

This diagram approximates the positions of gene expression in the embryo.

Head   Thorax   Abdomen
            (T1–T3)   (A1–A8)

19.19 Hox Genes in *Drosophila* *(Page 444)*

CHAPTER 20 Development and Evolutionary Change

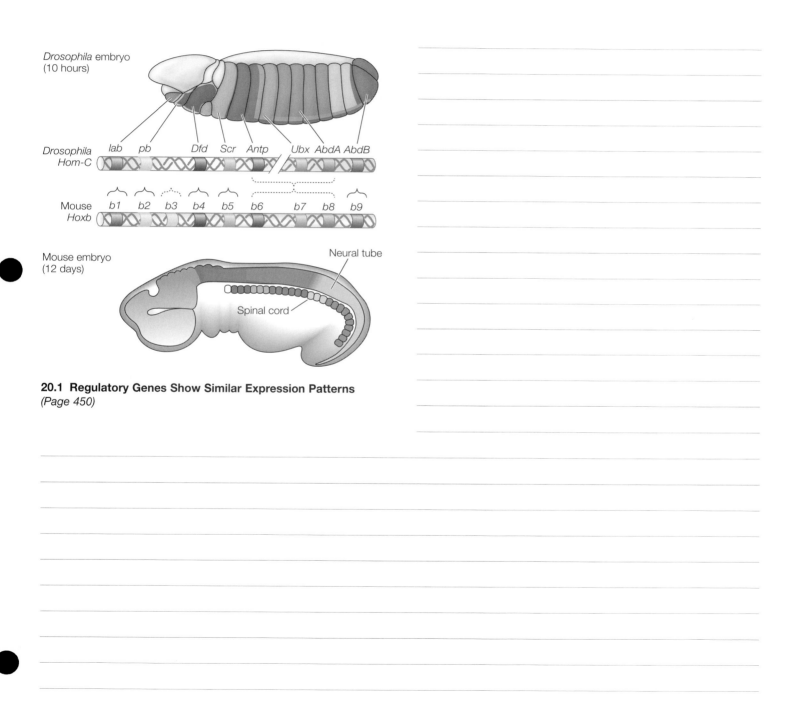

Drosophila embryo
(10 hours)

Drosophila
Hom-C

lab   pb   Dfd   Scr   Antp   Ubx   AbdA   AbdB

Mouse
Hoxb

b1   b2   b3   b4   b5   b6   b7   b8   b9

Mouse embryo
(12 days)

Neural tube

Spinal cord

20.1 Regulatory Genes Show Similar Expression Patterns
(Page 450)

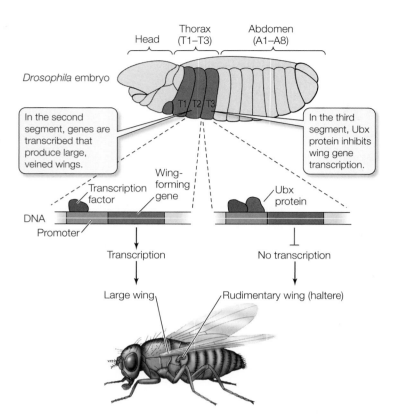

20.3 Segments Differentiate under Control of Genetic Switches
(Page 451)

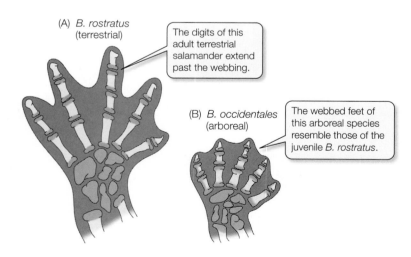

20.4 Heterochrony Created an Arboreal Salamander *(Page 452)*

EXPERIMENT

HYPOTHESIS: Adding Gremlin protein (a BMP4 inhibitor) to a developing chicken foot will transform it into a ducklike foot.

METHOD

Open chicken eggs and carefully add Gremlin-secreting beads to the webbing of embryonic chicken hindlimbs. Add beads that do not contain Gremlin to other hindlimbs (controls). Close the eggs and observe limb development.

RESULTS

In the hindlimbs in which Gremlin was secreted, the webbing does not undergo apoptosis, and the hindlimb resembles that of a duck. The control hindlimbs develop the normal chicken form.

Control   Gremlin added

 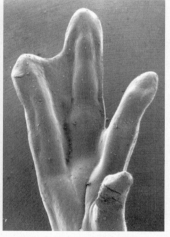

CONCLUSION: Differences in *gremlin* gene expression cause differences in morphology, allowing duck hindlimbs to retain their webbing.

20.6 Changing the Form of an Appendage *(Page 453)*

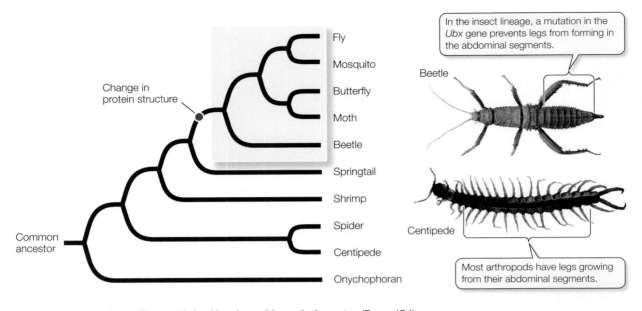

20.7 Mutation in a Hox Gene Changed the Number of Legs in Insects *(Page 454)*

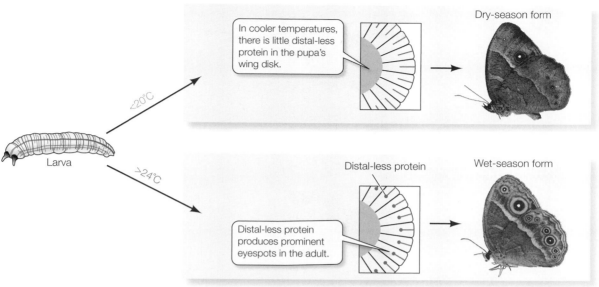

Dry-season form

In cooler temperatures, there is little distal-less protein in the pupa's wing disk.

Larva

Distal-less protein

Wet-season form

Distal-less protein produces prominent eyespots in the adult.

Bicyclus anynana

20.8 Eyespot Development Responds to Temperature *(Page 455)*

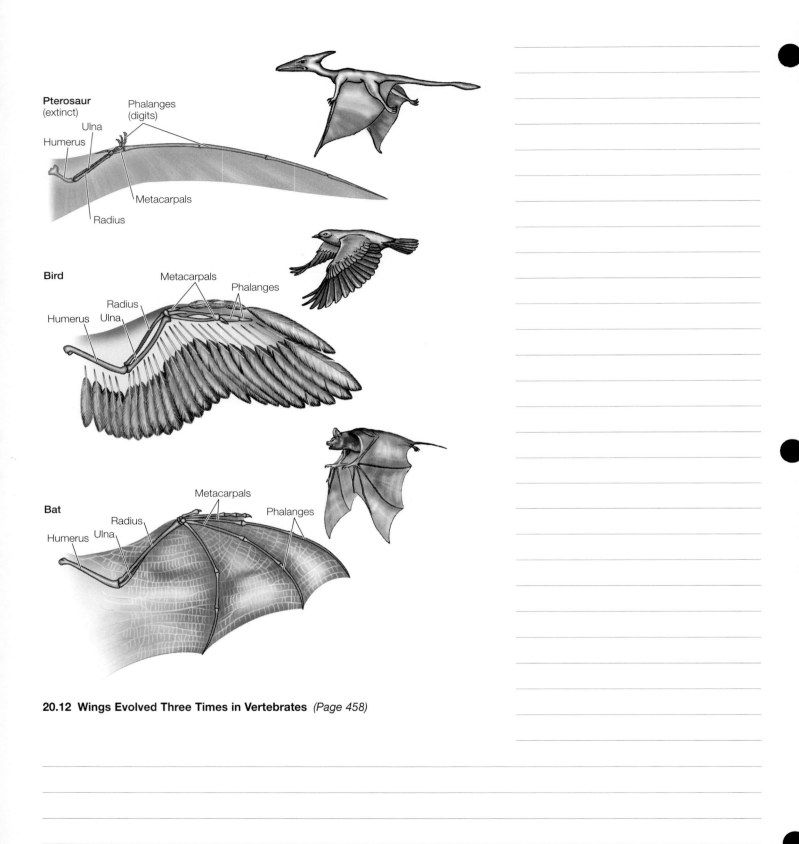

Pterosaur
(extinct)

Phalanges
(digits)

Ulna

Humerus

Metacarpals

Radius

Bird

Metacarpals

Phalanges

Radius

Ulna

Humerus

Bat

Metacarpals

Phalanges

Radius

Ulna

Humerus

20.12 Wings Evolved Three Times in Vertebrates *(Page 458)*

21 The History of Life on Earth

TABLE 21.1

Half-Lives of Some Radioisotopes

RADIOISOTOPE	HALF-LIFE
Phosphorus-32 (^{32}P)	14.3 days
Tritium (^{3}H)	12.3 years
Carbon-14 (^{14}C)	5,700 years
Potassium-40 (^{40}K)	1.3 billion years
Uranium-238 (^{238}U)	4.5 billion years

(Page 466)

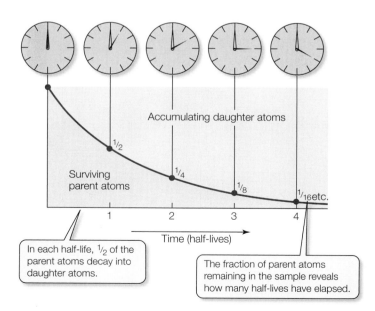

Accumulating daughter atoms

Surviving parent atoms

½

¼

⅛

$^{1}/_{16}$etc.

1   2   3   4

Time (half-lives)

In each half-life, ½ of the parent atoms decay into daughter atoms.

The fraction of parent atoms remaining in the sample reveals how many half-lives have elapsed.

21.1 Radioactive Isotopes Allow Us to Date Ancient Rocks
(Page 467)

TABLE 21.2

Earth's Geological History

RELATIVE TIME SPAN	ERA	PERIOD	ONSET	MAJOR PHYSICAL CHANGES ON EARTH
	Cenozoic	Quaternary	1.8 mya	Cold/dry climate; repeated glaciations
		Tertiary	65 mya	Continents near current positions; climate cools
	Mesozoic	Cretaceous	145 mya	Northern continents attached; Gondwana begins to drift apart; meteorite strikes Yucatán Peninsula
		Jurassic	200 mya	Two large continents form: Laurasia (north) and Gondwana (south); climate warm
		Triassic	251 mya	Pangaea begins to slowly drift apart; hot/humid climate
Precambrian	Paleozoic	Permian	297 mya	Continents aggregate into Pangaea; large glaciers form; dry climates form in interior of Pangaea
		Carboniferous	359 mya	Climate cools; marked latitudinal climate gradients
		Devonian	416 mya	Continents collide at end of period; meteorite probably strikes Earth
		Silurian	444 mya	Sea levels rise; two large continents form; hot/humid climate
		Ordovician	488 mya	Massive glaciation, sea level drops 50 meters
		Cambrian	542 mya	O_2 levels approach current levels
	Precambrian		600 mya	O_2 level at >5% of current level
			1.5 bya	O_2 level at >1% of current level
			3.8 bya	O_2 first appears in atmosphere
			4.5 bya	

[a]mya, million years ago; bya, billion years ago.

(Page 466)

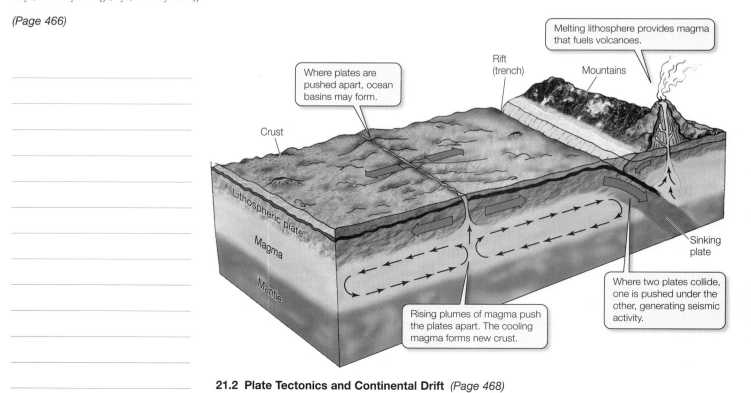

Where plates are pushed apart, ocean basins may form.

Rift (trench)

Mountains

Melting lithosphere provides magma that fuels volcanoes.

Crust

Lithospheric plate

Magma

Mantle

Sinking plate

Where two plates collide, one is pushed under the other, generating seismic activity.

Rising plumes of magma push the plates apart. The cooling magma forms new crust.

21.2 Plate Tectonics and Continental Drift *(Page 468)*

MAJOR EVENTS IN THE HISTORY OF LIFE

Humans evolve; many large mammals become extinct

Diversification of birds, mammals, flowering plants, and insects

Dinosaurs continue to diversify; flowering plants and mammals diversify; mass extinction at end of period (≈76% of species disappear)

Diverse dinosaurs; radiation of ray-finned fishes

Early dinosaurs; first mammals; marine invertebrates diversify; first flowering plants; mass extinction at end of period (≈65% of species disappear)

Reptiles diversify; amphibians decline; mass extinction at end of period (≈96% of species disappear)

Extensive "fern" forests; first reptiles; insects diversify

Fishes diversify; first insects and amphibians; mass extinction at end of period (≈75% of species disappear)

Jawless fishes diversify; first ray-finned fishes; plants and animals colonize land

Mass extinction at end of period (≈75% of species disappear)

Most animal phyla present; diverse photosynthetic protists

Ediacaran fauna

Eukaryotes evolve; several animal phyla appear

Origin of life; prokaryotes flourish

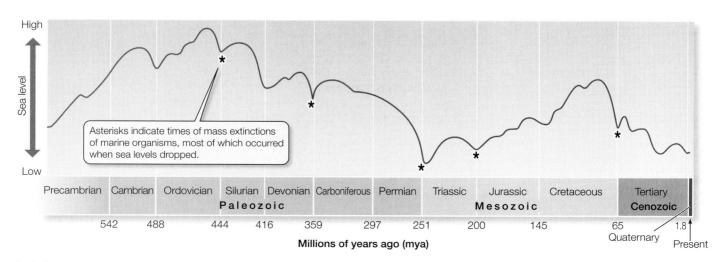

21.3 Sea Levels Have Changed Repeatedly *(Page 469)*

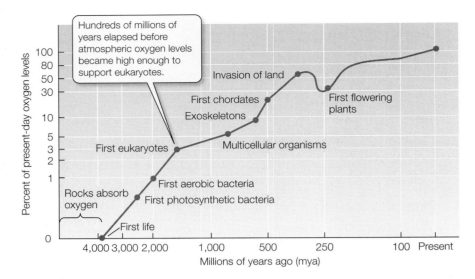

21.5 Larger Cells Need More Oxygen *(Page 470)*

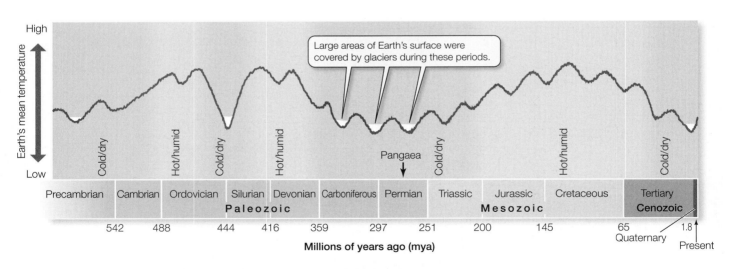

21.6 Hot/Humid and Cold/Dry Conditions Have Alternated over Earth's History *(Page 470)*

(A)

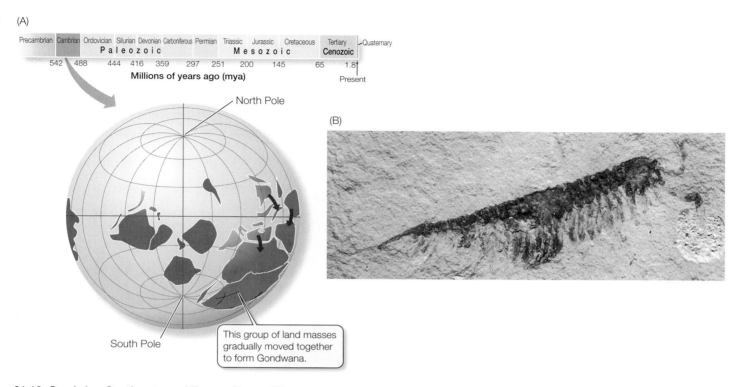

Precambrian	Cambrian	Ordovician	Silurian	Devonian	Carboniferous	Permian	Triassic	Jurassic	Cretaceous	Tertiary	Quaternary
			Paleozoic					Mesozoic		Cenozoic	

542   488   444   416   359   297   251   200   145   65   1.8

Millions of years ago (mya)

Present

North Pole

South Pole

This group of land masses gradually moved together to form Gondwana.

21.10 Cambrian Continents and Fauna *(Page 473)*

(B)

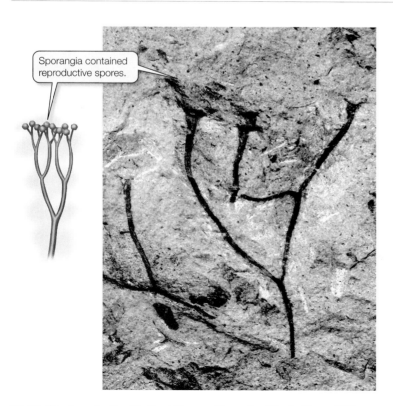

Sporangia contained reproductive spores.

21.11 *Cooksonia,* the Earliest Known Vascular Plant *(Page 474)*

(A)

| Precambrian | Cambrian | Ordovician | Silurian | Devonian | Carboniferous | Permian | Triassic | Jurassic | Cretaceous | Tertiary | Quaternary |

P a l e o z o i c   M e s o z o i c   Cenozoic

542 488 444 416 359 297 251 200 145 65 1.8

Millions of years ago (mya)   Present

During the Devonian period, the northern and southern continents were approaching one another.

(B)

Laurasia

Gondwana

21.12 Devonian Continents and Marine Communities *(Page 474)*

| Precambrian | Cambrian | Ordovician | Silurian | Devonian | Carboniferous | Permian | Triassic | Jurassic | Cretaceous | Tertiary | Quaternary |

P a l e o z o i c   M e s o z o i c   Cenozoic

542 488 444 416 359 297 251 200 145 65 1.8

Millions of years ago (mya)   Present

21.14 A Carboniferous "Crinoid Meadow" *(Page 475)*

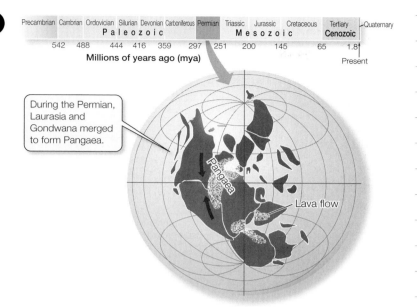

21.15 Pangaea Formed in the Permian Period *(Page 476)*

21.16 Jurassic Parkland *(Page 476)*

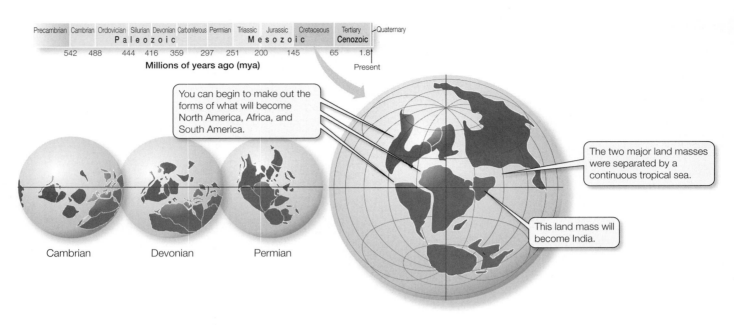

Precambrian Cambrian Ordovician Silurian Devonian Carboniferous Permian Triassic Jurassic Cretaceous Tertiary Quaternary
Paleozoic **Mesozoic** **Cenozoic**
542   488   444   416   359   297   251   200   145   65   1.8   Present
Millions of years ago (mya)

You can begin to make out the forms of what will become North America, Africa, and South America.

The two major land masses were separated by a continuous tropical sea.

This land mass will become India.

Cambrian   Devonian   Permian

21.17 Positions of the Continents during the Cretaceous Period *(Page 477)*

TABLE 21.3

Subdivisions of the Cenozoic Era

PERIOD	EPOCH	EPOCH ONSET (MYA)
Quaternary	Holocene[a]	0.01 (~10,000 years ago)
	Pleistocene	1.8
Tertiary	Pliocene	5.3
	Miocene	23
	Oligocene	34
	Eocene	55.8
	Paleocene	65

[a] The Holocene is also known as the Recent.

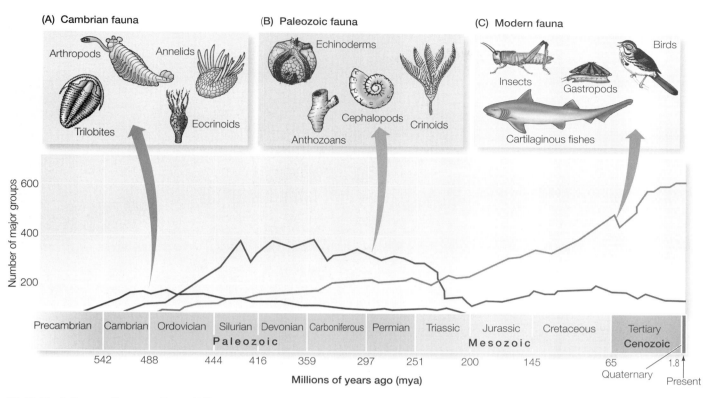

21.19 Evolutionary Faunas *(Page 478)*

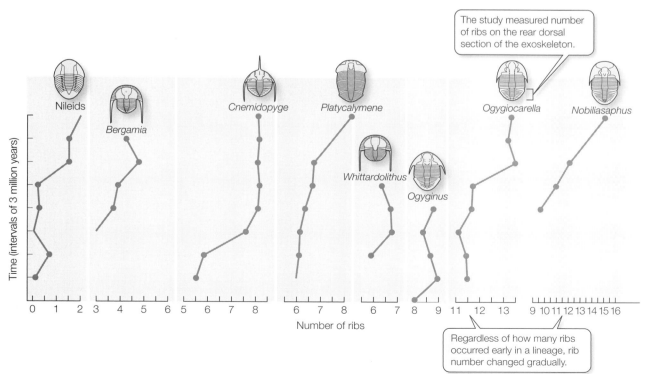

21.21 Rib Number Evolved Gradually in Trilobites *(Page 480)*

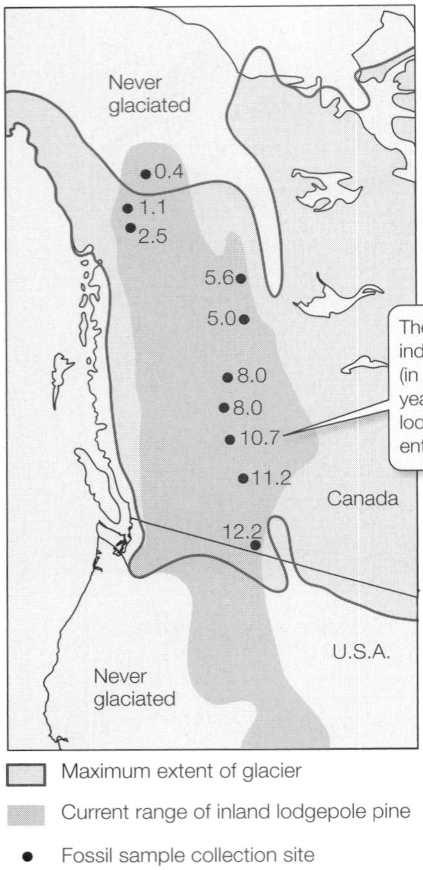

Never glaciated

●0.4
●1.1
●2.5

5.6●
5.0●

●8.0
●8.0
●10.7
●11.2

12.2●

The numbers indicate the time (in thousands of years BP) when lodgepole pine entered area.

Canada

U.S.A.

Never glaciated

Maximum extent of glacier

Current range of inland lodgepole pine

● Fossil sample collection site

21.22 Some Species Expanded Their Ranges as Continental Glaciers Retreated *(Page 481)*

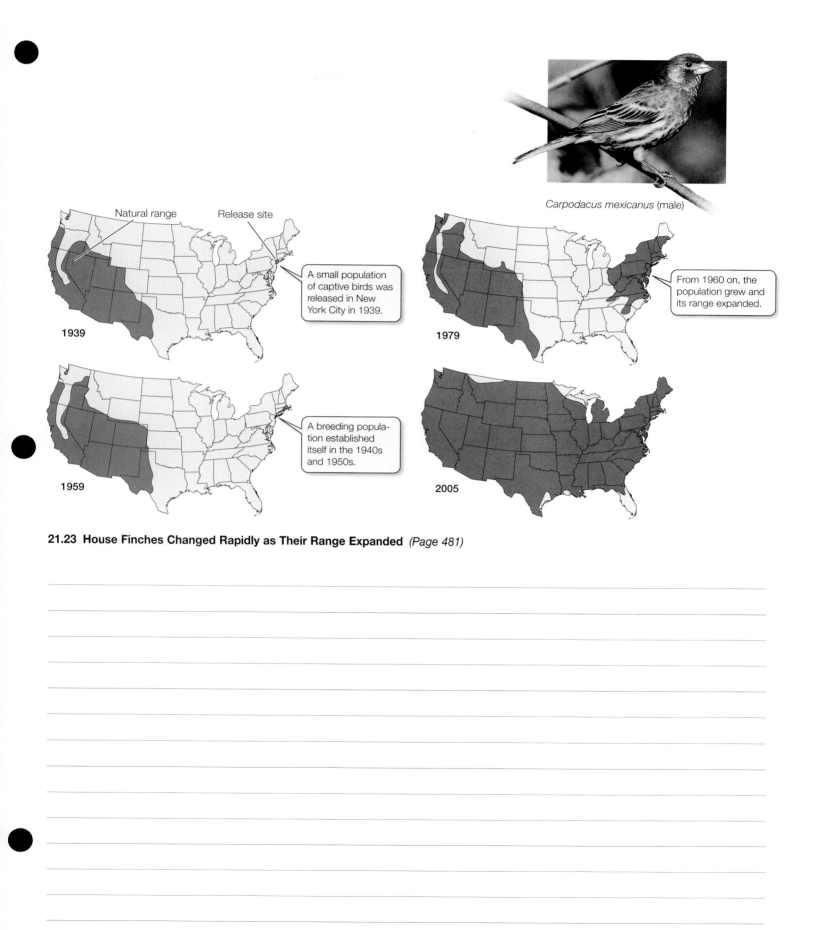

Carpodacus mexicanus (male)

Natural range   Release site

1939

A small population of captive birds was released in New York City in 1939.

1979

From 1960 on, the population grew and its range expanded.

1959

A breeding population established itself in the 1940s and 1950s.

2005

21.23 House Finches Changed Rapidly as Their Range Expanded *(Page 481)*

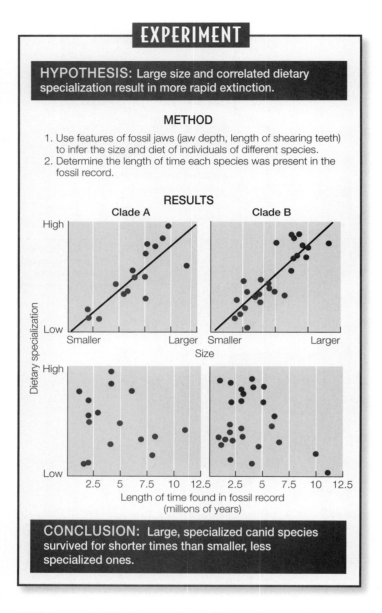

EXPERIMENT

HYPOTHESIS: Large size and correlated dietary specialization result in more rapid extinction.

METHOD

1. Use features of fossil jaws (jaw depth, length of shearing teeth) to infer the size and diet of individuals of different species.
2. Determine the length of time each species was present in the fossil record.

RESULTS

Clade A   Clade B

Dietary specialization

High

Low

Smaller   Larger   Smaller   Larger

Size

High

Low

2.5   5   7.5   10   12.5   2.5   5   7.5   10   12.5

Length of time found in fossil record
(millions of years)

CONCLUSION: Large, specialized canid species survived for shorter times than smaller, less specialized ones.

21.24 Larger Canids Lasted Shorter Times *(Page 482)*

CHAPTER 22 The Mechanisms of Evolution

(A)

(B)

Galápagos
Islands

| Pinta |
| Marchena |
| Genovesa |

Santiago

Bartolome´

Santa Cruz

Isabela

San Cristo´bal

Fernandina Tortuga

Floreana Española

22.1 Darwin and the Voyage of the *Beagle* *(Page 488)*

273

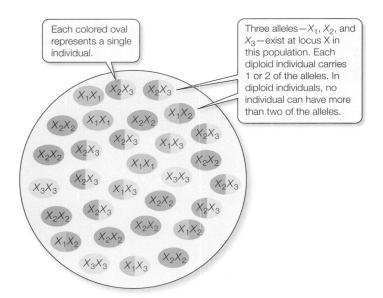

Each colored oval represents a single individual.

Three alleles—X_1, X_2, and X_3—exist at locus X in this population. Each diploid individual carries 1 or 2 of the alleles. In diploid individuals, no individual can have more than two of the alleles.

22.3 A Gene Pool *(Page 490)*

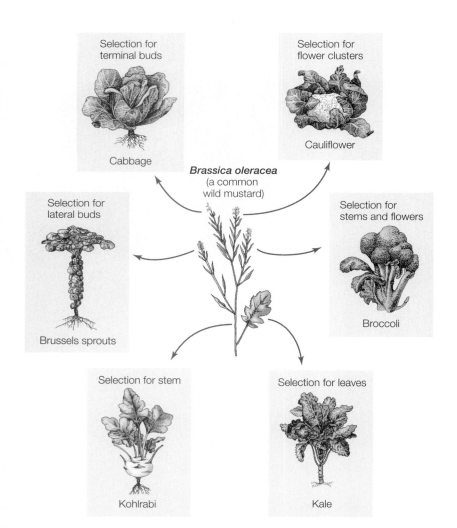

Selection for terminal buds

Cabbage

Selection for flower clusters

Cauliflower

Brassica oleracea (a common wild mustard)

Selection for lateral buds

Brussels sprouts

Selection for stems and flowers

Broccoli

Selection for stem

Kohlrabi

Selection for leaves

Kale

22.4 Many Vegetables from One Species *(Page 490)*

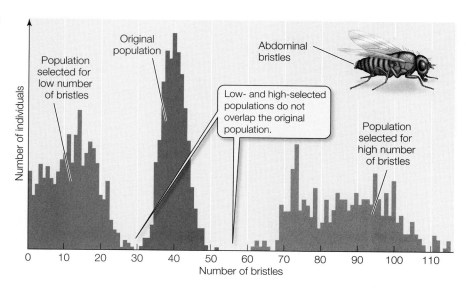

22.5 Artificial Selection Reveals Genetic Variation *(Page 491)*

RESEARCH METHOD

① Determine the allele frequencies in the population.

In any population:

$$\text{Frequency of allele } A = p = \frac{2N_{AA} + N_{Aa}}{2N} \qquad \text{Frequency of allele } a = q = \frac{2N_{aa} + N_{Aa}}{2N}$$

where N is the total number of individuals in the population.

② Compute allele frequencies for different populations.

**For population 1
(mostly homozygotes):**

$N_{AA} = 90$, $N_{Aa} = 40$, and $N_{aa} = 70$

so

$$p = \frac{180 + 40}{400} = 0.55$$

$$q = \frac{140 + 40}{400} = 0.45$$

**For population 2
(mostly heterozygotes):**

$N_{AA} = 45$, $N_{Aa} = 130$, and $N_{aa} = 25$

so

$$p = \frac{90 + 130}{400} = 0.55$$

$$q = \frac{50 + 130}{400} = 0.45$$

22.6 Calculating Allele Frequencies *(Page 492)*

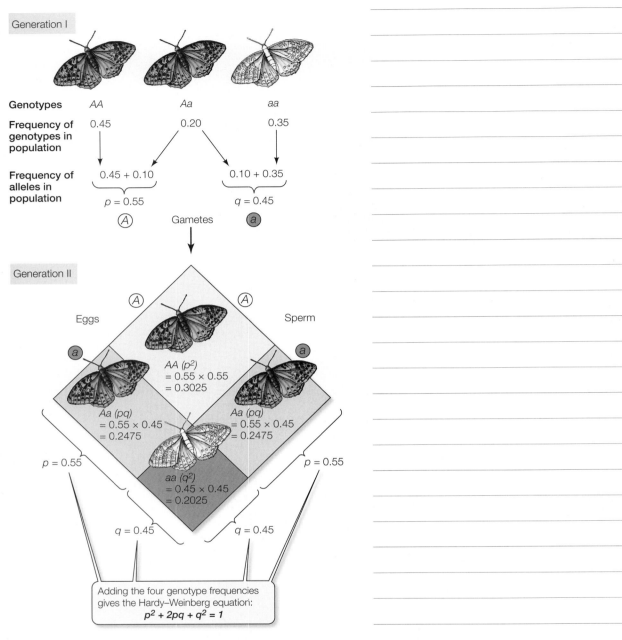

22.7 Calculating Hardy–Weinberg Genotype Frequencies
(Page 493)

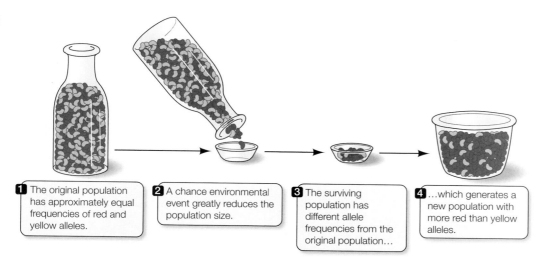

1 The original population has approximately equal frequencies of red and yellow alleles.

2 A chance environmental event greatly reduces the population size.

3 The surviving population has different allele frequencies from the original population…

4 …which generates a new population with more red than yellow alleles.

22.8 A Population Bottleneck *(Page 494)*

European populations of *D. subobscura* have 80 inversions.

These two populations of *D. subobscura* are very similar, but each has only 20 inversions.

22.10 A Founder Effect *(Page 496)*

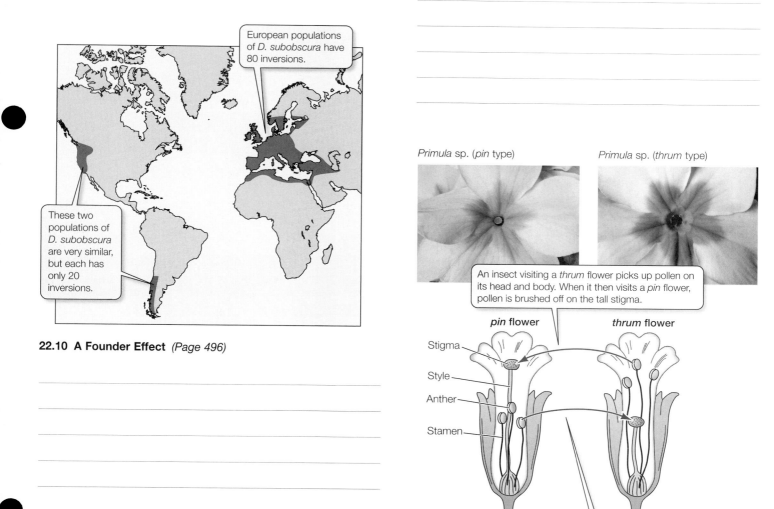

Primula sp. (*pin* type)

Primula sp. (*thrum* type)

An insect visiting a *thrum* flower picks up pollen on its head and body. When it then visits a *pin* flower, pollen is brushed off on the tall stigma.

pin flower

thrum flower

Stigma

Style

Anther

Stamen

An insect visiting a *pin* flower picks up pollen on its proboscis and head. When it then visits a *thrum* flower, it deposits pollen on the short stigma.

22.11 Flower Structure Fosters Nonrandom Mating *(Page 496)*

(A) Stabilizing selection

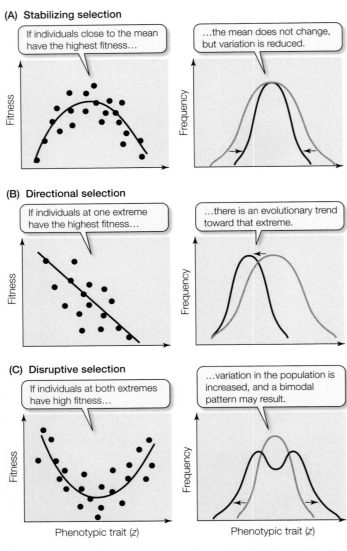

If individuals close to the mean have the highest fitness…

…the mean does not change, but variation is reduced.

Fitness

Frequency

(B) Directional selection

If individuals at one extreme have the highest fitness…

…there is an evolutionary trend toward that extreme.

Fitness

Frequency

(C) Disruptive selection

If individuals at both extremes have high fitness…

…variation in the population is increased, and a bimodal pattern may result.

Fitness

Frequency

Phenotypic trait (*z*)

Phenotypic trait (*z*)

22.12 Natural Selection Can Operate on Quantitative Variation in Several Ways *(Page 497)*

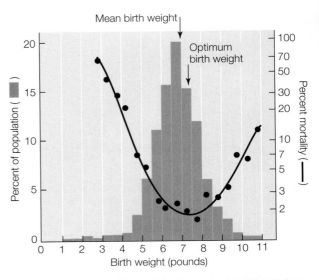

Mean birth weight

Optimum birth weight

Percent of population (▓)

Percent mortality (—)

Birth weight (pounds)

22.13 Human Birth Weight Is Influenced by Stabilizing Selection *(Page 498)*

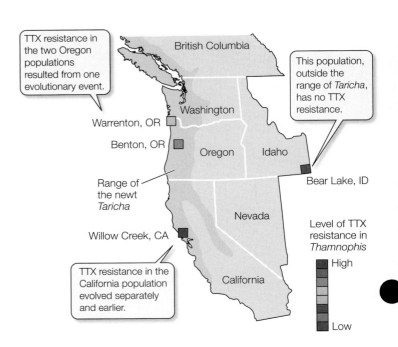

TTX resistance in the two Oregon populations resulted from one evolutionary event.

British Columbia

This population, outside the range of *Taricha*, has no TTX resistance.

Warrenton, OR

Washington

Benton, OR

Oregon   Idaho

Range of the newt *Taricha*

Bear Lake, ID

Nevada

Willow Creek, CA

Level of TTX resistance in *Thamnophis*

High

TTX resistance in the California population evolved separately and earlier.

California

Low

22.14 Resistance to TTX Is Associated with the Presence of Newts *(Page 498)*

22.15 Disruptive Selection Results in a Bimodal Distribution *(Page 499)*

EXPERIMENT

HYPOTHESIS: Sexual selection is responsible for the evolution of long tails in African long-tailed widowbirds.

METHOD

Capture males and artificially lengthen or shorten their tails by cutting feathers or gluing on feathers. Release males, then measure reproductive success by counting the nests with eggs or young on each male's territory.

RESULTS

Males whose tails were artificially lengthened attracted more females and had greater reproductive success…

…than males with normal or shortened tails, or the control group.

Average number of nests per male

Artificially lengthened · Normal · Control · Artificially shortened

CONCLUSION: Sexual selection in widowbirds favors long tails.

22.16 The Longer the Tail, the Better the Male *(Page 499)*

(A)

Taeniopygia guttata

22.17 Bright Bills Signal Good Health *(Page 500)*

(B)

EXPERIMENT

HYPOTHESIS: Having a bright red bill signals good health in a male zebra finch.

METHOD

Provide carotenoids in drinking water for experimental, but not for control males. Challenge all males immunologically and measure response.

RESULTS

Experimental males responded more strongly to the immunological challenge. They also developed brighter bills than control males.

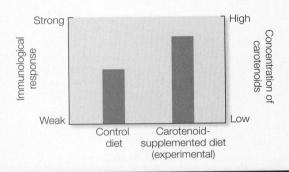

CONCLUSION: Males with high carotenoid levels have bright bills and are immunologically strong. Therefore, bill color is an indication of the health of a male zebra finch.

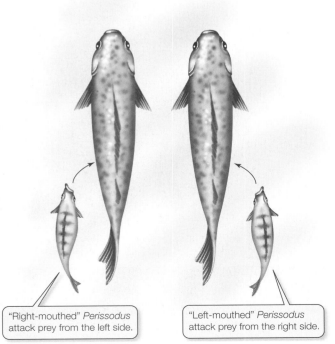

"Right-mouthed" *Perissodus* attack prey from the left side.

"Left-mouthed" *Perissodus* attack prey from the right side.

22.18 A Stable Polymorphism *(Page 502)*

EXPERIMENT

HYPOTHESIS: Heterozygous male *Colias* butterflies should have better mating sucess than homozygous males because they can fly farther under a broader range of temperatures.

METHOD

Capture butterflies in the field, transfer them to the laboratory, and determine their genotypes. Allow females to lay eggs, and determine the genotypes of the offspring (and thus paternity and mating success) of males.

RESULTS

For both species, the proportion of heterozygous males that mated successfully was higher than the proportion of all males seeking females ("flying").

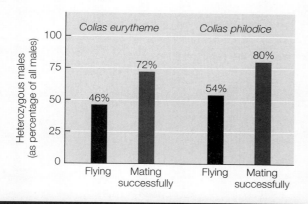

CONCLUSION: Heterozygous *Colias* males have a mating advantage over homozygous males.

22.19 A Heterozygote Mating Advantage *(Page 502)*

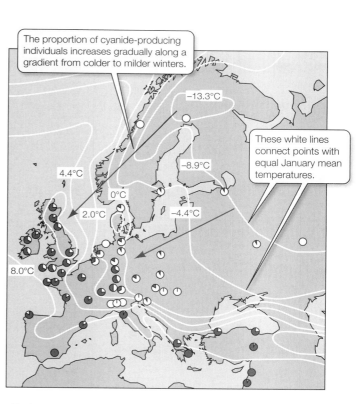

The proportion of cyanide-producing individuals increases gradually along a gradient from colder to milder winters.

These white lines connect points with equal January mean temperatures.

Red indicates proportion of plants producing cyanide — White indicates proportion of plants not producing cyanide

22.20 Geographic Variation in a Defensive Chemical *(Page 503)*

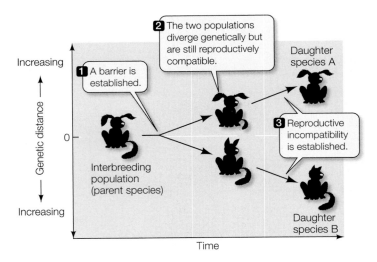

23.2 Speciation May Be a Gradual Process *(Page 510)*

23.3 Allopatric Speciation *(Page 511)*

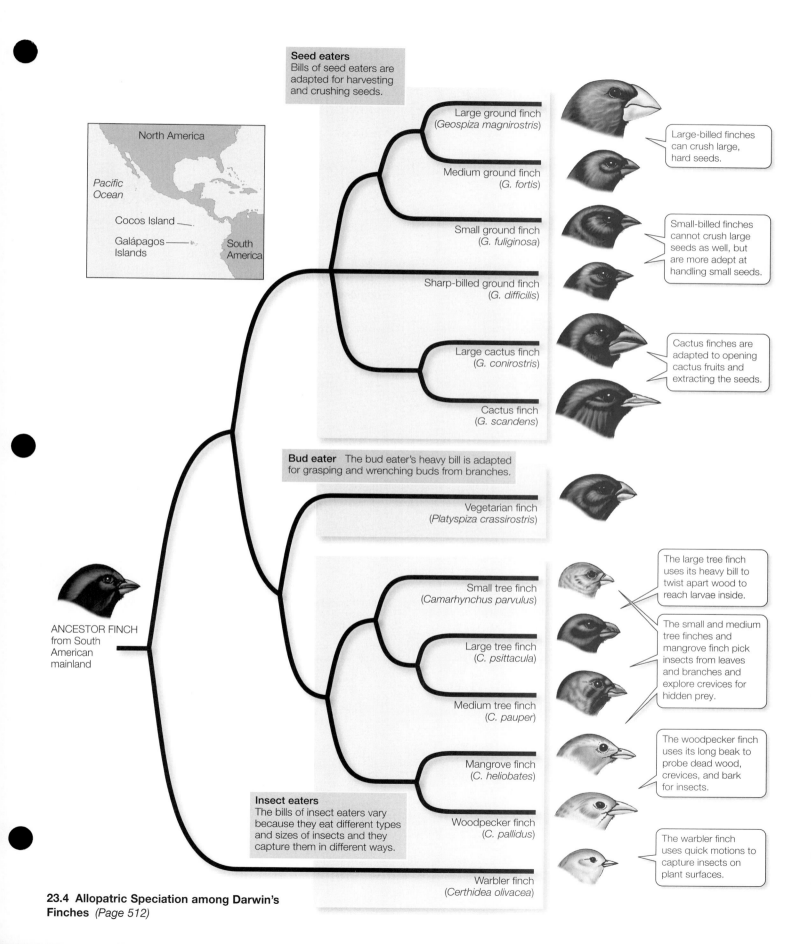

23.4 Allopatric Speciation among Darwin's Finches *(Page 512)*

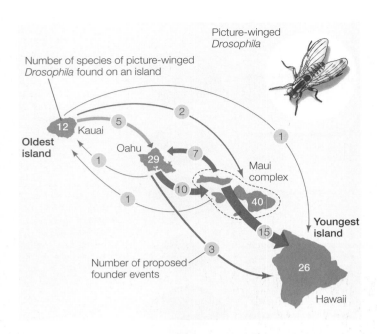

Picture-winged
Drosophila

Number of species of picture-winged
Drosophila found on an island

**Oldest
island**

Kauai 12

Oahu 5

29

7

2

1

Maui
complex

10

40

**Youngest
island**

15

Number of proposed
founder events 3

26

Hawaii

23.5 Founder Events Lead to Allopatric Speciation *(Page 513)*

The difference in allele frequencies is
associated with sympatric divergence.

Apple host

Hawthorn host

**23.6 Sympatric Speciation May Be Underway in *Rhagoletis
pomonella*** *(Page 513)*

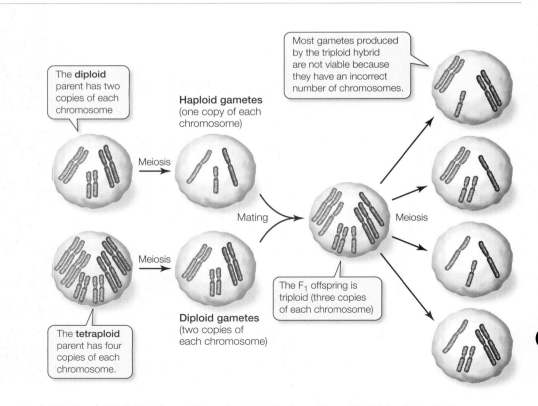

The **diploid**
parent has two
copies of each
chromosome

Haploid gametes
(one copy of each
chromosome)

Meiosis

Meiosis

Mating

The **tetraploid**
parent has four
copies of each
chromosome.

Diploid gametes
(two copies of
each chromosome)

The F_1 offspring is
triploid (three copies
of each chromosome)

Meiosis

Most gametes produced
by the triploid hybrid
are not viable because
they have an incorrect
number of chromosomes.

23.7 Tetraploids Are Soon Reproductively Isolated from Diploids *(Page 514)*

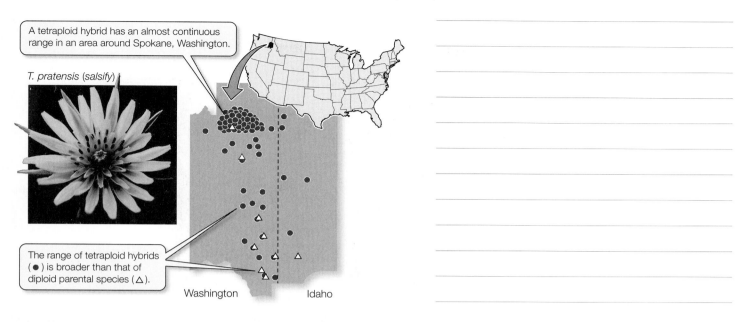

A tetraploid hybrid has an almost continuous range in an area around Spokane, Washington.

T. pratensis (salsify)

The range of tetraploid hybrids (●) is broader than that of diploid parental species (△).

Washington   Idaho

23.8 Polyploids May Outperform Their Parent Species
(Page 514)

(B) *Aquilegia pubescens*

(A) *Aquilegia formosa*

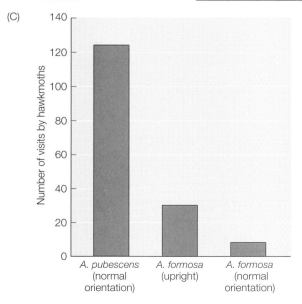

(C)

Number of visits by hawkmoths

A. pubescens (normal orientation)   A. formosa (upright)   A. formosa (normal orientation)

23.10 Hawkmoths Favor Flowers of One Columbine Species
(Page 516)

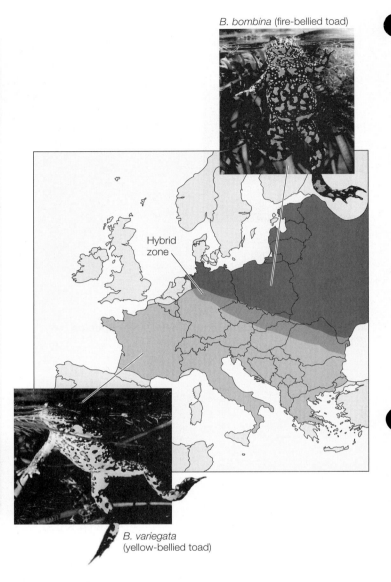

B. bombina (fire-bellied toad)

EXPERIMENT

HYPOTHESIS: *Phlox drummondii* has red flowers only where it is sympatric with pink-flowered *P. cuspidata* because having red flowers decreases interspecific hybridization.

METHOD

1. Introduce equal numbers of red- and pink-flowered *P. drummondii* individuals into an area with many pink-flowered *P. cuspidata*.

2. After the flowering season ends, assess the genetic composition of the seeds produced by *P. drummondii* plants of both colors.

RESULTS

Of the seeds produced by pink-flowered *P. drummondii*, 38% were hybrids with *P. cuspidata*. Only 13% of the seeds produced by red-flowered individuals were genetic hybrids.

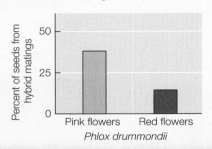

CONCLUSION: For *Phlox drummondii*, having flowers that differ in color from those of *P. cuspidata* reduces the amount of interspecific hybridization.

23.11 Prezygotic Reproductive Barriers *(Page 517)*

B. variegata (yellow-bellied toad)

23.12 Hybrid Zones May Be Long and Narrow *(Page 518)*

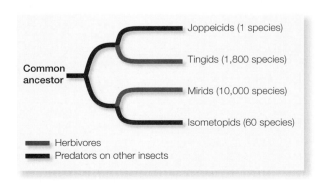

Joppeicids (1 species)
Tingids (1,800 species)
Mirids (10,000 species)
Isometopids (60 species)

Common ancestor

Herbivores
Predators on other insects

23.13 Dietary Shifts Can Promote Speciation *(Page 519)*

CHAPTER 24 The Evolution of Genes and Genomes

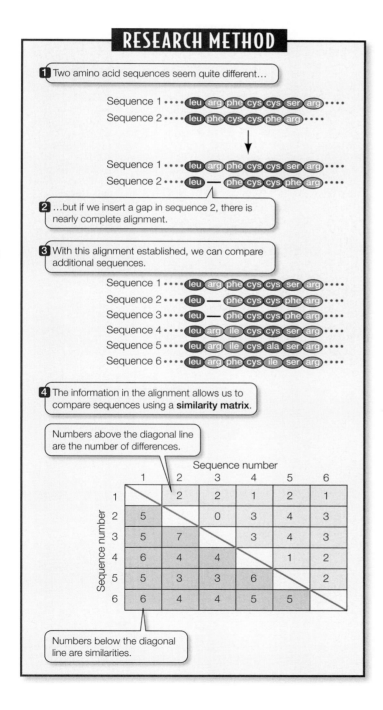

RESEARCH METHOD

1 Two amino acid sequences seem quite different...

Sequence 1 ···· leu arg phe cys cys ser arg ····
Sequence 2 ···· leu phe cys cys phe arg ····

Sequence 1 ···· leu arg phe cys cys ser arg ····
Sequence 2 ···· leu — phe cys cys phe arg ····

2 ...but if we insert a gap in sequence 2, there is nearly complete alignment.

3 With this alignment established, we can compare additional sequences.

Sequence 1 ···· leu arg phe cys cys ser arg ····
Sequence 2 ···· leu — phe cys cys phe arg ····
Sequence 3 ···· leu — phe cys cys phe arg ····
Sequence 4 ···· leu arg ile cys cys ser arg ····
Sequence 5 ···· leu arg ile cys ala ser arg ····
Sequence 6 ···· leu arg phe cys ile ser arg ····

4 The information in the alignment allows us to compare sequences using a **similarity matrix**.

Numbers above the diagonal line are the number of differences.

Sequence number

	1	2	3	4	5	6
1		2	2	1	2	1
2	5		0	3	4	3
3	5	7		3	4	3
4	6	4	4		1	2
5	5	3	3	6		2
6	6	4	4	5	5	

Sequence number

Numbers below the diagonal line are similarities.

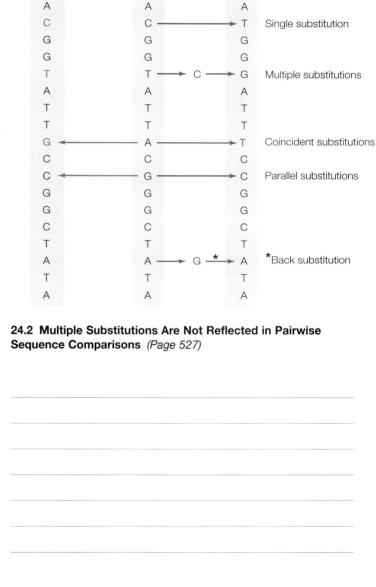

Descendant 1	Ancestral sequence	Descendant 2	
A	A	A	
C	C	T	Single substitution
G	G	G	
G	G	G	
T	T → C → G		Multiple substitutions
A	A	A	
T	T	T	
T	T	T	
G	A	T	Coincident substitutions
C	C	C	
C	G	C	Parallel substitutions
G	G	G	
G	G	G	
C	C	C	
T	T	T	
A	A → G *→ A		*Back substitution
T	T	T	
A	A	A	

24.2 Multiple Substitutions Are Not Reflected in Pairwise Sequence Comparisons *(Page 527)*

24.1 Amino Acid Sequence Alignment *(Page 526)*

Tuna

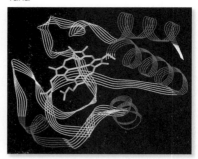

Rice

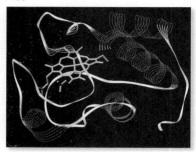

Acidic side chains
D Aspartic acid
E Glutamic acid

Basic side chains
H Histidine
K Lysine
R Arginine

Hydrophobic side chains
F Phenylalanine
I Isoleucine
L Leucine
M Methionine

V Valine
Y Tyrosine
W Tryptophan
A Alanine

Other
C Cysteine
P Proline
Q Glutamine
N Asparagine
S Serine
T Threonine
G Glycine

The number 1 indicates an invariant position in the cytochrome *c* molecule (i.e., all the organisms have the same amino acid in this position). Such a position is probably under strong stabilizing selection.

Amino acids at positions marked by red arrowheads have side chains that interact with the heme group.

Position in sequence	1			5					10					15					20					25					30				
Number of amino acids at the position	1	3	5	5	5	1	3	3	4	1	4	3	2	1	3	3	1	1	2	4	3	4	2	3	4	2	1	4	1	1	2	1	5	1
Human, chimpanzee	G	D	V	E	K	G	K	K	I	F	I	M	K	C	S	Q	C	H	T	V	E	K	G	G	K	H	K	T	G	P	N	L	H	G
Rhesus monkey	G	D	V	E	K	G	K	K	I	F	I	M	K	C	S	Q	C	H	T	V	E	K	G	G	K	H	K	T	G	P	N	L	H	G
Horse	G	D	V	E	K	G	K	K	I	F	V	Q	K	C	A	Q	C	H	T	V	E	K	G	G	K	H	K	T	G	P	N	L	H	G
Donkey	G	D	V	E	K	G	K	K	I	F	V	Q	K	C	A	Q	C	H	T	V	E	K	G	G	K	H	K	T	G	P	N	L	H	G
Cow, pig, sheep	G	D	V	E	K	G	K	K	I	F	V	Q	K	C	A	Q	C	H	T	V	E	K	G	G	K	H	K	T	G	P	N	L	H	G
Dog	G	D	V	E	K	G	K	K	I	F	V	Q	K	C	A	Q	C	H	T	V	E	K	G	G	K	H	K	T	G	P	N	L	H	G
Rabbit	G	D	V	E	K	G	K	K	I	F	V	Q	K	C	A	Q	C	H	T	V	E	K	G	G	K	H	K	T	G	P	N	L	H	G
Gray whale	G	D	V	E	K	G	K	K	I	F	V	Q	K	C	A	Q	C	H	T	V	E	K	G	G	K	H	K	T	G	P	N	L	H	G
Gray kangaroo	G	D	V	E	K	G	K	K	I	F	V	Q	K	C	A	Q	C	H	T	V	E	K	G	G	K	H	K	T	G	P	N	L	N	G
Chicken, turkey	G	D	I	E	K	G	K	K	I	F	V	Q	K	C	S	Q	C	H	T	V	E	K	G	G	K	H	K	T	G	P	N	L	H	G
Pigeon	G	D	I	E	K	G	K	K	I	F	V	Q	K	C	S	Q	C	H	T	V	E	K	G	G	K	H	K	T	G	P	N	L	H	G
Pekin duck	G	D	V	E	K	G	K	K	I	F	V	Q	K	C	S	Q	C	H	T	V	E	K	G	G	K	H	K	T	G	P	N	L	H	G
Snapping turtle	G	D	V	E	K	G	K	K	I	F	V	Q	K	C	A	Q	C	H	T	V	E	K	G	G	K	H	K	T	G	P	N	L	N	G
Rattlesnake	G	D	V	E	K	G	K	K	I	F	T	M	K	C	S	Q	C	H	T	V	E	K	G	G	K	H	K	T	G	P	N	L	H	G
Bullfrog	G	D	V	E	K	G	K	K	I	F	V	Q	K	C	A	Q	C	H	T	C	E	K	G	G	K	H	K	V	G	P	N	L	Y	G
Tuna	G	D	V	A	K	G	K	K	T	F	V	Q	K	C	A	Q	C	H	T	V	E	N	G	G	K	H	K	V	G	P	N	L	W	G
Dogfish	G	D	V	E	K	G	K	K	V	F	V	Q	K	C	A	Q	C	H	T	V	E	N	G	G	K	H	K	T	G	P	N	L	S	G
Samia cynthia (moth)	G	N	A	E	N	G	K	K	I	F	V	Q	R	C	A	Q	C	H	T	V	E	A	G	G	K	H	K	V	G	P	N	L	H	G
Tobacco hornworm moth	G	N	A	D	N	G	K	K	I	F	V	Q	R	C	A	Q	C	H	T	V	E	A	G	G	K	H	K	V	G	P	N	L	N	G
Screwworm fly	G	D	V	E	K	G	K	K	I	F	V	Q	R	C	A	Q	C	H	T	V	E	A	G	G	K	H	K	V	G	P	N	L	H	G
Drosophila (fruit fly)	G	D	V	E	K	G	K	K	L	F	V	Q	R	C	A	Q	C	H	T	V	E	A	G	G	K	H	K	V	G	P	N	L	H	G
Baker's yeast	G	S	A	K	K	G	A	T	L	F	K	T	R	C	E	L	C	H	T	V	E	K	G	G	P	H	K	V	G	P	N	L	H	G
Candida krusei (yeast)	G	S	A	K	K	G	A	T	L	F	K	T	R	C	A	E	C	H	T	I	E	A	G	G	P	H	K	V	G	P	N	L	H	G
Neurospora crassa (mold)	G	D	S	K	K	G	A	N	L	F	K	T	R	C	A	E	C	H	E				N	L	T	Q	K	I	G	P	A	L	H	G
Wheat	G	N	P	D	A	G	A	K	I	F	K	T	K	C	A	Q	C	H	T	V	D	A	G	A			H	K	Q	G	P	N	L	H	G
Sunflower	G	D	P	T	T	G	A	K	I	F	K	T	K	C	A	Q	C	H	T	V	E	K	G	A			H	K	Q	G	P	N	L	N	G
Mung bean	G	D	S	K	S	G	E	K	I	F	K	T	K	C	A	Q	C	H	T	V	D	K	G	A			H	K	Q	G	P	N	L	N	G
Rice	G	N	P	K	A	G	E	K	I	F	K	T	K	C	A	Q	C	H	T	V	D	K	G	A			H	K	Q	G	P	N	L	N	G
Sesame	G	D	V	K	S	G	E	K	I	F	K	T	K	C	A	Q	C	H	T	V	D	K	G	A			H	K	Q	G	P	N	L	N	G

Gaps indicate insertion and/or deletion events.

24.3 Amino Acid Sequences of Cytochrome *c* (Page 528)

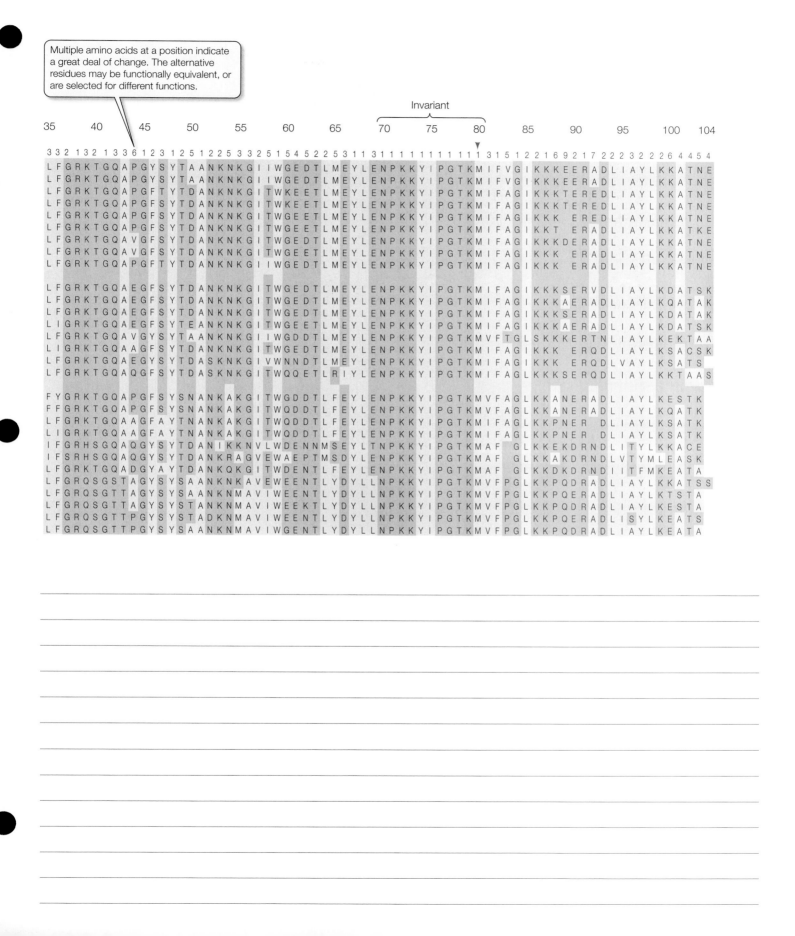

Multiple amino acids at a position indicate a great deal of change. The alternative residues may be functionally equivalent, or are selected for different functions.

EXPERIMENT

HYPOTHESIS: Heterogeneous environments lead to adaptive radiation, whereas homogeneous environments inhibit diversification.

METHOD

One colony of *Pseudomonas fluorescens* (all of a single genotype) is used to inoculate many replicate cultures.

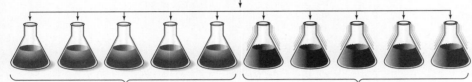

> Half of replicate cultures are kept **static**, so that many different local environments may develop.

> The other half of the cultures are **shaken**, to keep the environmental conditions uniform throughout the medium.

RESULTS

In the **shaken** flasks, the ancestral morphotype persists. But in the **static** flasks, two new morphotypes regularly evolve, each adapted to a different local environment. Molecular analysis reveals multiple genetic causes for the similar morphotypes.

Smooth morph
(ancestral)

"Wrinkly spreader"

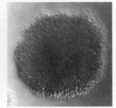
"Fuzzy spreader"

CONCLUSION: Heterogeneous environments promote diversification.

24.4 A Heterogeneous Environment Spurs Adaptive Radiation *(Page 530)*

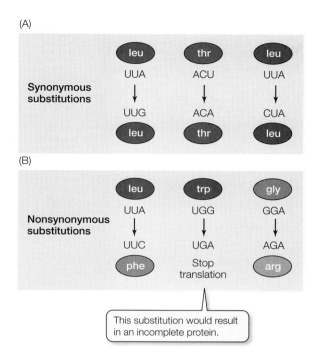

(A)

Synonymous substitutions

leu — UUA → UUG — leu

thr — ACU → ACA — thr

leu — UUA → CUA — leu

(B)

Nonsynonymous substitutions

leu — UUA → UUC — phe

trp — UGG → UGA — Stop translation

gly — GGA → AGA — arg

This substitution would result in an incomplete protein.

24.5 When One Nucleotide Does or Doesn't Make a Difference
(Page 531)

TABLE 24.1

Similarity Matrix for Lysozyme in Mammals

SPECIES	LANGUR	BABOON	HUMAN	RAT	CATTLE	HORSE
Langur*		14	18	38	32	65
Baboon	0		14	33	39	65
Human	0	1		37	41	64
Rat	0	1	0		55	64
Cattle*	5	0	0	0		71
Horse	0	0	0	0	1	

Shown above the diagonal line is the number of amino acid sequence *differences* between the two species being compared; below the line are the number of changes uniquely *shared* by the two species.

Asterisks (*) indicate foregut-fermenting species.

(Page 533)

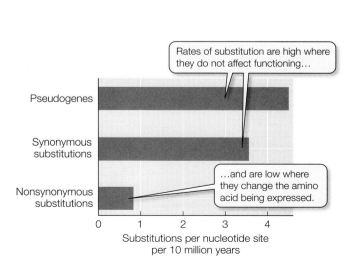

Rates of substitution are high where they do not affect functioning...

Pseudogenes

Synonymous substitutions

Nonsynonymous substitutions

...and are low where they change the amino acid being expressed.

Substitutions per nucleotide site per 10 million years

24.6 Rates of Substitution Differ *(Page 531)*

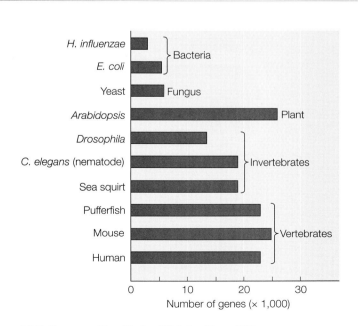

H. influenzae ⎤ Bacteria
E. coli ⎦

Yeast — Fungus

Arabidopsis — Plant

Drosophila ⎤
C. elegans (nematode) ⎬ Invertebrates
Sea squirt ⎦

Pufferfish ⎤
Mouse ⎬ Vertebrates
Human ⎦

Number of genes (× 1,000)

24.8 Genome Size Varies Widely *(Page 533)*

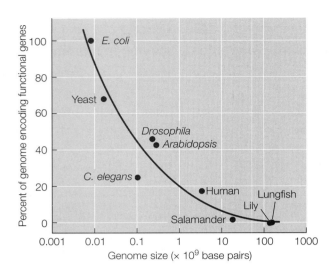

24.9 A Large Proportion of DNA Is Noncoding *(Page 534)*

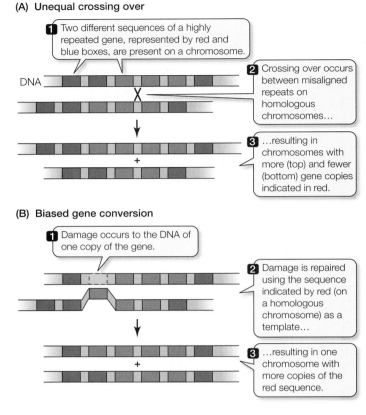

(A) Unequal crossing over

1. Two different sequences of a highly repeated gene, represented by red and blue boxes, are present on a chromosome.

DNA

2. Crossing over occurs between misaligned repeats on homologous chromosomes...

3. ...resulting in chromosomes with more (top) and fewer (bottom) gene copies indicated in red.

(B) Biased gene conversion

1. Damage occurs to the DNA of one copy of the gene.

2. Damage is repaired using the sequence indicated by red (on a homologous chromosome) as a template...

3. ...resulting in one chromosome with more copies of the red sequence.

24.11 Concerted Evolution *(Page 536)*

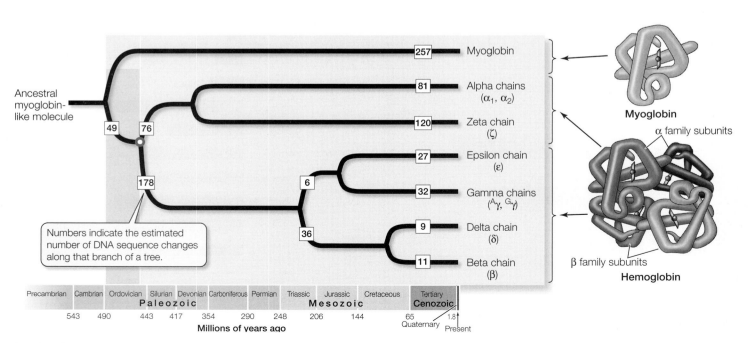

Ancestral myoglobin-like molecule

49

76

178

Numbers indicate the estimated number of DNA sequence changes along that branch of a tree.

6

36

257 Myoglobin

81 Alpha chains (α_1, α_2)

120 Zeta chain (ζ)

27 Epsilon chain (ϵ)

32 Gamma chains ($^A\gamma$, $^G\gamma$)

9 Delta chain (δ)

11 Beta chain (β)

Myoglobin

α family subunits

β family subunits

Hemoglobin

Precambrian	Cambrian	Ordovician	Silurian	Devonian	Carboniferous	Permian	Triassic	Jurassic	Cretaceous	Tertiary	
										Cenozoic	

Paleozoic   Mesozoic   Cenozoic

543   490   443   417   354   290   248   206   144   65   1.8

Quaternary   Present

Millions of years ago

24.10 A Globin Family Gene Tree *(Page 535)*

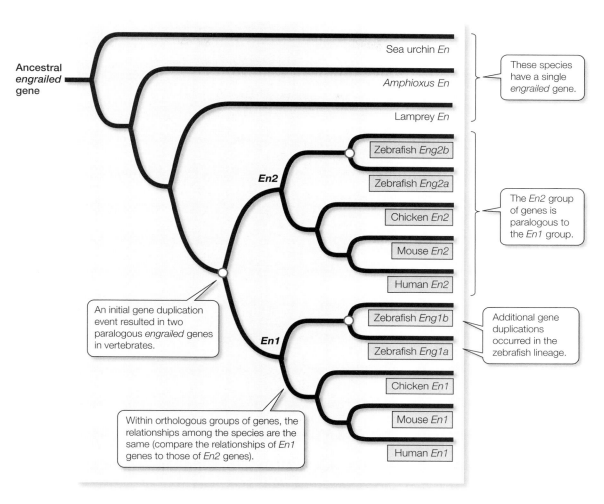

Ancestral *engrailed* gene

Sea urchin *En*

Amphioxus En

Lamprey *En*

These species have a single *engrailed* gene.

En2

Zebrafish *Eng2b*

Zebrafish *Eng2a*

Chicken *En2*

Mouse *En2*

Human *En2*

The *En2* group of genes is paralogous to the *En1* group.

An initial gene duplication event resulted in two paralogous *engrailed* genes in vertebrates.

En1

Zebrafish *Eng1b*

Zebrafish *Eng1a*

Chicken *En1*

Mouse *En1*

Human *En1*

Additional gene duplications occurred in the zebrafish lineage.

Within orthologous groups of genes, the relationships among the species are the same (compare the relationships of *En1* genes to those of *En2* genes).

24.12 Phylogeny of the *engrailed* Genes *(Page 537)*

(A)
PCR amplification introduces new variation into the DNA population.
DNA template
Transcription by T7 RNA polymerase
cDNA
RNA population
Reverse transcription
Selected RNA population
Select the RNA molecules with the highest ligation rates from the population.
(B)
Ligation rate (per hour)
10^1
10^0
10^–1
10^–2
10^–3
10^–4
10^–5
10^–6
0
2
4
6
8
10
Round

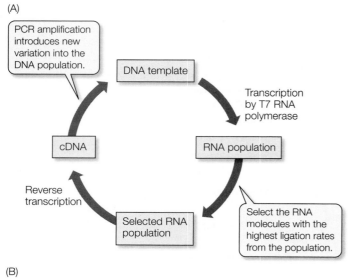

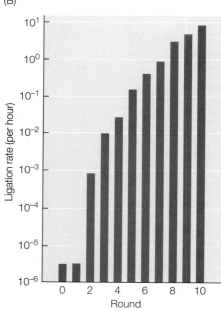

24.13 In Vitro Evolution of a Ribozyme *(Page 539)*

CHAPTER 25 Reconstructing and Using Phylogenies

(A)

The splits in branches are called **nodes**, and indicate a division of one lineage into two.

In this book, all phylogenetic trees show the common ancestor for the group on the left; this is called the **root** of the tree.

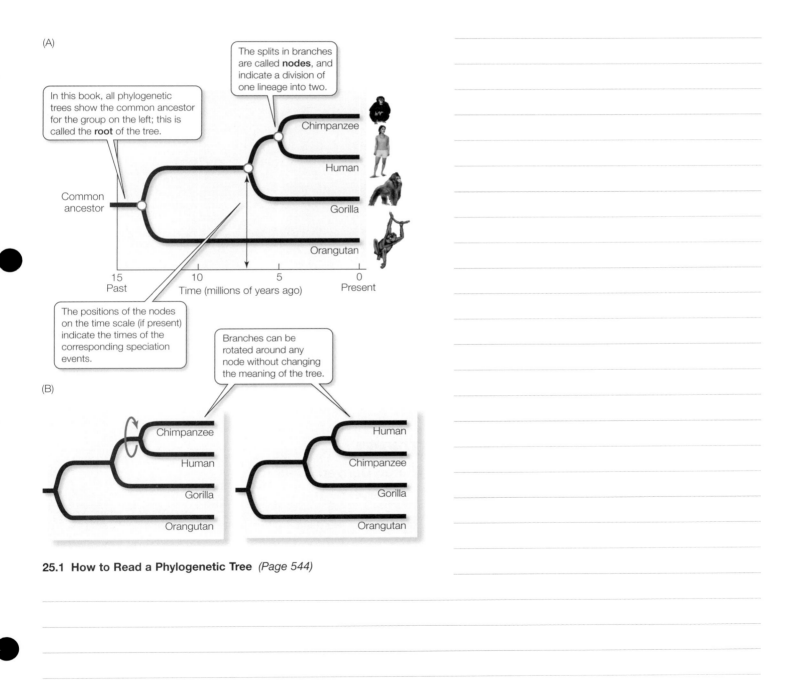

Common ancestor

Chimpanzee

Human

Gorilla

Orangutan

| 15 | 10 | 5 | 0 |
| Past | | | Present |

Time (millions of years ago)

The positions of the nodes on the time scale (if present) indicate the times of the corresponding speciation events.

Branches can be rotated around any node without changing the meaning of the tree.

(B)

Chimpanzee

Human

Gorilla

Orangutan

Human

Chimpanzee

Gorilla

Orangutan

25.1 How to Read a Phylogenetic Tree *(Page 544)*

Bat wing

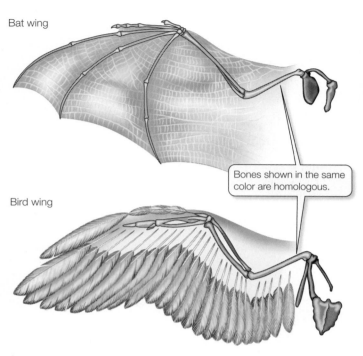

Bones shown in the same color are homologous.

Bird wing

25.2 The Bones Are Homologous; the Wings Are Not *(Page 545)*

TABLE 25.1

Eight Vertebrates Ordered According to Unique Shared Derived Traits

	DERIVED TRAIT[a]							
TAXON	JAWS	LUNGS	CLAWS OR NAILS	GIZZARD	FEATHERS	FUR	MAMMARY GLANDS	KERATINOUS SCALES
Lamprey (outgroup)	–	–	–	–	–	–	–	–
Perch	+	–	–	–	–	–	–	–
Salamander	+	+	–	–	–	–	–	–
Lizard	+	+	+	–	–	–	–	+
Crocodile	+	+	+	+	–	–	–	+
Pigeon	+	+	+	+	+	–	–	+
Mouse	+	+	+	–	–	+	+	–
Chimpanzee	+	+	+	–	–	+	+	–

[a]A plus sign indicates the trait is present, a minus sign that it is absent.

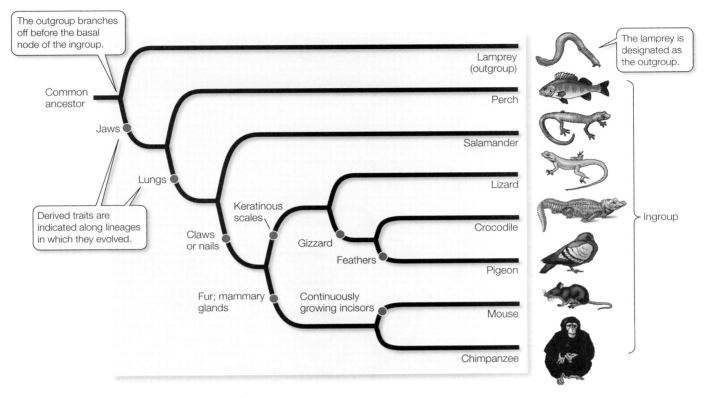

25.3 Inferring a Phylogenetic Tree *(Page 547)*

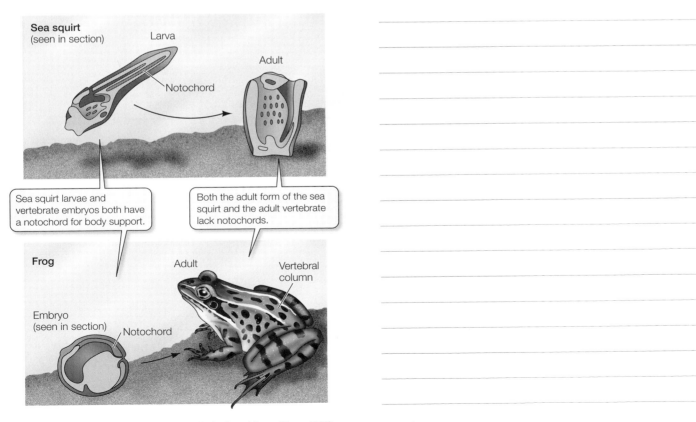

25.4 A Larva Reveals Evolutionary Relationships *(Page 548)*

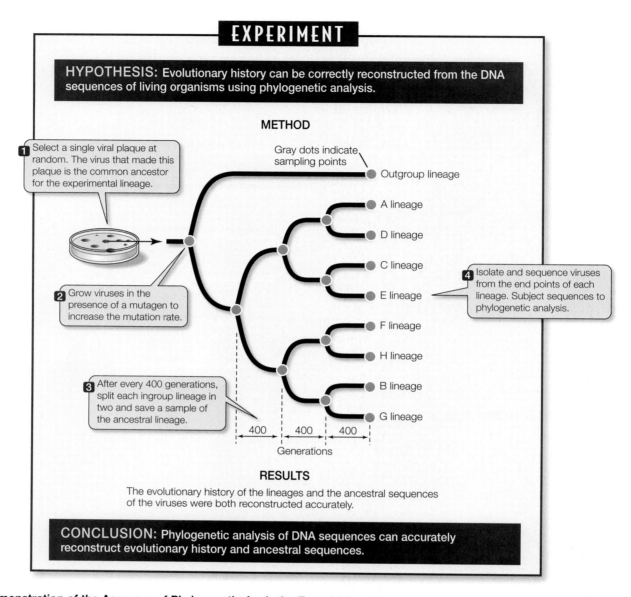

25.5 A Demonstration of the Accuracy of Phylogenetic Analysis *(Page 549)*

(A)

Harpagochromis sp.

Ptyochromis sp.

(B)

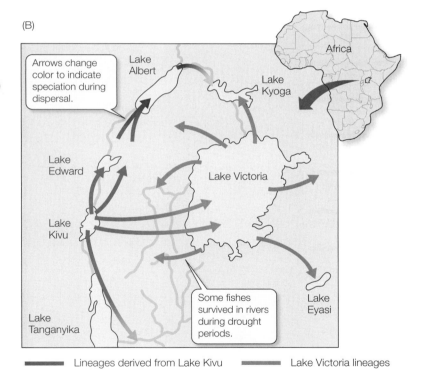

Lineages derived from Lake Kivu   Lake Victoria lineages

25.6 Origins of the Cichlid Fishes of Lake Victoria *(Page 550)*

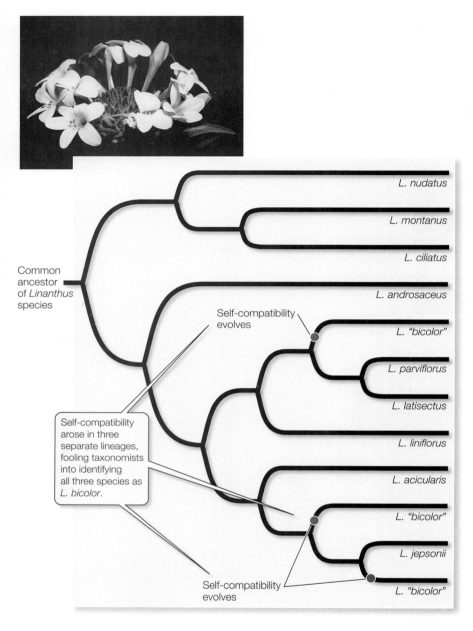

L. nudatus

L. montanus

L. ciliatus

L. androsaceus

Common
ancestor
of *Linanthus*
species

Self-compatibility
evolves

L. "bicolor"

L. parviflorus

L. latisectus

L. liniflorus

Self-compatibility
arose in three
separate lineages,
fooling taxonomists
into identifying
all three species as
L. bicolor.

L. acicularis

L. "bicolor"

L. jepsonii

Self-compatibility
evolves

L. "bicolor"

25.7 Phylogeny of a Section of the Plant Genus *Linanthus*
(Page 552)

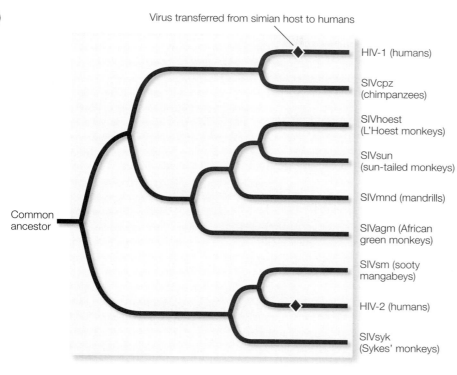

25.8 Phylogenetic Tree of Immunodeficiency Viruses *(Page 553)*

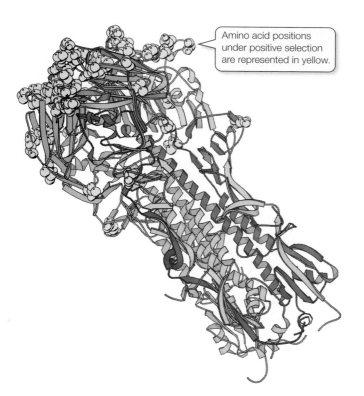

25.10 Model of Hemagglutinin, a Surface Protein of Influenza
(Page 553)

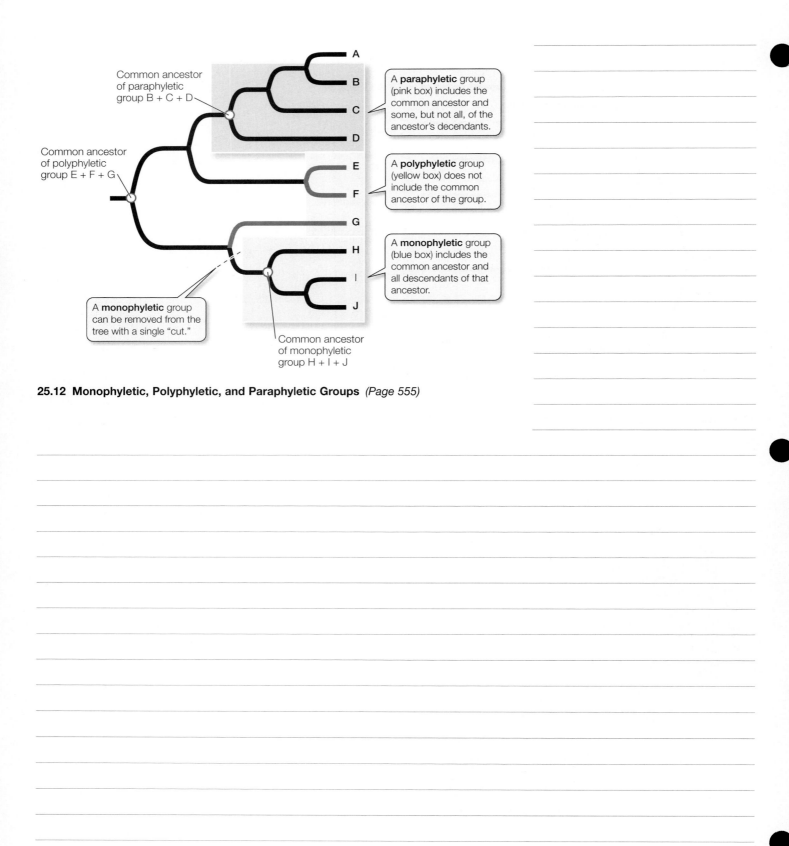

Common ancestor of paraphyletic group B + C + D

A **paraphyletic** group (pink box) includes the common ancestor and some, but not all, of the ancestor's decendants.

Common ancestor of polyphyletic group E + F + G

A **polyphyletic** group (yellow box) does not include the common ancestor of the group.

A **monophyletic** group (blue box) includes the common ancestor and all descendants of that ancestor.

A **monophyletic** group can be removed from the tree with a single "cut."

Common ancestor of monophyletic group H + I + J

25.12 Monophyletic, Polyphyletic, and Paraphyletic Groups *(Page 555)*

CHAPTER 26 Bacteria and Archaea: The Prokaryotic Domains

(Page 562)

TABLE 26.1			

The Three Domains of Life on Earth

CHARACTERISTIC	BACTERIA	DOMAIN ARCHAEA	EUKARYA
Membrane-enclosed nucleus	Absent	Absent	*Present*
Membrane-enclosed organelles	Absent	Absent	*Present*
Peptidoglycan in cell wall	*Present*	Absent	Absent
Membrane lipids	Ester-linked	*Ether-linked*	Ester-linked
	Unbranched	*Branched*	Unbranched
Ribosomes[a]	70S	70S	*80S*
Initiator tRNA	*Formylmethionine*	Methionine	Methionine
Operons	Yes	Yes	*No*
Plasmids	Yes	Yes	*Rare*
RNA polymerases	One	One[b]	*Three*
Ribosomes sensitive to chloramphenicol and streptomycin	*Yes*	No	No
Ribosomes sensitive to diphtheria toxin	*No*	Yes	Yes
Some are methanogens	No	*Yes*	No
Some fix nitrogen	Yes	Yes	*No*
Some conduct chlorophyll-based photosynthesis	Yes	*No*	Yes

[a]70S ribosomes are smaller than 80S ribosomes.
[b]Archaeal RNA polymerase is similar to eukaryotic polymerases.

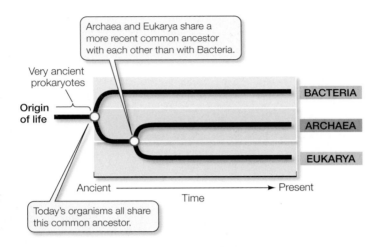

26.1 The Three Domains of the Living World *(Page 562)*

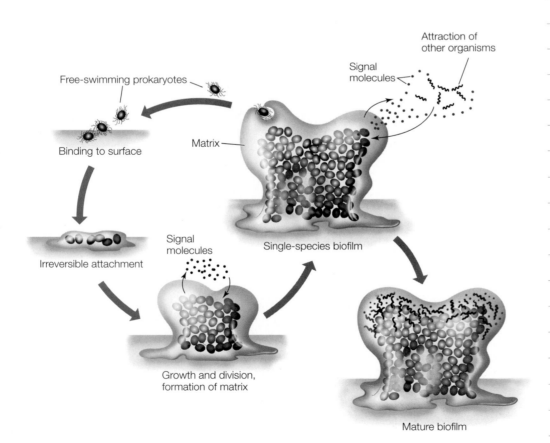

26.3 Forming a Biofilm *(Page 564)*

(A)

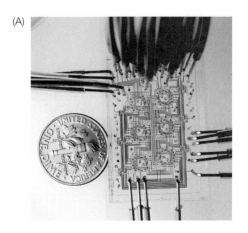

(B)

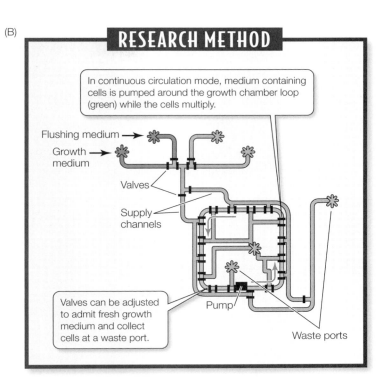

RESEARCH METHOD

In continuous circulation mode, medium containing cells is pumped around the growth chamber loop (green) while the cells multiply.

Flushing medium →

Growth → medium

Valves

Supply channels

Valves can be adjusted to admit fresh growth medium and collect cells at a waste port.

Pump

Waste ports

26.4 Microchemostats Allow Us to Study Microbial Dynamics *(Page 564)*

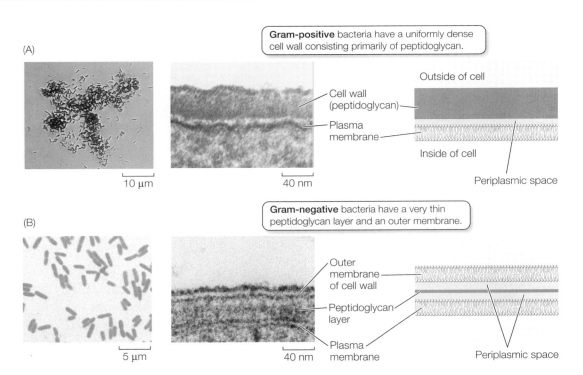

(A)

Gram-positive bacteria have a uniformly dense cell wall consisting primarily of peptidoglycan.

Outside of cell

Cell wall (peptidoglycan)

Plasma membrane

Inside of cell

Periplasmic space

10 µm

40 nm

(B)

Gram-negative bacteria have a very thin peptidoglycan layer and an outer membrane.

Outer membrane of cell wall

Peptidoglycan layer

Plasma membrane

Periplasmic space

5 µm

40 nm

26.5 The Gram Stain and the Bacterial Cell Wall *(Page 565)*

(A)

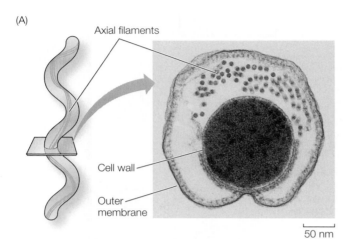

Axial filaments

Cell wall

Outer membrane

50 nm

(B)

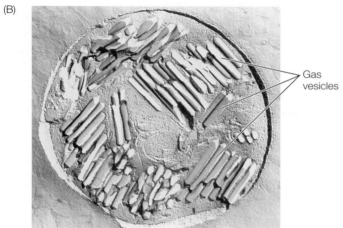

Gas vesicles

26.6 Structures Associated with Prokaryote Motility *(Page 566)*

TABLE 26.2

How Organisms Obtain Their Energy and Carbon

NUTRITIONAL CATEGORY	ENERGY SOURCE	CARBON SOURCE
Photoautotrophs (found in all three domains)	Light	Carbon dioxide
Photoheterotrophs (some bacteria)	Light	Organic compounds
Chemolithotrophs (some bacteria, many archaea)	Inorganic substances	Carbon dioxide
Chemoheterotrophs (found in all three domains)	Organic compounds	Organic compounds

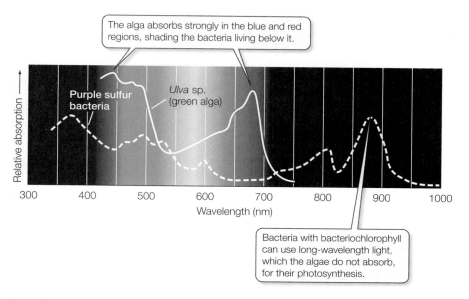

26.9 Bacteriochlorophyll Absorbs Long-Wavelength Light *(Page 568)*

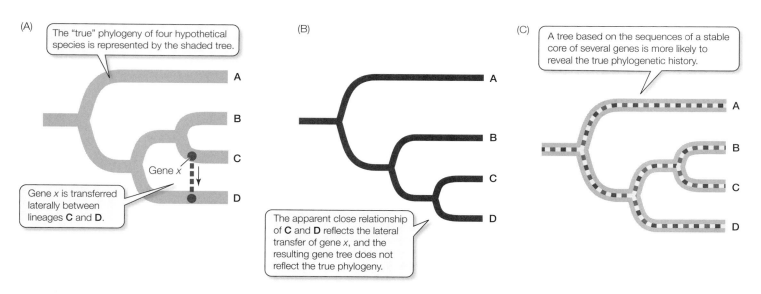

26.10 Lateral Gene Transfer Complicates Phylogenetic Relationships *(Page 570)*

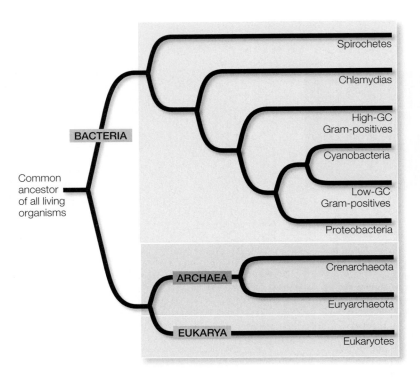

26.11 Two Domains: A Brief Overview *(Page 571)*

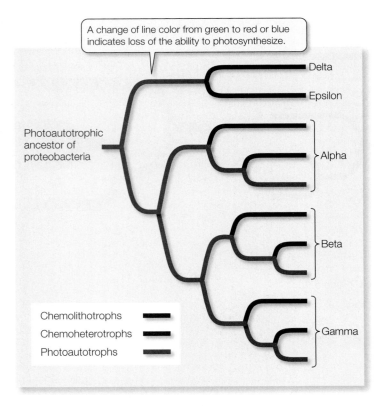

26.19 Modes of Nutrition in the Proteobacteria *(Page 575)*

EXPERIMENT

HYPOTHESIS: Some prokaryotes can grow and multiply at temperatures above 120°C.

METHOD

1. Seal samples of unidentified microorganisms taken from the vicinity of a thermal vent in tubes with medium containing Fe^{3+} as an electron acceptor. Control tubes contain the electron acceptor but no cells.

2. Hold experimental and control tubes for 10 hours in a sterilizer at 121°C. Reduction of the Fe^{3+} produces Fe^{2+} as magnetite, indicating the presence of living cells (left-hand photo).

3. In a second experiment, isolate and test for growth at various temperatures.

RESULTS

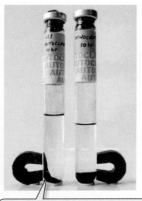

The iron-containing solids were attracted to a magnet only in those tubes that contained living cells.

Cells multiplied most rapidly at about 105°C but divided about once a day even at 121°C.

CONCLUSION: Some organisms in the sample multiplied at 121°C, the highest temperature yet known to allow growth of an organism.

26.21 What Is the Highest Temperature an Organism Can Tolerate? *(Page 576)*

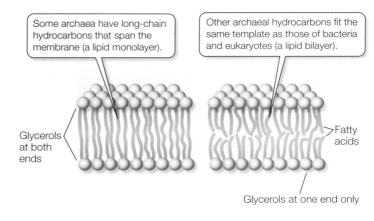

Some archaea have long-chain hydrocarbons that span the membrane (a lipid monolayer).

Other archaeal hydrocarbons fit the same template as those of bacteria and eukaryotes (a lipid bilayer).

Glycerols at both ends

Fatty acids

Glycerols at one end only

26.22 Membrane Architecture in Archaea *(Page 576)*

Page 576 In-Text Art (1)

Page 576 In-Text Art (2)

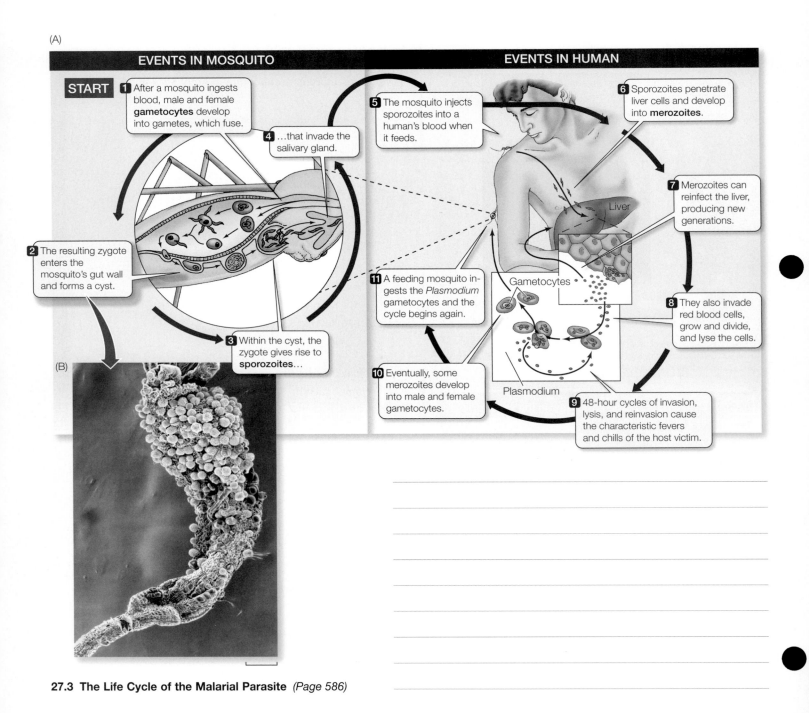

(A)

EVENTS IN MOSQUITO

EVENTS IN HUMAN

START

1 After a mosquito ingests blood, male and female **gametocytes** develop into gametes, which fuse.

4 ...that invade the salivary gland.

5 The mosquito injects sporozoites into a human's blood when it feeds.

6 Sporozoites penetrate liver cells and develop into **merozoites**.

2 The resulting zygote enters the mosquito's gut wall and forms a cyst.

7 Merozoites can reinfect the liver, producing new generations.

Liver

11 A feeding mosquito ingests the *Plasmodium* gametocytes and the cycle begins again.

Gametocytes

8 They also invade red blood cells, grow and divide, and lyse the cells.

3 Within the cyst, the zygote gives rise to **sporozoites**...

(B)

10 Eventually, some merozoites develop into male and female gametocytes.

Plasmodium

9 48-hour cycles of invasion, lysis, and reinvasion cause the characteristic fevers and chills of the host victim.

27.3 The Life Cycle of the Malarial Parasite *(Page 586)*

TABLE 27.1

Major Eukaryote Clades

CLADE	ATTRIBUTES	EXAMPLE (GENUS)
Chromalveolates		
Haptophytes	Unicellular, often with calcium carbonate scales	*Emiliania*
Alveolates	Sac-like structures beneath plasma membrane	
Apicomplexans	Apical complex for penetration of host	*Plasmodium*
Dinoflagellates	Pigments give golden-brown color	*Gonyaulax*
Ciliates	Cilia; two types of nuclei	*Paramecium*
Stramenopiles	Hairy and smooth flagella	
Brown algae	Multicellular; marine; photosynthetic	*Macrocystis*
Diatoms	Unicellular; photosynthetic; two-part cell walls	*Thalassiosira*
Oomycetes	Mostly coenocytic; heterotrophic	*Saprolegnia*
Plantae		
Glaucophytes	Peptidoglycan in chloroplasts	*Cyanophora*
Red algae	No flagella; chlorophyll *a* and *c*; phycoerythrin	*Chondrus*
Chlorophytes	Chlorophyll *a* and *b*	*Ulva*
*Land plants (Chs. 28–29)	Chlorophyll *a* and *b*; protected embryo	*Ginkgo*
Charophytes	Chlorophyll *a* and *b*; mitotic spindle oriented as in land plants	*Chara*
Excavates		
Diplomonads	No mitochondria; two nuclei; flagella	*Giardia*
Parabasalids	No mitochondria; flagella and undulating membrane	*Trichomonas*
Heteroloboseans	Can transform between amoeboid and flagellate stages	*Naegleria*
Euglenids	Flagella; spiral strips of protein support cell surface	*Euglena*
Kinetoplastids	Kinetoplast within mitochondrion	*Trypanosoma*
Rhizaria		
Cercozoans	Threadlike pseudopods	*Cercomonas*
Foraminiferans	Long, branched pseudopods; calcium carbonate shells	*Globigerina*
Radiolarians	Glassy endoskeleton; thin, stiff pseudopods	*Astrolithium*
Unikonts		
Opisthokonts	Single, posterior flagellum	
*Fungi (Ch. 30)	Heterotrophs that feed by absorption	*Penicillium*
Choanoflagellates	Resemble sponge cells; heterotrophic; with flagella	*Choanoeca*
*Animals (Chs. 31–33)	Heterotrophs that feed by ingestion	*Drosophila*
Amoebozoans	Amoebas with lobe-shaped pseudopods	
Loboseans	Feed individually	*Amoeba*
Plasmodial slime molds	Form coenocytic feeding bodies	*Physarum*
Cellular slime molds	Cells retain their identity in pseudoplasmodium	*Dictyostelium*

*Clades marked with an asterisk are made up of multicellular organisms and are discussed in the chapters indicated. All other groups listed are treated here as *microbial eukaryotes* (often known as *protists*).

(Page 584)

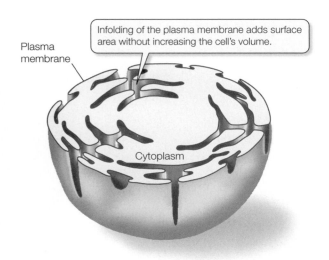

Plasma membrane

Infolding of the plasma membrane adds surface area without increasing the cell's volume.

Cytoplasm

27.6 Membrane Infolding *(Page 588)*

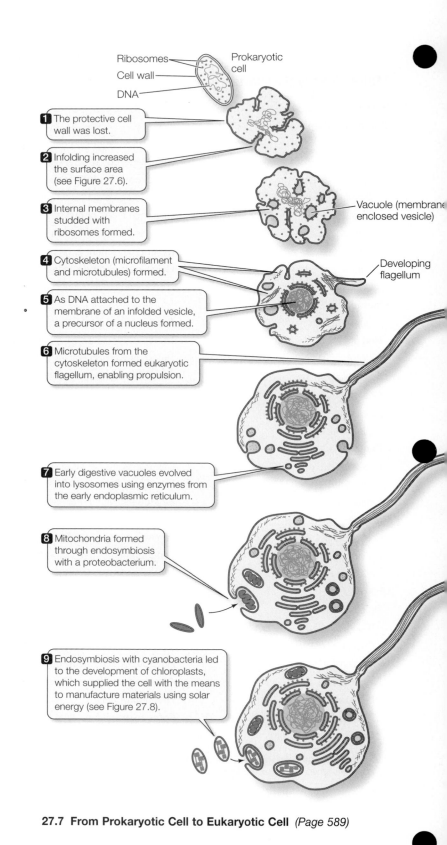

Ribosomes

Cell wall

DNA

Prokaryotic cell

1 The protective cell wall was lost.

2 Infolding increased the surface area (see Figure 27.6).

3 Internal membranes studded with ribosomes formed.

Vacuole (membrane enclosed vesicle)

4 Cytoskeleton (microfilament and microtubules) formed.

Developing flagellum

5 As DNA attached to the membrane of an infolded vesicle, a precursor of a nucleus formed.

6 Microtubules from the cytoskeleton formed eukaryotic flagellum, enabling propulsion.

7 Early digestive vacuoles evolved into lysosomes using enzymes from the early endoplasmic reticulum.

8 Mitochondria formed through endosymbiosis with a proteobacterium.

9 Endosymbiosis with cyanobacteria led to the development of chloroplasts, which supplied the cell with the means to manufacture materials using solar energy (see Figure 27.8).

27.7 From Prokaryotic Cell to Eukaryotic Cell *(Page 589)*

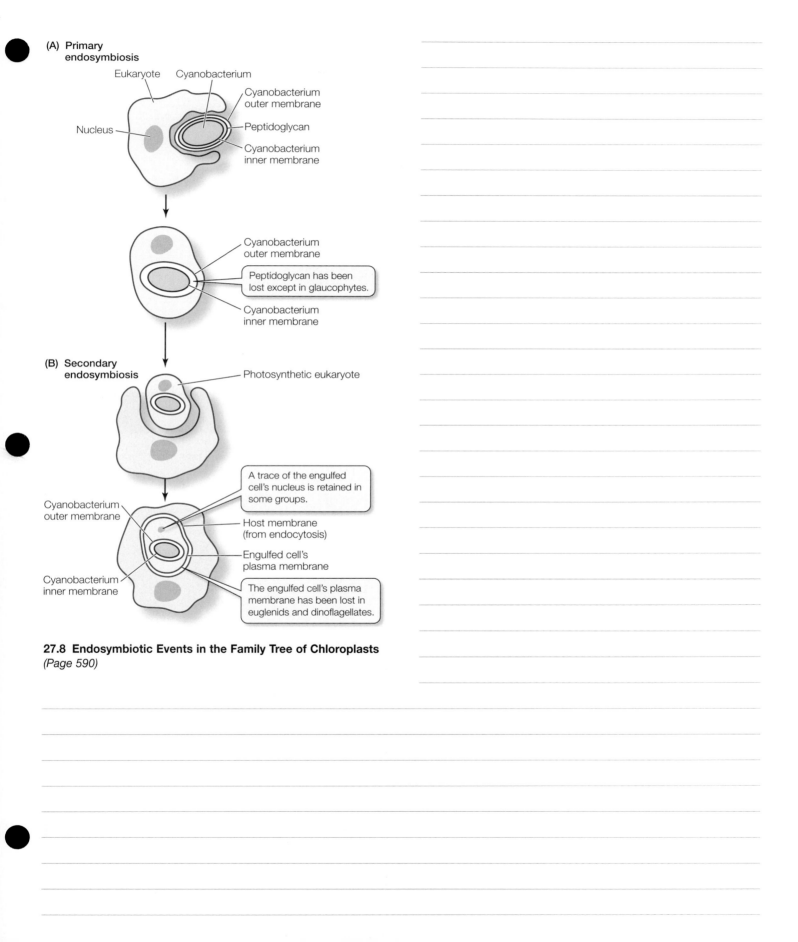

(A) Primary endosymbiosis

Eukaryote

Cyanobacterium

Cyanobacterium outer membrane

Peptidoglycan

Nucleus

Cyanobacterium inner membrane

Cyanobacterium outer membrane

Peptidoglycan has been lost except in glaucophytes.

Cyanobacterium inner membrane

(B) Secondary endosymbiosis

Photosynthetic eukaryote

A trace of the engulfed cell's nucleus is retained in some groups.

Cyanobacterium outer membrane

Host membrane (from endocytosis)

Engulfed cell's plasma membrane

Cyanobacterium inner membrane

The engulfed cell's plasma membrane has been lost in euglenids and dinoflagellates.

27.8 Endosymbiotic Events in the Family Tree of Chloroplasts
(Page 590)

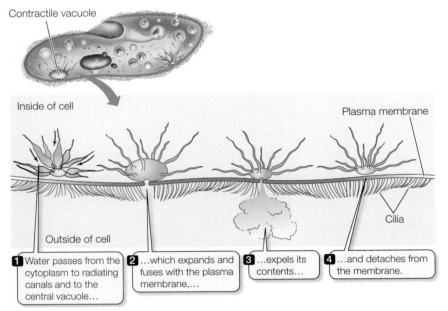

Contractile vacuole

Inside of cell

Plasma membrane

Outside of cell

Cilia

1 Water passes from the cytoplasm to radiating canals and to the central vacuole…

2 …which expands and fuses with the plasma membrane,…

3 …expels its contents…

4 …and detaches from the membrane.

27.10 Contractile Vacuoles Bail Out Excess Water
(Page 592)

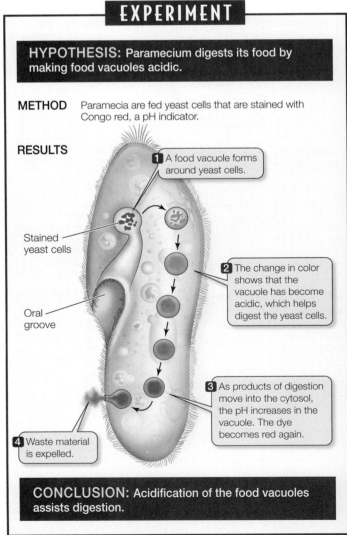

EXPERIMENT

HYPOTHESIS: Paramecium digests its food by making food vacuoles acidic.

METHOD   Paramecia are fed yeast cells that are stained with Congo red, a pH indicator.

RESULTS

1 A food vacuole forms around yeast cells.

Stained yeast cells

Oral groove

2 The change in color shows that the vacuole has become acidic, which helps digest the yeast cells.

3 As products of digestion move into the cytosol, the pH increases in the vacuole. The dye becomes red again.

4 Waste material is expelled.

CONCLUSION: Acidification of the food vacuoles assists digestion.

27.11 Food Vacuoles Handle Digestion and Excretion *(Page 592)*

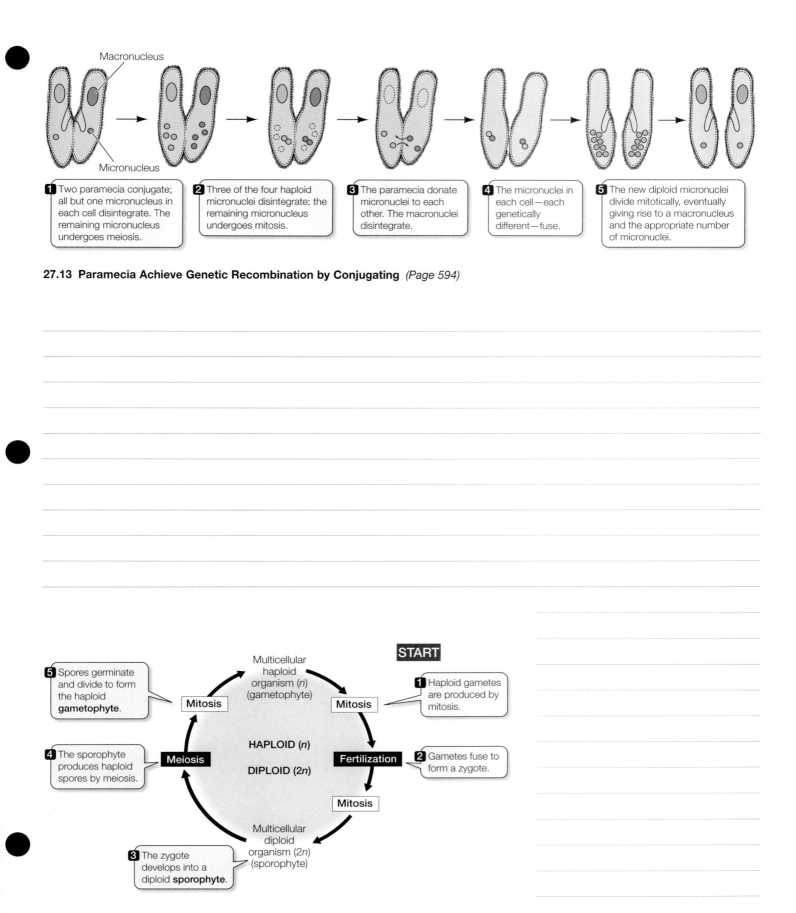

27.13 Paramecia Achieve Genetic Recombination by Conjugating *(Page 594)*

Macronucleus

Micronucleus

1 Two paramecia conjugate; all but one micronucleus in each cell disintegrate. The remaining micronucleus undergoes meiosis.

2 Three of the four haploid micronuclei disintegrate; the remaining micronucleus undergoes mitosis.

3 The paramecia donate micronuclei to each other. The macronuclei disintegrate.

4 The micronuclei in each cell—each genetically different—fuse.

5 The new diploid micronuclei divide mitotically, eventually giving rise to a macronucleus and the appropriate number of micronuclei.

START

5 Spores germinate and divide to form the haploid **gametophyte**.

Mitosis

Multicellular haploid organism (n) (gametophyte)

Mitosis

1 Haploid gametes are produced by mitosis.

4 The sporophyte produces haploid spores by meiosis.

Meiosis

HAPLOID (n)

DIPLOID (2n)

Fertilization

2 Gametes fuse to form a zygote.

Mitosis

Multicellular diploid organism (2n) (sporophyte)

3 The zygote develops into a diploid **sporophyte**.

27.14 Alternation of Generations *(Page 594)*

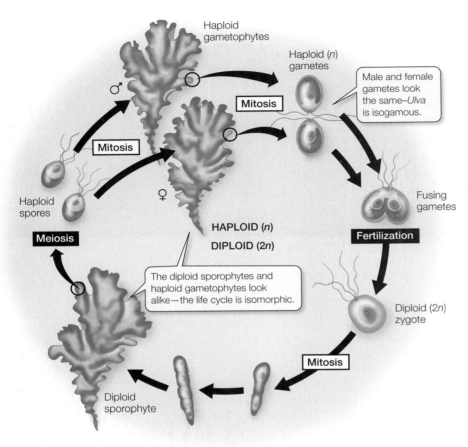

Haploid gametophytes

Haploid (*n*) gametes

Male and female gametes look the same—*Ulva* is isogamous.

Mitosis

Mitosis

♂

Mitosis

Haploid spores

♀

Fusing gametes

HAPLOID (*n*)

DIPLOID (2*n*)

Fertilization

Meiosis

The diploid sporophytes and haploid gametophytes look alike—the life cycle is isomorphic.

Diploid (2*n*) zygote

Diploid sporophyte

Mitosis

27.15 An Isomorphic Life Cycle *(Page 595)*

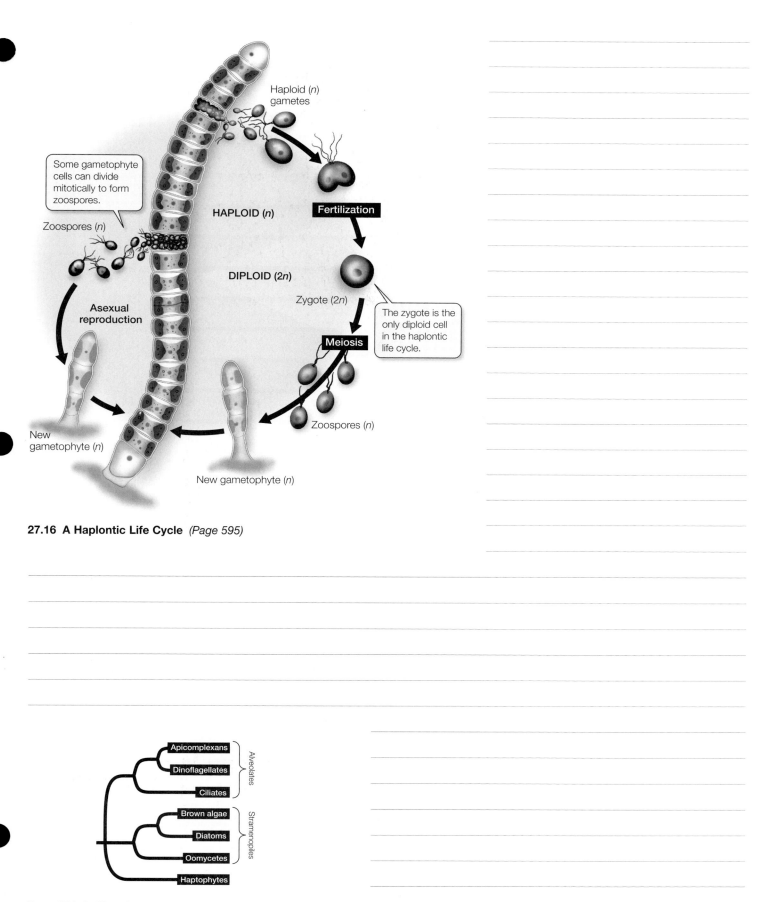

Haploid (*n*) gametes

Some gametophyte cells can divide mitotically to form zoospores.

Zoospores (*n*)

HAPLOID (*n*)

Fertilization

DIPLOID (2*n*)

Zygote (2*n*)

The zygote is the only diploid cell in the haplontic life cycle.

Asexual reproduction

Meiosis

New gametophyte (*n*)

New gametophyte (*n*)

Zoospores (*n*)

27.16 A Haplontic Life Cycle *(Page 595)*

Apicomplexans

Dinoflagellates

Ciliates

Alveolates

Brown algae

Diatoms

Oomycetes

Stramenopiles

Haptophytes

Page 596 In-Text Art

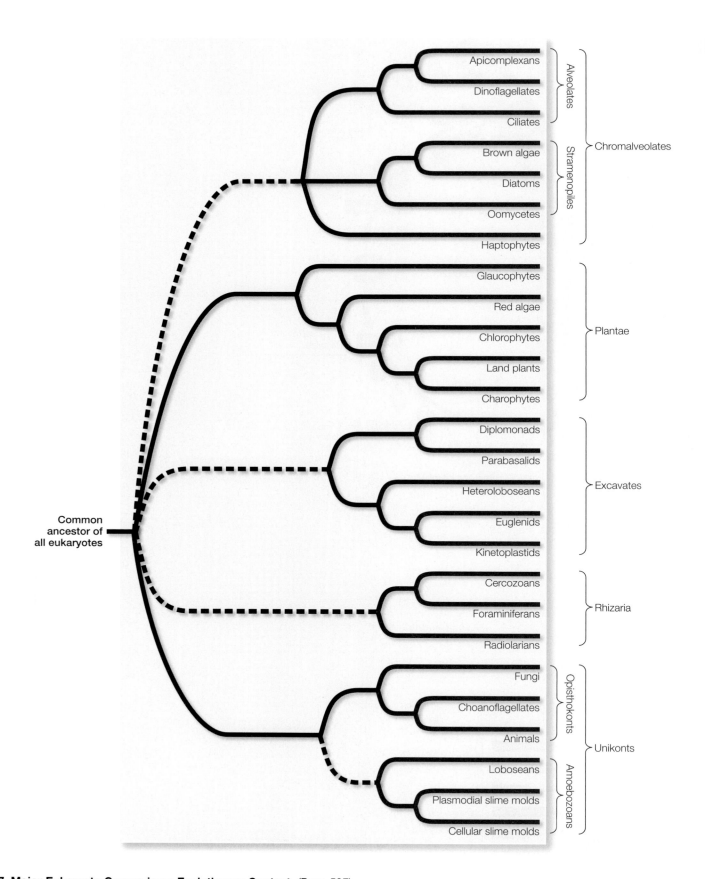

27.17 Major Eukaryote Groups in an Evolutionary Context *(Page 597)*

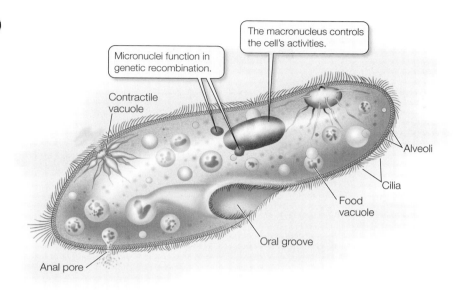

Micronuclei function in genetic recombination.

The macronucleus controls the cell's activities.

Contractile vacuole

Alveoli

Cilia

Food vacuole

Oral groove

Anal pore

27.20 Anatomy of *Paramecium* (Page 599)

EXPERIMENT

HYPOTHESIS: Copepods fail to flourish during a diatom bloom because of the toxicity of the diatoms.

METHOD

1. Female copepods are fed for 10 days on either the diatom *Skeletonema costatum* or a nontoxic diet of dinoflagellates (control).
2. Eggs and newly hatched larvae are counted each day.

RESULTS

Although almost all eggs of the control group hatched each day, fewer and fewer of the eggs from the diatom-fed females hatched as the experiment progressed.

Control females

Diatom-fed females

CONCLUSION: Toxicity is a plausible explanation for the failure of copepods to flourish during a diatom bloom.

27.22 Why Don't Copepods Flourish During Diatom Blooms?
(Page 600)

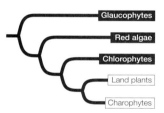

Glaucophytes
Red algae
Chlorophytes
Land plants
Charophytes

Page 600 In-Text Art

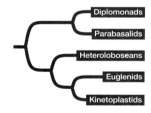

Diplomonads
Parabasalids
Heteroloboseans
Euglenids
Kinetoplastids

Page 602 In-Text Art

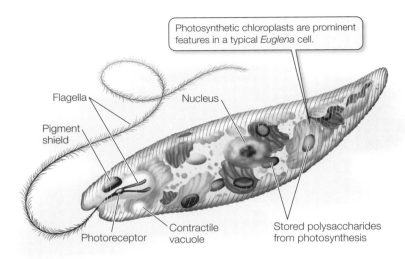

Flagella

Pigment shield

Nucleus

Photosynthetic chloroplasts are prominent features in a typical *Euglena* cell.

Photoreceptor

Contractile vacuole

Stored polysaccharides from photosynthesis

27.28 A Photosynthetic Euglenid *(Page 604)*

TABLE 27.2

A Comparison of Three Kinetoplastid Trypanosomes

	TRYPANOSOMA BRUCEI	*TRYPANOSOMA CRUZI*	*LEISHMANIA MAJOR*
Human disease	Sleeping sickness	Chagas' disease	Leishmaniasis
Insect vector	Tsetse fly	Assassin bug	Sand fly
Vaccine or effective cure	None	None	None
Strategy for survival	Changes surface recognition molecules frequently	Causes changes in surface recognition molecules on host cell	Reduces effectiveness of macrophage hosts
Site in human body	Bloodstream; attacks nerve tissue in final stages	Enters cells, especially muscle cells	Enters cells, primarily macrophages
Deaths per year	>50,000	43,000	60,000 (?)

(Page 604)

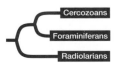

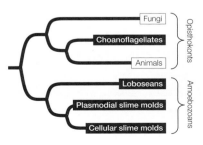

Page 604 In-Text Art   **Page 605 In-Text Art**

CHAPTER 28 Plants without Seeds: From Sea to Land

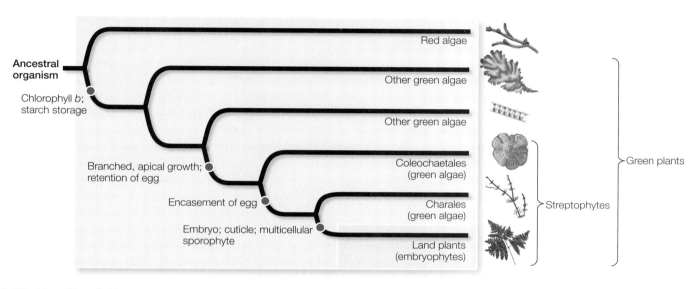

28.1 What Is a Plant? *(Page 612)*

TABLE 28.1

Classification of Land Plants

GROUP	COMMON NAME	CHARACTERISTICS
NONVASCULAR PLANTS		
Hepatophyta	Liverworts	No filamentous stage; gametophyte flat
Anthocerophyta	Hornworts	Embedded archegonia; sporophyte grows basally (from the ground)
Bryophyta	Mosses	Filamentous stage; sporophyte grows apically (from the tip)
VASCULAR PLANTS		
Lycophyta	Club mosses and allies	Microphylls in spirals; sporangia in leaf axils
Pteridophyta	Horsetails, whisk ferns, ferns	Differentiation between main stem and side branches (overtopping growth)
SEED PLANTS		
Gymnosperms		
Cycadophyta	Cycads	Compound leaves; swimming sperm; seeds on modified leaves
Ginkgophyta	Ginkgo	Deciduous; fan-shaped leaves; swimming sperm
Gnetophyta	Gnetophytes	Vessels in vascular tissue; opposite, simple leaves
Coniferophyta	Conifers	Seeds in cones; needle-like or scale-like leaves
Angiosperms	Flowering plants	Endosperm; carpels; gametophytes much reduced; seeds within fruit

Note: No extinct groups are included in this classification.

(Page 612)

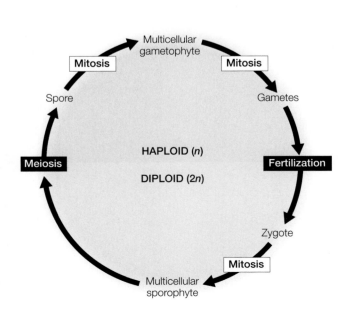

28.4 Alternation of Generations in Plants *(Page 614)*

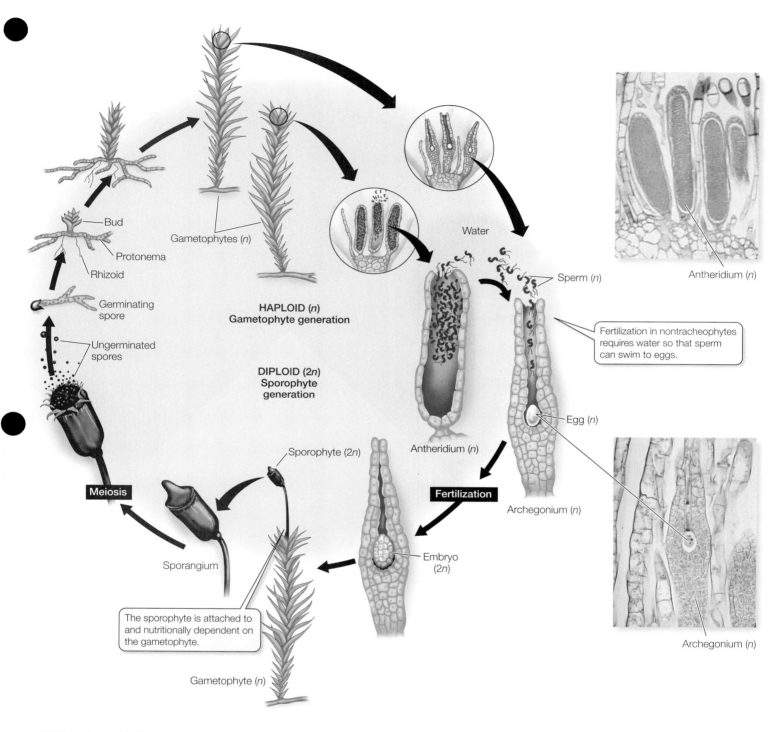

Bud

Protonema

Rhizoid

Germinating
spore

Ungerminated
spores

Gametophytes (n)

HAPLOID (n)
Gametophyte generation

DIPLOID (2n)
Sporophyte
generation

Water

Sperm (n)

Antheridium (n)

Fertilization in nontracheophytes
requires water so that sperm
can swim to eggs.

Egg (n)

Antheridium (n)

Archegonium (n)

Meiosis

Sporophyte (2n)

Fertilization

Sporangium

Embryo
(2n)

The sporophyte is attached to
and nutritionally dependent on
the gametophyte.

Archegonium (n)

Gametophyte (n)

28.5 A Moss Life Cycle *(Page 615)*

EXPERIMENT

HYPOTHESIS: Ancient microfossils, predating the earliest known nonvascular plant fossils, could be fragments of ancient liverworts.

METHOD

1. Investigators allowed liverworts to rot in soil or subjected them to high-temperature acid treatment, then examined the degraded material by light and scanning electron microscopy.

2. They compared images of the degraded material with those of microfossils from ancient rocks.

RESULTS

Sheets of decay-resistant cells from the lower surface of the degraded liverworts resembled some cell-sheet microfossils, showing rosette-like groupings of cells around "pores."

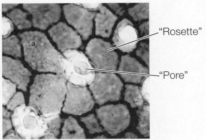

"Rosette"

"Pore"

Resistant fragments of liverwort rhizoids resembled some tubular microfossils.

CONCLUSION: The ancient microfossils may be fragments of liverworts.

28.6 Mimicking a Microfossil *(Page 616)*

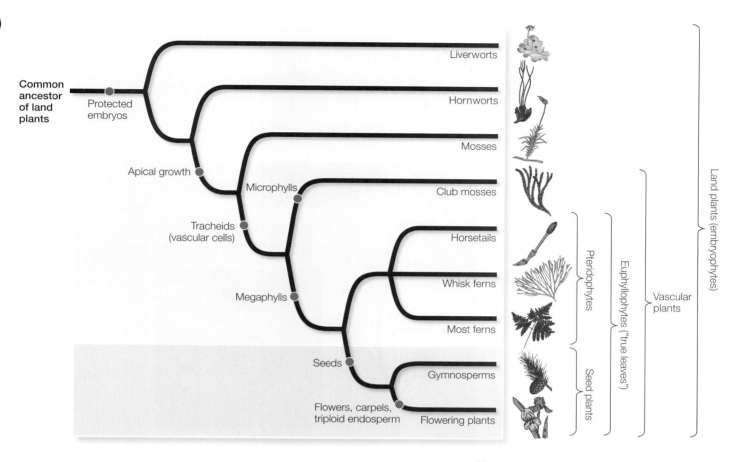

28.7 The Evolution of Today's Plants *(Page 617)*

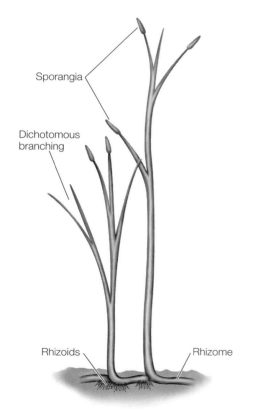

28.9 An Ancient Vascular Plant Relative *(Page 619)*

(A)

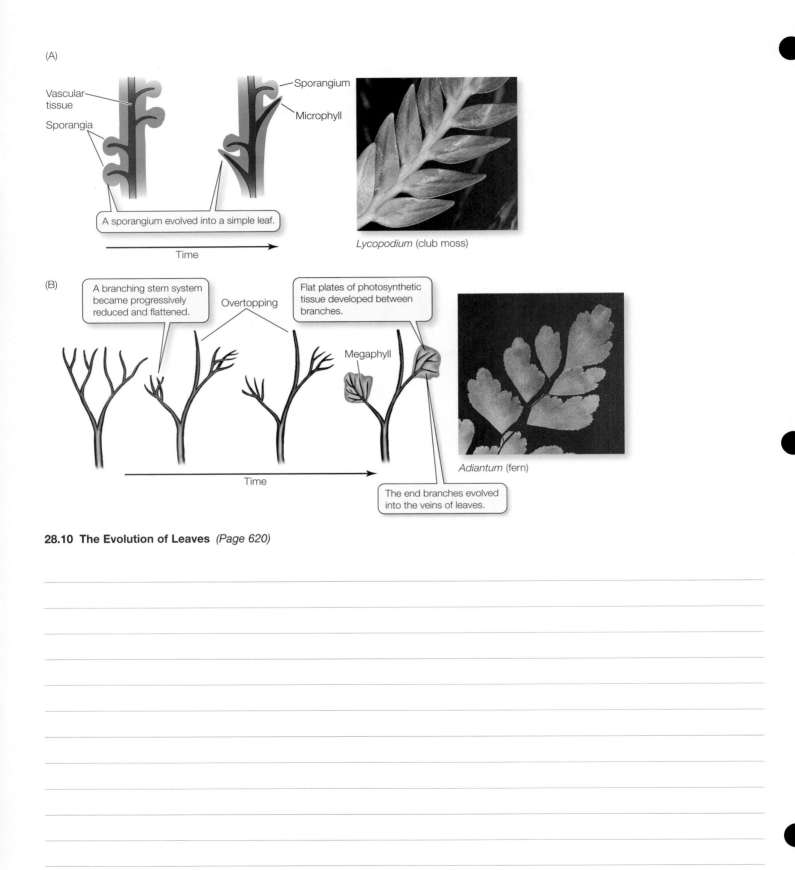

Vascular tissue

Sporangia

Sporangium

Microphyll

A sporangium evolved into a simple leaf.

Time

Lycopodium (club moss)

(B)

A branching stem system became progressively reduced and flattened.

Overtopping

Flat plates of photosynthetic tissue developed between branches.

Megaphyll

Time

The end branches evolved into the veins of leaves.

Adiantum (fern)

28.10 The Evolution of Leaves *(Page 620)*

EXPERIMENT

HYPOTHESIS: High concentrations of CO_2 in the Devonian atmosphere delayed the increase in leaf size.

METHOD

1. Scientists analyzed 300 plant fossils from the Devonian and Carboniferous periods and calculated the sizes of the leaves.

2. They compared the pattern of change in leaf size over time with that of change in atmosphere CO_2 concentrations.

RESULTS

Among these fossils, a rise in leaf size corresponded in time with a decline in the estimated concentration of CO_2 in the atmosphere.

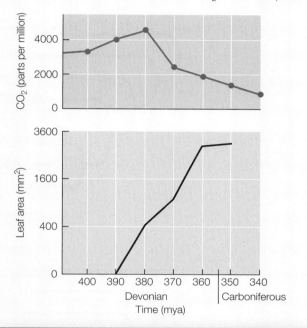

CONCLUSION: Leaf sizes increased as CO_2 concentrations decreased.

28.11 Decreasing CO_2 Levels and the Evolution of Megaphylls
(Page 621)

(A) Homospory

Homosporous plants produce a single type of spore.

The spores of homosporous plants produce a single type of gametophyte with both male and female reproductive organs.

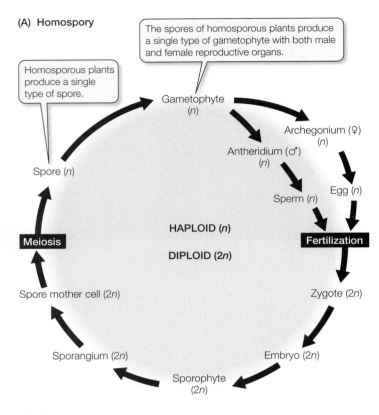

Gametophyte (*n*)

Archegonium (♀) (*n*)

Antheridium (♂) (*n*)

Spore (*n*)

Sperm (*n*)

Egg (*n*)

HAPLOID (*n*)

Meiosis

Fertilization

DIPLOID (2*n*)

Spore mother cell (2*n*)

Zygote (2*n*)

Sporangium (2*n*)

Embryo (2*n*)

Sporophyte (2*n*)

(B) Heterospory

Heterosporous plants produce two types of spores: a larger megaspore and a smaller microspore.

The spores of heterosporous plants produce male and female gametophytes.

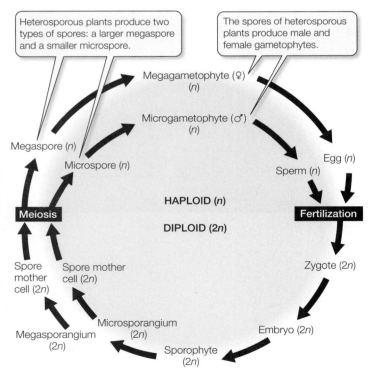

Megagametophyte (♀) (*n*)

Microgametophyte (♂) (*n*)

Megaspore (*n*)

Microspore (*n*)

Egg (*n*)

Sperm (*n*)

HAPLOID (*n*)

Meiosis

Fertilization

DIPLOID (2*n*)

Spore mother cell (2*n*)

Spore mother cell (2*n*)

Zygote (2*n*)

Megasporangium (2*n*)

Microsporangium (2*n*)

Embryo (2*n*)

Sporophyte (2*n*)

28.12 Homospory and Heterospory *(Page 622)*

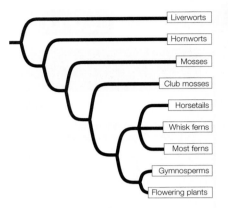

Liverworts
Hornworts
Mosses
Club mosses
Horsetails
Whisk ferns
Most ferns
Gymnosperms
Flowering plants

Page 622 In-Text Art

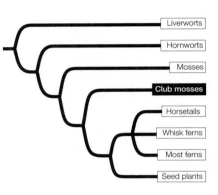

Liverworts
Hornworts
Mosses
Club mosses
Horsetails
Whisk ferns
Most ferns
Seed plants

Page 624 In-Text Art

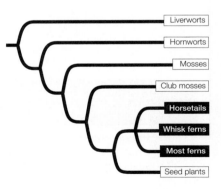

Liverworts
Hornworts
Mosses
Club mosses
Horsetails
Whisk ferns
Most ferns
Seed plants

Page 625 In-Text Art

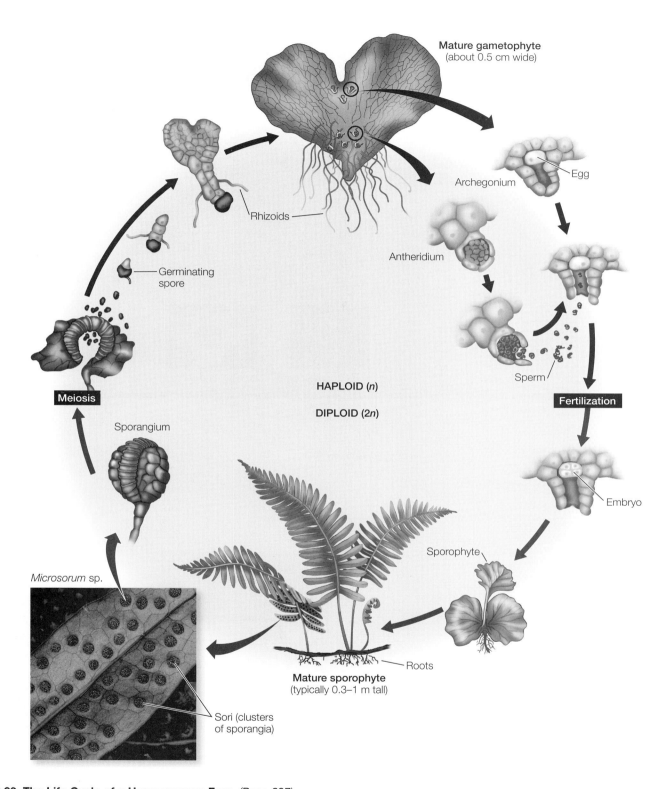

Mature gametophyte
(about 0.5 cm wide)

Archegonium

Egg

Antheridium

Rhizoids

Sperm

Germinating
spore

HAPLOID (*n*)

DIPLOID (2*n*)

Fertilization

Meiosis

Sporangium

Embryo

Microsorum sp.

Sporophyte

Roots

Mature sporophyte
(typically 0.3–1 m tall)

Sori (clusters
of sporangia)

28.20 The Life Cycle of a Homosporous Fern *(Page 627)*

CHAPTER 29 The Evolution of Seed Plants

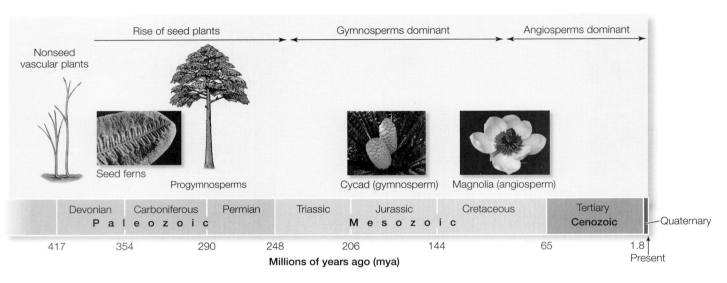

29.1 Highlights in the History of Seed Plants *(Page 632)*

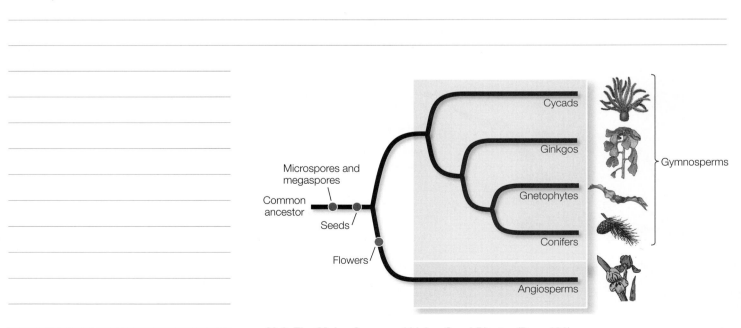

29.2 The Major Groups of Living Seed Plants *(Page 632)*

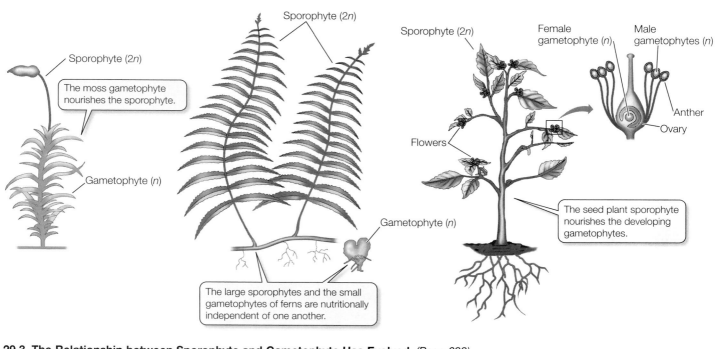

Sporophyte (2n)

The moss gametophyte nourishes the sporophyte.

Gametophyte (n)

Sporophyte (2n)

Gametophyte (n)

The large sporophytes and the small gametophytes of ferns are nutritionally independent of one another.

Sporophyte (2n)

Flowers

Female gametophyte (n)

Male gametophytes (n)

Anther

Ovary

The seed plant sporophyte nourishes the developing gametophytes.

29.3 The Relationship between Sporophyte and Gametophyte Has Evolved *(Page 633)*

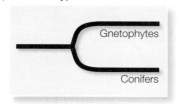

(A) Gnetifer hypothesis

Gnetophytes

Conifers

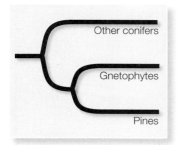

(B) Gnepine hypothesis

Other conifers

Gnetophytes

Pines

29.6 Two Interpretations of Conifer Phylogeny *(Page 636)*

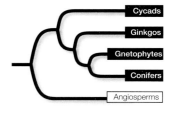

Cycads

Ginkgos

Gnetophytes

Conifers

Angiosperms

Page 634 In-Text Art

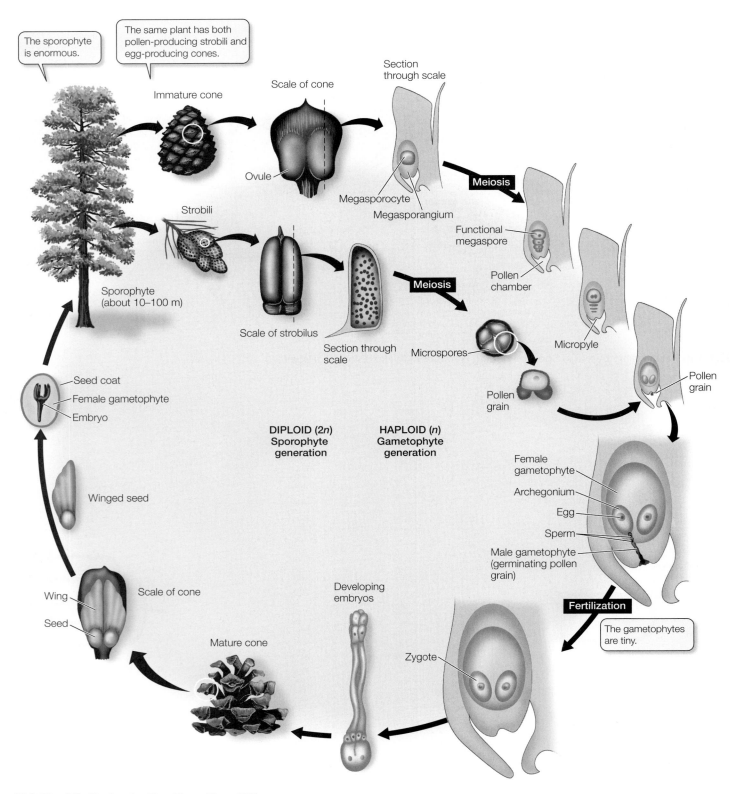

The sporophyte is enormous.

The same plant has both pollen-producing strobili and egg-producing cones.

Immature cone

Scale of cone

Section through scale

Ovule

Megasporocyte

Megasporangium

Meiosis

Functional megaspore

Pollen chamber

Strobili

Sporophyte (about 10–100 m)

Scale of strobilus

Section through scale

Meiosis

Microspores

Micropyle

Pollen grain

Pollen grain

Seed coat

Female gametophyte

Embryo

DIPLOID (2n)
Sporophyte
generation

HAPLOID (n)
Gametophyte
generation

Pollen grain

Female gametophyte

Archegonium

Egg

Sperm

Male gametophyte (germinating pollen grain)

Winged seed

Wing

Seed

Scale of cone

Mature cone

Developing embryos

Zygote

Fertilization

The gametophytes are tiny.

29.8 The Life Cycle of a Pine Tree *(Page 637)*

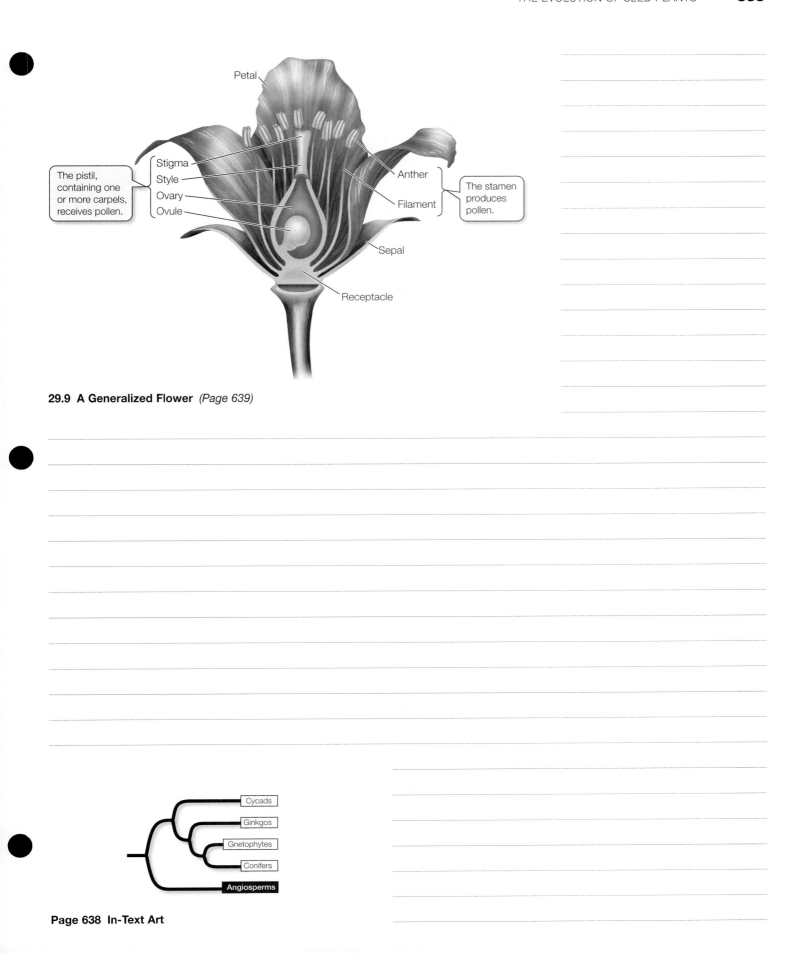

Petal

Stigma

The pistil,
containing one
or more carpels,
receives pollen.

Style

Ovary

Ovule

Anther

The stamen
produces
pollen.

Filament

Sepal

Receptacle

29.9 A Generalized Flower *(Page 639)*

Cycads

Ginkgos

Gnetophytes

Conifers

Angiosperms

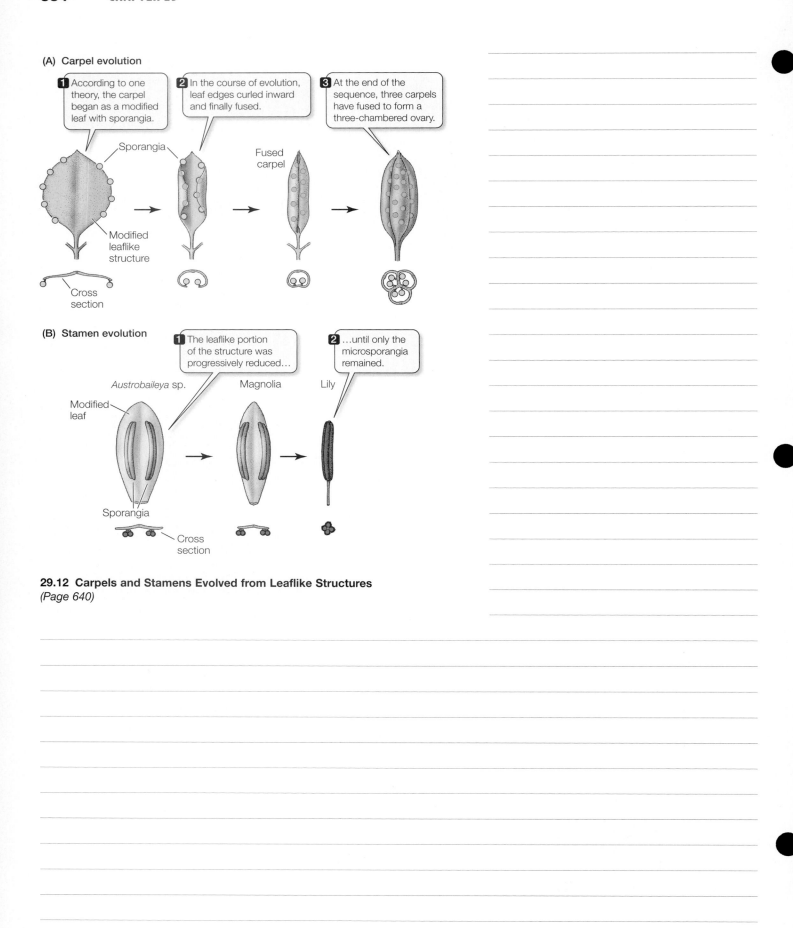

(A) Carpel evolution

1 According to one theory, the carpel began as a modified leaf with sporangia.

2 In the course of evolution, leaf edges curled inward and finally fused.

3 At the end of the sequence, three carpels have fused to form a three-chambered ovary.

Sporangia

Fused carpel

Modified leaflike structure

Cross section

(B) Stamen evolution

1 The leaflike portion of the structure was progressively reduced...

2 ...until only the microsporangia remained.

Austrobaileya sp.

Magnolia

Lily

Modified leaf

Sporangia

Cross section

29.12 Carpels and Stamens Evolved from Leaflike Structures
(Page 640)

(A)

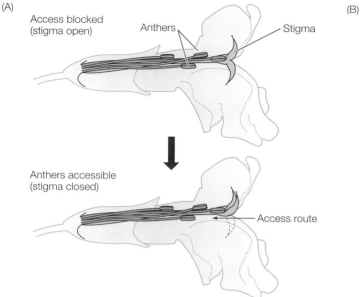

Access blocked
(stigma open)

Anthers

Stigma

Anthers accessible
(stigma closed)

Access route

29.13 Stigma Behavior Increases Pollen Export in Monkeyflowers
(Page 641)

(B)

EXPERIMENT

HYPOTHESIS: Stigma responses in the bush monkeyflower favor the export of pollen.

METHOD

1. Set up experimental arrays of monkeyflowers such that only one flower in each array can donate pollen.

2. Pollen donors in some arrays are normal controls (untouched open stigmas); stigmas of other donors are artificially closed; a third group of donor stigmas are permanently propped open.

3. After hummingbirds visit the arrays, count the grains from each donor on the stigma of the next flower visited.

RESULTS

Almost twice as much pollen was exported from control flowers (i.e., those whose stigmas functioned normally) as from those whose stigmas were experimentally propped open. Experimentally closing the stigmas resulted in even greater pollen dispersal.

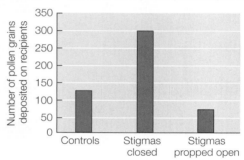

CONCLUSION: Stigma responses enhance the male function of the flower (pollen dispersal) once its female function (pollen deposition) has been performed.

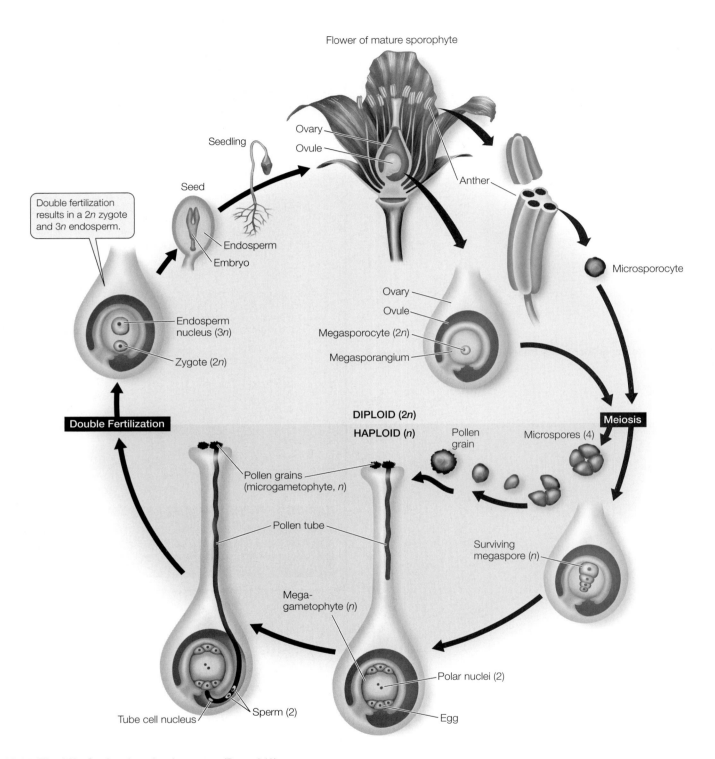

Flower of mature sporophyte

Ovary
Ovule

Seedling

Seed

Anther

Double fertilization results in a 2n zygote and 3n endosperm.

Endosperm
Embryo

Microsporocyte

Ovary
Ovule
Megasporocyte (2n)
Megasporangium

Endosperm nucleus (3n)

Zygote (2n)

DIPLOID (2n)

HAPLOID (n)

Pollen grain

Microspores (4)

Meiosis

Double Fertilization

Pollen grains (microgametophyte, n)

Surviving megaspore (n)

Pollen tube

Mega-gametophyte (n)

Polar nuclei (2)

Tube cell nucleus

Sperm (2)

Egg

29.14 The Life Cycle of an Angiosperm *(Page 642)*

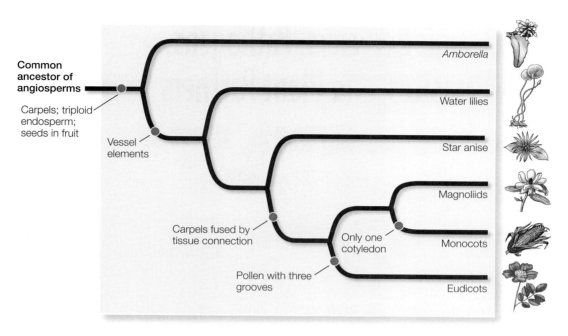

Common ancestor of angiosperms

Carpels; triploid endosperm; seeds in fruit

Vessel elements

Carpels fused by tissue connection

Pollen with three grooves

Only one cotyledon

Amborella

Water lilies

Star anise

Magnoliids

Monocots

Eudicots

29.16 Evolutionary Relationships among the Angiosperms *(Page 644)*

TABLE 29.1

Some Medicinal Plants and Their Products

PRODUCT	PLANT SOURCE	MEDICAL APPLICATION
Atropine	Belladonna	Dilating pupils for eye examination
Bromelain	Pineapple stem	Controlling tissue inflammation
Digitalin	Foxglove	Strengthening heart muscle contraction
Ephedrine	*Ephedra*	Easing nasal congestion
Menthol	Japanese mint	Relief of coughing
Morphine	Opium poppy	Relief of pain
Quinine	Cinchona bark	Treatment of malaria
Taxol	Pacific yew	Treatment of ovarian and breast cancers
Tubocurarine	Curare plant	As muscle relaxant in surgery
Vincristine	Periwinkle	Treatment of leukemia and lymphoma

30 Fungi: Recyclers, Pathogens, Parasites, and Plant Partners

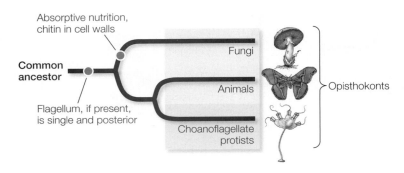

30.1 Fungi in Evolutionary Context *(Page 652)*

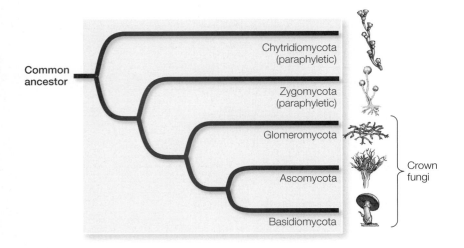

30.2 Phylogeny of the Fungi *(Page 652)*

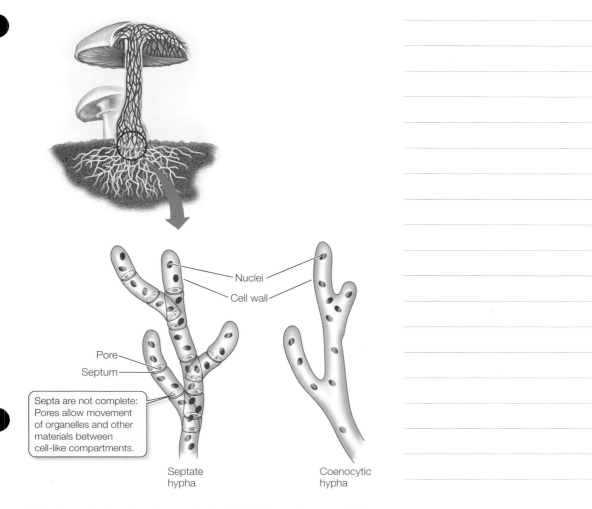

Nuclei

Cell wall

Pore

Septum

Septa are not complete: Pores allow movement of organelles and other materials between cell-like compartments.

Septate hypha

Coenocytic hypha

30.4 Most Hyphae Are Incompletely Divided into Separate Cells *(Page 653)*

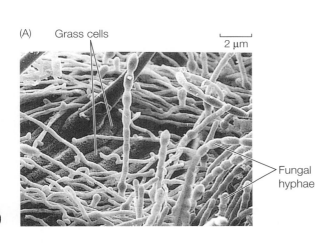

(A)   Grass cells

2 μm

Fungal hyphae

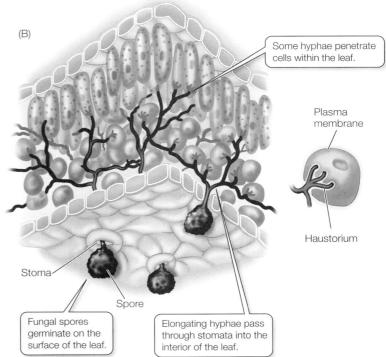

(B)

Some hyphae penetrate cells within the leaf.

Plasma membrane

Haustorium

Stoma

Spore

Fungal spores germinate on the surface of the leaf.

Elongating hyphae pass through stomata into the interior of the leaf.

30.5 Attacks on a Leaf *(Page 654)*

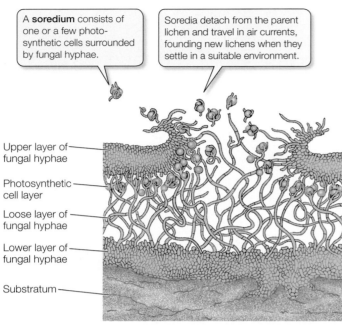

A **soredium** consists of one or a few photosynthetic cells surrounded by fungal hyphae.

Soredia detach from the parent lichen and travel in air currents, founding new lichens when they settle in a suitable environment.

Upper layer of fungal hyphae

Photosynthetic cell layer

Loose layer of fungal hyphae

Lower layer of fungal hyphae

Substratum

30.9 Lichen Anatomy *(Page 657)*

EXPERIMENT

HYPOTHESIS: Fungi, not ants, produce the substance that keeps "foreign" fungi from growing in the garden.

METHOD

1. Collect samples of distinct fungus gardens and ants from the field and establish colonies in the laboratory.

2. Some ants from each of two colonies (1 and 2) are forced to feed for 10 days on fungi obtained from the other colony. Other ants continue to feed on their "home" gardens.

3. At the beginning and end of the 10-day period, place fecal droplets of ants on fungi from their own garden or from the "alien" garden.

4. Test for incompatibility reactions (avoidance of droplet by fungus, toxicity to fungus).

RESULTS

Ants originally from:
Garden 1   Garden 2

Ants fed with **fungus** from:
Garden 1
Garden 2

Fecal droplet incompatibility

Before   After   Before   After
Force-feeding (10 days)

Fungus tested:
● Fungus 1
● Fungus 2

Incompatibility:
0 = Complete compatibility
5 = Complete incompatibility

CONCLUSION: Foreign fungi are deterred by substances produced by the fungus an ant eats, not by the ants themselves.

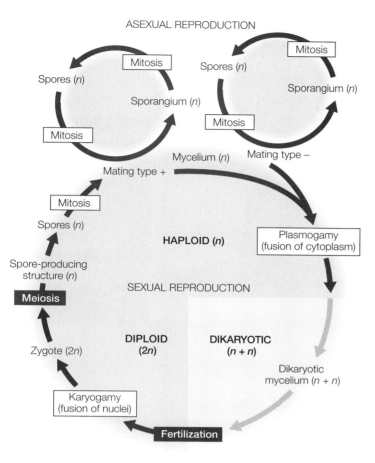

30.12 Asexual and Sexual Reproduction in a Fungal Life Cycle
(Page 659)

(A) Chytrids

The life cycle of the aquatic chytrids features alternation of generations. They have no dikaryotic stage.

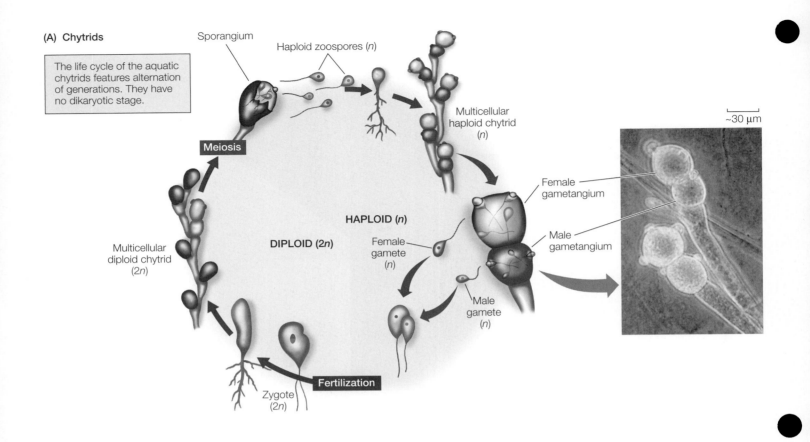

Sporangium

Haploid zoospores (n)

Multicellular haploid chytrid (n)

Meiosis

HAPLOID (n)

DIPLOID (2n)

Multicellular diploid chytrid (2n)

Female gamete (n)

Female gametangium

Male gametangium

Male gamete (n)

~30 μm

Zygote (2n)

Fertilization

(B) Zygomycetes

The zygomycete sporangium contains haploid nuclei that are incorporated into spores.

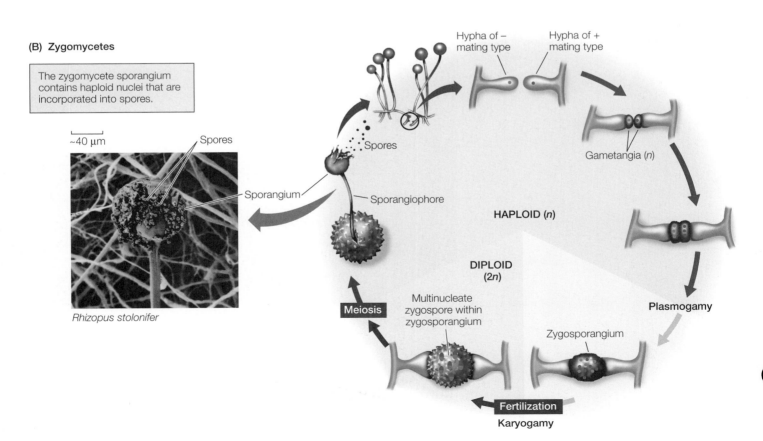

Hypha of – mating type

Hypha of + mating type

Spores

Gametangia (n)

Sporangium

Spores

Sporangiophore

HAPLOID (n)

DIPLOID (2n)

Plasmogamy

Meiosis

Multinucleate zygospore within zygosporangium

Zygosporangium

~40 μm

Spores

Rhizopus stolonifer

Fertilization

Karyogamy

30.13 Sexual Life Cycles Vary among Different Groups of Fungi *(Page 660)*

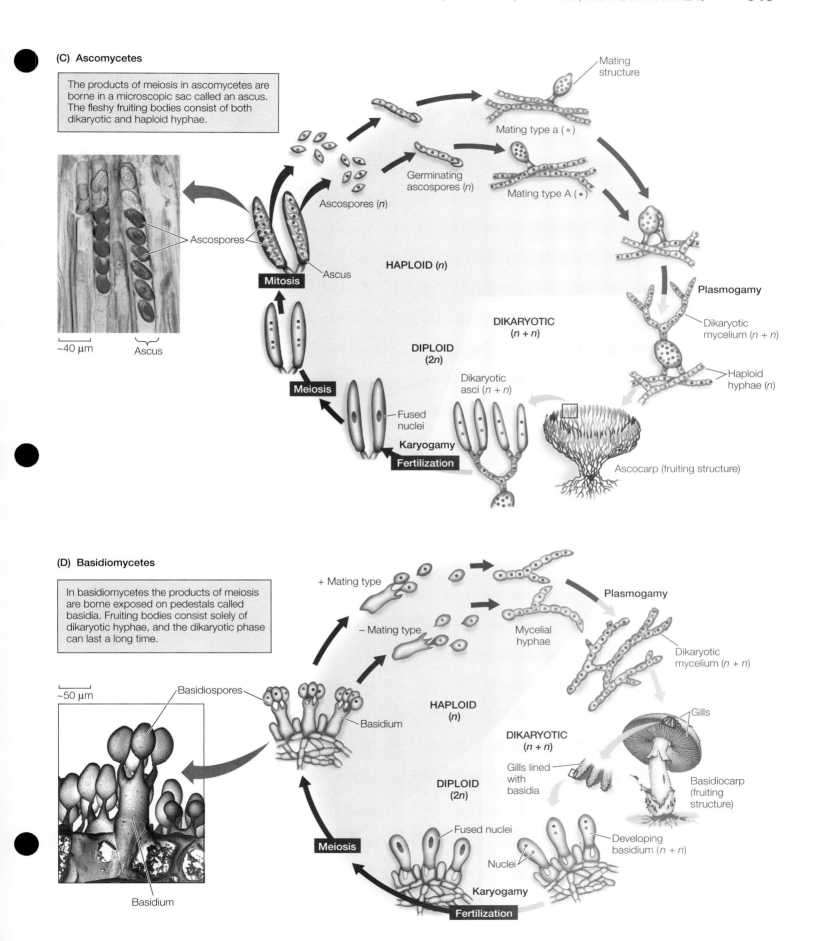

(C) Ascomycetes

The products of meiosis in ascomycetes are borne in a microscopic sac called an ascus. The fleshy fruiting bodies consist of both dikaryotic and haploid hyphae.

Mating structure

Mating type a ( )

Germinating ascospores (n)

Mating type A ( )

Ascospores (n)

Plasmogamy

Dikaryotic mycelium (n + n)

HAPLOID (n)

Mitosis

Ascus

DIKARYOTIC (n + n)

DIPLOID (2n)

Haploid hyphae (n)

Ascospores

~40 μm

Ascus

Meiosis

Dikaryotic asci (n + n)

Fused nuclei

Karyogamy

Fertilization

Ascocarp (fruiting structure)

(D) Basidiomycetes

In basidiomycetes the products of meiosis are borne exposed on pedestals called basidia. Fruiting bodies consist solely of dikaryotic hyphae, and the dikaryotic phase can last a long time.

+ Mating type

Plasmogamy

– Mating type

Mycelial hyphae

Dikaryotic mycelium (n + n)

~50 μm

Basidiospores

Basidium

HAPLOID (n)

Gills

DIKARYOTIC (n + n)

Gills lined with basidia

DIPLOID (2n)

Basidiocarp (fruiting structure)

Fused nuclei

Developing basidium (n + n)

Nuclei

Basidium

Meiosis

Karyogamy

Fertilization

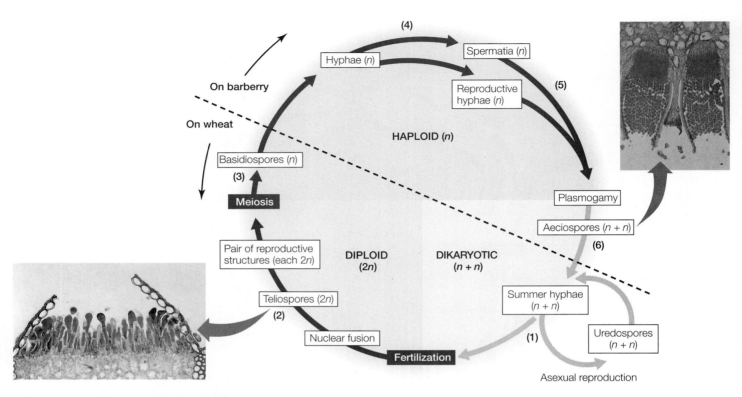

(4)

Hyphae (*n*)

Spermatia (*n*)

On barberry

Reproductive
hyphae (*n*)

(5)

On wheat

HAPLOID (*n*)

Basidiospores (*n*)

(3)

Meiosis

Plasmogamy

Aeciospores (*n* + *n*)

(6)

Pair of reproductive
structures (each 2*n*)

DIPLOID
(2*n*)

DIKARYOTIC
(*n* + *n*)

Summer hyphae
(*n* + *n*)

Teliospores (2*n*)

(2)

Uredospores
(*n* + *n*)

Nuclear fusion

Fertilization

(1)

Asexual reproduction

30.14 Fungal Life Cycles Can Be Very Complex *(Page 662)*

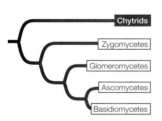

Chytrids

Zygomycetes

Glomeromycetes

Ascomycetes

Basidiomycetes

Page 663 In-Text Art

TABLE 30.1			
A Classification of the Fungi			
GROUP	**COMMON NAME**	**FEATURES**	**EXAMPLES**
Chytridiomycetes	Chytrids	Aquatic; zoospores have flagella	*Allomyces*
Zygomycetes	Conjugating fungi	Zygosporangium; no regularly occurring septa; usually no fleshy fruiting body	*Rhizopus*
Glomeromycetes	Mycorrhizal fungi	Form arbuscular mycorrhizae on plant roots	*Glomus*
Ascomycetes	Sac fungi	Ascus; perforated septa	*Neurospora*
Basidiomycetes	Club fungi	Basidium; perforated septa	*Armillariella*

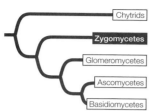

Page 664 In-Text Art

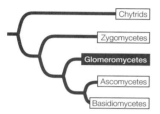

Page 665 In-Text Art (1)

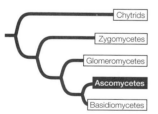

Page 665 In-Text Art (2)

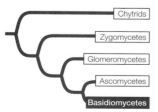

Page 667 In-Text Art

CHAPTER 31 Animal Origins and the Evolution of Body Plans

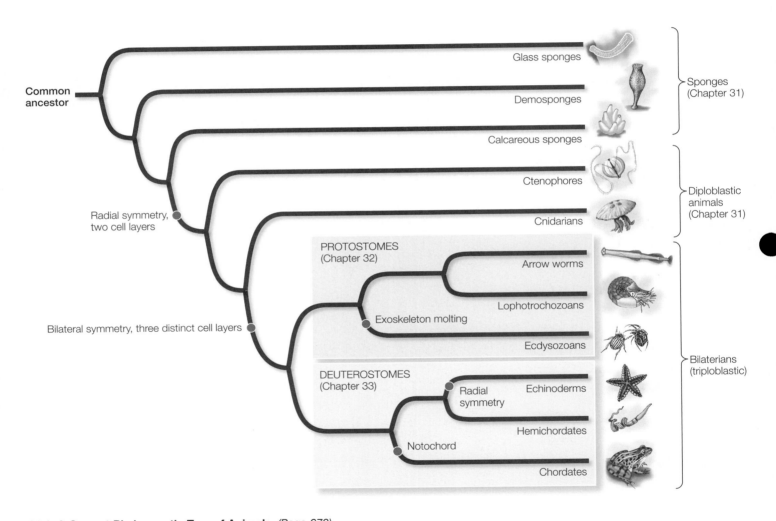

Common ancestor

Glass sponges

Demosponges

Calcareous sponges

Sponges (Chapter 31)

Ctenophores

Radial symmetry, two cell layers

Cnidarians

Diploblastic animals (Chapter 31)

PROTOSTOMES (Chapter 32)

Arrow worms

Lophotrochozoans

Exoskeleton molting

Bilateral symmetry, three distinct cell layers

Ecdysozoans

DEUTEROSTOMES (Chapter 33)

Radial symmetry

Echinoderms

Hemichordates

Notochord

Chordates

Bilaterians (triploblastic)

31.1 A Current Phylogenetic Tree of Animals *(Page 672)*

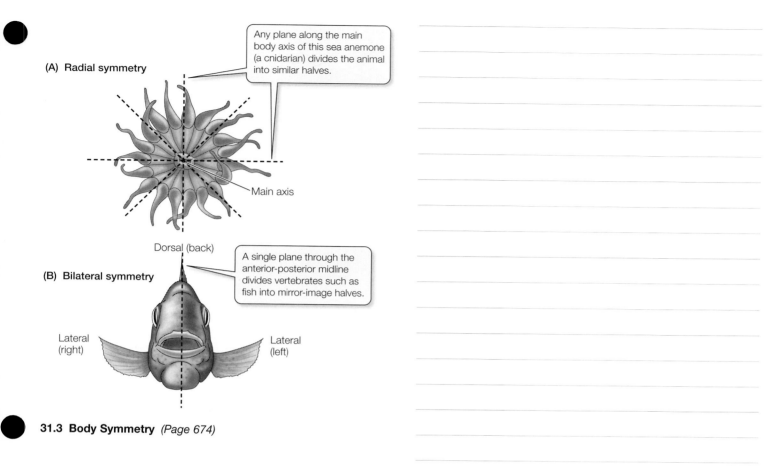

(A) Radial symmetry

Any plane along the main body axis of this sea anemone (a cnidarian) divides the animal into similar halves.

Main axis

Dorsal (back)

(B) Bilateral symmetry

A single plane through the anterior-posterior midline divides vertebrates such as fish into mirror-image halves.

Lateral (right)

Lateral (left)

31.3 Body Symmetry *(Page 674)*

(A) Acoelomate (flatworm)

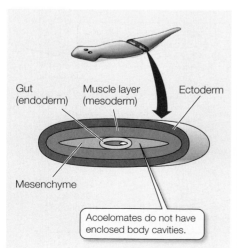

Gut (endoderm)

Muscle layer (mesoderm)

Ectoderm

Mesenchyme

Acoelomates do not have enclosed body cavities.

(B) Pseudocoelomate (roundworm)

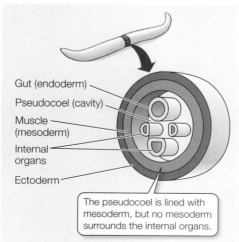

Gut (endoderm)

Pseudocoel (cavity)

Muscle (mesoderm)

Internal organs

Ectoderm

The pseudocoel is lined with mesoderm, but no mesoderm surrounds the internal organs.

(C) Coelomate (earthworm)

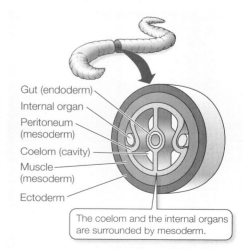

Gut (endoderm)

Internal organ

Peritoneum (mesoderm)

Coelom (cavity)

Muscle (mesoderm)

Ectoderm

The coelom and the internal organs are surrounded by mesoderm.

31.4 Animal Body Cavities *(Page 675)*

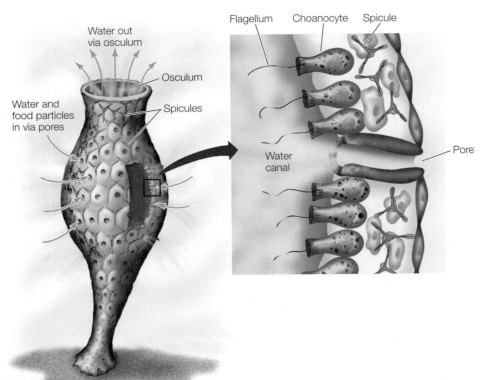

31.7 Even Sessile Filter Feeders Expend Energy
(Page 677)

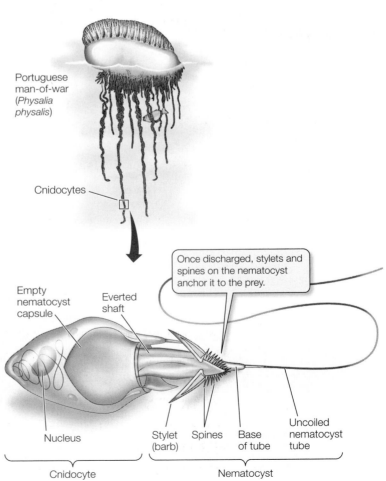

31.10 Nematocysts Are Potent Weapons *(Page 679)*

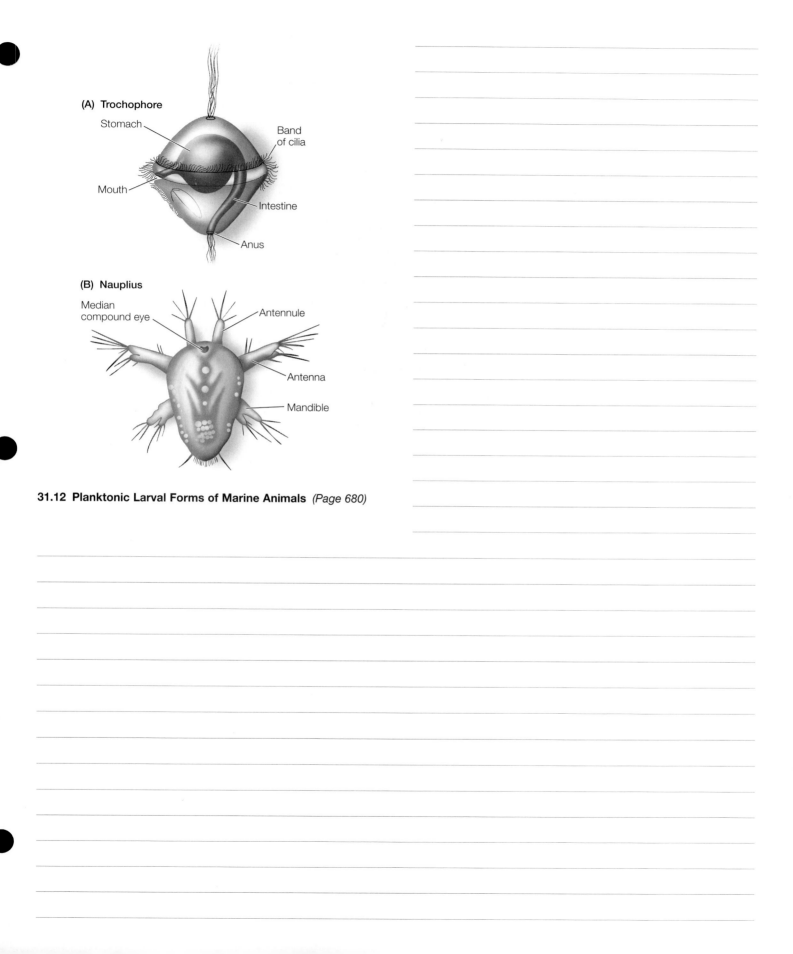

(A) Trochophore

Stomach

Band of cilia

Mouth

Intestine

Anus

(B) Nauplius

Median compound eye

Antennule

Antenna

Mandible

31.12 Planktonic Larval Forms of Marine Animals *(Page 680)*

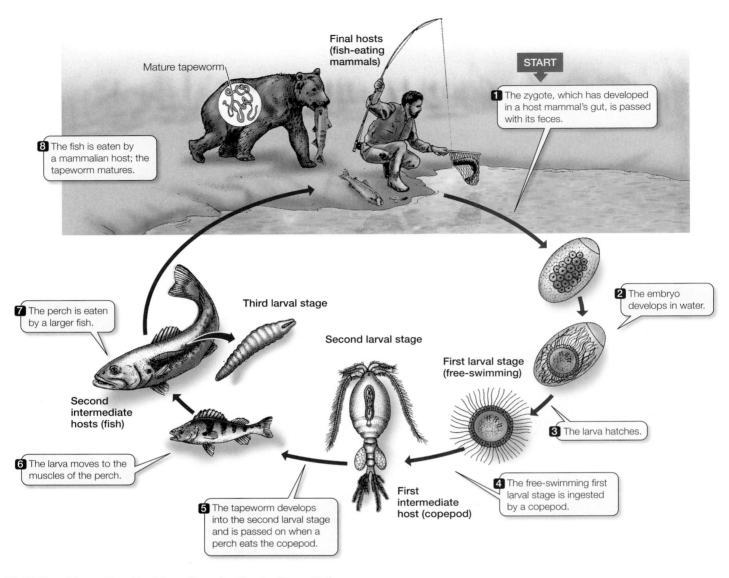

Mature tapeworm

Final hosts (fish-eating mammals)

START

1 The zygote, which has developed in a host mammal's gut, is passed with its feces.

8 The fish is eaten by a mammalian host; the tapeworm matures.

2 The embryo develops in water.

Third larval stage

7 The perch is eaten by a larger fish.

Second larval stage

First larval stage (free-swimming)

3 The larva hatches.

Second intermediate hosts (fish)

6 The larva moves to the muscles of the perch.

First intermediate host (copepod)

4 The free-swimming first larval stage is ingested by a copepod.

5 The tapeworm develops into the second larval stage and is passed on when a perch eats the copepod.

31.15 Reaching a New Host by a Complex Route *(Page 682)*

TABLE 31.1

Summary of Living Members of the Major Groups of Animals

	APPROXIMATE NUMBER OF LIVING SPECIES DESCRIBED	MAJOR GROUPS		APPROXIMATE NUMBER OF LIVING SPECIES DESCRIBED	MAJOR GROUPS
Glass sponges	500		**Ecdysozoans**		
Demosponges	7,000		Kinorhynchs	150	
Calcareous sponges	500		Loriciferans	100	
Ctenophores	100		Priapulids	16	
Cnidarians	11,000	Anthozoans: Corals, sea anemones	Horsehair worms	320	
		Hydrozoans: Hydras and hydroids	Nematodes	25,000	
		Scyphozoans: Jellyfishes	Onychophorans	150	
			Tardigrades	800	
			Arthropods:		
PROTOSTOMES			Crustaceans	50,000	Crabs, shrimps, lobsters, barnacles, copepods
Arrow worms	100				
Lophotrochozoans			Hexapods	1,000,000	Insects and relatives
Ectoprocts	4,500		Myriapods	14,000	Millipedes, centipedes
Flatworms	25,000	Free-living flatworms; flukes and tapeworms (all parasitic); monogeneans (ectoparasites of fishes)	Chelicerates	89,000	Horseshoe crabs, arachnids (scorpions, harvestmen, spiders, mites, ticks)
Rotifers	1,800		**DEUTEROSTOMES**		
Ribbon worms	1,000		Echinoderms	7,000	Crinoids (sea lilies and feather stars); brittle stars; sea stars; sea daisies; sea urchins; sea cucumbers
Phoronids	20				
Brachiopods	335		Hemichordates	95	Acorn worms and pterobranchs
Annelids	16,500	Polychaetes (all marine)	Chordates	55,000	Urochordates: Sea squirts
		Clitellates: Earthworms, freshwater worms, leeches			Cephalochordates: Lancelets
Mollusks	95,000	Monoplacophorans			Agnathans: Lampreys, hagfishes
		Chitons			Cartilaginous fishes
		Bivalves: Clams, oysters, mussels			Ray-finned fishes
		Gastropods: Snails, slugs, limpets			Lobe-finned fishes
		Cephalopods: Squids, octopuses, nautiloids			Amphibians
					Reptiles (including birds)
					Mammals

(Page 683)

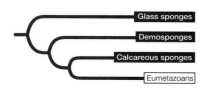

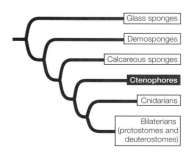

Page 683 In-Text Art

Page 684 In-Text Art

(A)

Gut

Tentacle
sheath

Ctenes

Pharynx

Mouth

Tentacle

Prey adhere to the sticky cells
that cover the tentacles.

(B) *Mnemiopsis* sp.

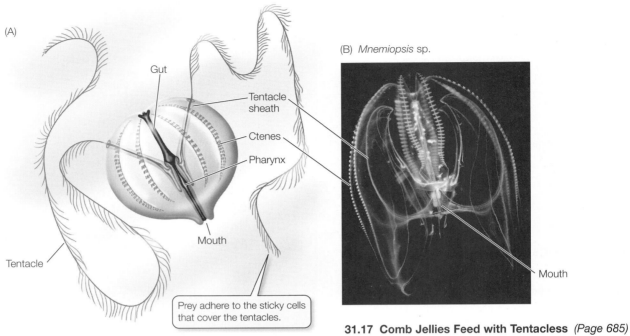

Mouth

31.17 Comb Jellies Feed with Tentacless *(Page 685)*

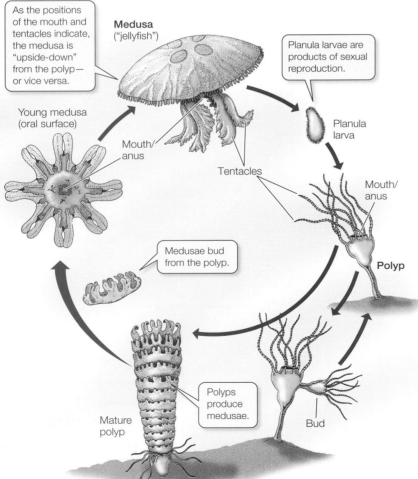

As the positions
of the mouth and
tentacles indicate,
the medusa is
"upside-down"
from the polyp—
or vice versa.

Medusa
("jellyfish")

Planula larvae are
products of sexual
reproduction.

Young medusa
(oral surface)

Planula
larva

Mouth/
anus

Tentacles

Mouth/
anus

Medusae bud
from the polyp.

Polyp

Mature
polyp

Polyps
produce
medusae.

Bud

31.18 The Cnidarian Life Cycle Has Two Stages *(Page 685)*

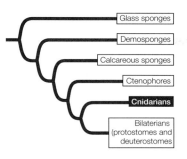

Page 685 In-Text Art

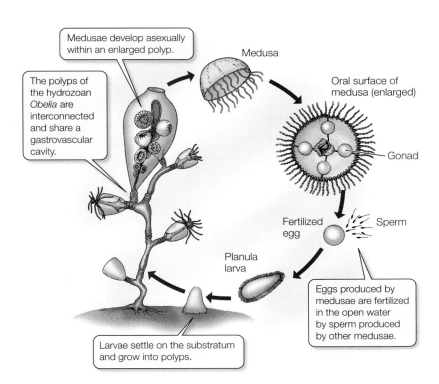

Medusae develop asexually within an enlarged polyp.

Medusa

The polyps of the hydrozoan *Obelia* are interconnected and share a gastrovascular cavity.

Oral surface of medusa (enlarged)

Gonad

Fertilized egg

Sperm

Planula larva

Eggs produced by medusae are fertilized in the open water by sperm produced by other medusae.

Larvae settle on the substratum and grow into polyps.

31.21 Hydrozoans Often Have Colonial Polyps *(Page 687)*

32 Protostome Animals

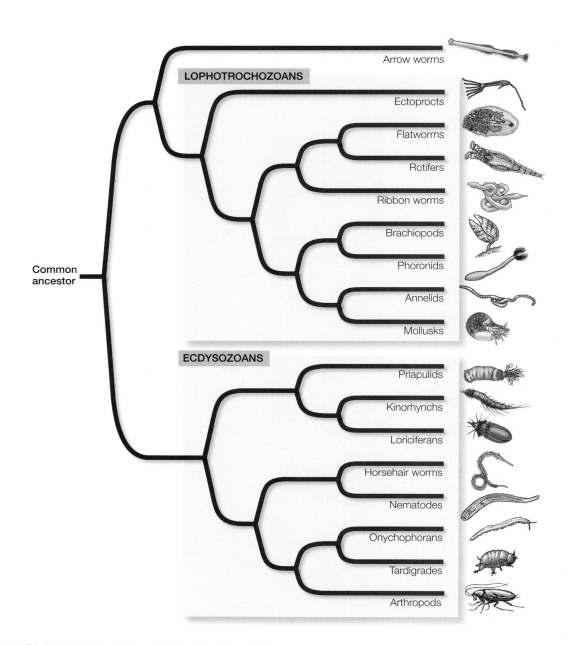

32.1 A Current Phylogenetic Tree of Protostomes *(Page 692)*

Common ancestor

LOPHOTROCHOZOANS

Arrow worms
Ectoprocts
Flatworms
Rotifers
Ribbon worms
Brachiopods
Phoronids
Annelids
Mollusks

ECDYSOZOANS

Priapulids
Kinorhynchs
Loriciferans
Horsehair worms
Nematodes
Onychophorans
Tardigrades
Arthropods

TABLE 32.1

Anatomical Characteristics of Some Major Protostome Groups[a]

GROUP	BODY CAVITY	DIGESTIVE TRACT	CIRCULATORY SYSTEM
Arrow worms	Coelom	Complete	None
LOPHOTROCHOZOANS			
Flatworms	None	Dead-end sac	None
Rotifers	Pseudocoelom	Complete	None
Ectoprocts	Coelom	Complete	None
Brachiopods	Coelom	Complete in most	Open
Phoronids	Coelom	Complete	Closed
Ribbon worms	Coelom	Complete	Closed
Annelids	Coelom	Complete	Closed or open
Mollusks	Reduced coelom	Complete	Open except in cephalopods
ECDYSOZOANS			
Horsehair worms	Pseudocoelom	Greatly reduced	None
Nematodes	Pseudoceolom	Complete	None
Arthropods	Hemocoel	Complete	Open

[a]Note that all protostomes have bilateral symmetry.

(Page 692)

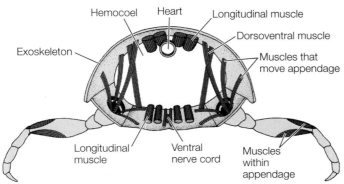

32.4 Arthropod Skeletons Are Rigid and Jointed *(Page 694)*

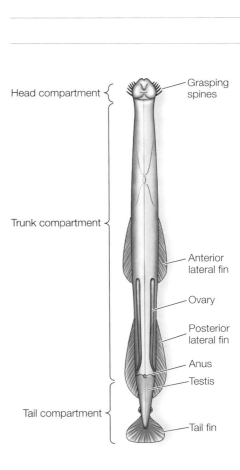

32.5 An Arrow Worm *(Page 694)*

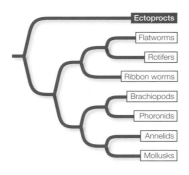

Page 695 In-Text Art (1)

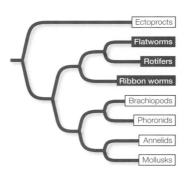

Page 695 In-Text Art (2)

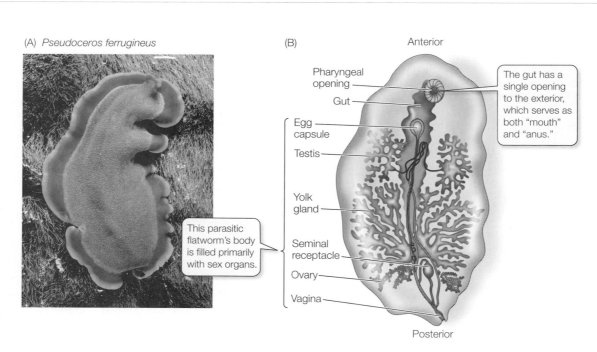

(A) *Pseudoceros ferrugineus*

This parasitic flatworm's body is filled primarily with sex organs.

(B)

Anterior

Pharyngeal opening

Gut

The gut has a single opening to the exterior, which serves as both "mouth" and "anus."

Egg capsule

Testis

Yolk gland

Seminal receptacle

Ovary

Vagina

Posterior

32.7 Flatworms May Live Freely or Parasitically *(Page 696)*

(A) *Philodina roseola*

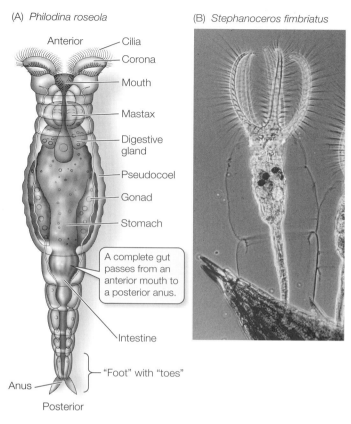

Anterior

Cilia

Corona

Mouth

Mastax

Digestive gland

Pseudocoel

Gonad

Stomach

A complete gut passes from an anterior mouth to a posterior anus.

Intestine

"Foot" with "toes"

Anus

Posterior

32.8 Rotifers *(Page 696)*

(B) *Stephanoceros fimbriatus*

(A)

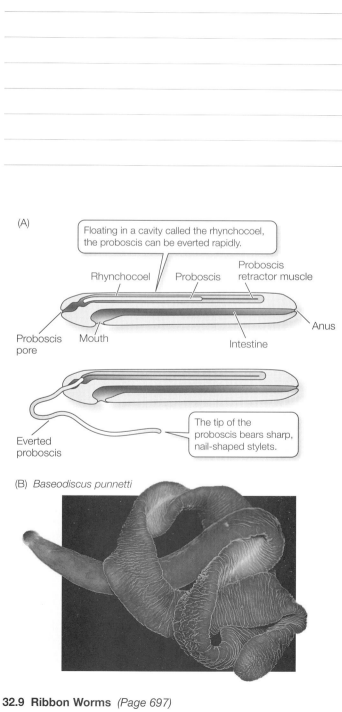

Floating in a cavity called the rhynchocoel, the proboscis can be everted rapidly.

Rhynchocoel

Proboscis

Proboscis retractor muscle

Proboscis pore

Mouth

Intestine

Anus

Everted proboscis

The tip of the proboscis bears sharp, nail-shaped stylets.

(B) *Baseodiscus punnetti*

32.9 Ribbon Worms *(Page 697)*

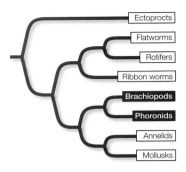

Page 697 In-Text Art

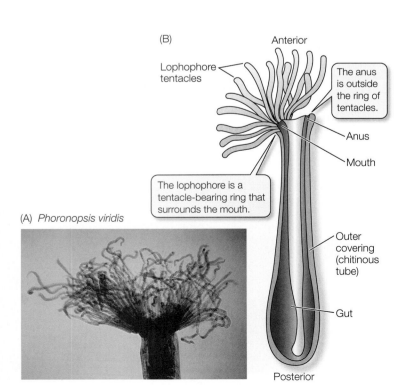

(B)

Anterior

Lophophore
tentacles

The anus
is outside
the ring of
tentacles.

Anus

Mouth

The lophophore is a
tentacle-bearing ring that
surrounds the mouth.

(A) *Phoronopsis viridis*

Outer
covering
(chitinous
tube)

Gut

Posterior

32.10 Phoronids Have Impressive Lophophores *(Page 697)*

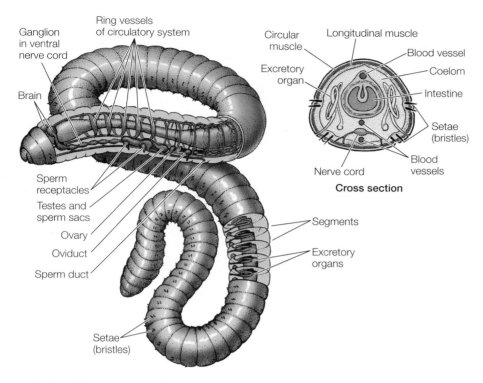

Ganglion in ventral nerve cord

Ring vessels of circulatory system

Brain

Sperm receptacles

Testes and sperm sacs

Ovary

Oviduct

Sperm duct

Setae (bristles)

Segments

Excretory organs

Circular muscle

Longitudinal muscle

Excretory organ

Blood vessel

Coelom

Intestine

Setae (bristles)

Blood vessels

Nerve cord

Cross section

32.12 Annelids Have Many Body Segments *(Page 698)*

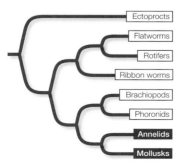

Ectoprocts

Flatworms

Rotifers

Ribbon worms

Brachiopods

Phoronids

Annelids

Mollusks

Page 698 In-Text Art

Generalized molluscan body plan

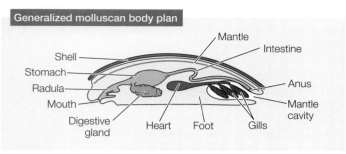

Mantle
Shell
Intestine
Stomach
Radula
Anus
Mouth
Mantle cavity
Digestive gland
Heart
Foot
Gills

Chitons

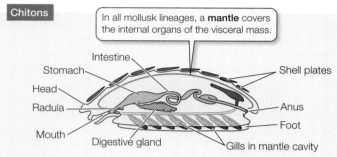

In all mollusk lineages, a **mantle** covers the internal organs of the visceral mass.

Intestine
Stomach
Shell plates
Head
Radula
Anus
Mouth
Foot
Digestive gland
Gills in mantle cavity

Gastropods

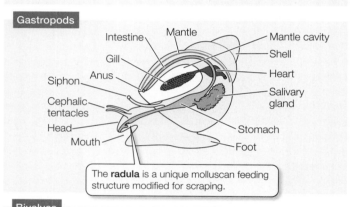

Intestine
Mantle
Mantle cavity
Gill
Shell
Anus
Heart
Siphon
Cephalic tentacles
Salivary gland
Head
Stomach
Mouth
Foot

The **radula** is a unique molluscan feeding structure modified for scraping.

Bivalves

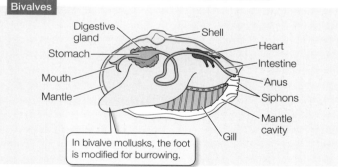

Digestive gland
Shell
Stomach
Heart
Mouth
Intestine
Mantle
Anus
Siphons
Mantle cavity
Gill

In bivalve mollusks, the foot is modified for burrowing.

Cephalopods

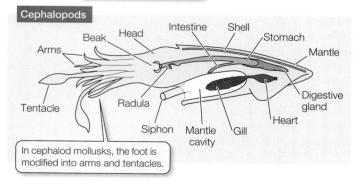

Intestine
Shell
Stomach
Beak
Head
Mantle
Arms
Tentacle
Digestive gland
Radula
Siphon
Mantle cavity
Gill
Heart

In cephalod mollusks, the foot is modified into arms and tentacles.

32.14 Molluscan Body Plans *(Page 700)*

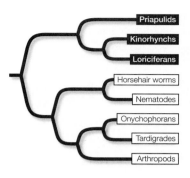

Priapulids
Kinorhynchs
Loriciferans
Horsehair worms
Nematodes
Onychophorans
Tardigrades
Arthropods

Page 703 In-Text Art (1)

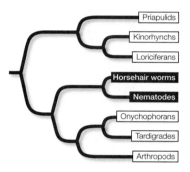

Priapulids
Kinorhynchs
Loriciferans
Horsehair worms
Nematodes
Onychophorans
Tardigrades
Arthropods

Page 703 In-Text Art (2)

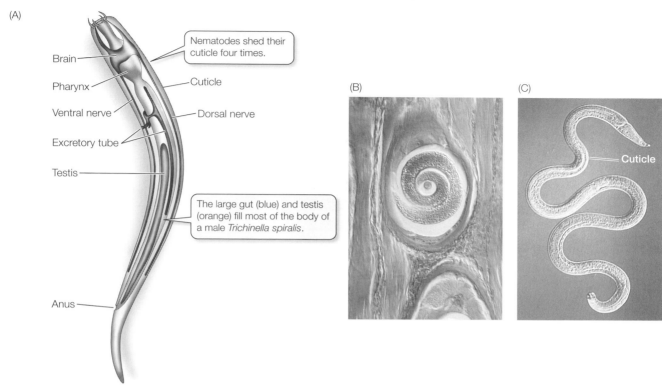

(A)

Brain

Pharynx

Ventral nerve

Excretory tube

Testis

Anus

Nematodes shed their cuticle four times.

Cuticle

Dorsal nerve

The large gut (blue) and testis (orange) fill most of the body of a male *Trichinella spiralis*.

(B)

(C)

Cuticle

32.18 Nematodes *(Page 704)*

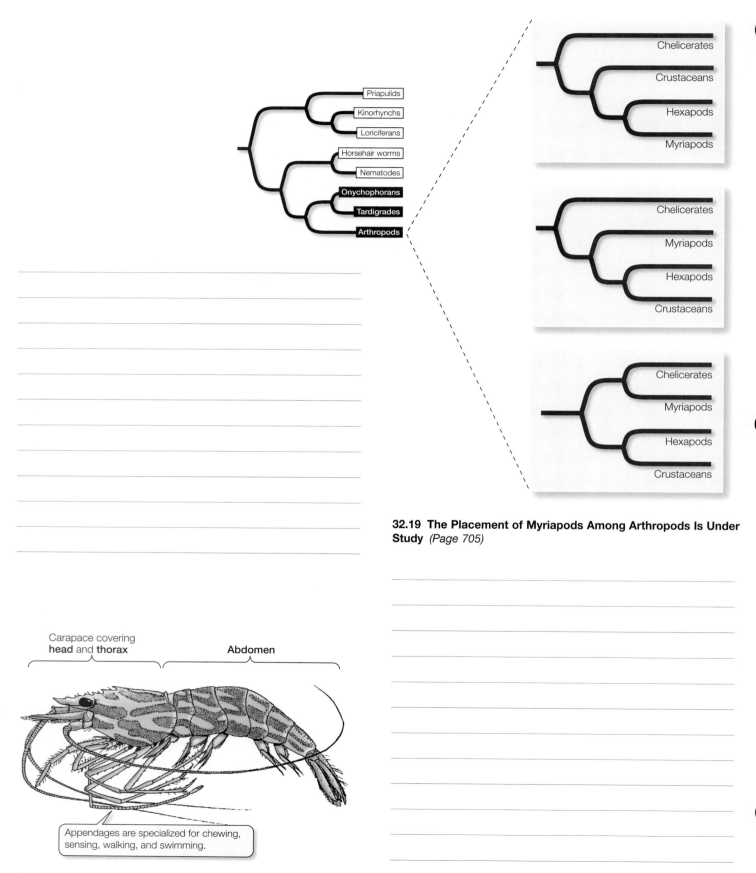

Priapulids

Kinorhynchs

Loriciferans

Horsehair worms

Nematodes

Onychophorans

Tardigrades

Arthropods

Chelicerates

Crustaceans

Hexapods

Myriapods

Chelicerates

Myriapods

Hexapods

Crustaceans

Chelicerates

Myriapods

Hexapods

Crustaceans

32.19 The Placement of Myriapods Among Arthropods Is Under Study *(Page 705)*

Carapace covering
head and **thorax**

Abdomen

Appendages are specialized for chewing, sensing, walking, and swimming.

32.23 Crustacean Structure *(Page 707)*

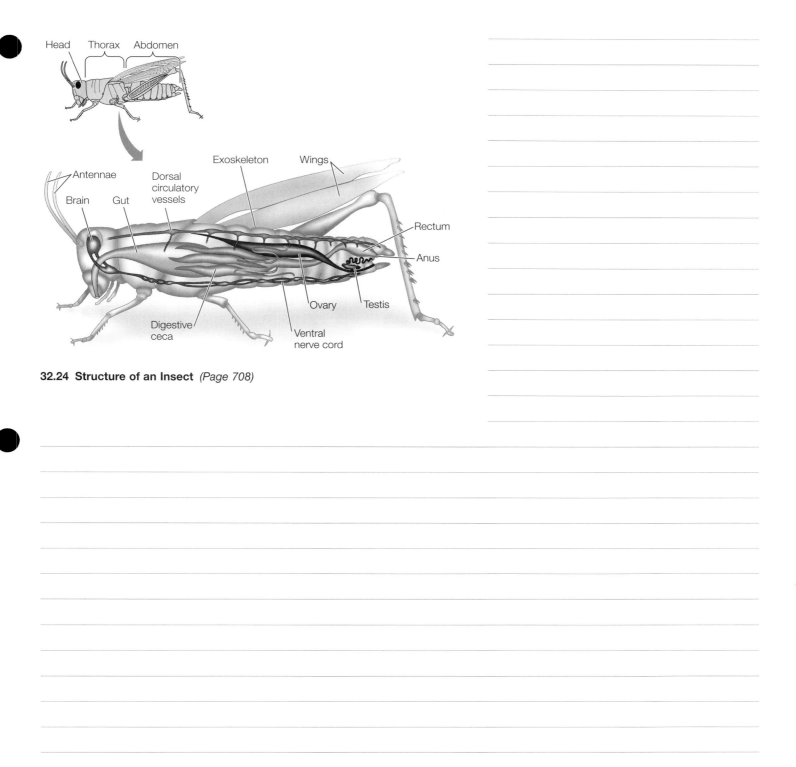

32.24 Structure of an Insect *(Page 708)*

TABLE 32.2

The major insect groups[a]

GROUP	APPROXIMATE NUMBER OF DESCRIBED LIVING SPECIES
Jumping bristletails (*Archaeognatha*)	300
Silverfish (*Thysanura*)	370
PTERYGOTE (WINGED) INSECTS (*PTERYGOTA*)	
Mayflies (*Ephemeroptera*)	2,000
Dragonflies and damselflies (*Odonata*)	5,000
Neopterans (*Neoptera*) [b]	
Ice-crawlers (*Grylloblattodea*)	25
Gladiators (*Mantophasmatodea*)	15
Stoneflies (*Plecoptera*)	1,700
Webspinners (*Embioptera*)	300
Angel insects (*Zoraptera*)	30
Earwigs (*Dermaptera*)	1,800
Grasshoppers and crickets (*Orthoptera*)	20,000
Stick insects (*Phasmida*)	3,000
Cockroaches (*Blattodea*)	3,500
Termites (*Isoptera*)	2,750
Mantids (*Mantodea*)	2,300
Booklice and barklice (*Psocoptera*)	3,000
Thrips (*Thysanoptera*)	5,000
Lice (*Phthiraptera*)	3,100
True bugs, cicadas, aphids, leafhoppers (*Hemiptera*)	80,000
Holometabolous neopterans (*Holometabola*) [c]	
Ants, bees, wasps (*Hymenoptera*)	125,000
Beetles (*Coleoptera*)	375,000
Strepsipterans (*Strepsiptera*)	600
Lacewings, ant lions, dobsonflies (*Neuropterida*)	4,700
Scorpionflies (*Mecoptera*)	600
Fleas (*Siphonaptera*)	2,400
True flies (*Diptera*)	120,000
Caddisflies (*Trichoptera*)	5,000
Butterflies and moths (*Lepidoptera*)	250,000

[a] The hexapod relatives of insects include the springtails (*Collembola*; 3,000 spp.), two-pronged bristletails (*Diplura*; 600 spp.), and proturans (*Protura*; 10 spp.). All are wingless and have internal mouthparts.

[b] Neopteran insects can tuck their wings close to their bodies

[c] Holometabolous insects are neopterans that undergo complete metamorphosis.

(Page 709)

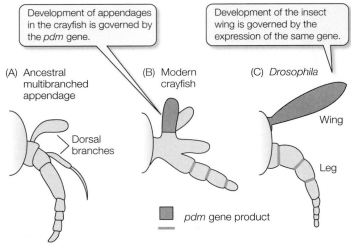

32.28 **The Origin of Insect Wings?** *(Page 710)*

CHAPTER 33 Deuterostome Animals

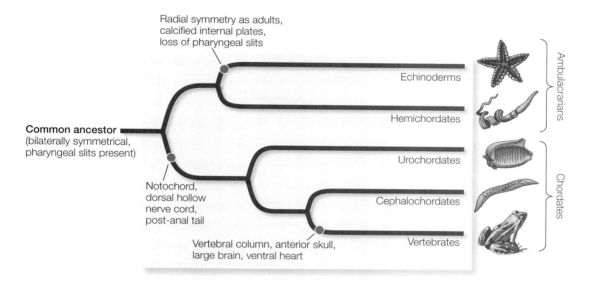

Radial symmetry as adults,
calcified internal plates,
loss of pharyngeal slits

Echinoderms

Hemichordates

Ambulacrarians

Common ancestor
(bilaterally symmetrical,
pharyngeal slits present)

Urochordates

Cephalochordates

Chordates

Notochord,
dorsal hollow
nerve cord,
post-anal tail

Vertebral column, anterior skull,
large brain, ventral heart

Vertebrates

33.1 A Current Phylogenetic Tree of the Deuterostomes *(Page 718)*

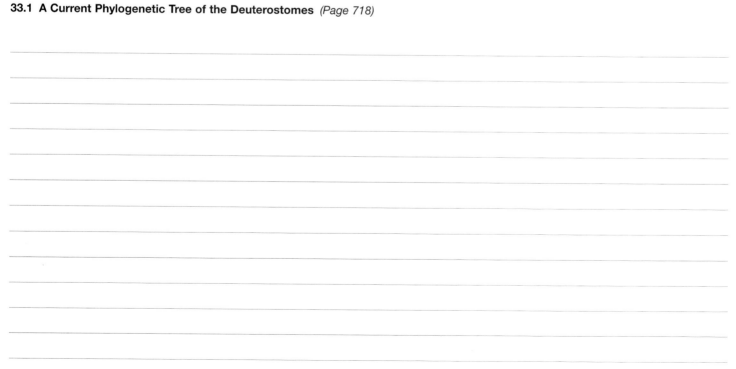

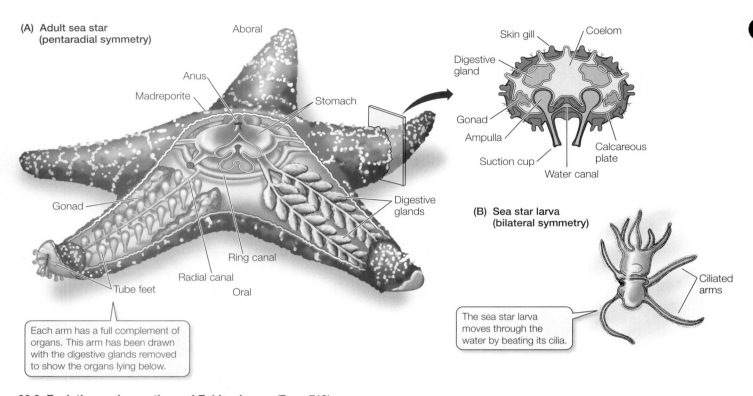

(A) Adult sea star (pentaradial symmetry)

Aboral

Anus

Madreporite

Stomach

Gonad

Digestive glands

Ring canal

Radial canal

Oral

Tube feet

Each arm has a full complement of organs. This arm has been drawn with the digestive glands removed to show the organs lying below.

Skin gill

Coelom

Digestive gland

Gonad

Ampulla

Suction cup

Calcareous plate

Water canal

(B) Sea star larva (bilateral symmetry)

Ciliated arms

The sea star larva moves through the water by beating its cilia.

33.3 Evolutionary Innovations of Echinoderms *(Page 719)*

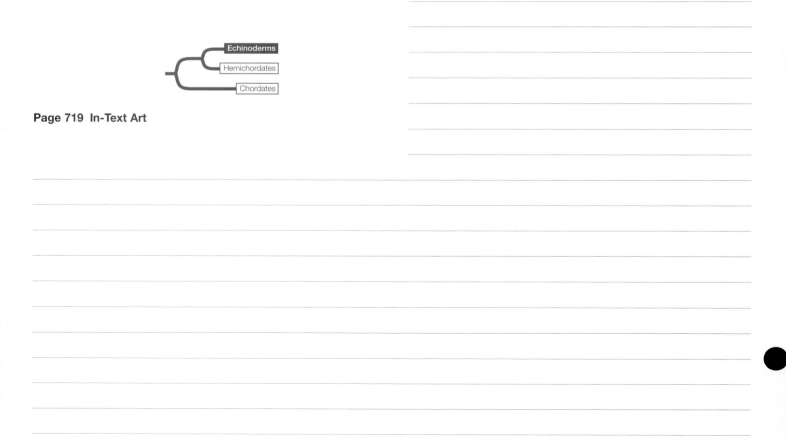

Echinoderms

Hemichordates

Chordates

Page 719 In-Text Art

(A)

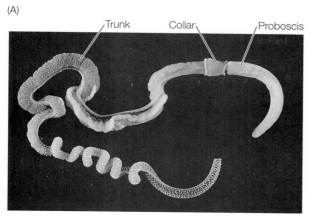

Trunk   Collar   Proboscis

Saccoglossus kowalevskii

(B)

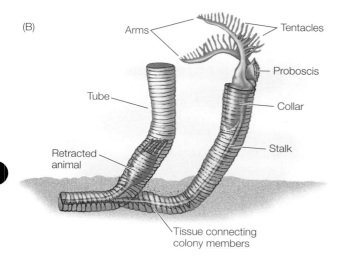

Arms   Tentacles

Proboscis

Tube

Collar

Stalk

Retracted animal

Tissue connecting colony members

33.5 Hemichordates *(Page 721)*

Echinoderms

Hemichordates

Chordates

Page 721 In-Text Art

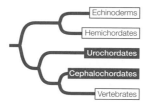

Echinoderms

Hemichordates

Urochordates

Cephalochordates

Vertebrates

Page 722 In-Text Art

(A)

A sea squirt larva has all the chordate features, some of which are lost in the adult.

Dorsal hollow nerve cord

Notochord

Tail

Pharyngeal slits

(B)

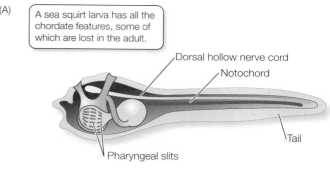

Dorsal hollow nerve cord

Notochord

Tail

Pharyngeal basket

An adult lancelet retains all the chordate features including a much enlarged pharyngeal basket.

Gill slits

Branchiostoma sp.

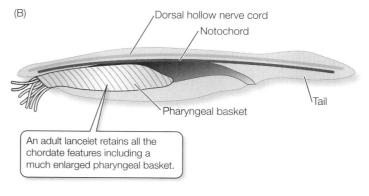

33.6 The Key Features of Chordates Are Most Apparent in Early Developmental Stages *(Page 722)*

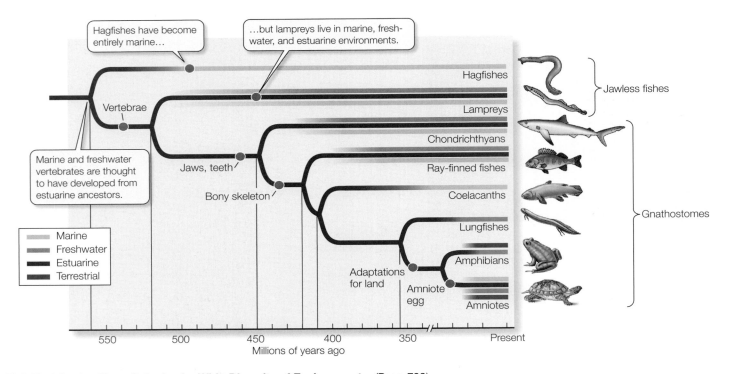

Hagfishes have become entirely marine…

…but lampreys live in marine, fresh-water, and estuarine environments.

Vertebrae

Marine and freshwater vertebrates are thought to have developed from estuarine ancestors.

Jaws, teeth

Bony skeleton

Marine
Freshwater
Estuarine
Terrestrial

Adaptations for land

Amniote egg

Hagfishes

Lampreys

Chondrichthyans

Ray-finned fishes

Coelacanths

Lungfishes

Amphibians

Amniotes

Jawless fishes

Gnathostomes

550   500   450   400   350   Present
Millions of years ago

33.8 Vertebrates Have Colonized a Wide Diversity of Environments *(Page 723)*

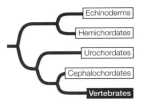

Echinoderms
Hemichordates
Urochordates
Cephalochordates
Vertebrates

Page 724 In-Text Art

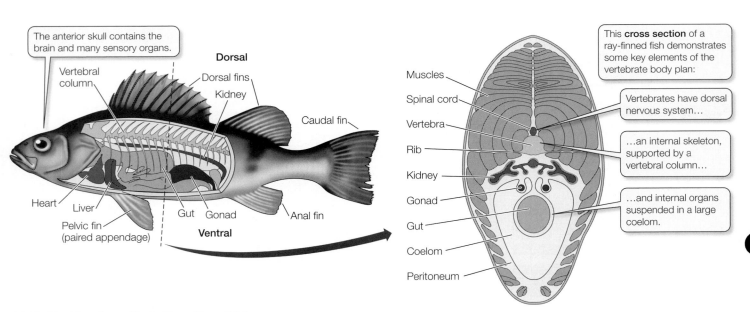

The anterior skull contains the brain and many sensory organs.

Dorsal

Vertebral column

Dorsal fins

Kidney

Caudal fin

Heart   Liver

Gut   Gonad

Anal fin

Pelvic fin (paired appendage)

Ventral

Muscles

Spinal cord

Vertebra

Rib

Kidney

Gonad

Gut

Coelom

Peritoneum

This **cross section** of a ray-finned fish demonstrates some key elements of the vertebrate body plan:

Vertebrates have dorsal nervous system…

…an internal skeleton, supported by a vertebral column…

…and internal organs suspended in a large coelom.

33.10 The Vertebrate Body Plan *(Page 725)*

(A) **Jawless fishes**

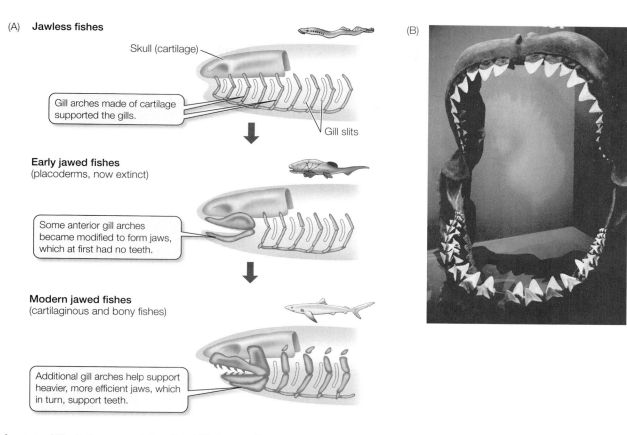

Skull (cartilage)

Gill arches made of cartilage supported the gills.

Gill slits

Early jawed fishes
(placoderms, now extinct)

Some anterior gill arches became modified to form jaws, which at first had no teeth.

Modern jawed fishes
(cartilaginous and bony fishes)

Additional gill arches help support heavier, more efficient jaws, which in turn, support teeth.

(B)

33.11 Jaws and Teeth Increased Feeding Efficiency *(Page 725)*

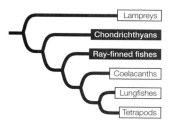

Lampreys

Chondrichthyans

Ray-finned fishes

Coelacanths

Lungfishes

Tetrapods

Page 726 In-Text Art

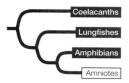

Coelacanths

Lungfishes

Amphibians

Amniotes

Page 728 In-Text Art

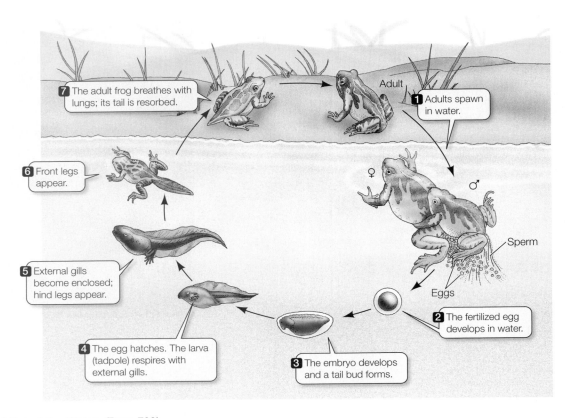

33.15 In and Out of the Water *(Page 729)*

Shell — Embryo — Amniotic cavity

Amnion ⎫
Chorion ⎬ Extraembryonic
Allantois ⎪ membranes
Yolk sac ⎭

Yolk

33.17 An Egg for Dry Places *(Page 730)*

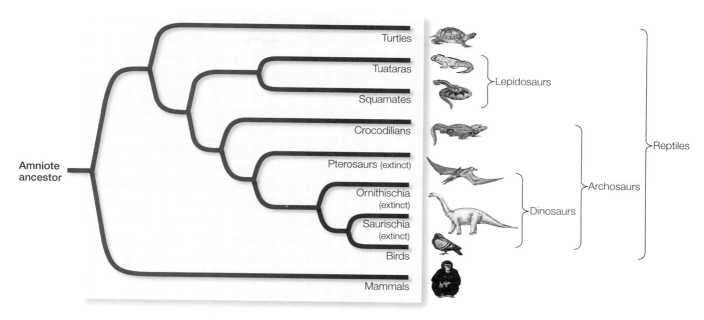

33.18 A Current Phylogenetic Tree of Amniotes *(Page 731)*

Page 731 In-Text Art

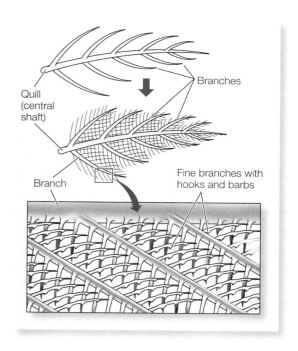

33.22 A Major Evolutionary Breakthrough *(Page 734)*

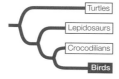

Page 733 In-Text Art

Page 734 In-Text Art

TABLE 33.1

Major groups of living eutherian mammals

GROUP	APPROXIMATE NUMBER OF LIVING SPECIES	EXAMPLES
Gnawing mammals (*Rodentia*)	>2,000	Rats, mice, squirrels, woodchucks, ground squirrels, beaver, capybara
Flying mammals (*Chiroptera*)	1,000	Bats
Soricomorph insectivores (*Soricomorpha*)	430	Shrews, moles
Even-toed hoofed mammals and cetaceans (*Cetartiodactyla*)	300	Deer, sheep, goats, cattle, antelope, giraffes, camels, swine, hippopotamus, cetaceans (whales, dolphins)
Carnivores (*Carnivora*)	280	Wolves, dogs, bears, cats, weasels, pinnipeds (seals, sea lions, walruses)
Primates (*Primates*)	235	Lemurs, monkeys, apes, humans
Lagomorphs (*Lagomorpha*)	80	Rabbits, hares, pikas
African insectivores (*Afrosoricida*)	30	Tenrecs and golden moles
Spiny insectivores (*Erinaceomorpha*)	24	Hedgehogs
Armored mammals (*Cingulata*)	21	Armadillos
Tree shrews (*Scandentia*)	20	Tree shrews
Odd-toed hoofed mammals (*Perissodactyla*)	20	Horses, zebras, tapirs, rhinoceros
Long-nosed insectivores (*Macroscelidea*)	15	Elephant shrews
Pilosans (*Pilosa*)	10	Anteaters and sloths
Pholidotans (*Pholidota*)	8	Pangolins
Sirenians (*Sirenia*)	5	Manatees, dugongs
Hyracoids (*Hyracoidea*)	4	Hyraxes, dassies
Elephants (*Proboscidea*)	3	African and Indian elephants
Dermopterans (*Dermoptera*)	2	Flying lemurs
Aardvark (*Tubulidentata*)	1	Aardvark

(Page 736)

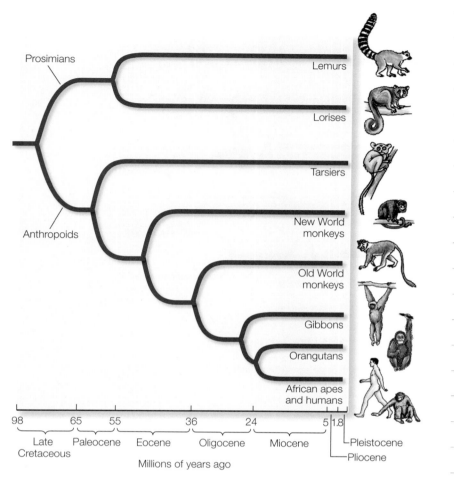

Prosimians

Anthropoids

Lemurs

Lorises

Tarsiers

New World
monkeys

Old World
monkeys

Gibbons

Orangutans

African apes
and humans

98	65	55	36	24	5	1.8

Late
Cretaceous

Paleocene

Eocene

Oligocene

Miocene

Pleistocene

Pliocene

Millions of years ago

33.27 A Current Phylogenetic Tree of the Primates *(Page 738)*

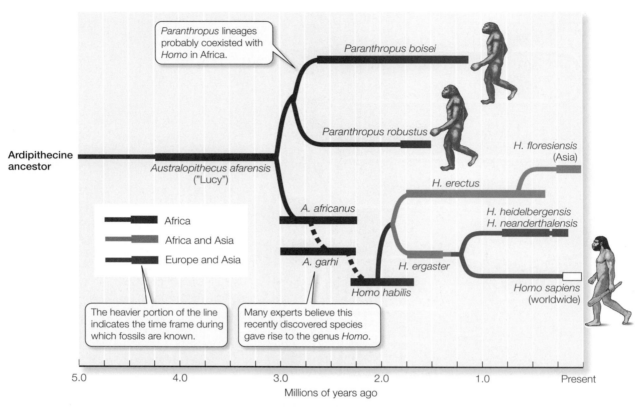

33.31 A Current Phylogenetic Tree of *Homo sapiens* and Our Close Extinct Relatives *(Page 740)*

_____

_____

_____

_____

_____

_____

_____

_____

_____

_____

_____

_____

_____

_____

_____

_____

_____

CHAPTER 34 The Plant Body

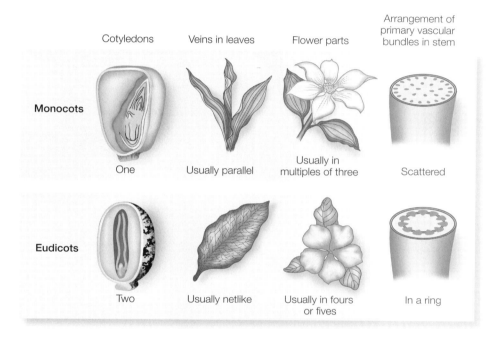

	Cotyledons	Veins in leaves	Flower parts	Arrangement of primary vascular bundles in stem
Monocots	One	Usually parallel	Usually in multiples of three	Scattered
Eudicots	Two	Usually netlike	Usually in fours or fives	In a ring

34.1 Monocots versus Eudicots *(Page 746)*

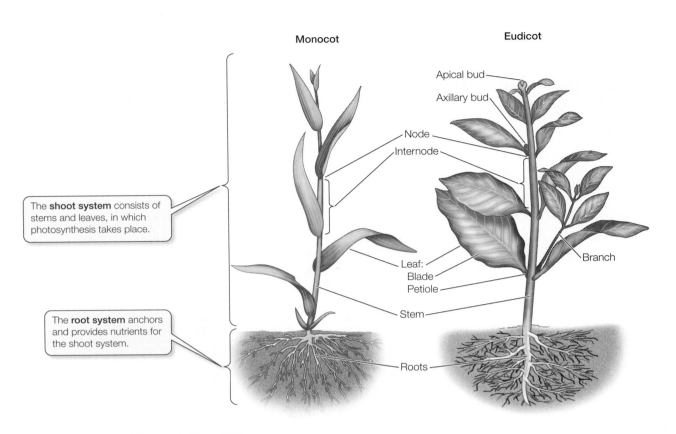

Monocot

Eudicot

Apical bud

Axillary bud

Node

Internode

The **shoot system** consists of stems and leaves, in which photosynthesis takes place.

Leaf:
Blade
Petiole

Branch

The **root system** anchors and provides nutrients for the shoot system.

Stem

Roots

34.2 Vegetative Organs and Systems *(Page 746)*

Leaflets

Axillary bud

Simple

Compound

Doubly compound

34.5 Simple and Compound Leaves *(Page 748)*

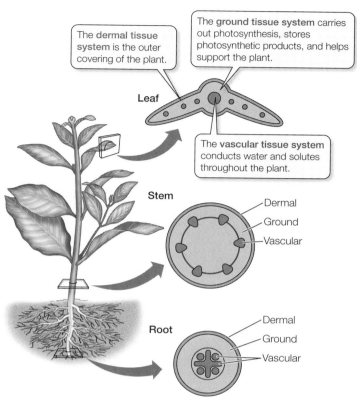

The **dermal tissue system** is the outer covering of the plant.

The **ground tissue system** carries out photosynthesis, stores photosynthetic products, and helps support the plant.

Leaf

The **vascular tissue system** conducts water and solutes throughout the plant.

Stem

Dermal
Ground
Vascular

Root

Dermal
Ground
Vascular

34.6 Three Tissue Systems Extend Throughout the Plant Body
(Page 748)

(A) Primary cell wall   Plasma membrane

Plant cell

The cell plate is the first barrier to form.

(B) Middle lamella

Each daughter cell deposits a primary wall.

The cells expand.

(C)

Secondary wall

(D)

The primary cell wall thins.

After the cells stop expanding, they may deposit more layers, forming secondary walls.

34.7 Cell Wall Formation *(Page 749)*

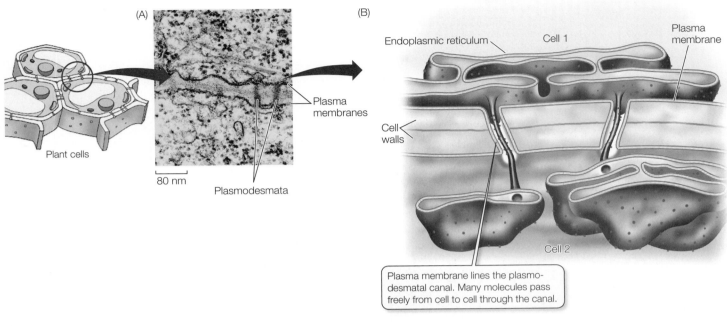

(A)

Plant cells

80 nm

Plasmodesmata

Plasma membranes

(B)

Endoplasmic reticulum

Cell 1

Plasma membrane

Cell walls

Cell 2

Plasma membrane lines the plasmo-desmatal canal. Many molecules pass freely from cell to cell through the canal.

34.8 Plasmodesmata *(Page 750)*

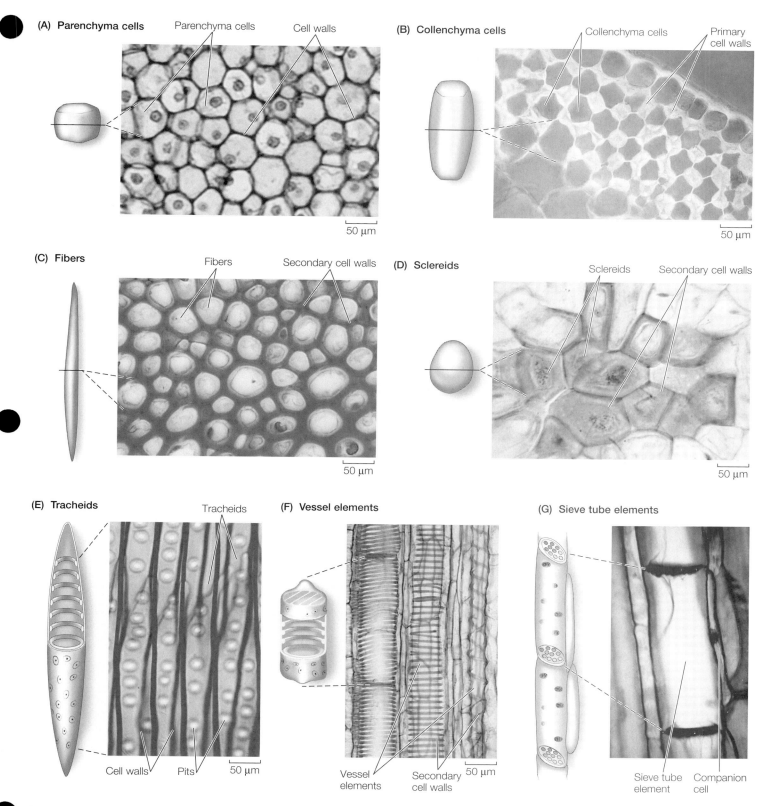

(A) Parenchyma cells
Parenchyma cells
Cell walls
50 μm

(B) Collenchyma cells
Collenchyma cells
Primary cell walls
50 μm

(C) Fibers
Fibers
Secondary cell walls
50 μm

(D) Sclereids
Sclereids
Secondary cell walls
50 μm

(E) Tracheids
Tracheids
Cell walls
Pits
50 μm

(F) Vessel elements
Vessel elements
Secondary cell walls
50 μm

(G) Sieve tube elements
Sieve tube element
Companion cell

34.9 Plant Cell Types *(Page 751)*

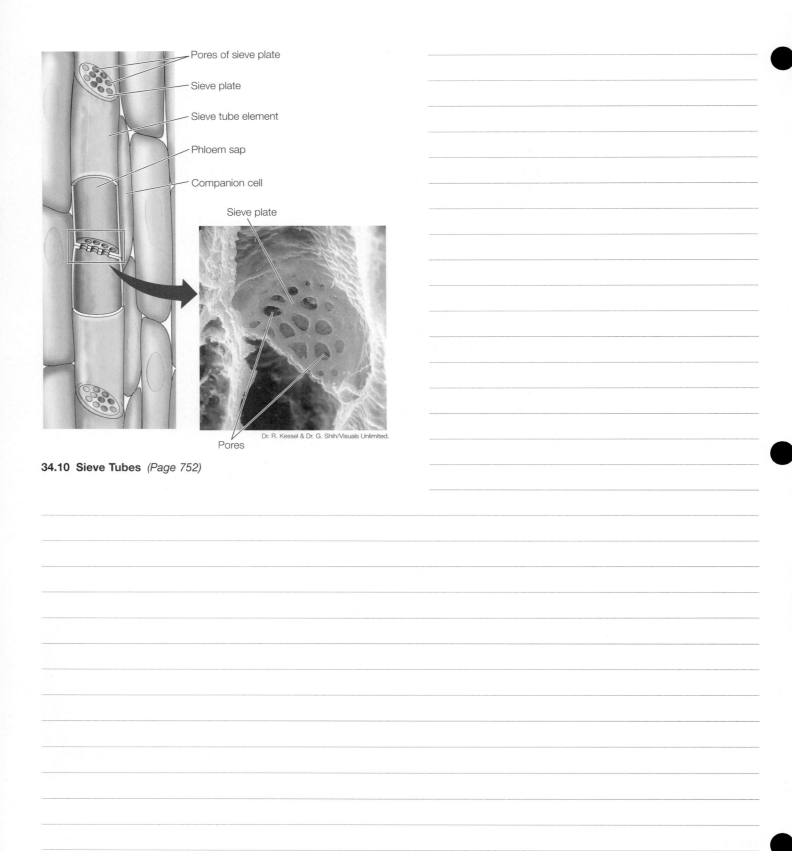

Pores of sieve plate

Sieve plate

Sieve tube element

Phloem sap

Companion cell

Sieve plate

Dr. R. Kessel & Dr. G. Shih/Visuals Unlimited.

Pores

34.10 Sieve Tubes *(Page 752)*

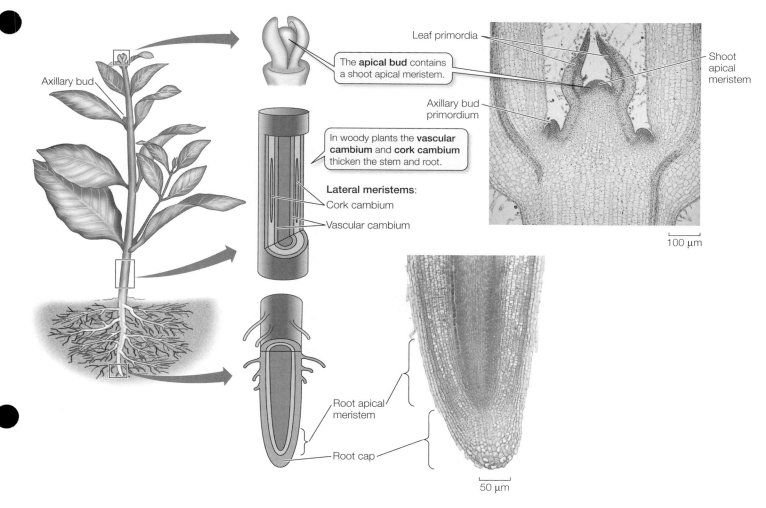

Axillary bud

Leaf primordia

The **apical bud** contains a shoot apical meristem.

Shoot apical meristem

Axillary bud primordium

In woody plants the **vascular cambium** and **cork cambium** thicken the stem and root.

Lateral meristems:
Cork cambium
Vascular cambium

100 μm

Root apical meristem

Root cap

50 μm

34.11 Apical and Lateral Meristems *(Page 753)*

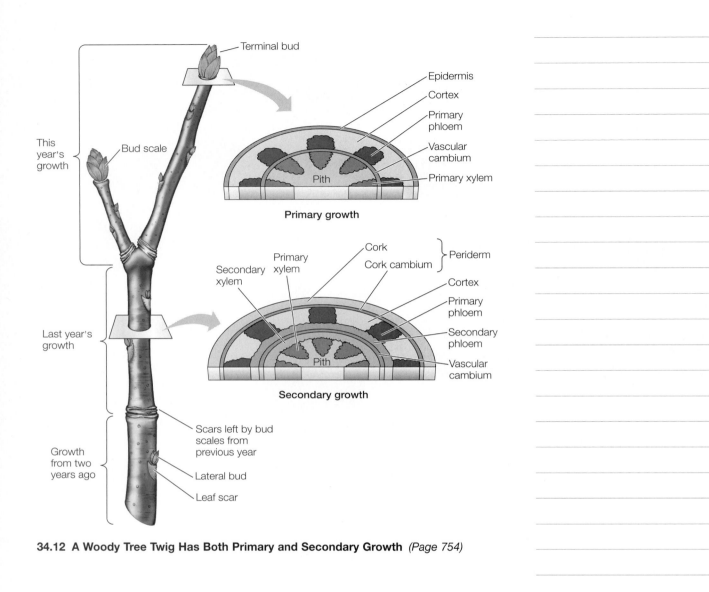

Terminal bud

This year's growth

Bud scale

Epidermis
Cortex
Primary phloem
Vascular cambium
Primary xylem
Pith

Primary growth

Last year's growth

Secondary xylem
Primary xylem
Cork
Cork cambium } Periderm
Cortex
Primary phloem
Secondary phloem
Vascular cambium
Pith

Secondary growth

Growth from two years ago

Scars left by bud scales from previous year
Lateral bud
Leaf scar

34.12 A Woody Tree Twig Has Both Primary and Secondary Growth *(Page 754)*

Apical meristems	→	Primary meristems	→	Tissue systems

Root or shoot apical meristem	Protoderm	→	Dermal tissue system
	Ground meristem	→	Ground tissue system
	Procambium	→	Vascular tissue system

Page 754 In-Text Art

(A)

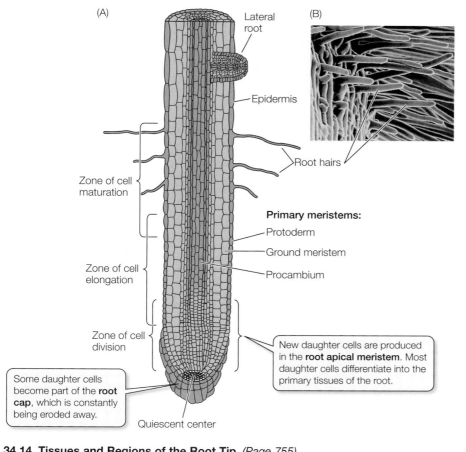

Lateral root

Epidermis

Root hairs

(B)

Zone of cell maturation

Primary meristems:

Protoderm

Ground meristem

Procambium

Zone of cell elongation

Zone of cell division

New daughter cells are produced in the **root apical meristem**. Most daughter cells differentiate into the primary tissues of the root.

Some daughter cells become part of the **root cap**, which is constantly being eroded away.

Quiescent center

34.14 Tissues and Regions of the Root Tip *(Page 755)*

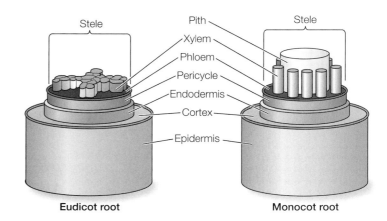

Stele

Pith

Xylem

Phloem

Pericycle

Endodermis

Cortex

Epidermis

Stele

Eudicot root

Monocot root

34.15 Products of the Root's Primary Meristems *(Page 756)*

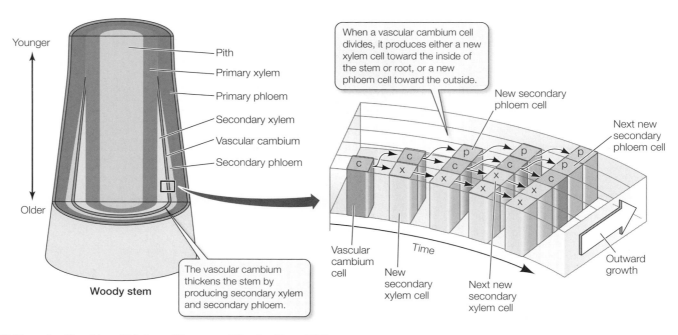

Younger

Older

Pith

Primary xylem

Primary phloem

Secondary xylem

Vascular cambium

Secondary phloem

Woody stem

The vascular cambium thickens the stem by producing secondary xylem and secondary phloem.

When a vascular cambium cell divides, it produces either a new xylem cell toward the inside of the stem or root, or a new phloem cell toward the outside.

New secondary phloem cell

Next new secondary phloem cell

Vascular cambium cell

New secondary xylem cell

Next new secondary xylem cell

Time

Outward growth

34.18 Vascular Cambium Thickens Stems and Roots *(Page 758)*

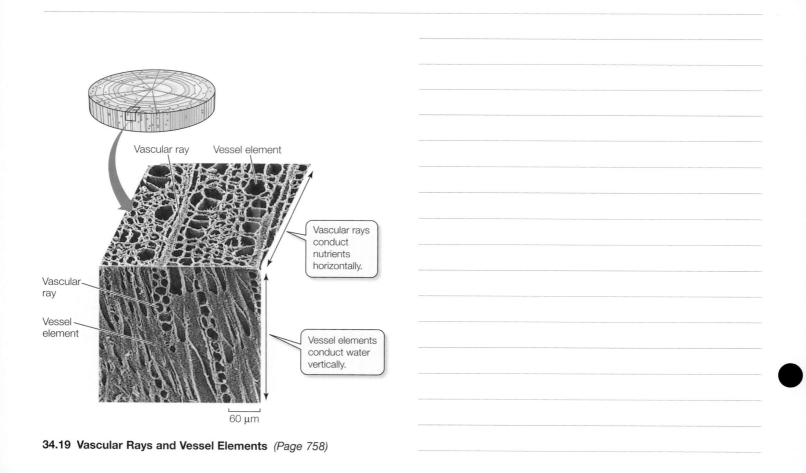

Vascular ray

Vessel element

Vascular rays conduct nutrients horizontally.

Vascular ray

Vessel element

Vessel elements conduct water vertically.

60 μm

34.19 Vascular Rays and Vessel Elements *(Page 758)*

EXPERIMENT

HYPOTHESIS: A needle of a ponderosa pine tree connects with the xylem laid down in the year the needle was formed.

METHOD

1. Immerse the basal ends of 2-cm-long segments of young branches in a dye solution.
2. Clip the tip of one needle in the segment, and apply vacuum to the cut needle.
3. After 5 minutes of the vacuum treatment, cut the segment several mm above the base. Observe which annual ring(s) contain the dye.

RESULTS

When 1-year-old needles were tested, the dye was always found in the annual ring formed 1 year before the experiment (y–1). When 2-year-old needles were tested, the dye was always found in the annual ring formed 2 years before the experiment (y–2), and sometimes also in the ring formed the following year.

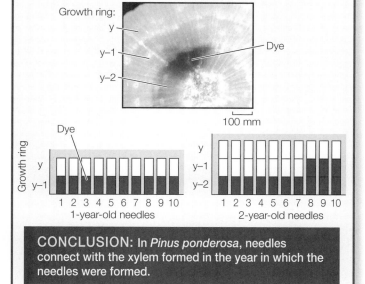

CONCLUSION: In *Pinus ponderosa*, needles connect with the xylem formed in the year in which the needles were formed.

34.21 How Do Leaves Relate to Annual Rings? *(Page 759)*

(A)

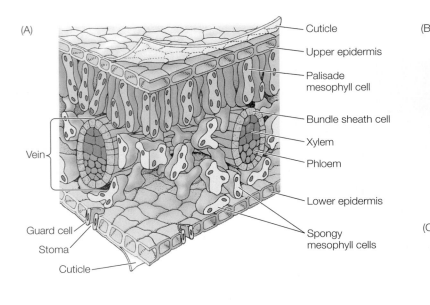

Cuticle

Upper epidermis

Palisade mesophyll cell

Bundle sheath cell

Xylem

Phloem

Lower epidermis

Spongy mesophyll cells

Vein

Guard cell

Stoma

Cuticle

(B)

Guard cells   Stoma

(C)

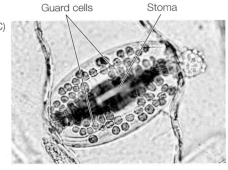

34.23 The Eudicot Leaf *(Page 761)*

35 Transport in Plants

H$_2$O, carbohydrates, etc.

O$_2$

CO$_2$

H$_2$O

H$_2$O and dissolved minerals

35.1 The Pathways of Water and Solutes in the Plant
(Page 766)

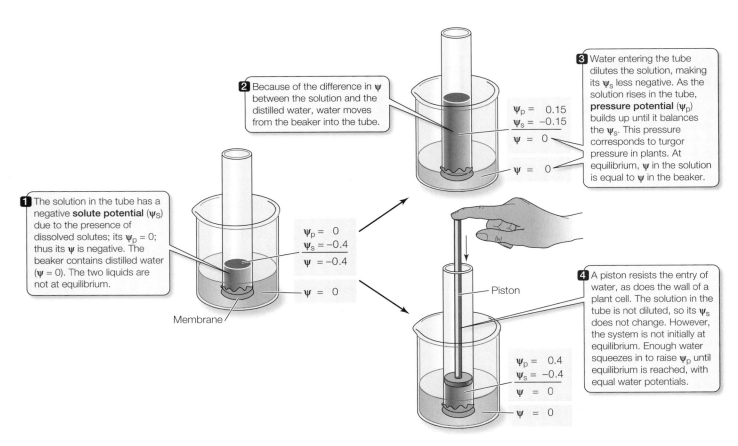

2 Because of the difference in ψ between the solution and the distilled water, water moves from the beaker into the tube.

3 Water entering the tube dilutes the solution, making its ψ_s less negative. As the solution rises in the tube, **pressure potential** (ψ_p) builds up until it balances the ψ_s. This pressure corresponds to turgor pressure in plants. At equilibrium, ψ in the solution is equal to ψ in the beaker.

$$\psi_p = 0.15$$
$$\psi_s = -0.15$$
$$\psi = 0$$

$$\psi = 0$$

1 The solution in the tube has a negative **solute potential** (ψ_s) due to the presence of dissolved solutes; its $\psi_p = 0$; thus its ψ is negative. The beaker contains distilled water (ψ = 0). The two liquids are not at equilibrium.

$$\psi_p = 0$$
$$\psi_s = -0.4$$
$$\psi = -0.4$$

$$\psi = 0$$

Membrane

Piston

4 A piston resists the entry of water, as does the wall of a plant cell. The solution in the tube is not diluted, so its ψ_s does not change. However, the system is not initially at equilibrium. Enough water squeezes in to raise ψ_p until equilibrium is reached, with equal water potentials.

$$\psi_p = 0.4$$
$$\psi_s = -0.4$$
$$\psi = 0$$

$$\psi = 0$$

35.2 Water Potential, Solute Potential, and Pressure Potential *(Page 766)*

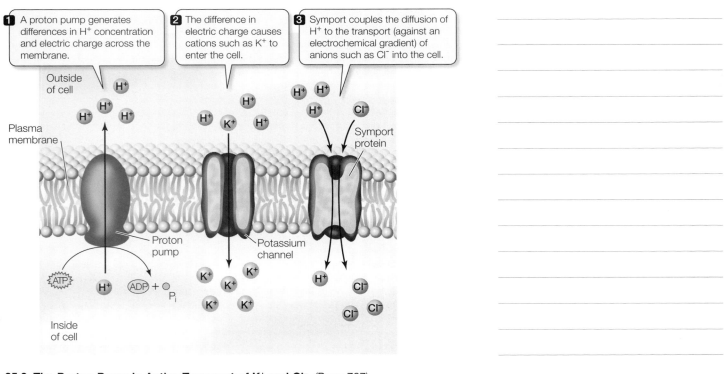

1 A proton pump generates differences in H^+ concentration and electric charge across the membrane.

2 The difference in electric charge causes cations such as K^+ to enter the cell.

3 Symport couples the diffusion of H^+ to the transport (against an electrochemical gradient) of anions such as Cl^- into the cell.

Outside of cell

Plasma membrane

Symport protein

Proton pump

Potassium channel

ATP

ADP + P_i

Inside of cell

35.3 The Proton Pump in Active Transport of K^+ and Cl^- *(Page 767)*

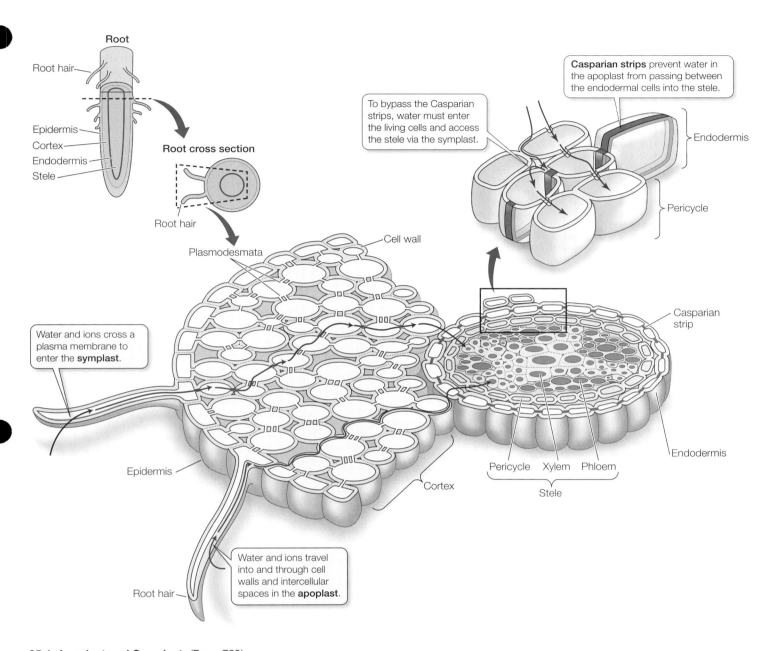

Root

Root hair

Epidermis
Cortex
Endodermis
Stele

Root cross section

Root hair

Casparian strips prevent water in the apoplast from passing between the endodermal cells into the stele.

To bypass the Casparian strips, water must enter the living cells and access the stele via the symplast.

Endodermis

Pericycle

Plasmodesmata

Cell wall

Water and ions cross a plasma membrane to enter the **symplast**.

Casparian strip

Epidermis

Cortex

Endodermis

Pericycle   Xylem   Phloem

Stele

Root hair

Water and ions travel into and through cell walls and intercellular spaces in the **apoplast**.

35.4 Apoplast and Symplast *(Page 769)*

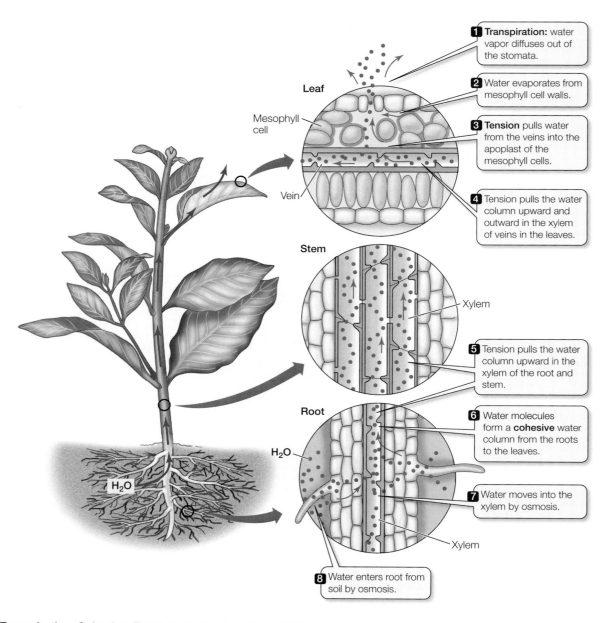

Leaf

Mesophyll cell

Vein

1 **Transpiration:** water vapor diffuses out of the stomata.

2 Water evaporates from mesophyll cell walls.

3 **Tension** pulls water from the veins into the apoplast of the mesophyll cells.

4 Tension pulls the water column upward and outward in the xylem of veins in the leaves.

Stem

Xylem

5 Tension pulls the water column upward in the xylem of the root and stem.

Root

H₂O

Xylem

6 Water molecules form a **cohesive** water column from the roots to the leaves.

7 Water moves into the xylem by osmosis.

8 Water enters root from soil by osmosis.

H₂O

35.6 The Transpiration–Cohesion–Tension Mechanism *(Page 771)*

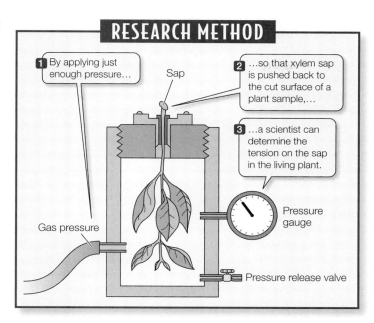

RESEARCH METHOD

1 By applying just enough pressure…

Sap

2 …so that xylem sap is pushed back to the cut surface of a plant sample,…

3 …a scientist can determine the tension on the sap in the living plant.

Gas pressure

Pressure gauge

Pressure release valve

35.7 A Pressure Chamber *(Page 772)*

EXPERIMENT

HYPOTHESIS: K^+ affects the xylem flow rate.

METHOD

Two xylem-containing flaps were cut along the stem of a tobacco plant. One was connected to a source of pure water (the control). The other flap could be connected to either pure water or to a solution containing a known concentration of K^+.

RESULTS

The addition of the K^+ solution dramatically increased the flow rate. The rate returned to the control level when the K^+ solution was replaced by pure water.

H_2O (control)

K^+ solution or H_2O

The experimental trace spiked immediately after the injection of K^+ solution.

2

Return to H_2O alone

K^+ solution injected

Relative flow rate

1

H_2O (control)

The control trace (H_2O only) showed little variation in flow rate.

0   1000   2000   3000   4000

Time (seconds)

CONCLUSION: K^+ increases the rate of flow in the xylem.

35.8 Potassium Ions Speed Transport in the Xylem *(Page 772)*

(A)

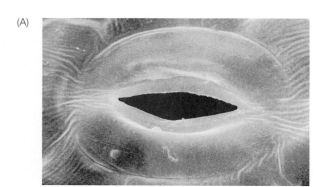

(B)

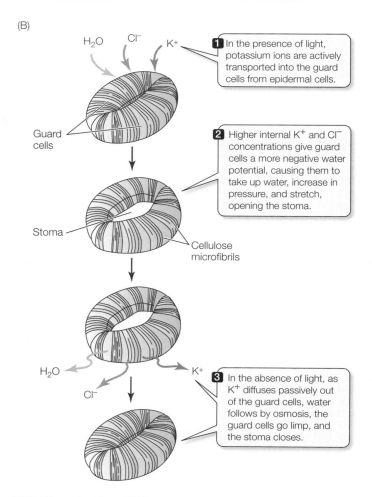

H_2O Cl^- K^+

1 In the presence of light, potassium ions are actively transported into the guard cells from epidermal cells.

Guard cells

2 Higher internal K^+ and Cl^- concentrations give guard cells a more negative water potential, causing them to take up water, increase in pressure, and stretch, opening the stoma.

Stoma

Cellulose microfibrils

H_2O K^+

Cl^-

3 In the absence of light, as K^+ diffuses passively out of the guard cells, water follows by osmosis, the guard cells go limp, and the stoma closes.

35.9 Stomata *(Page 773)*

EXPERIMENT

HYPOTHESIS: Guard cells of open stomata contain more potassium ions than do those of closed stomata.

METHOD

1. Peel strips of epidermis from leaves of broad beans in the dark (closed stomata) and in the light (open stomata).
2. Examine the strips to locate stomata.
3. Scan across guard cells with the electron probe microanalyzer set to measure K^+ concentration.

RESULTS

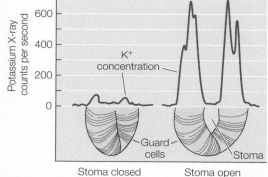

K^+ concentration

Guard cells

Stoma

Stoma closed   Stoma open

CONCLUSION: K^+ concentration within the guard cells surrounding an open stoma was much greater than that in the guard cells surrounding a closed stoma.

35.10 Measuring Tiny Amounts of Potassium in Guard Cells
(Page 774)

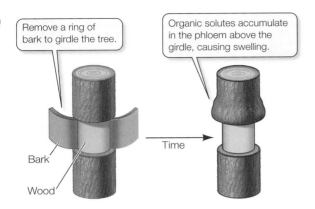

35.11 Girdling Blocks Translocation in the Phloem
(Page 775)

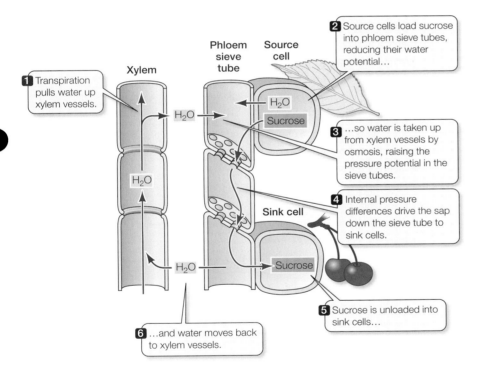

35.13 The Pressure Flow Model *(Page 776)*

TABLE 35.1

Mechanisms of Sap Flow in Plant Vascular Tissues

	XYLEM	PHLOEM
Driving force for bulk flow	Transpiration from leaves	Active transport of sucrose at source
Site of bulk flow	Non-living vessel elements and tracheids (cohesion)	Living sieve tube elements
Pressure potential in sap	Negative (pull from top; tension)	Positive (push from source; pressure)

TABLE 36.1

Mineral Elements Required by Plants

ELEMENT	ABSORBED FORM	MAJOR FUNCTIONS
MACRONUTRIENTS		
Nitrogen (N)	NO_3^- and NH_4^+	In proteins, nucleic acids, etc.
Phosphorus (P)	$H_2PO_4^-$ and HPO_4^{2-}	In nucleic acids, ATP, phospholipids, etc.
Potassium (K)	K^+	Enzyme activation; water balance; ion balance; stomatal opening
Sulfur (S)	SO_4^{2-}	In proteins and coenzymes
Calcium (Ca)	Ca^{2+}	Affects the cytoskeleton, membranes, and many enzymes; second messenger
Magnesium (Mg)	Mg^{2+}	In chlorophyll; required by many enzymes; stabilizes ribosomes
MICRONUTRIENTS		
Iron (Fe)	Fe^{2+} and Fe^{3+}	In active site of many redox enzymes and electron carriers; chlorophyll synthesis
Chlorine (Cl)	Cl^-	Photosynthesis; ion balance
Manganese (Mn)	Mn^{2+}	Activation of many enzymes
Boron (B)	$B(OH)_3$	Possibly carbohydrate transport (poorly understood)
Zinc (Zn)	Zn^{2+}	Enzyme activation; auxin synthesis
Copper (Cu)	Cu^{2+}	In active site of many redox enzymes and electron carriers
Nickel (Ni)	Ni^{2+}	Activation of one enzyme
Molybdenum (Mo)	MoO_4^{2-}	Nitrate reduction

(Page 782)

TABLE 36.2

Some Mineral Deficiencies in Plants

DEFICIENCY	SYMPTOMS
Calcium	Growing points die back; young leaves are yellow and crinkly
Iron	Young leaves are white or yellow
Magnesium	Older leaves have yellow in stripes between veins
Manganese	Younger leaves are pale with green veins
Nitrogen	Oldest leaves turn yellow and die prematurely; plant is stunted
Phosphorus	Plant is dark green with purple veins and is stunted
Potassium	Older leaves have dead edges
Sulfur	Young leaves are yellow to white with yellow veins
Zinc	Young leaves are abnormally small; older leaves have many dead spots

(Page 783)

EXPERIMENT

HYPOTHESIS: Nickel is an essential element for a plant to complete its life cycle.

METHOD

1. Grow barley plants for 3 generations in nutrient solutions containing 0, 0.6, and 1.0 μM $NiSO_4$.
2. Harvest seeds from 5–6 third-generation plants in each of the groups.
3. Determine the nickel concentration in seeds from each plant.
4. Germinate other seeds from the same plants and plot the success of germination against nickel concentration.

RESULTS

There was a positive correlation between seed germination and seed nickel concentration. There was no germination at the lowest nickel concentration.

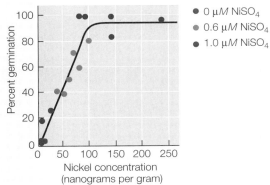

CONCLUSION: Barley seeds from nickel-free plants require nickel in order to germinate, and thereby complete the life cycle.

36.3 Is Nickel an Essential Element for Plant Growth? *(Page 785)*

RESEARCH METHOD

METHOD

Grow seedlings in a medium that lacks the element in question (in this case, nitrogen)

Seedling grown in a complete growth medium.

Seedling grown in a medium lacking nitrogen.

RESULTS

Growth is normal.

Growth is abnormal, and plant cannot complete its life cycle.

36.2 Identifying Essential Elements for Plants *(Page 784)*

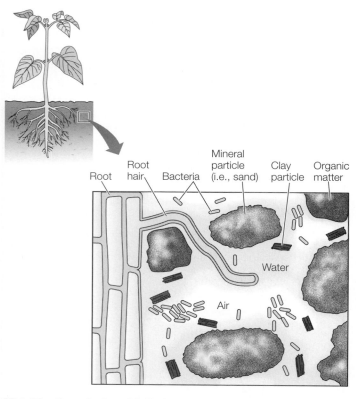

Root · Root hair · Bacteria · Mineral particle (i.e., sand) · Clay particle · Organic matter

Water

Air

36.4 The Complexity of Soil *(Page 785)*

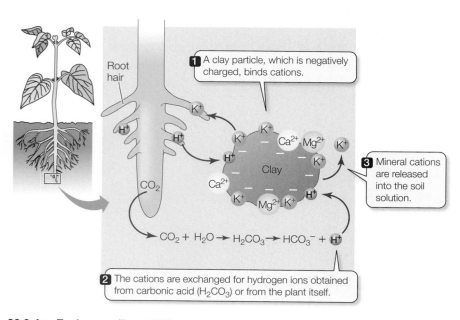

Root hair

1 A clay particle, which is negatively charged, binds cations.

K⁺

H⁺

H⁺

CO_2

K⁺ · K⁺ · Ca²⁺ · Mg²⁺ · K⁺

H⁺ · Clay · K⁺

Ca²⁺ · K⁺ · Mg²⁺ · K⁺ · H⁺

3 Mineral cations are released into the soil solution.

$CO_2 + H_2O \rightarrow H_2CO_3 \rightarrow HCO_3^- + H^+$

2 The cations are exchanged for hydrogen ions obtained from carbonic acid (H_2CO_3) or from the plant itself.

36.6 Ion Exchange *(Page 787)*

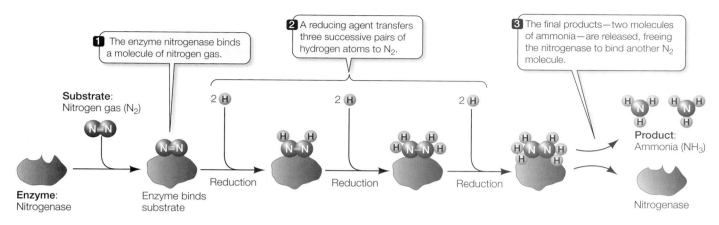

36.8 Nitrogenase Fixes Nitrogen *(Page 789)*

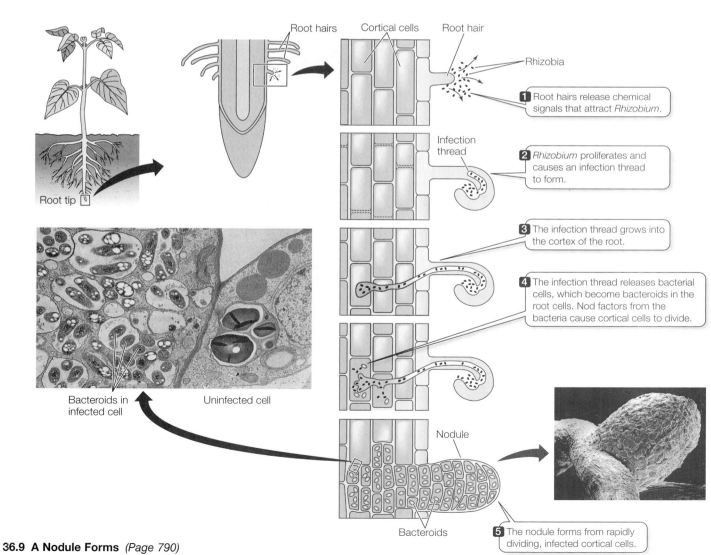

36.9 A Nodule Forms *(Page 790)*

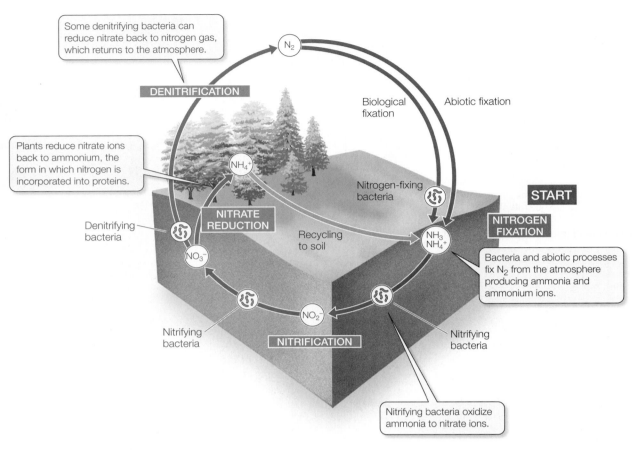

36.10 The Nitrogen Cycle *(Page 791)*

CHAPTER 37 Regulation of Plant Growth

TABLE 37.1

Plant Growth Hormones

HORMONE	TYPICAL ACTIVITIES
Abscisic acid	Maintains seed dormancy and winter dormancy; closes stomata
Auxins	Promote stem elongation, adventitious root initiation, and fruit growth; inhibit axillary bud outgrowth and leaf abscission
Brassinosteroids	Promote stem and pollen tube elongation; promote vascular tissue differentiation
Cytokinins	Inhibit leaf senescence; promote cell division and axillary bud outgrowth; affect root growth
Ethylene	Promotes fruit ripening and leaf abscission; inhibits stem elongation and gravitropism
Gibberellins	Promote seed germination, stem growth, and fruit development; break winter dormancy; mobilize nutrient reserves in grass seeds

(Page 798)

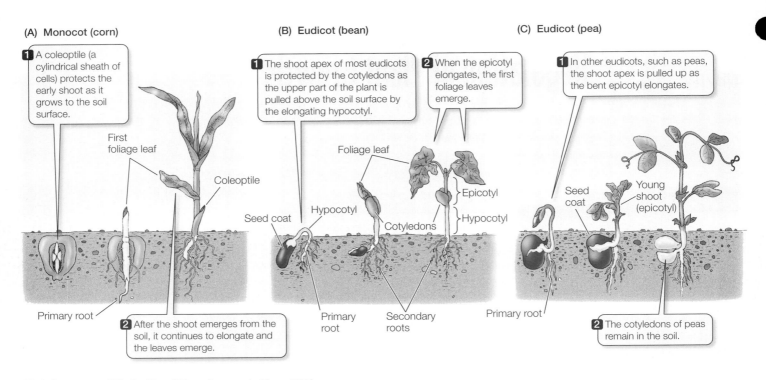

(A) Monocot (corn)

1 A coleoptile (a cylindrical sheath of cells) protects the early shoot as it grows to the soil surface.

First foliage leaf

Coleoptile

Primary root

2 After the shoot emerges from the soil, it continues to elongate and the leaves emerge.

(B) Eudicot (bean)

1 The shoot apex of most eudicots is protected by the cotyledons as the upper part of the plant is pulled above the soil surface by the elongating hypocotyl.

2 When the epicotyl elongates, the first foliage leaves emerge.

Foliage leaf

Seed coat

Hypocotyl

Cotyledons

Epicotyl

Hypocotyl

Primary root

Secondary roots

(C) Eudicot (pea)

1 In other eudicots, such as peas, the shoot apex is pulled up as the bent epicotyl elongates.

Seed coat

Young shoot (epicotyl)

Primary root

2 The cotyledons of peas remain in the soil.

37.1 Patterns of Early Shoot Development *(Page 799)*

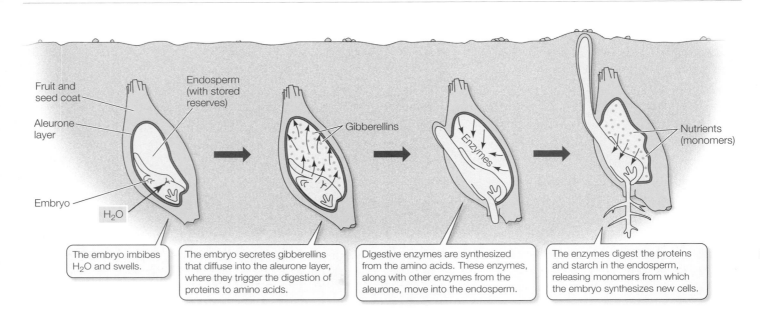

Fruit and seed coat

Endosperm (with stored reserves)

Aleurone layer

Gibberellins

Embryo

H₂O

Enzymes

Nutrients (monomers)

The embryo imbibes H_2O and swells.

The embryo secretes gibberellins that diffuse into the aleurone layer, where they trigger the digestion of proteins to amino acids.

Digestive enzymes are synthesized from the amino acids. These enzymes, along with other enzymes from the aleurone, move into the endosperm.

The enzymes digest the proteins and starch in the endosperm, releasing monomers from which the embryo synthesizes new cells.

37.4 Embryos Mobilize Their Reserves *(Page 801)*

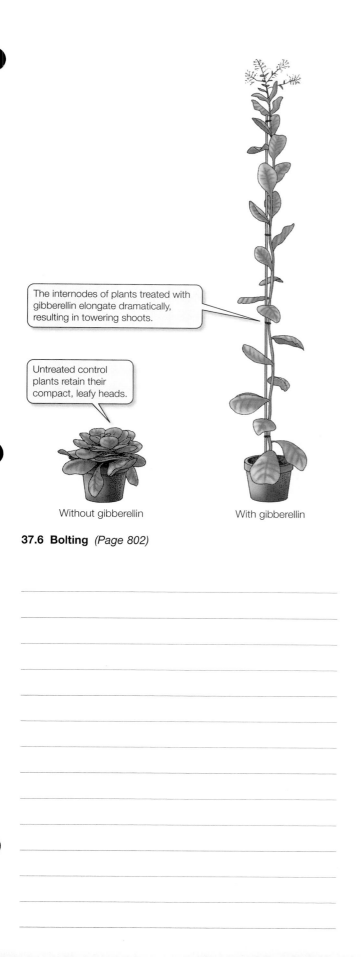

The internodes of plants treated with gibberellin elongate dramatically, resulting in towering shoots.

Untreated control plants retain their compact, leafy heads.

Without gibberellin

With gibberellin

37.6 Bolting *(Page 802)*

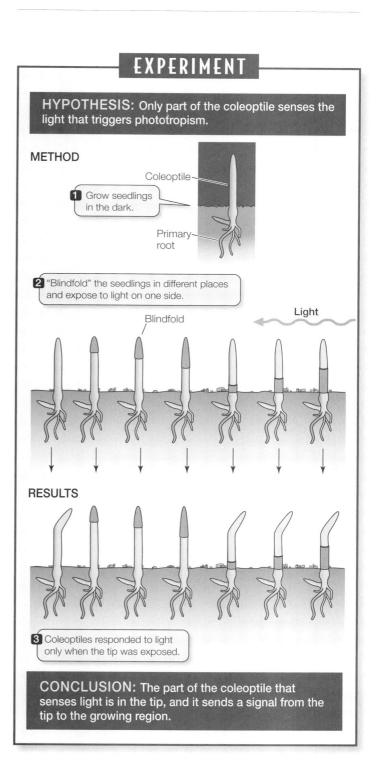

EXPERIMENT

HYPOTHESIS: Only part of the coleoptile senses the light that triggers phototropism.

METHOD

Coleoptile

1 Grow seedlings in the dark.

Primary root

2 "Blindfold" the seedlings in different places and expose to light on one side.

Blindfold

Light

RESULTS

3 Coleoptiles responded to light only when the tip was exposed.

CONCLUSION: The part of the coleoptile that senses light is in the tip, and it sends a signal from the tip to the growing region.

37.7 The Darwins' Phototropism Experiment *(Page 804)*

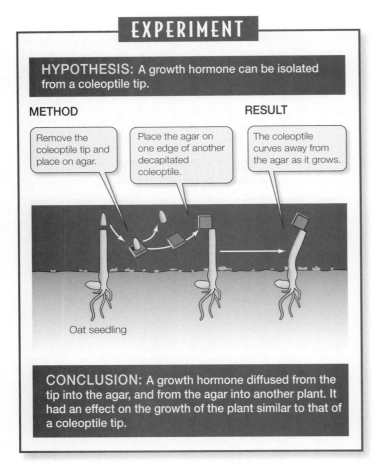

37.8 Went's Experiment *(Page 804)*

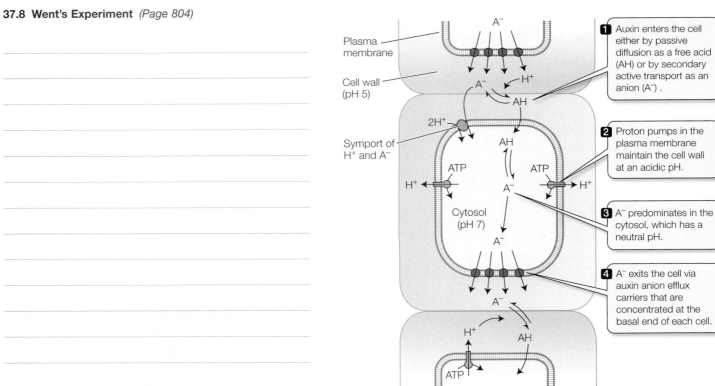

37.9 Polar Transport of Auxin *(Page 805)*

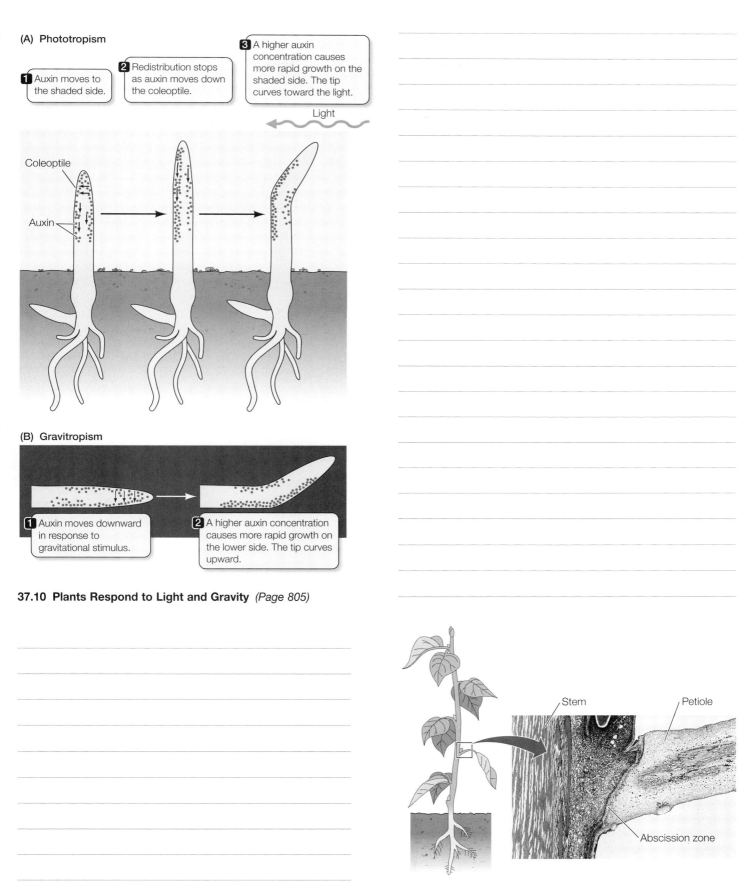

(A) Phototropism

1 Auxin moves to the shaded side.

2 Redistribution stops as auxin moves down the coleoptile.

3 A higher auxin concentration causes more rapid growth on the shaded side. The tip curves toward the light.

Light

Coleoptile

Auxin

(B) Gravitropism

1 Auxin moves downward in response to gravitational stimulus.

2 A higher auxin concentration causes more rapid growth on the lower side. The tip curves upward.

37.10 Plants Respond to Light and Gravity *(Page 805)*

Stem

Petiole

Abscission zone

37.11 Changes Occur when a Leaf Is About to Fall *(Page 806)*

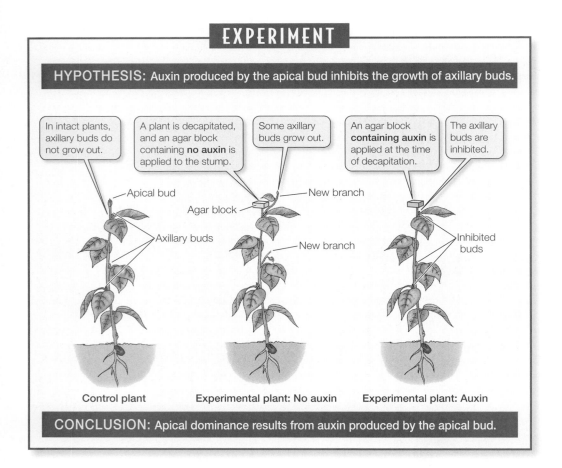

37.12 Auxin and Apical Dominance *(Page 806)*

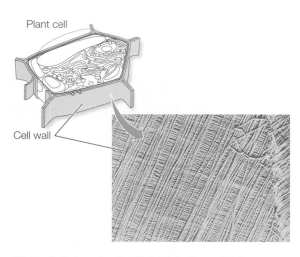

37.13 Cellulose in the Cell Wall *(Page 807)*

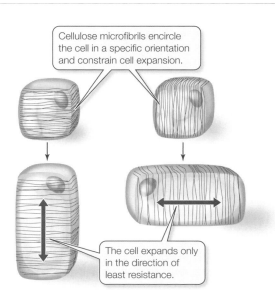

37.14 Plant Cells Expand *(Page 807)*

REGULATION OF PLANT GROWTH

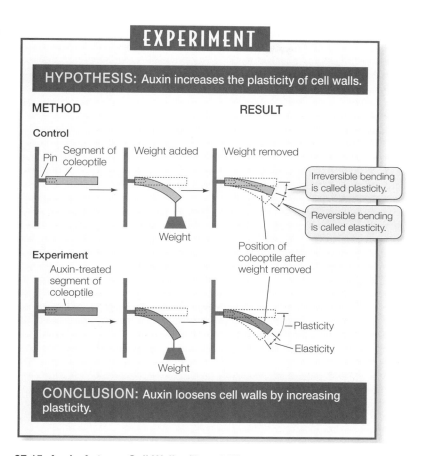

37.15 Auxin Acts on Cell Walls *(Page 808)*

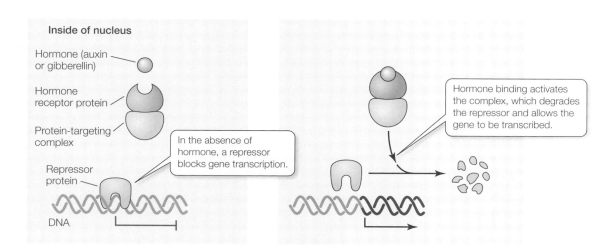

37.16 Similar Signal Transduction Pathways Are Used by Auxin and Gibberellins *(Page 808)*

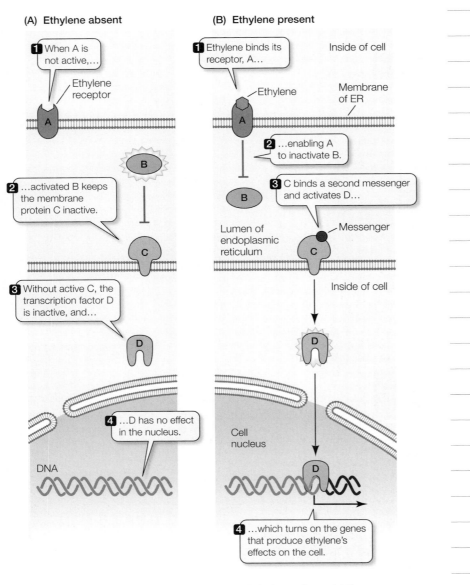

37.18 The Signal Transduction Pathway for Ethylene *(Page 811)*

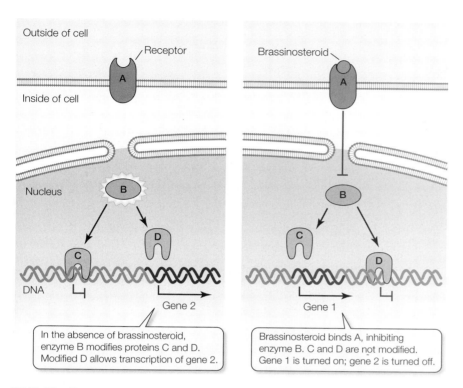

In the absence of brassinosteroid, enzyme B modifies proteins C and D. Modified D allows transcription of gene 2.

Brassinosteroid binds A, inhibiting enzyme B. C and D are not modified. Gene 1 is turned on; gene 2 is turned off.

37.19 The Brassinosteroid Signal Transduction Pathway Begins at the Plasma Membrane *(Page 812)*

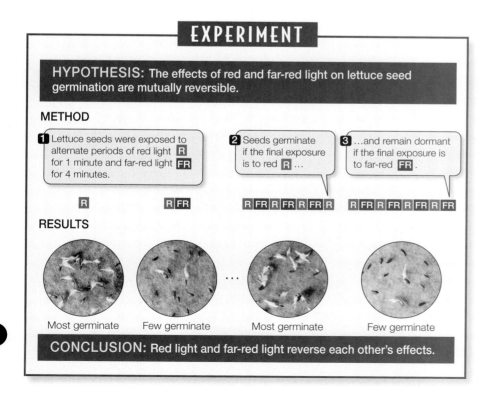

EXPERIMENT

HYPOTHESIS: The effects of red and far-red light on lettuce seed germination are mutually reversible.

METHOD

1 Lettuce seeds were exposed to alternate periods of red light **R** for 1 minute and far-red light **FR** for 4 minutes.

2 Seeds germinate if the final exposure is to red **R** …

3 …and remain dormant if the final exposure is to far-red **FR** .

R   **R** **FR**   **R** **FR** **R** **FR** **R** **FR** **R**   **R** **FR** **R** **FR** **R** **FR** **R** **FR**

RESULTS

Most germinate   Few germinate   … Most germinate   Few germinate

CONCLUSION: Red light and far-red light reverse each other's effects.

37.20 Sensitivity of Seeds to Red and Far-Red Light *(Page 813)*

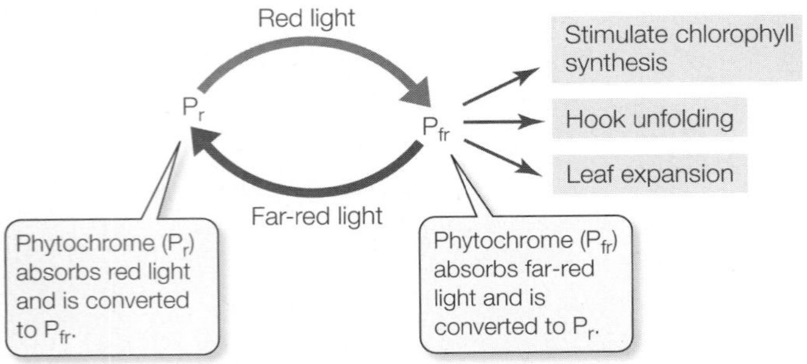

Red light

P_r

P_{fr}

Far-red light

Stimulate chlorophyll synthesis

Hook unfolding

Leaf expansion

Phytochrome (P_r) absorbs red light and is converted to P_{fr}.

Phytochrome (P_{fr}) absorbs far-red light and is converted to P_r.

Page 813 In-Text Art

38 Reproduction in Flowering Plants

38.1 Development of Gametophytes and Nuclear Fusion *(Page 820)*

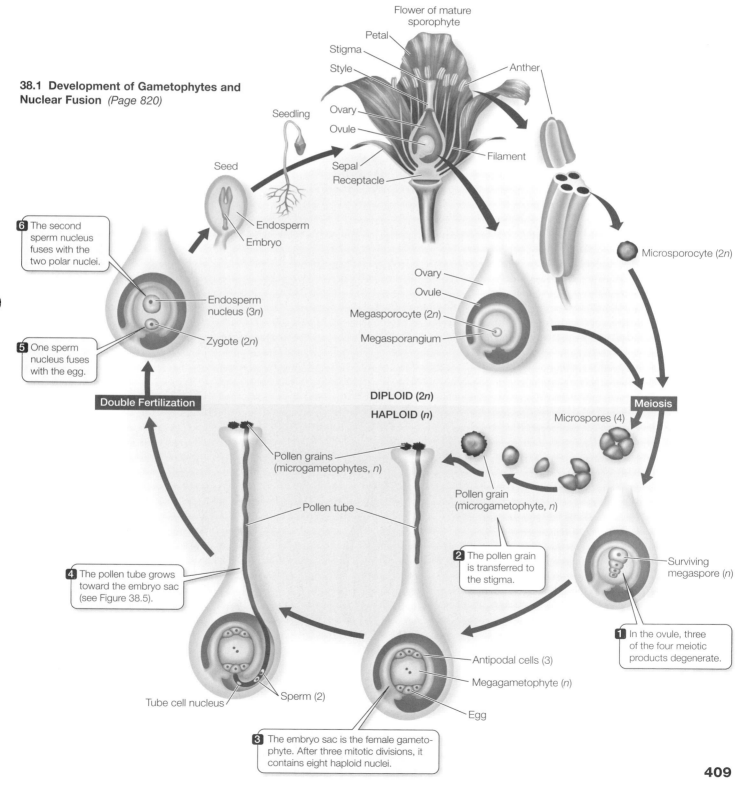

Flower of mature sporophyte

Petal
Stigma
Style
Anther
Ovary
Ovule
Filament
Sepal
Receptacle

Seedling

Seed

Endosperm
Embryo

6 The second sperm nucleus fuses with the two polar nuclei.

Endosperm nucleus (3*n*)

5 One sperm nucleus fuses with the egg.

Zygote (2*n*)

Double Fertilization

Pollen grains (microgametophytes, *n*)

Pollen tube

4 The pollen tube grows toward the embryo sac (see Figure 38.5).

Tube cell nucleus

Sperm (2)

Ovary
Ovule
Megasporocyte (2*n*)
Megasporangium

DIPLOID (2*n*)
HAPLOID (*n*)

Microsporocyte (2*n*)

Meiosis

Microspores (4)

Pollen grain (microgametophyte, *n*)

2 The pollen grain is transferred to the stigma.

Surviving megaspore (*n*)

1 In the ovule, three of the four meiotic products degenerate.

Antipodal cells (3)

Megagametophyte (*n*)

Egg

3 The embryo sac is the female gametophyte. After three mitotic divisions, it contains eight haploid nuclei.

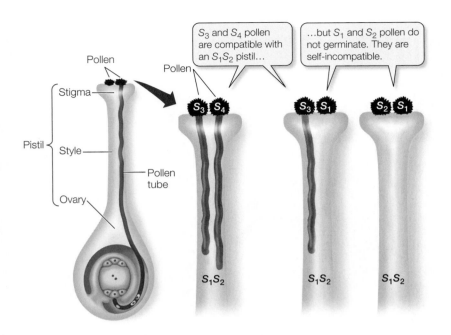

38.3 Self-Incompatibility *(Page 822)*

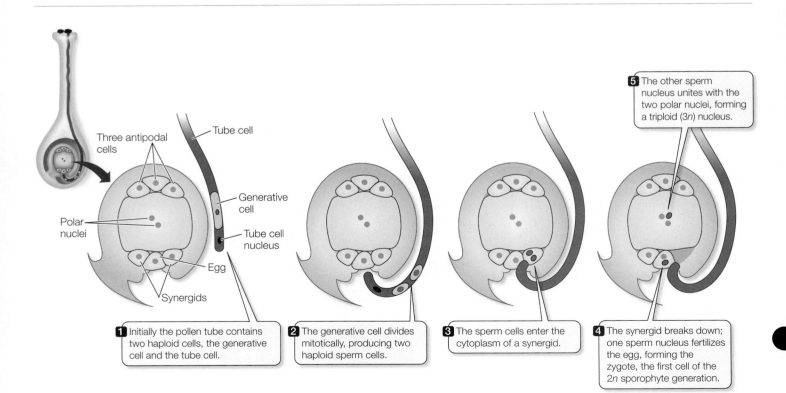

38.5 Double Fertilization *(Page 823)*

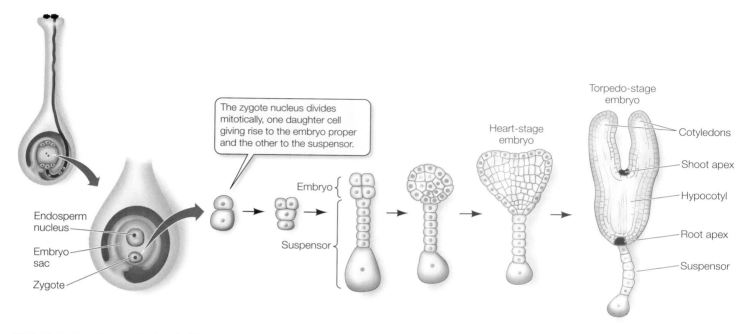

The zygote nucleus divides mitotically, one daughter cell giving rise to the embryo proper and the other to the suspensor.

Endosperm nucleus

Embryo sac

Zygote

Embryo

Suspensor

Heart-stage embryo

Torpedo-stage embryo

Cotyledons

Shoot apex

Hypocotyl

Root apex

Suspensor

38.6 Early Development of a Eudicot *(Page 823)*

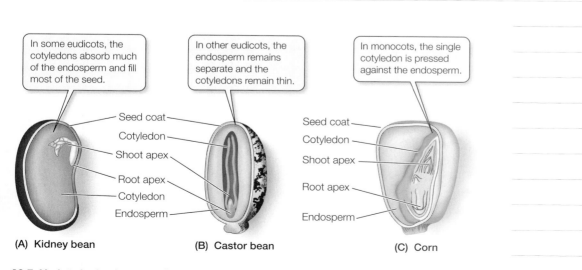

In some eudicots, the cotyledons absorb much of the endosperm and fill most of the seed.

In other eudicots, the endosperm remains separate and the cotyledons remain thin.

In monocots, the single cotyledon is pressed against the endosperm.

Seed coat
Cotyledon
Shoot apex
Root apex
Cotyledon
Endosperm

Seed coat
Cotyledon
Shoot apex
Root apex
Endosperm

(A) Kidney bean

(B) Castor bean

(C) Corn

38.7 Variety in Angiosperm Seeds *(Page 824)*

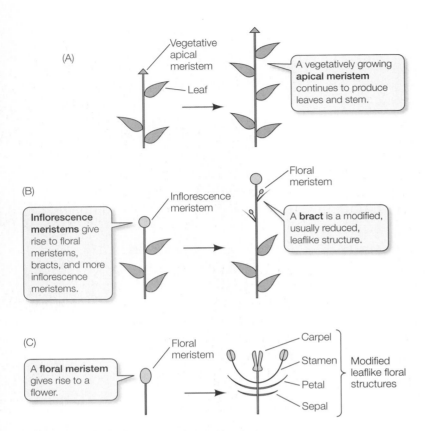

(A)

Vegetative apical meristem

Leaf

A vegetatively growing **apical meristem** continues to produce leaves and stem.

(B)

Inflorescence meristem

Floral meristem

Inflorescence meristems give rise to floral meristems, bracts, and more inflorescence meristems.

A **bract** is a modified, usually reduced, leaflike structure.

(C)

Floral meristem

Carpel

Stamen

Petal

Sepal

Modified leaflike floral structures

A **floral meristem** gives rise to a flower.

38.9 Flowering and the Apical Meristem *(Page 825)*

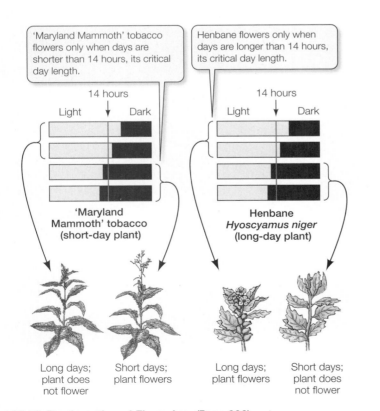

'Maryland Mammoth' tobacco flowers only when days are shorter than 14 hours, its critical day length.

Henbane flowers only when days are longer than 14 hours, its critical day length.

14 hours

Light   Dark

14 hours

Light   Dark

'Maryland Mammoth' tobacco (short-day plant)

Henbane *Hyoscyamus niger* (long-day plant)

Long days; plant does not flower

Short days; plant flowers

Long days; plant flowers

Short days; plant does not flower

38.10 Day Length and Flowering *(Page 826)*

EXPERIMENT

HYPOTHESIS: Short-day plants measure day length.

METHOD

Move plants between light and dark rooms for specified numbers of hours.

RESULTS

Light constant — Darkness varied

| 16 | 6 |
| 16 | 7 | No flowering
| 16 | 8 |

| 16 | 9 |
| 16 | 10 | Only plants given 9 or more hours of dark flowered.
| 16 | 11 |

Light varied — 8 or 10 hours of darkness

| 8 | 10 |
| 10 | 10 | Only plants given 10 hours of dark flowered.
| 12 | 10 |

| 8 | 8 |
| 10 | 8 | No flowering
| 12 | 8 |

Time (hours)

CONCLUSION: The data do not support the hypothesis. Short-day plants measure the length of the night and thus could more accurately be called long-night plants.

38.11 Night Length and Flowering *(Page 827)*

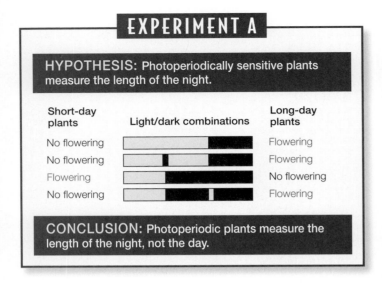

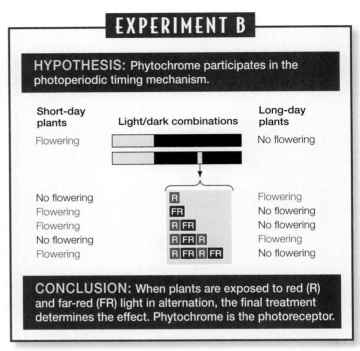

38.12 The Effect of Interrupted Days and Nights *(Page 828)*

38.13 Features of Circadian Rhythms *(Page 828)*

EXPERIMENT

HYPOTHESIS: Plants fix more carbon photo-synthetically when their circadian clock matches the environment's light-dark cycle.

METHOD

1. Select two mutant strains of *Arabidopsis thaliana*: one (*ztl-1*) with a long-cycle rhythm, and the other (*toc1-1*) with a short-cycle rhythm.

2. Grow plants of both mutant strains under either a 28-hour light-dark cycle or a 20-hour cycle instead of the normal 24-hour cycle.

3. Determine photosynthetic C fixation for each of the four groups.

RESULTS

Long-cycle mutants fixed more carbon per hour when grown on a longer-than-24-hour cycle, while short-cycle mutants performed better on a shorter-than-24-hour cycle.

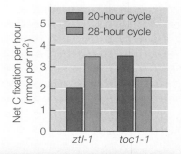

CONCLUSION: Each mutant performed best under the cycle that corresponded to its genetically determined circadian rhythm.

38.14 Does the Circadian Clock Help a Plant Interact with Its Environment? *(Page 829)*

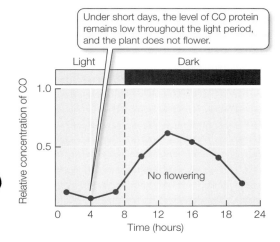

Under short days, the level of CO protein remains low throughout the light period, and the plant does not flower.

No flowering

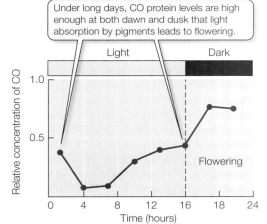

Under long days, CO protein levels are high enough at both dawn and dusk that light absorption by pigments leads to flowering.

Flowering

38.15 Photoreceptors and the Biological Clock Interact in Photoperiodic Plants *(Page 830)*

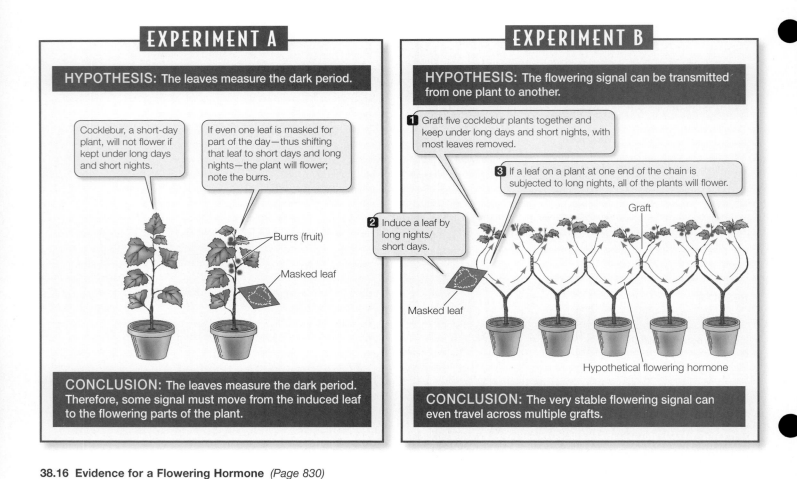

38.16 Evidence for a Flowering Hormone *(Page 830)*

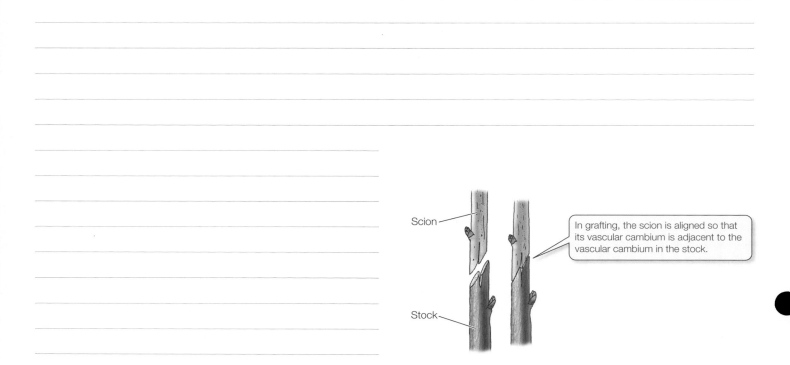

38.19 Grafting *(Page 833)*

39 Plant Responses to Environmental Challenges

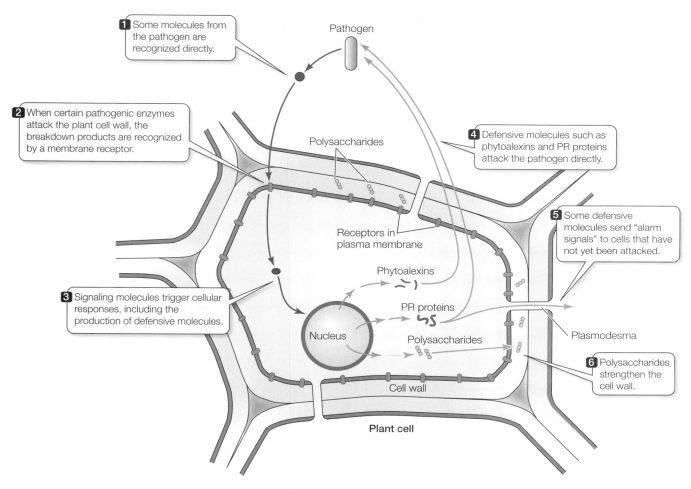

1 Some molecules from the pathogen are recognized directly.

Pathogen

2 When certain pathogenic enzymes attack the plant cell wall, the breakdown products are recognized by a membrane receptor.

Polysaccharides

4 Defensive molecules such as phytoalexins and PR proteins attack the pathogen directly.

Receptors in plasma membrane

5 Some defensive molecules send "alarm signals" to cells that have not yet been attacked.

Phytoalexins

3 Signaling molecules trigger cellular responses, including the production of defensive molecules.

PR proteins

Nucleus

Polysaccharides

Plasmodesma

6 Polysaccharides strengthen the cell wall.

Cell wall

Plant cell

39.1 Signaling between Plants and Pathogens *(Page 838)*

TABLE 39.1

Secondary Plant Metabolites Used in Defense

CLASS	TYPE	ROLE	EXAMPLE
Nitrogen-containing	Alkaloids	Affect herbivore nervous system	Nicotine in tobacco
	Glycosides	Release cyanide or sulfur compounds	Dhurrin in sorghum
	Nonprotein amino acids	Disrupt herbivore protein structure	Canavanine in jack bean
Phenolics	Flavonoids	Phytoalexins	Capsidol in peppers
	Quinones	Inhibit competing plants	Juglone in walnut
	Tannins	Deter herbivores and microbes	Many woods, such as oak
Terpenes	Monoterpenes	Insecticides	Pyrethroids in chrysanthemums
	Sesquiterpenes	Phytoalexins; deter herbivores	Gossypol in cotton
	Steroids	Mimic insect hormones and disrupt insect life cycles	α-Ecdysone in ferns
	Polyterpenes	Feeding deterrent?	Latex in rubber tree

(Page 839)

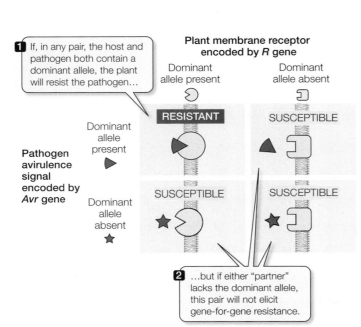

1 If, in any pair, the host and pathogen both contain a dominant allele, the plant will resist the pathogen...

Plant membrane receptor encoded by *R* gene

Dominant allele present

Dominant allele absent

Pathogen avirulence signal encoded by *Avr* gene

Dominant allele present

Dominant allele absent

RESISTANT

SUSCEPTIBLE

SUSCEPTIBLE

SUSCEPTIBLE

2 ...but if either "partner" lacks the dominant allele, this pair will not elicit gene-for-gene resistance.

39.3 Gene-for-Gene Resistance *(Page 840)*

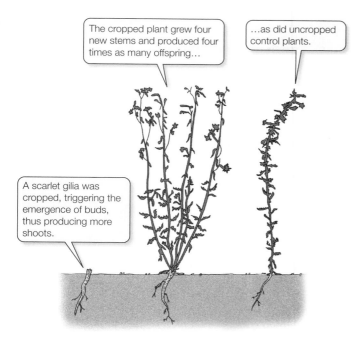

39.4 Overcompensation for Being Eaten *(Page 841)*

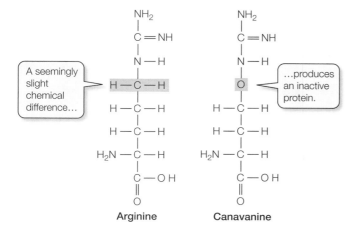

Arginine   Canavanine

Page 841 In-Text Art

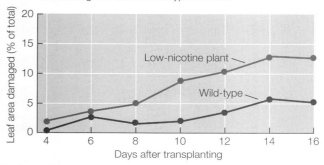

39.5 Some Plants Use Nicotine to Reduce Insect Attacks
(Page 842)

EXPERIMENT

HYPOTHESIS: Corn roots attacked by beetle larvae attract nematodes that will attack the larvae.

METHOD

1. Construct a test system with 6 arms radiating from a central chamber. Add soil, connecting all parts of the system.

2. Three of the control chambers contain only soil. The other three chambers each contain a single corn plant: One healthy plant, one with roots damaged by beetle larvae, and one with roots damaged by stabbing with a metal tool.

3. After three days, add ~2,000 nematodes to the central chamber.

4. After 24 hours count the nematodes in each connecting arm.

RESULTS

Nematodes moved into each of the arms, but by far the most moved into the arm leading to the larvae-damaged plant.

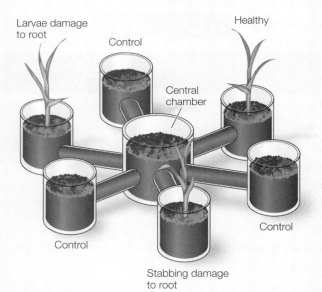

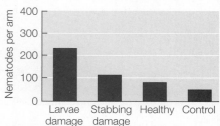

CONCLUSION: The nematodes were attracted to the roots that had been attacked by the beetle larvae.

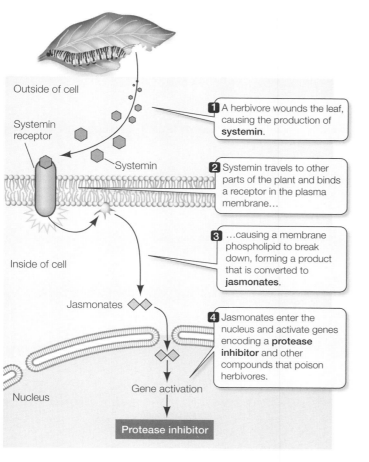

Outside of cell

Systemin receptor

Systemin

Inside of cell

Jasmonates

Nucleus

Gene activation

Protease inhibitor

1 A herbivore wounds the leaf, causing the production of **systemin**.

2 Systemin travels to other parts of the plant and binds a receptor in the plasma membrane…

3 …causing a membrane phospholipid to break down, forming a product that is converted to **jasmonates**.

4 Jasmonates enter the nucleus and activate genes encoding a **protease inhibitor** and other compounds that poison herbivores.

39.7 A Signaling Pathway for Synthesis of a Defensive Secondary Metabolite *(Page 843)*

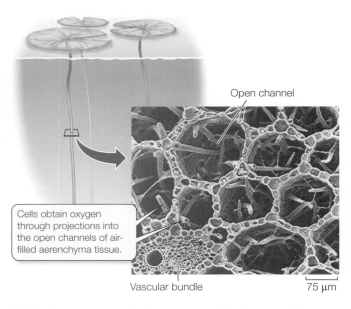

39.14 Aerenchyma Lets Oxygen Reach Submerged Tissues
(Page 847)

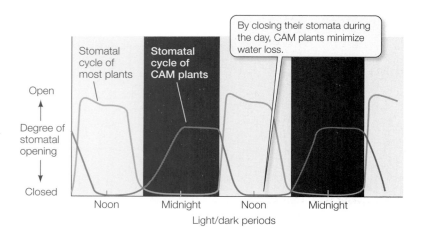

39.16 Stomatal Cycle *(Page 849)*

40 Physiology, Homeostasis, and Temperature Regulation

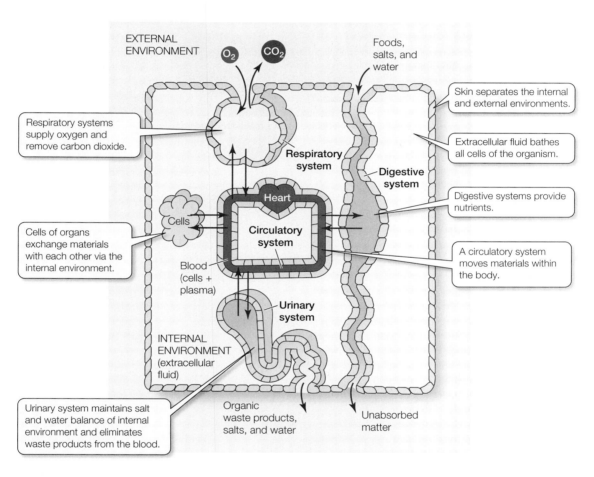

EXTERNAL ENVIRONMENT

O_2 CO_2

Foods, salts, and water

Respiratory systems supply oxygen and remove carbon dioxide.

Skin separates the internal and external environments.

Extracellular fluid bathes all cells of the organism.

Respiratory system

Digestive system

Cells of organs exchange materials with each other via the internal environment.

Cells

Heart

Circulatory system

Digestive systems provide nutrients.

Blood (cells + plasma)

A circulatory system moves materials within the body.

Urinary system

INTERNAL ENVIRONMENT (extracellular fluid)

Urinary system maintains salt and water balance of internal environment and eliminates waste products from the blood.

Organic waste products, salts, and water

Unabsorbed matter

40.1 Maintaining Internal Stability *(Page 856)*

40.2 Control, Regulation, and Feedback *(Page 857)*

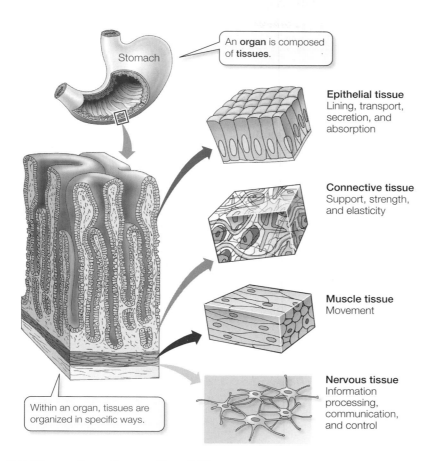

An **organ** is composed of **tissues**.

Epithelial tissue
Lining, transport, secretion, and absorption

Connective tissue
Support, strength, and elasticity

Muscle tissue
Movement

Nervous tissue
Information processing, communication, and control

Within an organ, tissues are organized in specific ways.

40.7 Tissues Form Organs *(Page 860)*

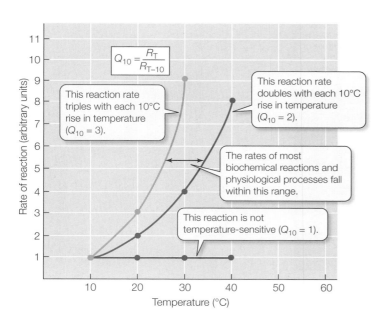

$$Q_{10} = \frac{R_T}{R_{T-10}}$$

This reaction rate triples with each 10°C rise in temperature ($Q_{10} = 3$).

This reaction rate doubles with each 10°C rise in temperature ($Q_{10} = 2$).

The rates of most biochemical reactions and physiological processes fall within this range.

This reaction is not temperature-sensitive ($Q_{10} = 1$).

40.8 Q₁₀ and Reaction Rate *(Page 861)*

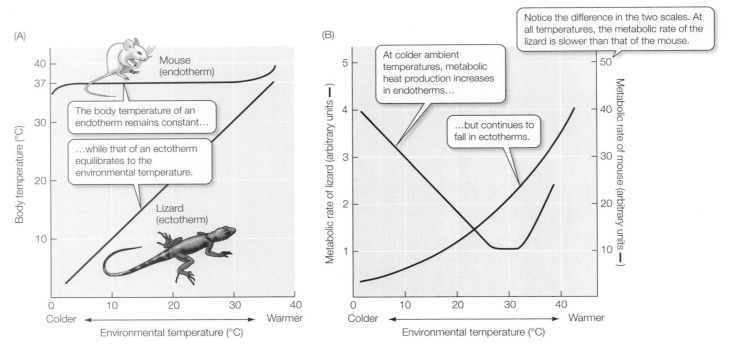

40.9 Ectotherms and Endotherms React Differently to Environmental Temperatures *(Page 862)*

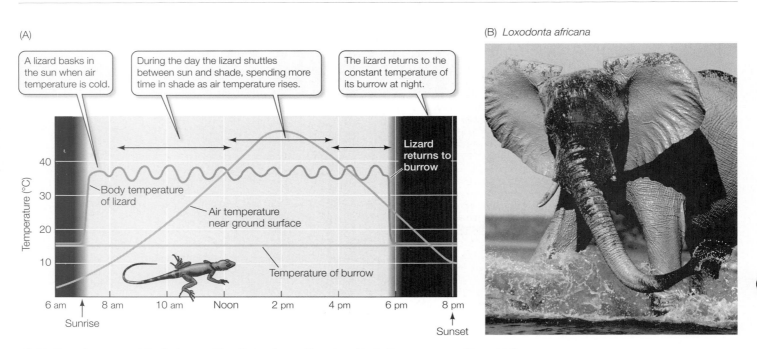

40.10 Ectotherms and Endotherms Use Behavior to Regulate Body Temperature *(Page 863)*

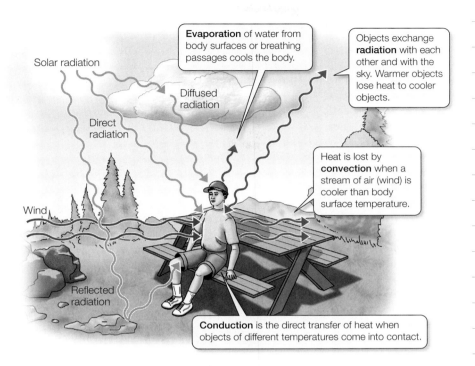

40.11 Animals Exchange Heat with the Environment *(Page 863)*

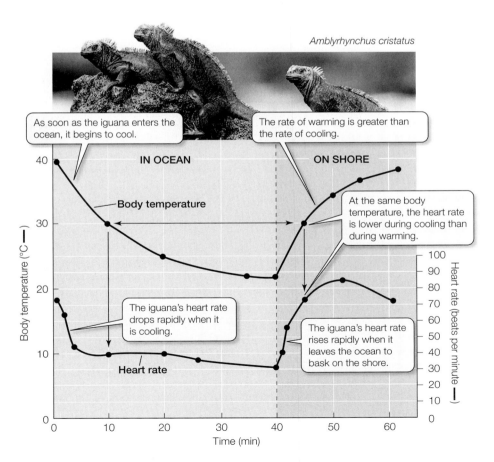

40.12 Some Ectotherms Regulate Blood Flow to the Skin *(Page 864)*

(A) "Cold" fish

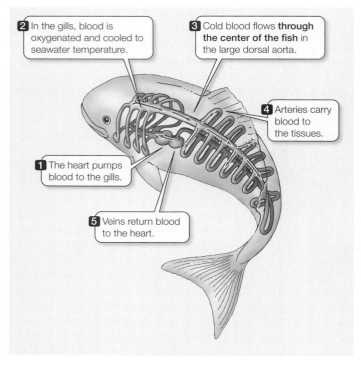

2 In the gills, blood is oxygenated and cooled to seawater temperature.

3 Cold blood flows **through the center of the fish** in the large dorsal aorta.

4 Arteries carry blood to the tissues.

1 The heart pumps blood to the gills.

5 Veins return blood to the heart.

40.13 "Cold" and "Hot" Fish *(Page 865)*

(B) "Hot" fish

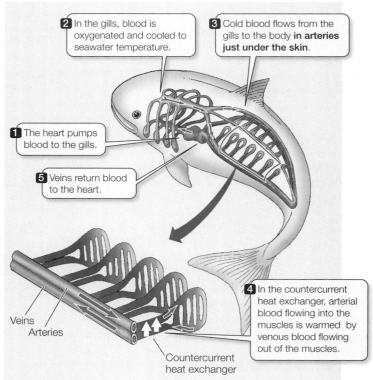

2 In the gills, blood is oxygenated and cooled to seawater temperature.

3 Cold blood flows from the gills to the body **in arteries just under the skin**.

1 The heart pumps blood to the gills.

5 Veins return blood to the heart.

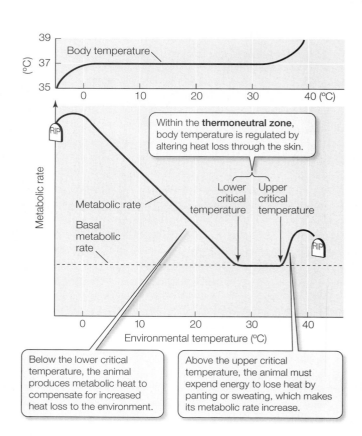

Veins
Arteries

4 In the countercurrent heat exchanger, arterial blood flowing into the muscles is warmed by venous blood flowing out of the muscles.

Countercurrent heat exchanger

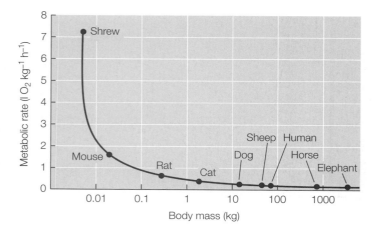

40.15 The Mouse-to-Elephant Curve *(Page 866)*

Body temperature

Within the **thermoneutral zone**, body temperature is regulated by altering heat loss through the skin.

Lower critical temperature

Upper critical temperature

Metabolic rate

Basal metabolic rate

Environmental temperature (°C)

Below the lower critical temperature, the animal produces metabolic heat to compensate for increased heat loss to the environment.

Above the upper critical temperature, the animal must expend energy to lose heat by panting or sweating, which makes its metabolic rate increase.

40.16 Environmental Temperature and Mammalian Metabolic Rates *(Page 867)*

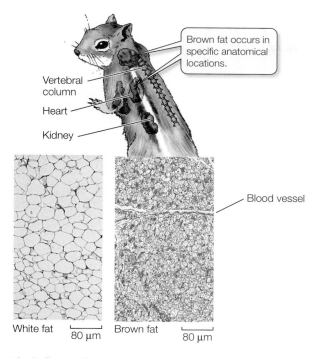

Brown fat occurs in specific anatomical locations.

Vertebral column

Heart

Kidney

Blood vessel

White fat   80 µm   Brown fat   80 µm

40.17 Brown Fat *(Page 867)*

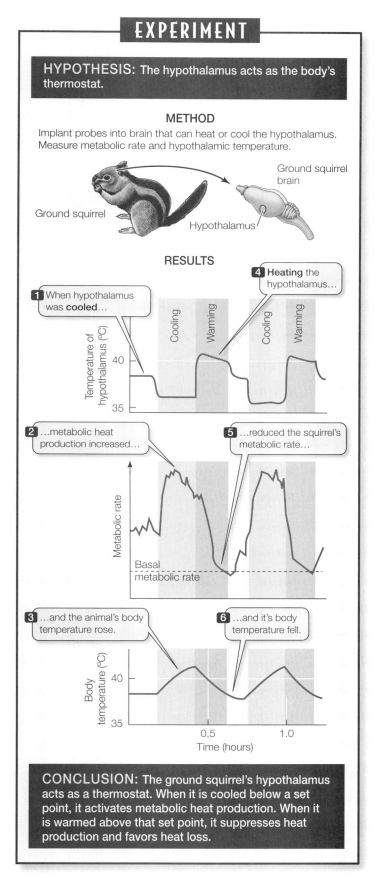

EXPERIMENT

HYPOTHESIS: The hypothalamus acts as the body's thermostat.

METHOD
Implant probes into brain that can heat or cool the hypothalamus. Measure metabolic rate and hypothalamic temperature.

Ground squirrel brain

Ground squirrel

Hypothalamus

RESULTS

4 **Heating** the hypothalamus…

1 When hypothalamus was **cooled**…

Cooling   Warming   Cooling   Warming

Temperature of hypothalamus (°C)   40   35

2 …metabolic heat production increased…

5 …reduced the squirrel's metabolic rate…

Metabolic rate

Basal metabolic rate

3 …and the animal's body temperature rose.

6 …and it's body temperature fell.

Body temperature (°C)   40   35

0.5   1.0
Time (hours)

CONCLUSION: The ground squirrel's hypothalamus acts as a thermostat. When it is cooled below a set point, it activates metabolic heat production. When it is warmed above that set point, it suppresses heat production and favors heat loss.

40.19 The Hypothalamus Regulates Body Temperature *(Page 869)*

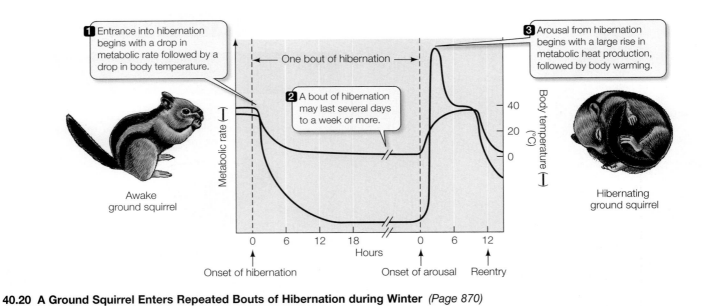

40.20 A Ground Squirrel Enters Repeated Bouts of Hibernation during Winter *(Page 870)*

CHAPTER 41 Animal Hormones

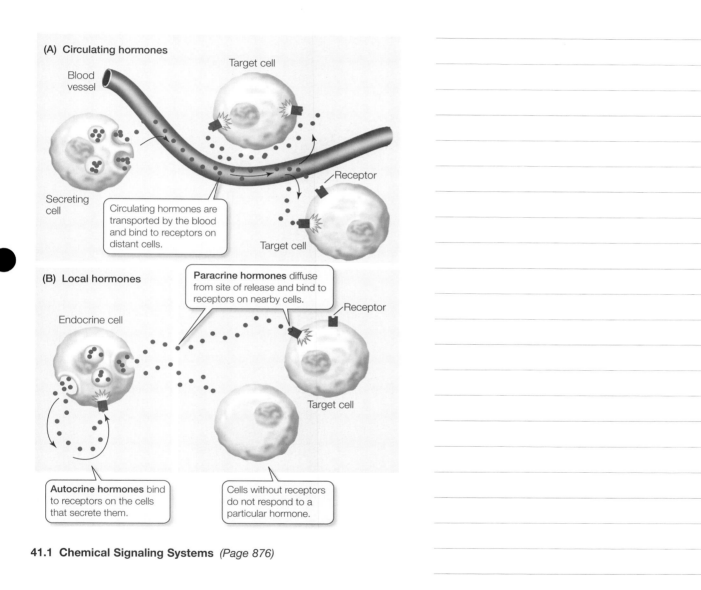

(A) Circulating hormones

Blood vessel

Target cell

Secreting cell

Circulating hormones are transported by the blood and bind to receptors on distant cells.

Receptor

Target cell

(B) Local hormones

Paracrine hormones diffuse from site of release and bind to receptors on nearby cells.

Receptor

Endocrine cell

Target cell

Autocrine hormones bind to receptors on the cells that secrete them.

Cells without receptors do not respond to a particular hormone.

41.1 Chemical Signaling Systems *(Page 876)*

EXPERIMENT

HYPOTHESIS: Some substance diffusing from the head of *Rhodnius prolixus* (a blood-sucking bug) controls molting.

Experiment 1

METHOD   Decapitate juvenile bugs at different times after a blood meal.

Juvenile bug (third instar)

Decapitation 1 hour after blood meal   Decapitation 1 week after blood meal

RESULTS

Does not molt (remains a juvenile)   Molts into an adult

CONCLUSION: Whether a decapitated *Rhodnius* will molt depends on the interval between a blood meal and the decapitation, which supports the idea that a substance must pass from head to body.

Experiment 2

METHOD   Decapitate juvenile bugs at different times after a blood meal.

Decapitation 1 hour after blood meal   Decapitation 1 week after blood meal

Join bugs with glass tube

RESULTS

Both bugs molt into adults

CONCLUSION: A diffusible substance is necessary for molting.

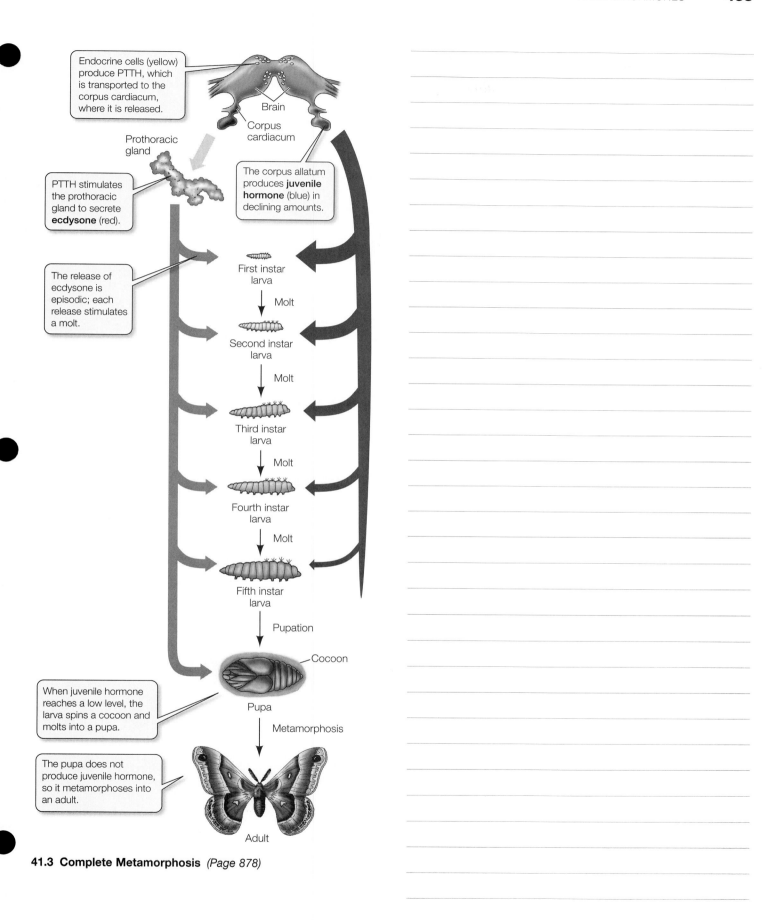

41.3 Complete Metamorphosis *(Page 878)*

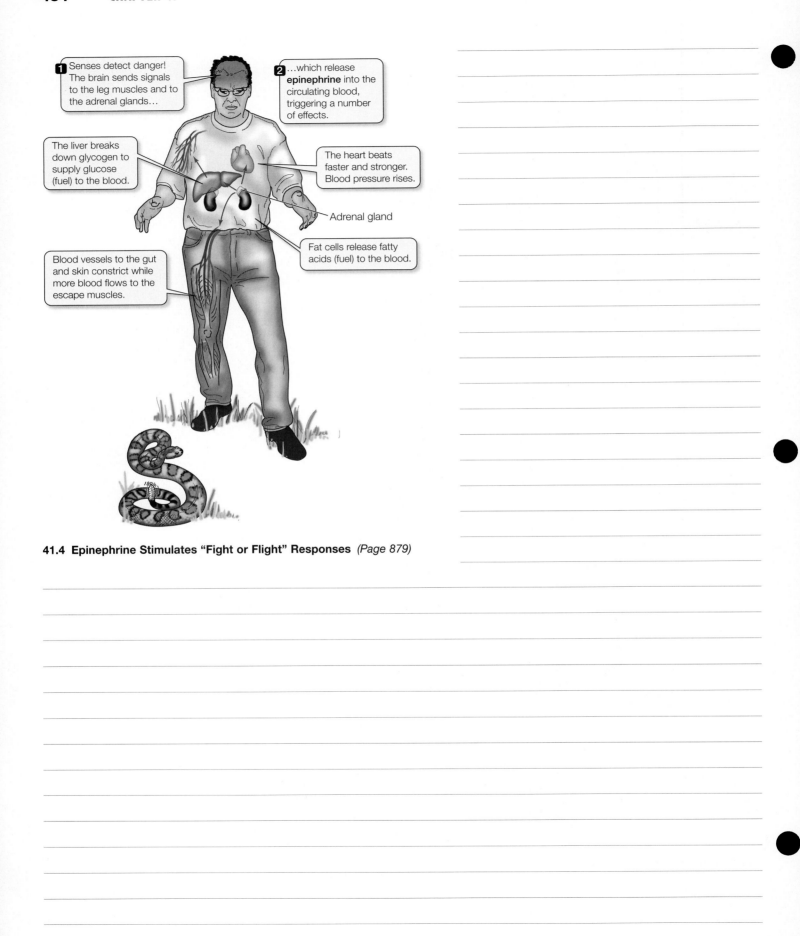

1 Senses detect danger! The brain sends signals to the leg muscles and to the adrenal glands…

2 …which release **epinephrine** into the circulating blood, triggering a number of effects.

The liver breaks down glycogen to supply glucose (fuel) to the blood.

The heart beats faster and stronger. Blood pressure rises.

Adrenal gland

Fat cells release fatty acids (fuel) to the blood.

Blood vessels to the gut and skin constrict while more blood flows to the escape muscles.

41.4 Epinephrine Stimulates "Fight or Flight" Responses *(Page 879)*

Hypothalamus (see Figure 41.6)
Release and release-inhibiting hormones control the anterior pituitary
ADH and *oxytocin* are transported to and released from the posterior pituitary

Anterior pituitary (see Figure 41.7)
Thyroid stimulating hormone (TSH): activates the thyroid gland
Follicle stimulating hormone (FSH): in females, stimulates maturation of ovarian follicles; in males, stimulates spermatogenesis
Luteinizing hormone (LH): in females, triggers ovulation and ovarian production of estrogens and progesterone; in males, stimulates production of testosterone
Adrenocorticotropic hormone (ACTH): stimulates adrenal cortex to secrete cortisol
Growth hormone: stimulates protein synthesis and growth
Prolactin: stimulates milk production

Posterior pituitary (see Figure 41.6)
Receives and releases two hypothalamic hormones:
Oxytocin: stimulates contraction of uterus, stimulates flow of milk.
Antidiuretic hormone (ADH): promotes water conservation by kidneys

Thymus gland (disappears in adults)
Thymosin: activates immune system T cells

Pancreas (islets of Langerhans)
Insulin: stimulates cells to take up and use glucose
Glucagon: stimulates liver to release glucose
Somatostatin: slows digestive tract functions including release of insulin and glucagon

Pineal gland
Melatonin: helps to regulate circadian rhythms

Thyroid gland (see Figures 41.9 and 41.10)
Thyroxine (T_3 and T_4): stimulates cell metabolism
Calcitonin: stimulates incorporation of calcium into bone

Parathyroid glands (on posterior surface of thyroid; see Figure 41.10)
Parathormone (PTH): stimulates release of calcium from bone

Adrenal gland (see Figure 41.11)
Cortex
Cortisol: mediates metabolic responses to stress
Aldosterone: involved in salt and water balance

Medulla
Epinephrine (adrenaline) and *norepinephrine* (noradrenaline): stimulate immediate fight or flight reactions

Gonads (see Chapter 42)
Ovaries (female)
Estrogens: development and maintenance of female sexual characteristics
Progesterone: supports pregnancy

Testes (male)
Testosterone: development and maintenance of male sexual characteristics

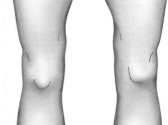

Other organs include cells that produce and secrete hormones

Organ	Hormone
Adipose tissue	Leptin
Heart	Atrial natriuretic peptide
Kidney	Erythropoietin
Stomach	Gastrin
Intestine	Secretin, cholecystokinin
Skin	Vitamin D (cholecalciferol)

41.5 The Endocrine System of Humans *(Page 880)*

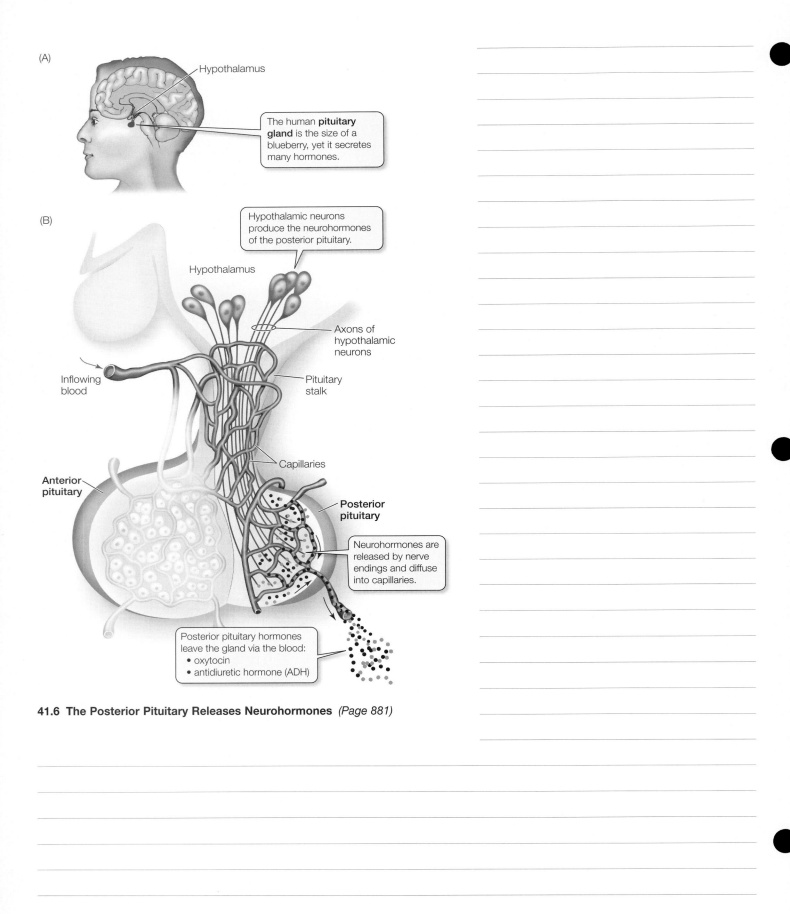

(A)

Hypothalamus

The human **pituitary gland** is the size of a blueberry, yet it secretes many hormones.

(B)

Hypothalamic neurons produce the neurohormones of the posterior pituitary.

Hypothalamus

Axons of hypothalamic neurons

Inflowing blood

Pituitary stalk

Capillaries

Anterior pituitary

Posterior pituitary

Neurohormones are released by nerve endings and diffuse into capillaries.

Posterior pituitary hormones leave the gland via the blood:
• oxytocin
• antidiuretic hormone (ADH)

41.6 The Posterior Pituitary Releases Neurohormones *(Page 881)*

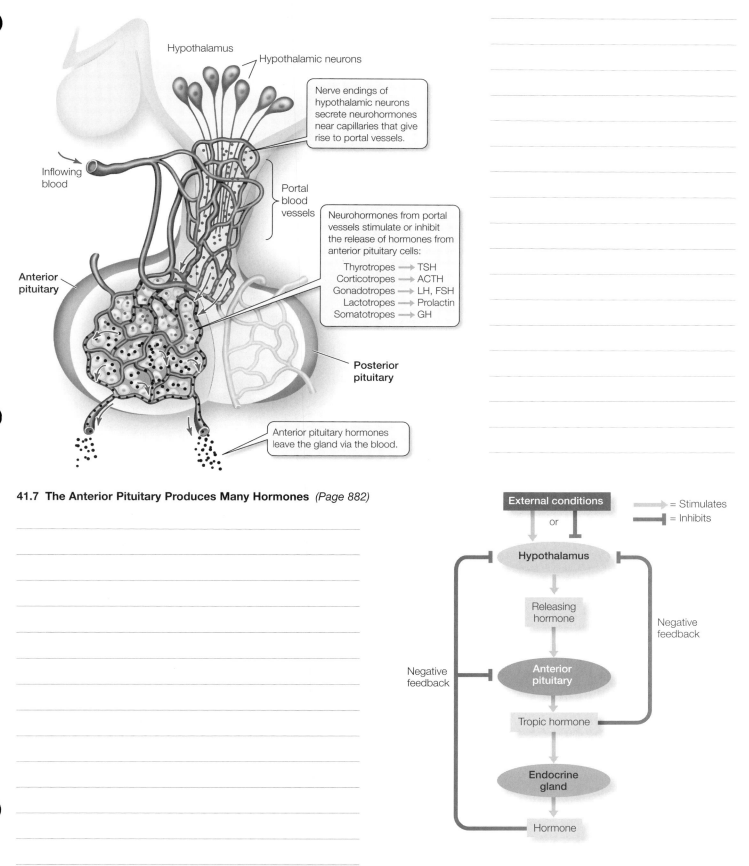

Hypothalamus

Hypothalamic neurons

Nerve endings of hypothalamic neurons secrete neurohormones near capillaries that give rise to portal vessels.

Inflowing blood

Portal blood vessels

Neurohormones from portal vessels stimulate or inhibit the release of hormones from anterior pituitary cells:

Thyrotropes ⟶ TSH
Corticotropes ⟶ ACTH
Gonadotropes ⟶ LH, FSH
Lactotropes ⟶ Prolactin
Somatotropes ⟶ GH

Anterior pituitary

Posterior pituitary

Anterior pituitary hormones leave the gland via the blood.

41.7 The Anterior Pituitary Produces Many Hormones *(Page 882)*

External conditions

⟶ = Stimulates
⊣ = Inhibits

or

Hypothalamus

Releasing hormone

Negative feedback

Anterior pituitary

Negative feedback

Tropic hormone

Endocrine gland

Hormone

41.8 Multiple Feedback Loops Control Hormone Secretion *(Page 883)*

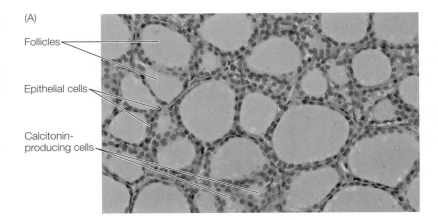

Thyroxine (T$_4$)

Page 883 In-Text Art

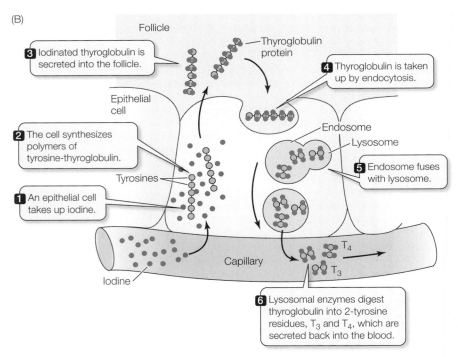

(A)

Follicles

Epithelial cells

Calcitonin-producing cells

(B)

Follicle

3 Iodinated thyroglobulin is secreted into the follicle.

Thyroglobulin protein

4 Thyroglobulin is taken up by endocytosis.

Epithelial cell

Endosome

Lysosome

2 The cell synthesizes polymers of tyrosine-thyroglobulin.

5 Endosome fuses with lysosome.

Tyrosines

1 An epithelial cell takes up iodine.

T$_4$

Capillary

T$_3$

Iodine

6 Lysosomal enzymes digest thyroglobulin into 2-tyrosine residues, T$_3$ and T$_4$, which are secreted back into the blood.

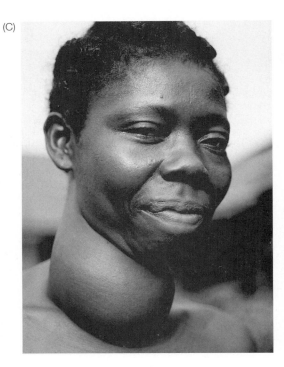

(C)

41.9 The Thyroid Gland Consists of Many Follicles *(Page 884)*

Triiodothyronine (T$_3$)

Page 884 In-Text Art

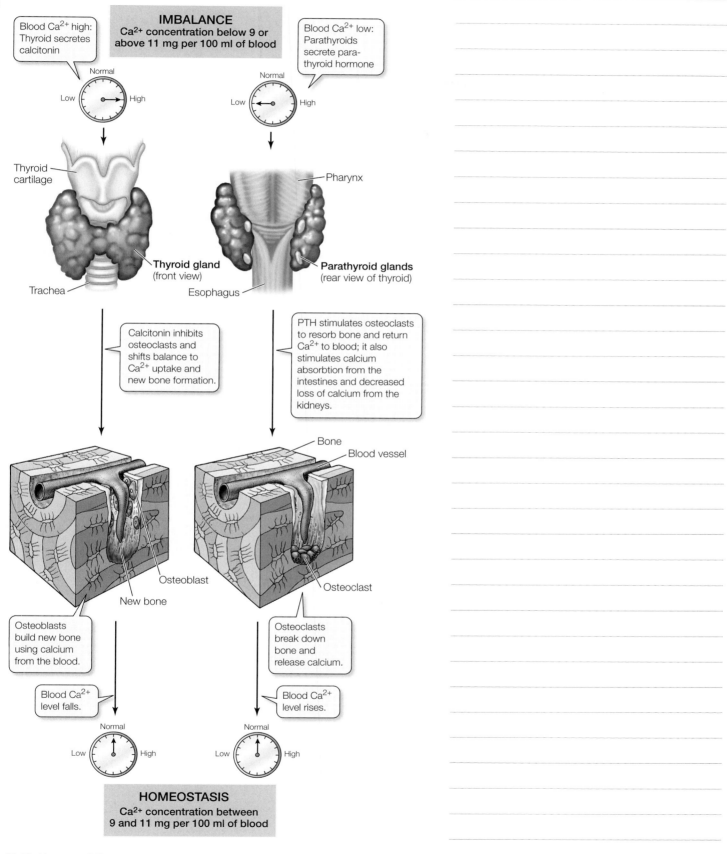

IMBALANCE
Ca²⁺ concentration below 9 or above 11 mg per 100 ml of blood

Blood Ca²⁺ high: Thyroid secretes calcitonin

Blood Ca²⁺ low: Parathyroids secrete para-thyroid hormone

Normal
Low — High

Normal
Low — High

Thyroid cartilage

Pharynx

Thyroid gland
(front view)

Parathyroid glands
(rear view of thyroid)

Trachea

Esophagus

Calcitonin inhibits osteoclasts and shifts balance to Ca²⁺ uptake and new bone formation.

PTH stimulates osteoclasts to resorb bone and return Ca²⁺ to blood; it also stimulates calcium absorbtion from the intestines and decreased loss of calcium from the kidneys.

Bone
Blood vessel

Osteoblast

New bone

Osteoclast

Osteoblasts build new bone using calcium from the blood.

Osteoclasts break down bone and release calcium.

Blood Ca²⁺ level falls.

Blood Ca²⁺ level rises.

Normal
Low — High

Normal
Low — High

HOMEOSTASIS
Ca²⁺ concentration between 9 and 11 mg per 100 ml of blood

41.10 Hormonal Regulation of Calcium (Page 886)

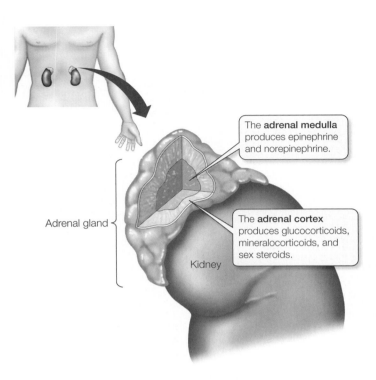

The **adrenal medulla** produces epinephrine and norepinephrine.

Adrenal gland

The **adrenal cortex** produces glucocorticoids, mineralocorticoids, and sex steroids.

Kidney

41.11 The Adrenal Gland Has an Outer and an Inner Portion
(Page 888)

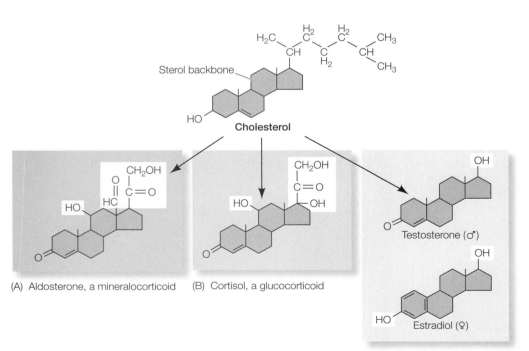

Sterol backbone

Cholesterol

(A) Aldosterone, a mineralocorticoid

(B) Cortisol, a glucocorticoid

Testosterone (♂)

Estradiol (♀)

(C) Sex steroids

41.12 The Corticosteroid Hormones are Built from Cholesterol *(Page 888)*

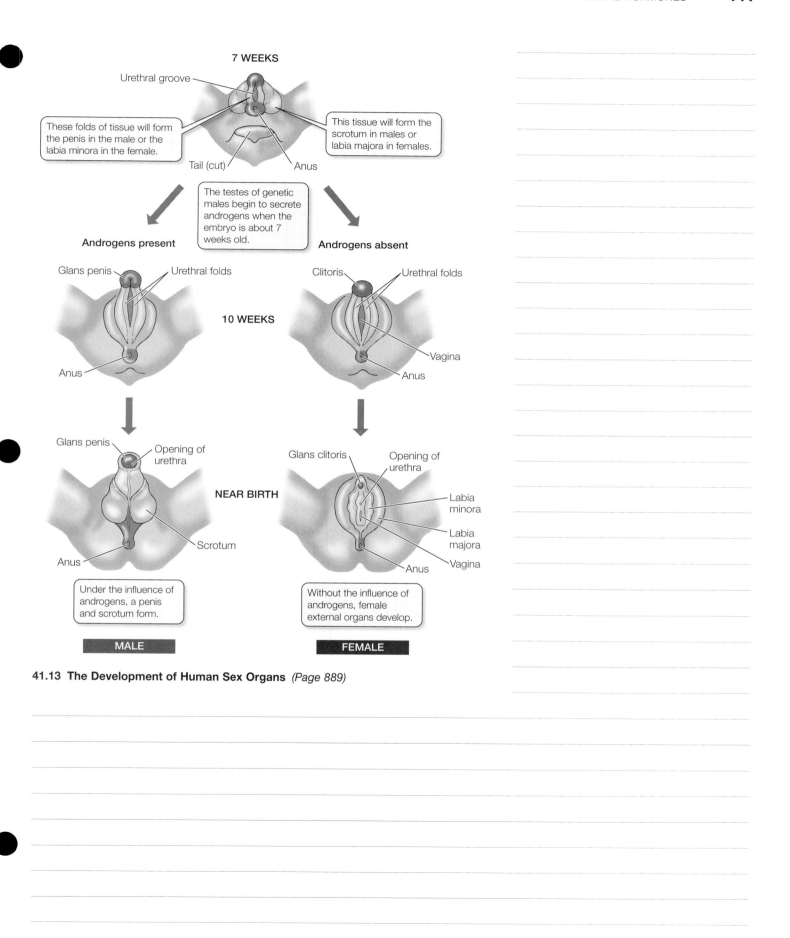

7 WEEKS

Urethral groove

These folds of tissue will form the penis in the male or the labia minora in the female.

This tissue will form the scrotum in males or labia majora in females.

Tail (cut)   Anus

The testes of genetic males begin to secrete androgens when the embryo is about 7 weeks old.

Androgens present   **Androgens absent**

Glans penis   Urethral folds   Clitoris   Urethral folds

10 WEEKS

Anus   Vagina

Anus

Glans penis   Opening of urethra   Glans clitoris   Opening of urethra

NEAR BIRTH

Labia minora

Anus   Scrotum   Labia majora

Vagina

Anus

Under the influence of androgens, a penis and scrotum form.

Without the influence of androgens, female external organs develop.

MALE   **FEMALE**

41.13 The Development of Human Sex Organs *(Page 889)*

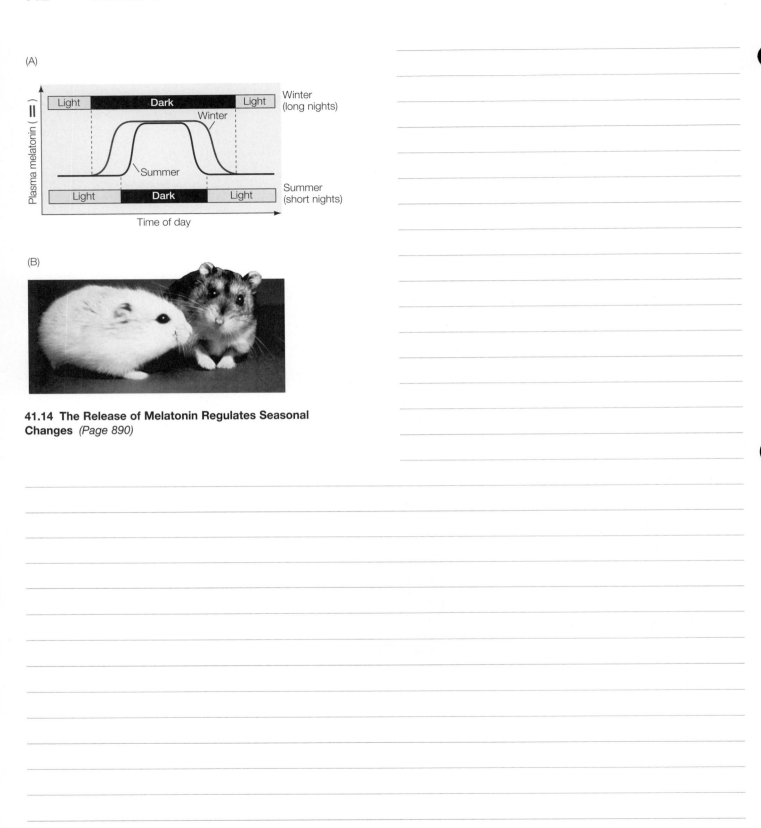

41.14 The Release of Melatonin Regulates Seasonal Changes *(Page 890)*

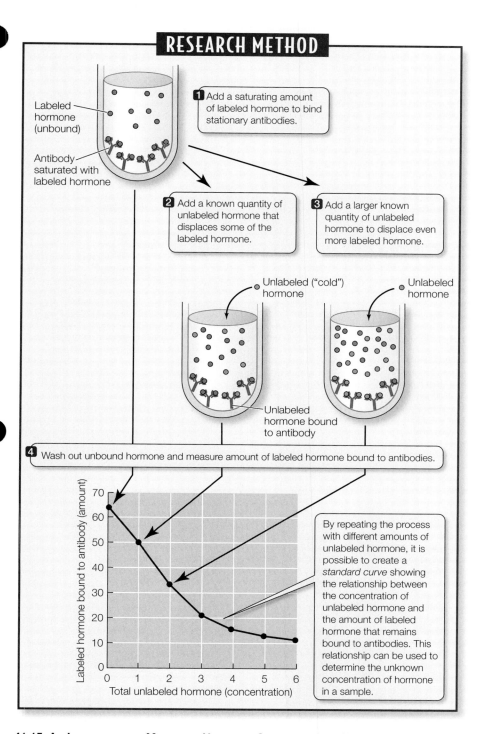

RESEARCH METHOD

Labeled hormone (unbound)

Antibody saturated with labeled hormone

1 Add a saturating amount of labeled hormone to bind stationary antibodies.

2 Add a known quantity of unlabeled hormone that displaces some of the labeled hormone.

3 Add a larger known quantity of unlabeled hormone to displace even more labeled hormone.

Unlabeled ("cold") hormone

Unlabeled hormone

Unlabeled hormone bound to antibody

4 Wash out unbound hormone and measure amount of labeled hormone bound to antibodies.

By repeating the process with different amounts of unlabeled hormone, it is possible to create a *standard curve* showing the relationship between the concentration of unlabeled hormone and the amount of labeled hormone that remains bound to antibodies. This relationship can be used to determine the unknown concentration of hormone in a sample.

Labeled hormone bound to antibody (amount)

Total unlabeled hormone (concentration)

41.15 An Immunoassay Measures Hormone Concentration *(Page 891)*

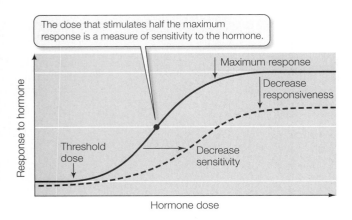

41.16 Dose–Response Curves Quantify Response to a Hormone
(Page 892)

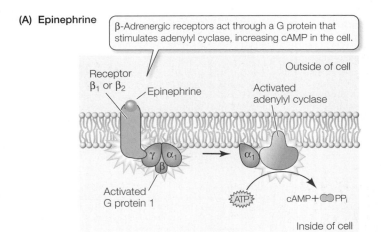

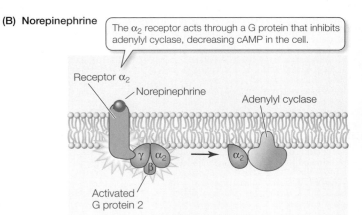

41.17 Some Hormones Can Activate a Variety of Signal Transduction Pathways *(Page 892)*

42 **Animal Reproduction**

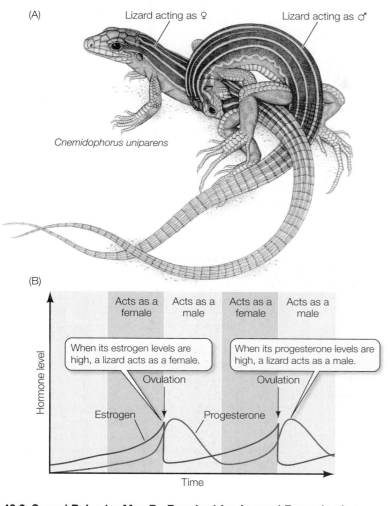

(A)

Lizard acting as ♀   Lizard acting as ♂

Cnemidophorus uniparens

(B)

| Acts as a female | Acts as a male | Acts as a female | Acts as a male |

When its estrogen levels are high, a lizard acts as a female.

When its progesterone levels are high, a lizard acts as a male.

Ovulation   Ovulation

Hormone level

Estrogen   Progesterone

Time

42.2 Sexual Behavior May Be Required for Asexual Reproduction
(Page 898)

(A) **SPERMATOGENESIS**

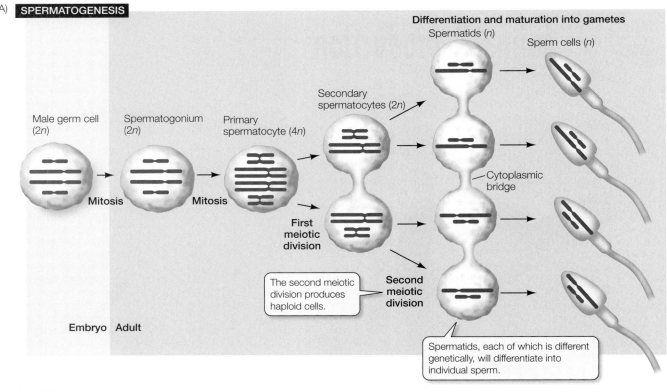

Differentiation and maturation into gametes
Spermatids (n)
Sperm cells (n)

Secondary spermatocytes (2n)

Male germ cell (2n)

Spermatogonium (2n)

Primary spermatocyte (4n)

Mitosis

Mitosis

First meiotic division

Cytoplasmic bridge

The second meiotic division produces haploid cells.

Second meiotic division

Embryo Adult

Spermatids, each of which is different genetically, will differentiate into individual sperm.

(B) **OOGENESIS**

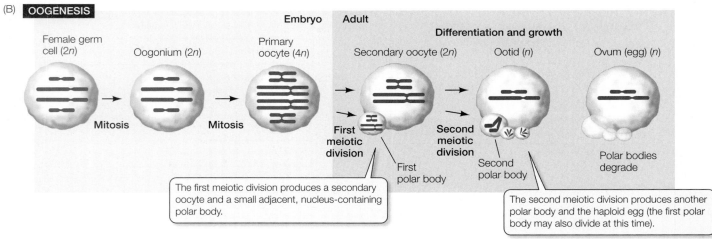

Embryo Adult

Differentiation and growth

Female germ cell (2n)

Oogonium (2n)

Primary oocyte (4n)

Secondary oocyte (2n)

Ootid (n)

Ovum (egg) (n)

Mitosis

Mitosis

First meiotic division

Second meiotic division

First polar body

Second polar body

Polar bodies degrade

The first meiotic division produces a secondary oocyte and a small adjacent, nucleus-containing polar body.

The second meiotic division produces another polar body and the haploid egg (the first polar body may also divide at this time).

42.3 Gametogenesis *(Page 900)*

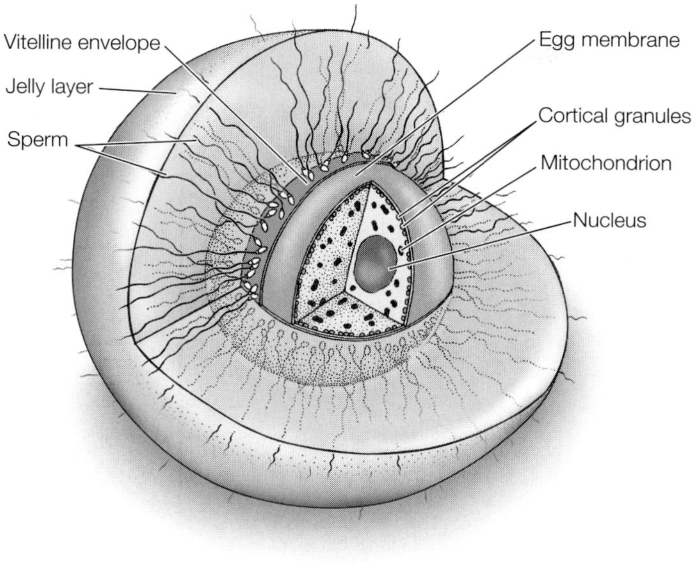

Vitelline envelope
Jelly layer
Sperm
Egg membrane
Cortical granules
Mitochondrion
Nucleus

42.4 The Sea Urchin Egg *(Page 901)*

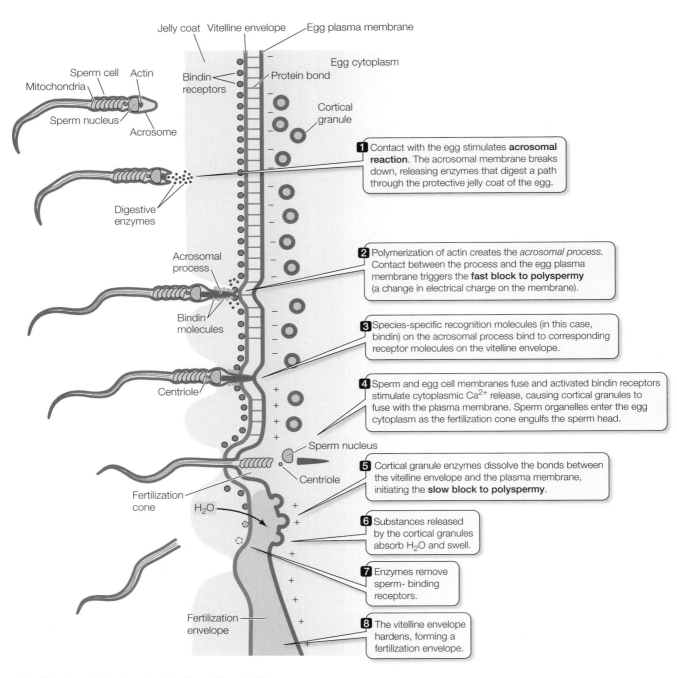

Jelly coat   Vitelline envelope   Egg plasma membrane

Egg cytoplasm

Sperm cell   Actin

Mitochondria

Bindin receptors   Protein bond

Cortical granule

Sperm nucleus

Acrosome

Digestive enzymes

1 Contact with the egg stimulates **acrosomal reaction**. The acrosomal membrane breaks down, releasing enzymes that digest a path through the protective jelly coat of the egg.

Acrosomal process

2 Polymerization of actin creates the *acrosomal process*. Contact between the process and the egg plasma membrane triggers the **fast block to polyspermy** (a change in electrical charge on the membrane).

Bindin molecules

3 Species-specific recognition molecules (in this case, bindin) on the acrosomal process bind to corresponding receptor molecules on the vitelline envelope.

Centriole

4 Sperm and egg cell membranes fuse and activated bindin receptors stimulate cytoplasmic Ca^{2+} release, causing cortical granules to fuse with the plasma membrane. Sperm organelles enter the egg cytoplasm as the fertilization cone engulfs the sperm head.

Sperm nucleus

Centriole

5 Cortical granule enzymes dissolve the bonds between the vitelline envelope and the plasma membrane, initiating the **slow block to polyspermy**.

Fertilization cone

H_2O

6 Substances released by the cortical granules absorb H_2O and swell.

7 Enzymes remove sperm- binding receptors.

Fertilization envelope

8 The vitelline envelope hardens, forming a fertilization envelope.

42.5 Fertilization of the Sea Urchin Egg *(Page 902)*

(A) Posterior view

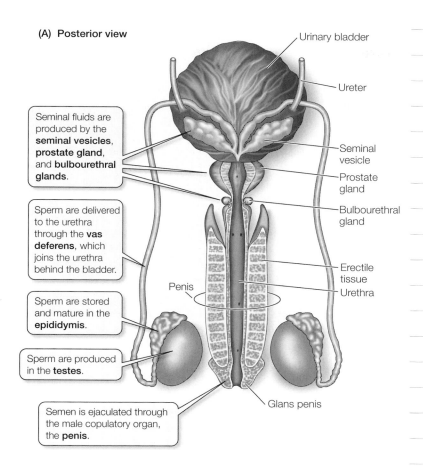

Urinary bladder

Ureter

Seminal fluids are produced by the **seminal vesicles**, **prostate gland**, and **bulbourethral glands**.

Seminal vesicle

Prostate gland

Sperm are delivered to the urethra through the **vas deferens**, which joins the urethra behind the bladder.

Bulbourethral gland

Penis

Erectile tissue

Urethra

Sperm are stored and mature in the **epididymis**.

Sperm are produced in the **testes**.

Glans penis

Semen is ejaculated through the male copulatory organ, the **penis**.

(B) Side view

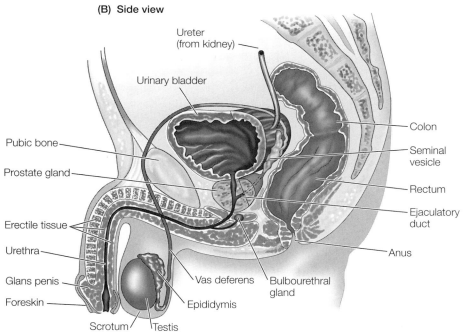

Ureter (from kidney)

Urinary bladder

Colon

Pubic bone

Seminal vesicle

Prostate gland

Rectum

Ejaculatory duct

Erectile tissue

Urethra

Anus

Glans penis

Vas deferens

Bulbourethral gland

Foreskin

Epididymis

Scrotum

Testis

42.9 The Reproductive Tract of the Human Male *(Page 906)*

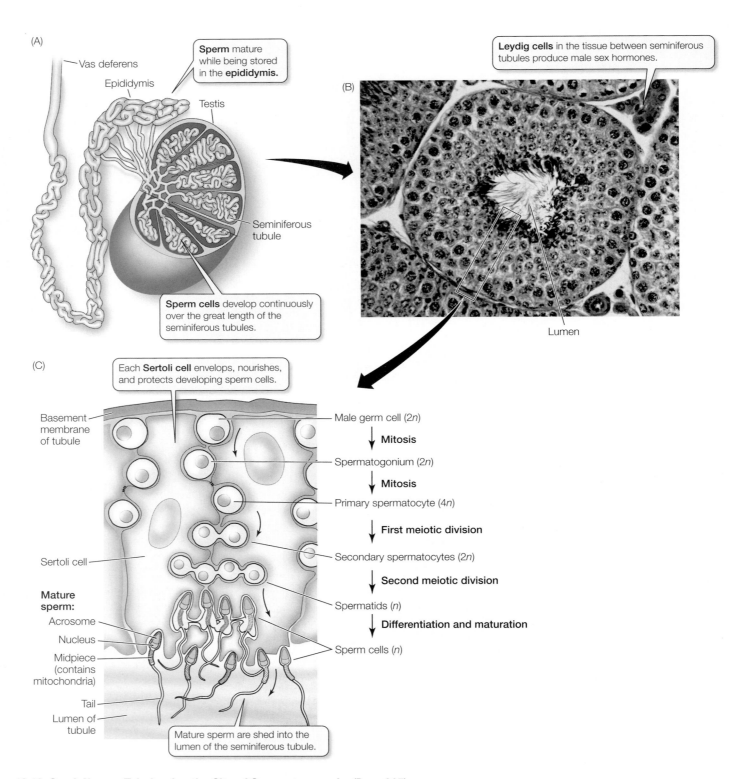

(A)

Vas deferens

Epididymis

Testis

Sperm mature while being stored in the **epididymis.**

Seminiferous tubule

Sperm cells develop continuously over the great length of the seminiferous tubules.

(B)

Leydig cells in the tissue between seminiferous tubules produce male sex hormones.

Lumen

(C)

Each **Sertoli cell** envelops, nourishes, and protects developing sperm cells.

Basement membrane of tubule

Sertoli cell

Mature sperm:

Acrosome

Nucleus

Midpiece (contains mitochondria)

Tail

Lumen of tubule

Mature sperm are shed into the lumen of the seminiferous tubule.

Male germ cell (2n)

↓ **Mitosis**

Spermatogonium (2n)

↓ **Mitosis**

Primary spermatocyte (4n)

↓ **First meiotic division**

Secondary spermatocytes (2n)

↓ **Second meiotic division**

Spermatids (n)

↓ **Differentiation and maturation**

Sperm cells (n)

42.10 Seminiferous Tubules Are the Site of Spermatogenesis *(Page 907)*

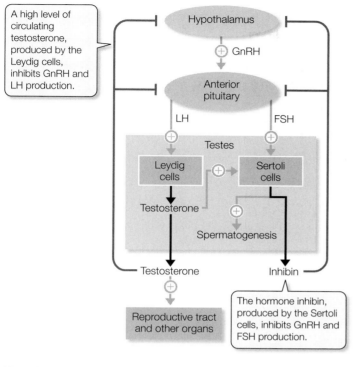

A high level of circulating testosterone, produced by the Leydig cells, inhibits GnRH and LH production.

The hormone inhibin, produced by the Sertoli cells, inhibits GnRH and FSH production.

42.11 Hormones Control the Male Reproductive System *(Page 908)*

(A) Frontal view

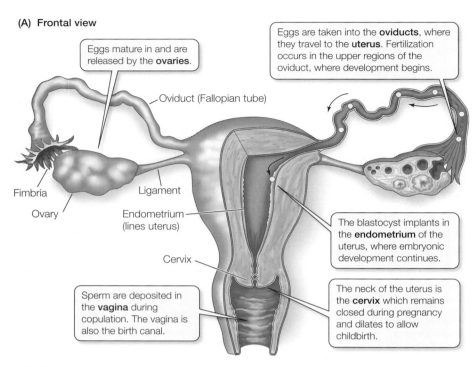

Eggs mature in and are released by the **ovaries**.

Eggs are taken into the **oviducts**, where they travel to the **uterus**. Fertilization occurs in the upper regions of the oviduct, where development begins.

Oviduct (Fallopian tube)

Fimbria

Ligament

Ovary

Endometrium (lines uterus)

The blastocyst implants in the **endometrium** of the uterus, where embryonic development continues.

Cervix

Sperm are deposited in the **vagina** during copulation. The vagina is also the birth canal.

The neck of the uterus is the **cervix** which remains closed during pregnancy and dilates to allow childbirth.

(B) Side view

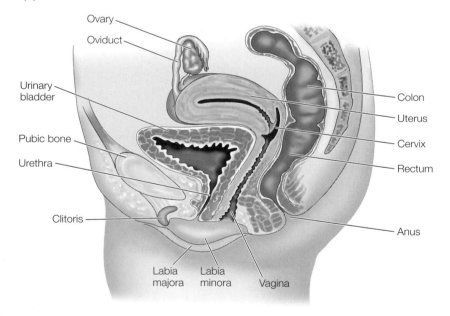

Ovary

Oviduct

Urinary bladder

Pubic bone

Urethra

Clitoris

Colon

Uterus

Cervix

Rectum

Anus

Labia majora

Labia minora

Vagina

42.12 The Reproductive Tract of the Human Female *(Page 909)*

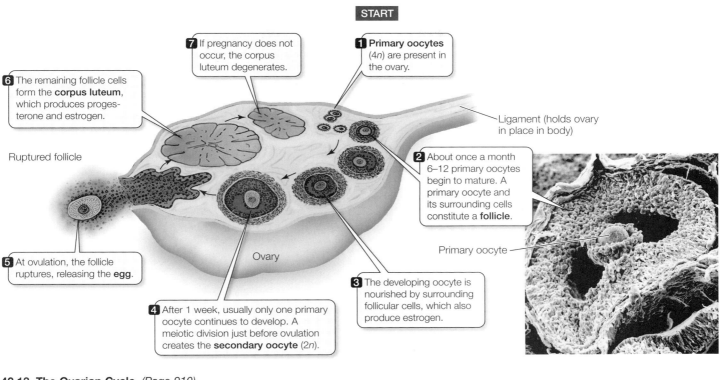

START

1 **Primary oocytes** (4n) are present in the ovary.

2 About once a month 6–12 primary oocytes begin to mature. A primary oocyte and its surrounding cells constitute a **follicle**.

3 The developing oocyte is nourished by surrounding follicular cells, which also produce estrogen.

4 After 1 week, usually only one primary oocyte continues to develop. A meiotic division just before ovulation creates the **secondary oocyte** (2n).

5 At ovulation, the follicle ruptures, releasing the **egg**.

6 The remaining follicle cells form the **corpus luteum**, which produces progesterone and estrogen.

7 If pregnancy does not occur, the corpus luteum degenerates.

Ruptured follicle

Ovary

Ligament (holds ovary in place in body)

Primary oocyte

42.13 The Ovarian Cycle *(Page 910)*

(A) Gonadotropins (from anterior pituitary)

FSH and LH are under control of GnRH from the hypothalamus and the ovarian hormones estrogen and progesterone.

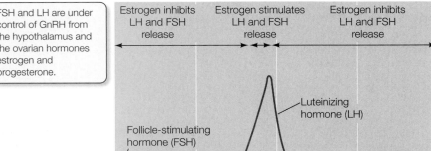

Estrogen inhibits LH and FSH release | Estrogen stimulates LH and FSH release | Estrogen inhibits LH and FSH release

Luteinizing hormone (LH)

Follicle-stimulating hormone (FSH)

(B) Events in ovary (ovarian cycle)

FSH stimulates the development of follicles; the LH surge causes ovulation and then the development of the corpus luteum.

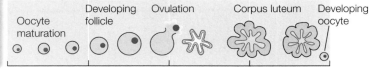

Oocyte maturation   Developing follicle   Ovulation   Corpus luteum   Developing oocyte

(C) Ovarian hormones and the uterine cycle

Estrogen and progesterone stimulate the development of the endometrium in preparation for pregnancy.

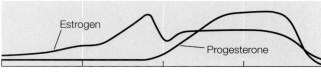

Estrogen

Progesterone

(D) Endometrium

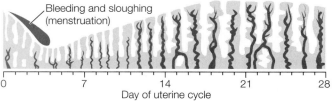

Bleeding and sloughing (menstruation)

0   7   14   21   28

Day of uterine cycle

42.14 The Ovarian and Uterine Cycles *(Page 911)*

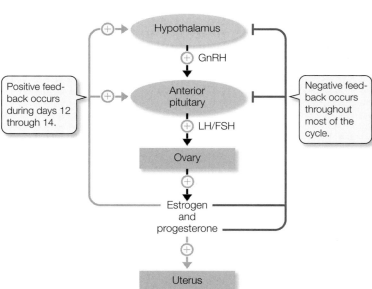

Positive feedback occurs during days 12 through 14.

Hypothalamus

⊕ GnRH

Anterior pituitary

⊕ LH/FSH

Ovary

⊕

Estrogen and progesterone

⊕

Uterus

Negative feedback occurs throughout most of the cycle.

42.15 Hormones Control the Ovarian and Uterine Cycles *(Page 912)*

(A)

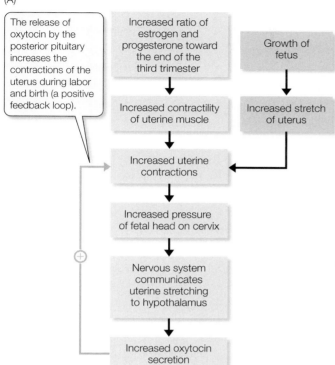

The release of oxytocin by the posterior pituitary increases the contractions of the uterus during labor and birth (a positive feedback loop).

Increased ratio of estrogen and progesterone toward the end of the third trimester

Growth of fetus

Increased contractility of uterine muscle

Increased stretch of uterus

⊕

Increased uterine contractions

Increased pressure of fetal head on cervix

Nervous system communicates uterine stretching to hypothalamus

Increased oxytocin secretion

(B)

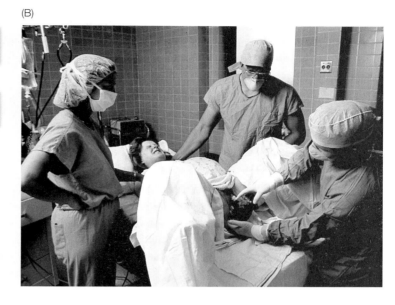

42.16 Control of Uterine Contractions and Childbirth *(Page 913)*

TABLE 42.1

Methods of Contraception

METHOD	MODE OF ACTION	FAILURE RATE[a]
Unprotected	No form of birth control	85
Douche	Supposedly flushes sperm from vagina	80
Periodic abstinence	Abstinence near time of ovulation	15–35
Coitus interruptus (withdrawal prior to ejaculation)	Prevents sperm from reaching egg	10–40
Vaginal jelly or foam	Kills sperm; blocks sperm movement	3–30
Diaphragm/jelly	Prevents sperm from entering uterus; kills sperm	3–25
Condom	Prevents sperm from entering vagina	3–20
Intrauterine device (IUD)	Prevents implantation of fertilized egg	0.5–6
RU-486	Prevents development of fertilized egg	0–15
Birth control pill	Prevents ovulation	0–3
Vasectomy	Prevents release of sperm	0–0.15
Tubal ligation	Prevents egg from entering uterus	0–0.05

[a]Number of pregnancies per 100 women per year.

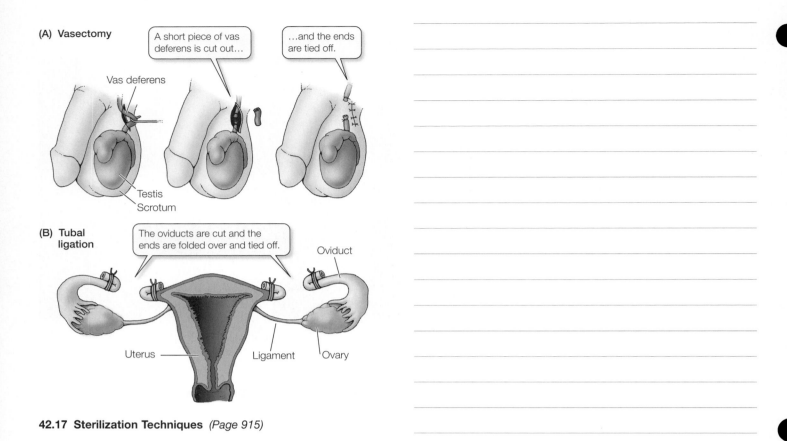

(A) Vasectomy

A short piece of vas deferens is cut out...

...and the ends are tied off.

Vas deferens

Testis
Scrotum

(B) Tubal ligation

The oviducts are cut and the ends are folded over and tied off.

Oviduct

Uterus   Ligament   Ovary

42.17 Sterilization Techniques *(Page 915)*

TABLE 42.2

Some Sexually Transmitted Diseases

DISEASE	INCIDENCE IN UNITED STATES	SYMPTOMS
Syphilis	80,000 new cases/yr	Primary stage (weeks): skin lesion (chancre) at site of infection
		Secondary stage (months): skin rash and flu-like symptoms; may be followed by a latent period
		Tertiary stage (years): deterioration of the cardiovascular and central nervous systems; death
Gonorrhea	800,000 new cases/yr	Pus-filled discharge from penis or vagina; burning urination. Infection can also start in throat or rectum
Chlamydia	>4,000,000 new cases/yr	Symptoms similar to gonorrhea, although often there are no obvious symptoms. Can lead to pelvic inflammatory disease in females
Genital herpes	500,000 new cases/yr	Small blisters that can cause itching or burning sensations are accompanied by inflammation and by secondary infections
Genital warts	10% of adults infected	Small growths on genital tissues. Increases risk of cervical cancer in women
Hepatitis B	5–20% of population	Fatigue, fever, nausea, loss of appetite, jaundice, abdominal pain, muscle and joint pain. Can lead to destruction of liver or liver cancer
HIV/AIDS	Approximately 900,000 cases[a]	Failure of the immune system (see Section 18.6)

[a]HIV/AIDS is widespread in other parts of the world, most notably in the southern part of the African continent. The infection is spreading rapidly in Southeast Asia and India. Estimated number of people infected with HIV worldwide in 2006 was 40 million.

(Page 917)

CHAPTER 43 Animal Development: From Genes to Organisms

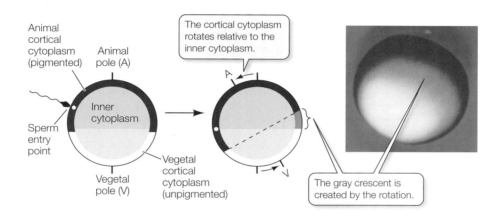

Animal cortical cytoplasm (pigmented)

Animal pole (A)

Inner cytoplasm

Sperm entry point

Vegetal pole (V)

Vegetal cortical cytoplasm (unpigmented)

The cortical cytoplasm rotates relative to the inner cytoplasm.

A

V

The gray crescent is created by the rotation.

43.1 The Gray Crescent *(Page 922)*

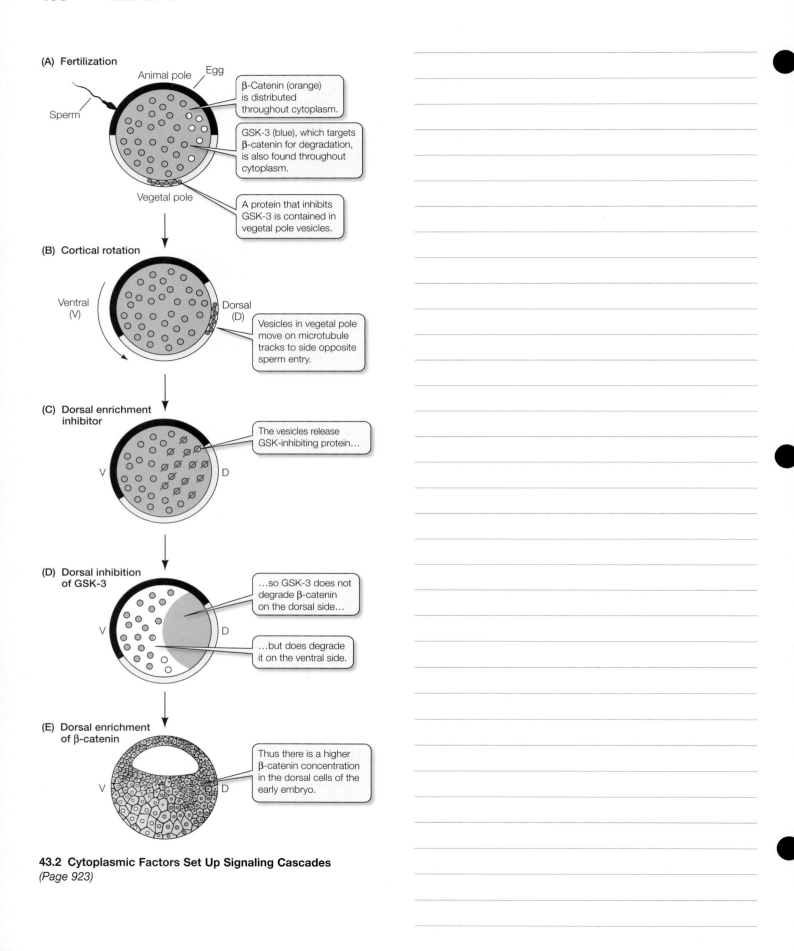

(A) Fertilization

Animal pole — Egg

Sperm

β-Catenin (orange) is distributed throughout cytoplasm.

GSK-3 (blue), which targets β-catenin for degradation, is also found throughout cytoplasm.

Vegetal pole

A protein that inhibits GSK-3 is contained in vegetal pole vesicles.

(B) Cortical rotation

Ventral (V)

Dorsal (D)

Vesicles in vegetal pole move on microtubule tracks to side opposite sperm entry.

(C) Dorsal enrichment inhibitor

V   D

The vesicles release GSK-inhibiting protein…

(D) Dorsal inhibition of GSK-3

V   D

…so GSK-3 does not degrade β-catenin on the dorsal side…

…but does degrade it on the ventral side.

(E) Dorsal enrichment of β-catenin

V   D

Thus there is a higher β-catenin concentration in the dorsal cells of the early embryo.

43.2 Cytoplasmic Factors Set Up Signaling Cascades
(Page 923)

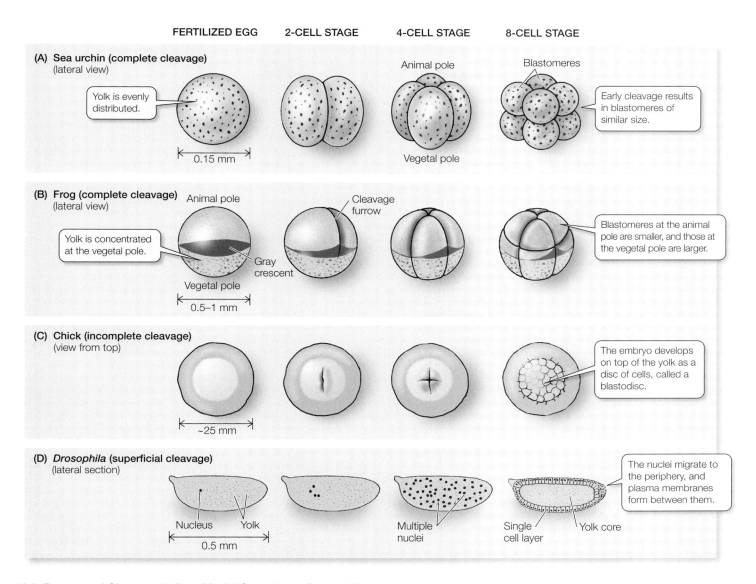

43.3 Patterns of Cleavage in Four Model Organisms *(Page 924)*

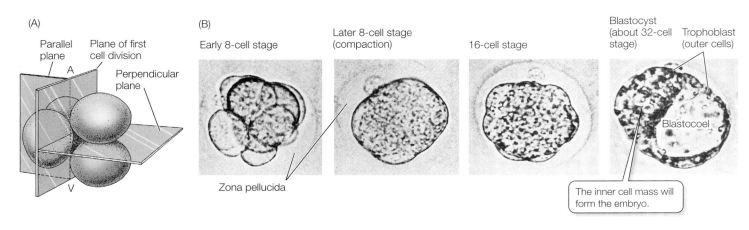

43.4 The Mammalian Zygote Becomes a Blastocyst *(Page 924)*

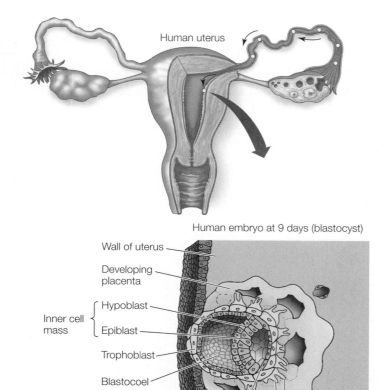

Human uterus

Human embryo at 9 days (blastocyst)

Wall of uterus

Developing placenta

Inner cell mass { Hypoblast

Epiblast

Trophoblast

Blastocoel

Endometrium

Amnion   Chorionic villi   Blood vessel

43.5 A Human Blastocyst at Implantation *(Page 925)*

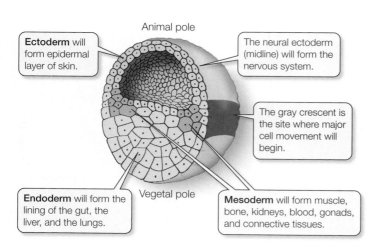

Animal pole

Ectoderm will form epidermal layer of skin.

The neural ectoderm (midline) will form the nervous system.

The gray crescent is the site where major cell movement will begin.

Endoderm will form the lining of the gut, the liver, and the lungs.

Vegetal pole

Mesoderm will form muscle, bone, kidneys, blood, gonads, and connective tissues.

43.6 Fate Map of a Frog Blastula *(Page 926)*

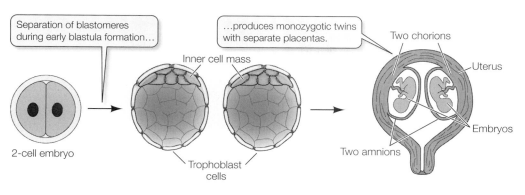

Separation of blastomeres during early blastula formation...

...produces monozygotic twins with separate placentas.

Inner cell mass

Two chorions

Uterus

Embryos

Two amnions

2-cell embryo

Trophoblast cells

43.7 Twinning in Humans *(Page 926)*

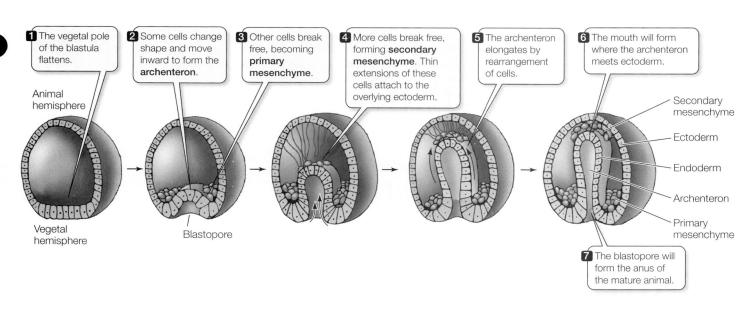

1 The vegetal pole of the blastula flattens.

2 Some cells change shape and move inward to form the **archenteron**.

3 Other cells break free, becoming **primary mesenchyme**.

4 More cells break free, forming **secondary mesenchyme**. Thin extensions of these cells attach to the overlying ectoderm.

5 The archenteron elongates by rearrangement of cells.

6 The mouth will form where the archenteron meets ectoderm.

Animal hemisphere

Vegetal hemisphere

Blastopore

Secondary mesenchyme

Ectoderm

Endoderm

Archenteron

Primary mesenchyme

7 The blastopore will form the anus of the mature animal.

43.8 Gastrulation in Sea Urchins *(Page 927)*

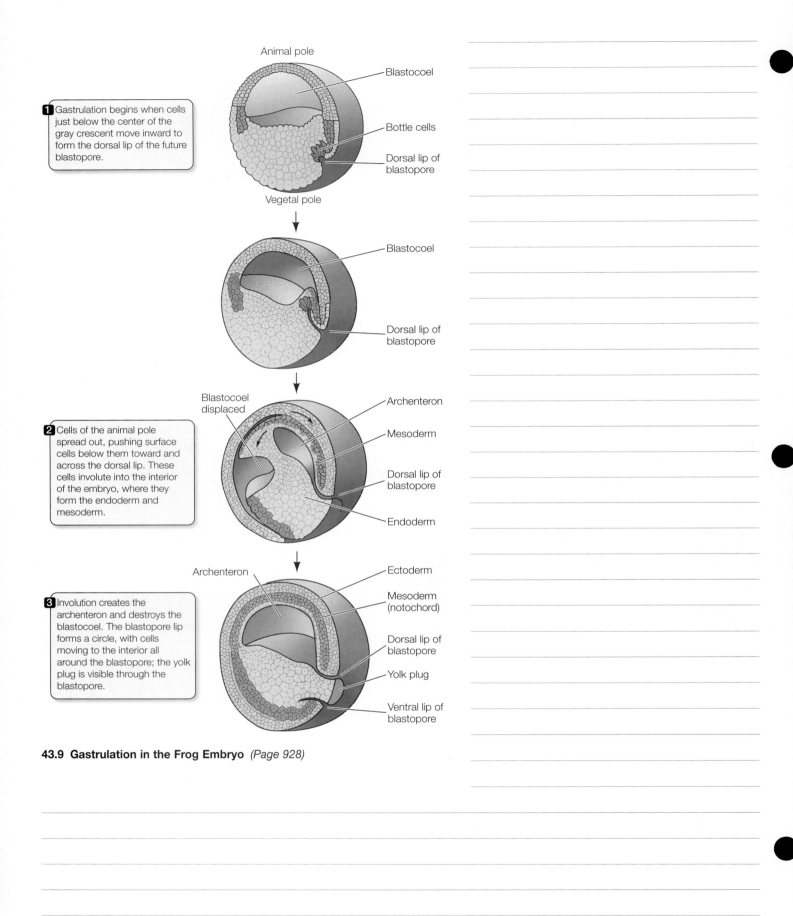

Animal pole

Blastocoel

1 Gastrulation begins when cells just below the center of the gray crescent move inward to form the dorsal lip of the future blastopore.

Bottle cells

Dorsal lip of blastopore

Vegetal pole

Blastocoel

Dorsal lip of blastopore

Blastocoel displaced

Archenteron

2 Cells of the animal pole spread out, pushing surface cells below them toward and across the dorsal lip. These cells involute into the interior of the embryo, where they form the endoderm and mesoderm.

Mesoderm

Dorsal lip of blastopore

Endoderm

Archenteron

Ectoderm

Mesoderm (notochord)

3 Involution creates the archenteron and destroys the blastocoel. The blastopore lip forms a circle, with cells moving to the interior all around the blastopore; the yolk plug is visible through the blastopore.

Dorsal lip of blastopore

Yolk plug

Ventral lip of blastopore

43.9 Gastrulation in the Frog Embryo *(Page 928)*

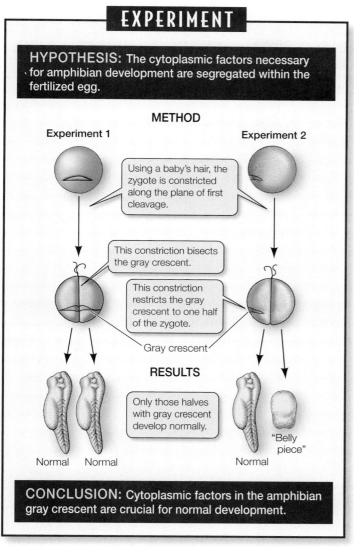

EXPERIMENT

HYPOTHESIS: The cytoplasmic factors necessary for amphibian development are segregated within the fertilized egg.

METHOD

Experiment 1

Experiment 2

Using a baby's hair, the zygote is constricted along the plane of first cleavage.

This constriction bisects the gray crescent.

This constriction restricts the gray crescent to one half of the zygote.

Gray crescent

RESULTS

Only those halves with gray crescent develop normally.

Normal   Normal

Normal

"Belly piece"

CONCLUSION: Cytoplasmic factors in the amphibian gray crescent are crucial for normal development.

43.10 Spemann's Experiment *(Page 929)*

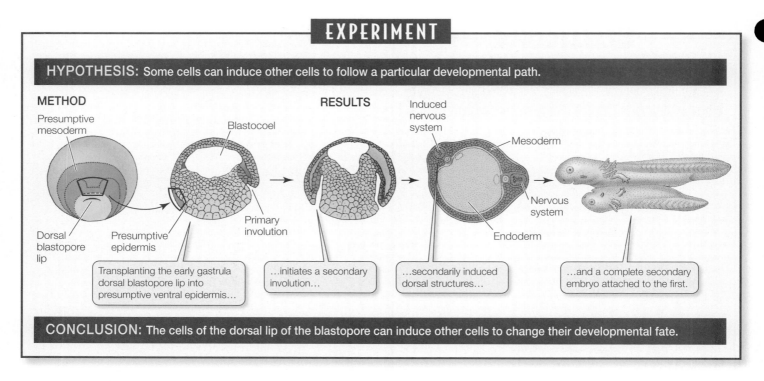

43.11 The Dorsal Lip Induces Embryonic Organization *(Page 930)*

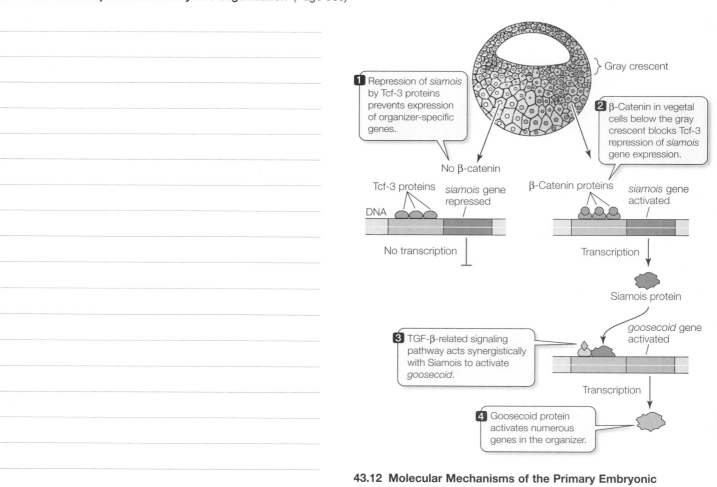

43.12 Molecular Mechanisms of the Primary Embryonic Organizer *(Page 930)*

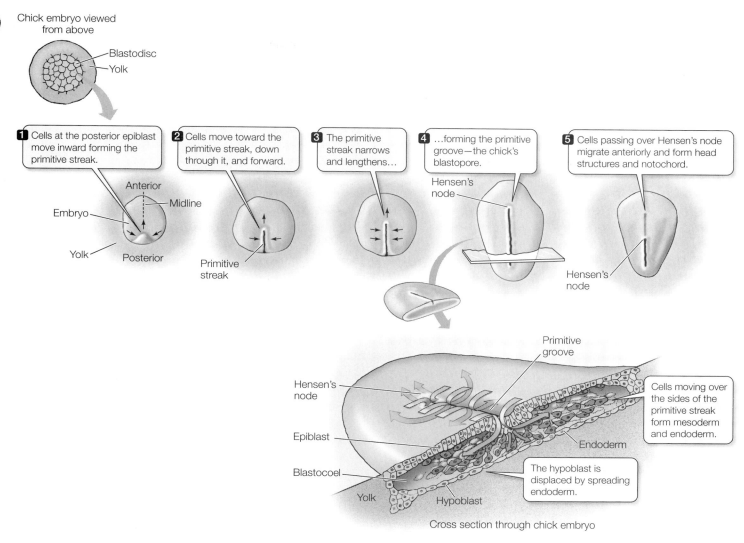

Chick embryo viewed
from above

Blastodisc
Yolk

1 Cells at the posterior epiblast move inward forming the primitive streak.

Anterior
Midline
Embryo
Yolk
Posterior

2 Cells move toward the primitive streak, down through it, and forward.

Primitive
streak

3 The primitive streak narrows and lengthens...

4 ...forming the primitive groove—the chick's blastopore.

Hensen's
node

5 Cells passing over Hensen's node migrate anteriorly and form head structures and notochord.

Hensen's
node

Primitive
groove

Hensen's
node

Epiblast

Blastocoel

Yolk   Hypoblast

Cells moving over the sides of the primitive streak form mesoderm and endoderm.

Endoderm

The hypoblast is displaced by spreading endoderm.

Cross section through chick embryo

43.13 Gastrulation in Birds *(Page 931)*

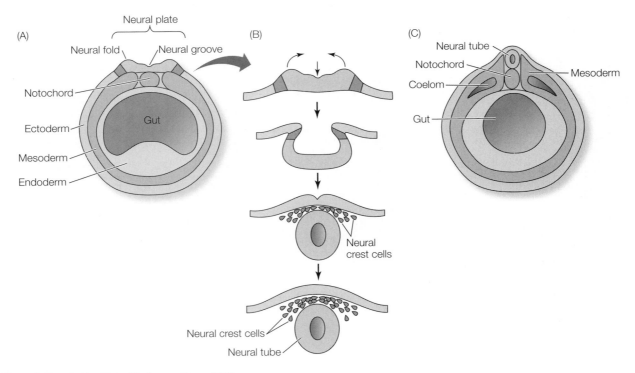

43.14 Neurulation in the Frog Embryo *(Page 933)*

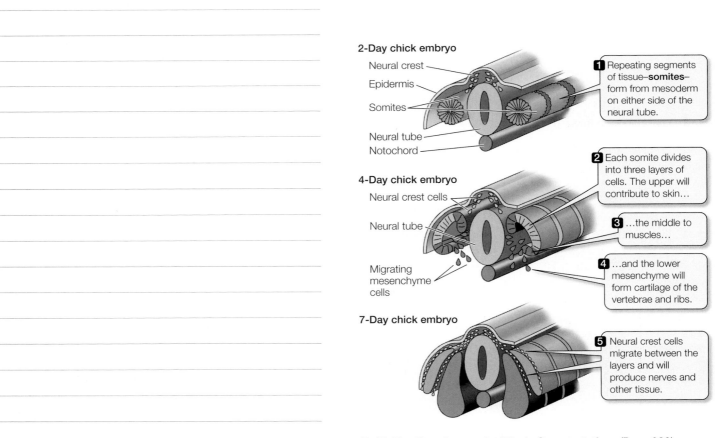

2-Day chick embryo

Neural crest
Epidermis
Somites
Neural tube
Notochord

1 Repeating segments of tissue—**somites**—form from mesoderm on either side of the neural tube.

4-Day chick embryo

Neural crest cells
Neural tube
Migrating mesenchyme cells

2 Each somite divides into three layers of cells. The upper will contribute to skin...

3 ...the middle to muscles...

4 ...and the lower mesenchyme will form cartilage of the vertebrae and ribs.

7-Day chick embryo

5 Neural crest cells migrate between the layers and will produce nerves and other tissue.

43.15 The Development of Body Segmentation *(Page 933)*

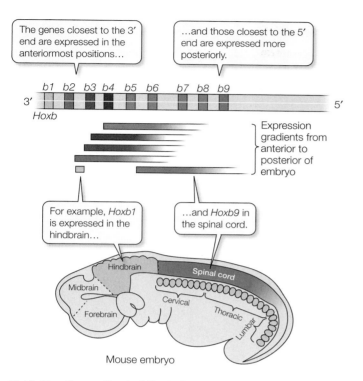

The genes closest to the 3' end are expressed in the anteriormost positions...

...and those closest to the 5' end are expressed more posteriorly.

b1 b2 b3 b4 b5 b6 b7 b8 b9

3'   5'

Hoxb

Expression gradients from anterior to posterior of embryo

For example, *Hoxb1* is expressed in the hindbrain...

...and *Hoxb9* in the spinal cord.

Hindbrain

Spinal cord

Midbrain

Forebrain

Cervical

Thoracic

Lumbar

Mouse embryo

43.16 Hox Genes Control Body Segmentation *(Page 934)*

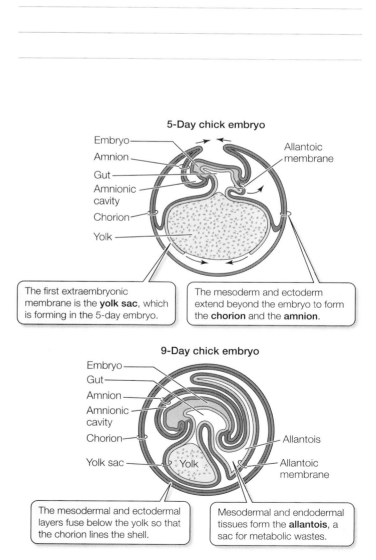

5-Day chick embryo

Embryo

Amnion

Gut

Amnionic cavity

Chorion

Yolk

Allantoic membrane

The first extraembryonic membrane is the **yolk sac**, which is forming in the 5-day embryo.

The mesoderm and ectoderm extend beyond the embryo to form the **chorion** and the **amnion**.

9-Day chick embryo

Embryo

Gut

Amnion

Amnionic cavity

Chorion

Yolk sac

Yolk

Allantois

Allantoic membrane

The mesodermal and ectodermal layers fuse below the yolk so that the chorion lines the shell.

Mesodermal and endodermal tissues form the **allantois**, a sac for metabolic wastes.

43.17 The Extraembryonic Membranes *(Page 935)*

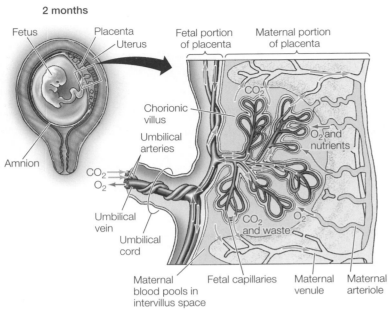

2 months

Fetus
Placenta
Uterus
Amnion

Fetal portion of placenta
Maternal portion of placenta

Chorionic villus
Umbilical arteries
CO_2
O_2
Umbilical vein
Umbilical cord

CO_2
O_2 and nutrients
CO_2 and waste
O_2

Maternal blood pools in intervillus space
Fetal capillaries
Maternal venule
Maternal arteriole

43.18 The Mammalian Placenta *(Page 935)*

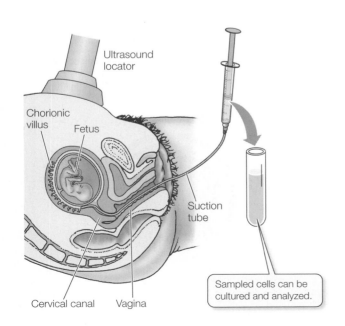

Ultrasound locator

Chorionic villus
Fetus

Suction tube

Sampled cells can be cultured and analyzed.

Cervical canal
Vagina

43.19 Chorionic Villus Sampling *(Page 936)*

CHAPTER 44 Neurons and Nervous Systems

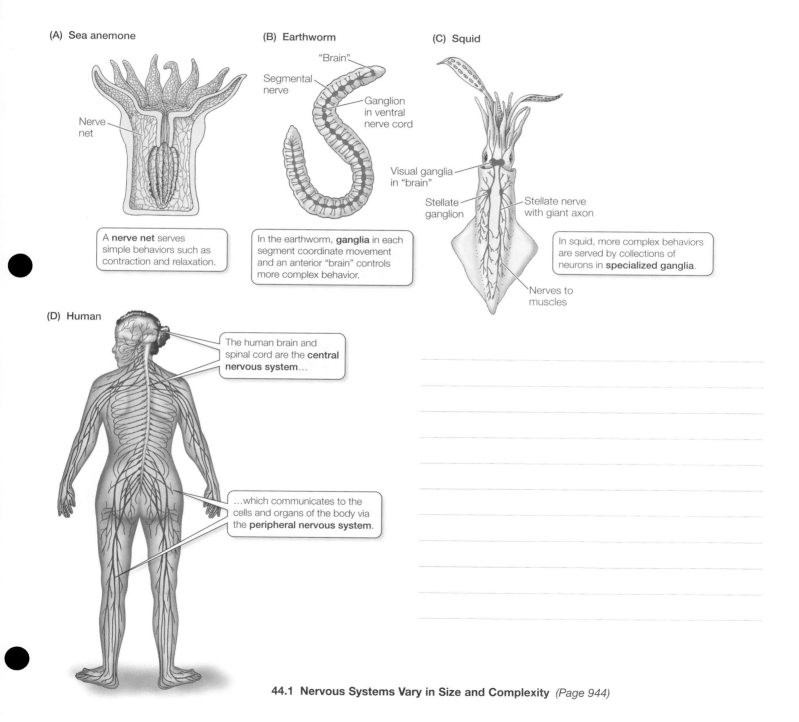

(A) Sea anemone

Nerve net

A **nerve net** serves simple behaviors such as contraction and relaxation.

(B) Earthworm

"Brain"

Segmental nerve

Ganglion in ventral nerve cord

In the earthworm, **ganglia** in each segment coordinate movement and an anterior "brain" controls more complex behavior.

(C) Squid

Visual ganglia in "brain"

Stellate ganglion

Stellate nerve with giant axon

Nerves to muscles

In squid, more complex behaviors are served by collections of neurons in **specialized ganglia**.

(D) Human

The human brain and spinal cord are the **central nervous system**...

...which communicates to the cells and organs of the body via the **peripheral nervous system**.

44.1 Nervous Systems Vary in Size and Complexity *(Page 944)*

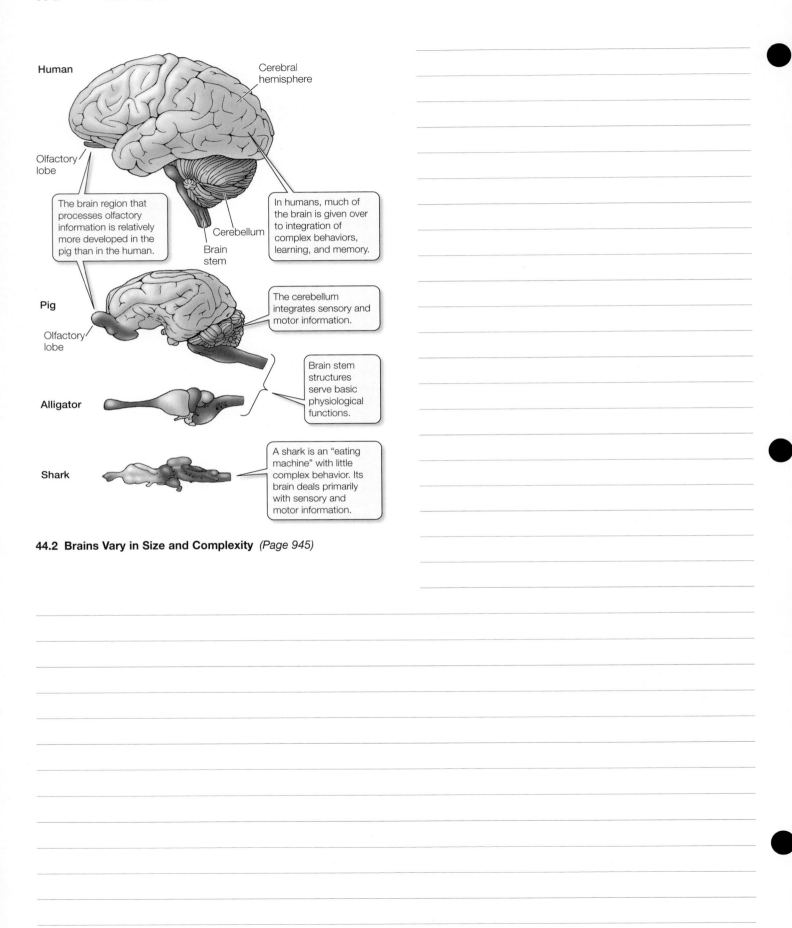

Human

Cerebral hemisphere

Olfactory lobe

The brain region that processes olfactory information is relatively more developed in the pig than in the human.

Cerebellum

Brain stem

In humans, much of the brain is given over to integration of complex behaviors, learning, and memory.

Pig

Olfactory lobe

The cerebellum integrates sensory and motor information.

Alligator

Brain stem structures serve basic physiological functions.

Shark

A shark is an "eating machine" with little complex behavior. Its brain deals primarily with sensory and motor information.

44.2 Brains Vary in Size and Complexity *(Page 945)*

(A) Generalized neuronal anatomy

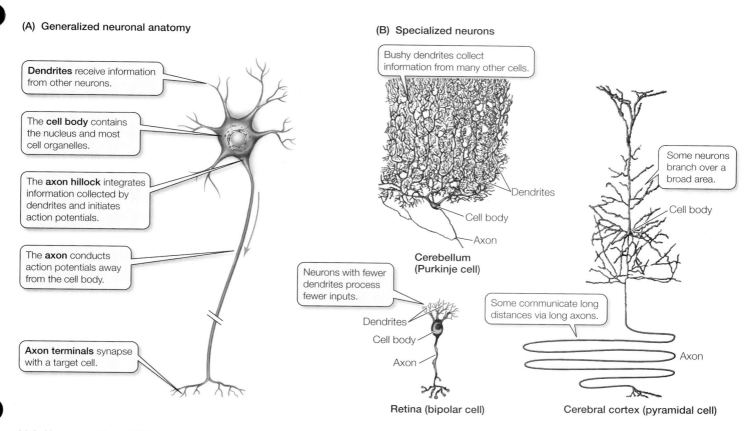

Dendrites receive information from other neurons.

The **cell body** contains the nucleus and most cell organelles.

The **axon hillock** integrates information collected by dendrites and initiates action potentials.

The **axon** conducts action potentials away from the cell body.

Axon terminals synapse with a target cell.

(B) Specialized neurons

Bushy dendrites collect information from many other cells.

Dendrites

Cell body

Axon

Cerebellum (Purkinje cell)

Neurons with fewer dendrites process fewer inputs.

Dendrites

Cell body

Axon

Retina (bipolar cell)

Some neurons branch over a broad area.

Cell body

Some communicate long distances via long axons.

Axon

Cerebral cortex (pyramidal cell)

44.3 Neurons *(Page 945)*

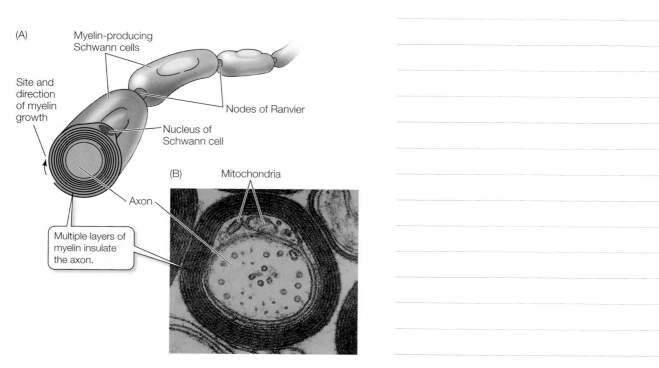

(A)

Myelin-producing Schwann cells

Site and direction of myelin growth

Nodes of Ranvier

Nucleus of Schwann cell

Axon

Multiple layers of myelin insulate the axon.

(B) Mitochondria

44.4 Wrapping Up an Axon *(Page 946)*

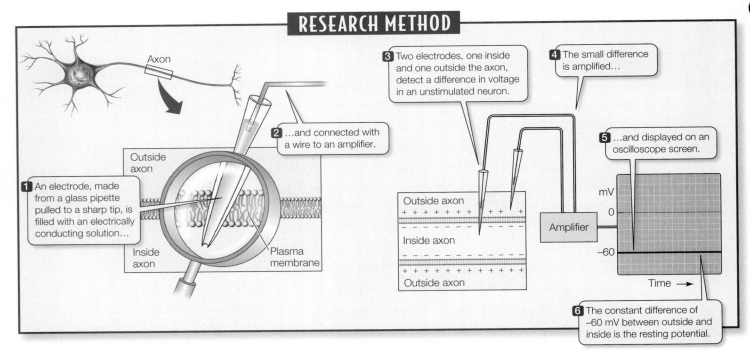

RESEARCH METHOD

1 An electrode, made from a glass pipette pulled to a sharp tip, is filled with an electrically conducting solution…

Outside axon

Inside axon

Plasma membrane

2 …and connected with a wire to an amplifier.

3 Two electrodes, one inside and one outside the axon, detect a difference in voltage in an unstimulated neuron.

4 The small difference is amplified…

5 …and displayed on an oscilloscope screen.

Outside axon
+ + + + + + + + + + +
– – – – – – – – – – –
Inside axon
– – – – – – – – – – –
+ + + + + + + + + + +
Outside axon

Amplifier

mV
0
–60
Time →

6 The constant difference of –60 mV between outside and inside is the resting potential.

44.5 Measuring the Resting Potential *(Page 947)*

(A) Na⁺ – K⁺ pump (ATPase)

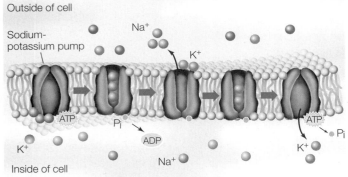

Outside of cell

Sodium-potassium pump

Na^+

K^+

ATP

P_i

ADP

ATP

P_i

K^+

Na^+

K^+

Inside of cell

(B) Na⁺ – K⁺ channels

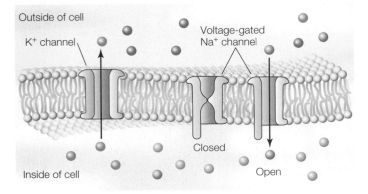

Outside of cell

K^+ channel

Voltage-gated Na^+ channel

Closed

Open

Inside of cell

44.6 Ion Pumps and Channels *(Page 948)*

EXPERIMENT

HYPOTHESIS: The resting potential of neurons is due to permeability of the membrane to potassium ions.

METHOD

1. Measure concentrations of ions inside and outside of a neuron.

To measure the concentration of ions in a neuron, the neuron (and its axon) must be big. Squid have giant neurons that control their escape response. It is possible to sample the cytoplasm of these axons, which are about 1 mm in diameter.

Squid axon
1 mm

Plasma membrane
Cytoplasm
Electrode

2. Use the Nernst equation to calculate what the membrane potential would be if it were permeable to each of these ions: Na^+, K^+, Ca^{2+}, and Cl^-.

Recall the Nernst equation from Section 5.3. It predicts the membrane potential resulting from membrane permeability to a single type of ion that differs in concentration on the two sides of the membrane. The equation is written

$$E_{ion} = 2.3 \frac{RT}{zF} \log \frac{[ion]_o}{[ion]_i}$$

where E is the equilibrium (resting) membrane potential (the voltage across the membrane in mV), R is the universal gas constant, T is the absolute temperature, z is the charge on the ion (+1, +1, +2, or −1, respectively, for the ions used here), and F is the Faraday constant. The subscripts o and i indicate the ion concentrations outside and inside the cell, respectively.

At this point you could just "plug and play," but do you understand this equation?

A concentration difference of ions across a membrane creates a *chemical* force that pushes the ions across the membrane; however, the resulting unbalanced *electrical charges* will pull the ions back the other way. At *equilibrium*, the work done moving ions in each direction will be the same.

The *chemical* work pushing the ions will equal 2.3 RT log $[ion]_o/[ion]_i$
The *electrical* work pulling the ions will equal zEF. So, at equilibrium:

$$zEF = 2.3\ RT\ \log \frac{[ion]_o}{[ion]_i}$$

Rearranging the equation to solve for E, we get the Nernst equation:

$$E_{ion} = 2.3 \frac{RT}{zF} \log \frac{[ion]_o}{[ion]_i}$$

We can simplify the equation by picking a temperature—let's use "room temperature," or 20°C—and solving for 2.3 RT/F. At 20°C, 2.3 RT/F equals 58. Thus:

$$E_{ion} = 58/z \log \frac{[ion]_o}{[ion]_i}$$

3. Measure the membrane potential across the squid giant axon and compare with calculated values for each ion.

RESULTS

1. Measuring ion concentrations in squid giant axon cytoplasm and in seawater, then solving the Nernst equation for each ion, we find:

Ion	Ion concentration (n*M*) in squid axon	in seawater	Predicted membrane potential (mV)
K^+	400	20	−75
Na^+	50	460	+56
Ca^{2+}	0.5	10	+38
Cl^-	50	560	−60

2. Using the method shown in Figure 44.5, the actual resting membrane potential of a squid giant axon is recorded to be −66 mV.

CONCLUSION: The resting potential of the squid giant axon can be due to permeability to K^+, but there is probably some permeability to another ion as well.

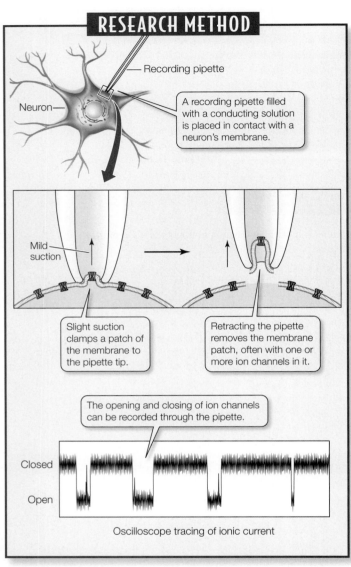

RESEARCH METHOD

Recording pipette

Neuron

A recording pipette filled with a conducting solution is placed in contact with a neuron's membrane.

Mild suction

Slight suction clamps a patch of the membrane to the pipette tip.

Retracting the pipette removes the membrane patch, often with one or more ion channels in it.

The opening and closing of ion channels can be recorded through the pipette.

Closed

Open

Oscilloscope tracing of ionic current

44.8 Patch Clamping *(Page 950)*

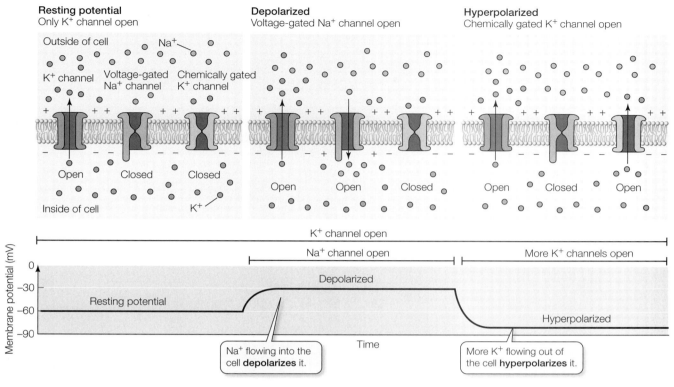

Resting potential
Only K⁺ channel open

Outside of cell

K⁺ channel   Voltage-gated   Chemically gated
             Na⁺ channel   K⁺ channel

Open   Closed   Closed

Inside of cell

Depolarized
Voltage-gated Na⁺ channel open

Open   Open   Closed

Hyperpolarized
Chemically gated K⁺ channel open

Open   Closed   Open

K⁺ channel open

Na⁺ channel open   More K⁺ channels open

Depolarized

Resting potential

Hyperpolarized

Time

Na⁺ flowing into the
cell **depolarizes** it.

More K⁺ flowing out of
the cell **hyperpolarizes** it.

44.9 Membranes Can Be Depolarized or Hyperpolarized *(Page 951)*

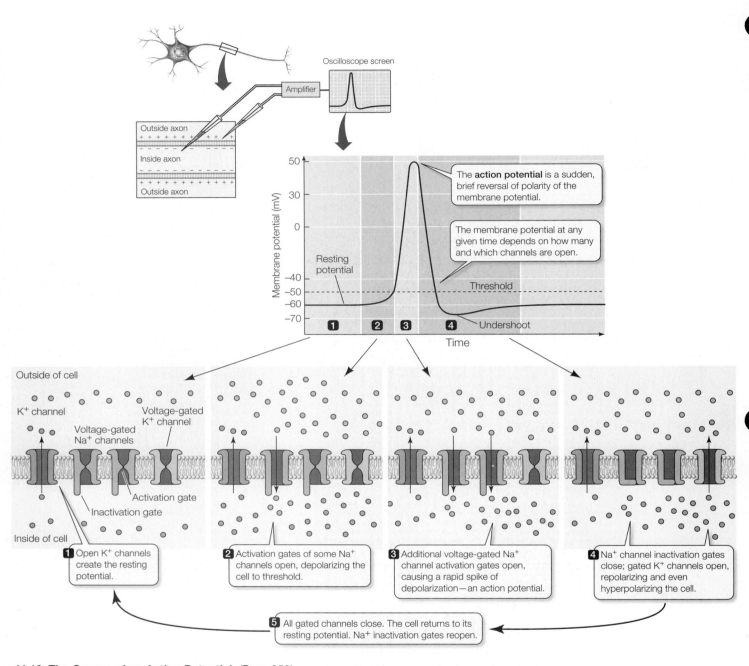

44.10 The Course of an Action Potential *(Page 952)*

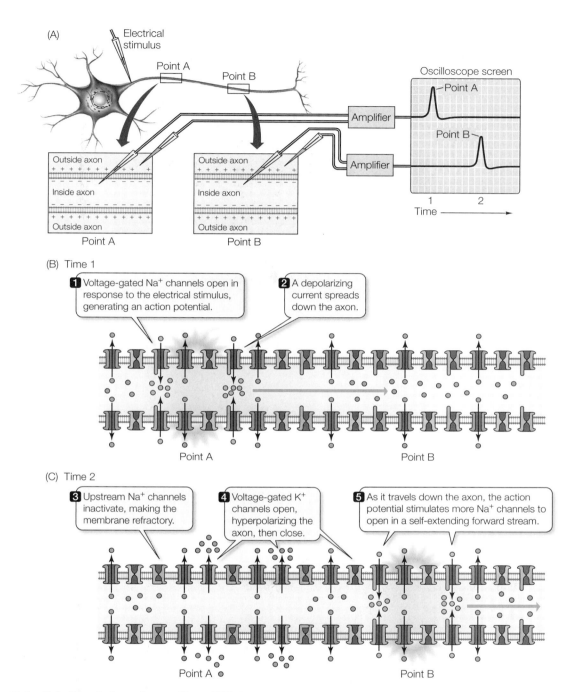

44.11 Action Potentials Travel along Axons *(Page 953)*

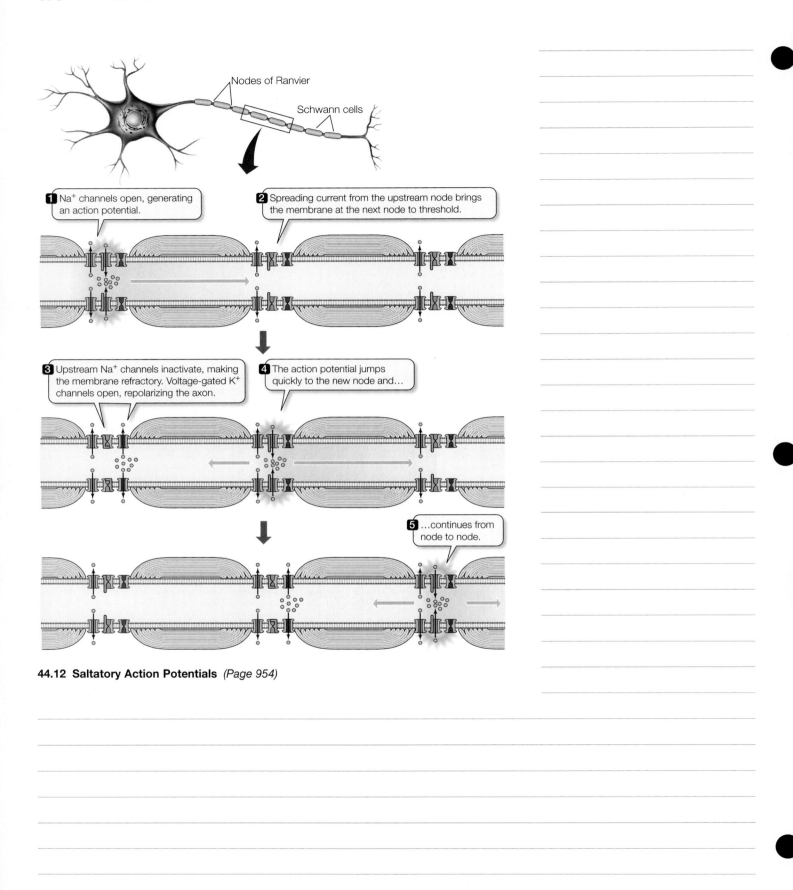

44.12 Saltatory Action Potentials *(Page 954)*

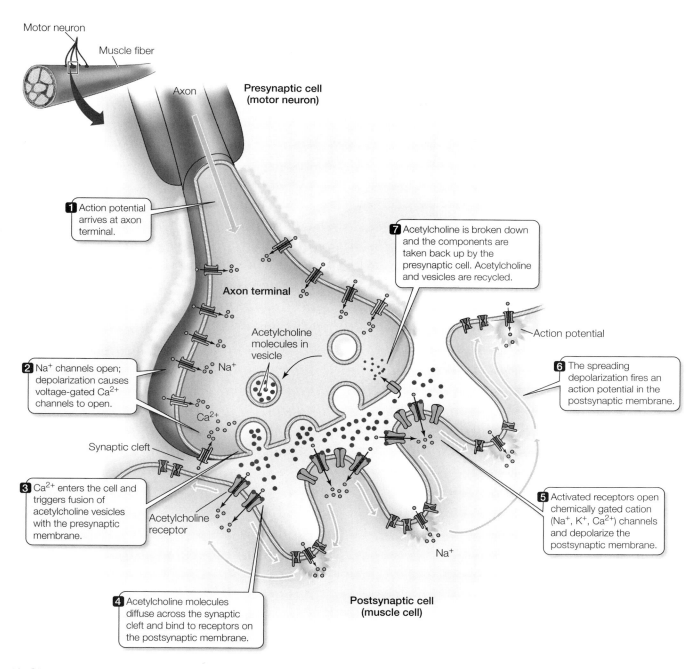

Motor neuron

Muscle fiber

Axon

Presynaptic cell (motor neuron)

1 Action potential arrives at axon terminal.

Axon terminal

7 Acetylcholine is broken down and the components are taken back up by the presynaptic cell. Acetylcholine and vesicles are recycled.

Acetylcholine molecules in vesicle

Action potential

2 Na^+ channels open; depolarization causes voltage-gated Ca^{2+} channels to open.

Na^+

Ca^{2+}

6 The spreading depolarization fires an action potential in the postsynaptic membrane.

Synaptic cleft

3 Ca^{2+} enters the cell and triggers fusion of acetylcholine vesicles with the presynaptic membrane.

Acetylcholine receptor

5 Activated receptors open chemically gated cation (Na^+, K^+, Ca^{2+}) channels and depolarize the postsynaptic membrane.

Na^+

4 Acetylcholine molecules diffuse across the synaptic cleft and bind to receptors on the postsynaptic membrane.

Postsynaptic cell (muscle cell)

44.13 Chemical Synaptic Transmission Begins with the Arrival of an Action Potential *(Page 956)*

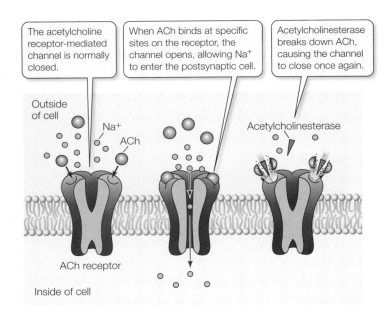

The acetylcholine receptor-mediated channel is normally closed.

When ACh binds at specific sites on the receptor, the channel opens, allowing Na⁺ to enter the postsynaptic cell.

Acetylcholinesterase breaks down ACh, causing the channel to close once again.

Outside of cell

Na⁺

ACh

Acetylcholinesterase

ACh receptor

Inside of cell

44.14 The Acetylcholine Receptor Is a Chemically Gated Channel *(Page 957)*

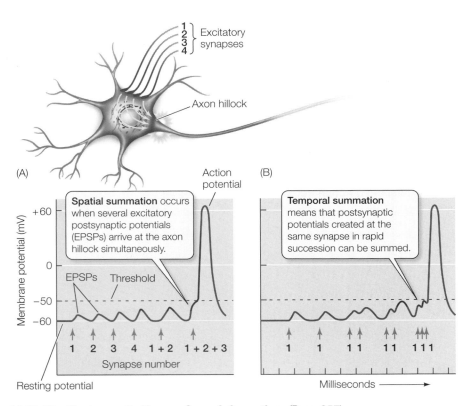

1
2
3
4
} Excitatory synapses

Axon hillock

(A)

Action potential

(B)

Spatial summation occurs when several excitatory postsynaptic potentials (EPSPs) arrive at the axon hillock simultaneously.

Temporal summation means that postsynaptic potentials created at the same synapse in rapid succession can be summed.

Membrane potential (mV)

+60

0

−50

−60

EPSPs   Threshold

1   2   3   4   1 + 2   1 + 2 + 3

Synapse number

Resting potential

1   1   1 1   1 1   1 1 1

Milliseconds ⟶

44.15 The Postsynaptic Neuron Sums Information *(Page 957)*

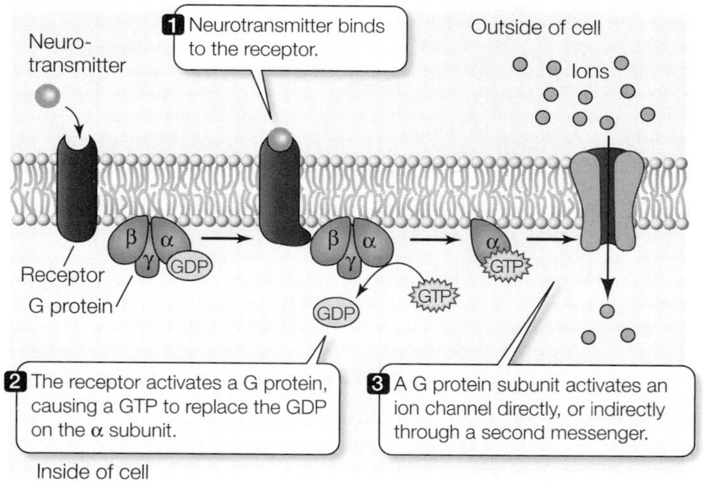

1 Neurotransmitter binds to the receptor.
Outside of cell
Neuro-transmitter
Ions
Receptor
G protein
β
α
γ
GDP
GTP
2 The receptor activates a G protein, causing a GTP to replace the GDP on the α subunit.
3 A G protein subunit activates an ion channel directly, or indirectly through a second messenger.
Inside of cell

44.16 Metabotropic Receptors Act through G Proteins
(Page 958)

TABLE 44.1

Some Well-Known Neurotransmitters

NEUROTRANSMITTER	ACTIONS	COMMENTS
Acetylcholine	The neurotransmitter of vertebrate motor neurons and of some neural pathways in the brain	Broken down in the synapse by acetylcholinesterase; blockers of this enzyme are powerful poisons
MONOAMINES		
Norepinephrine	Used in certain neural pathways in the brain. Also found in the peripheral nervous system, where it causes gut muscles to relax and the heart to beat faster	Related to epinephrine and acts at some of the same receptors
Dopamine	A neurotransmitter of the central nervous system	Involved in schizophrenia. Loss of dopamine neurons is the cause of Parkinson's disease
Histamine	A minor neurotransmitter in the brain	Involved in maintaining wakefulness
Serotonin	A neurotransmitter of the central nervous system that is involved in many systems, including pain control, sleep/wake control, and mood	Certain medications that elevate mood and counter anxiety act by inhibiting the reuptake of serotonin (so it remains active longer)
PURINES		
ATP	Co-released with many neurotransmitters	Large family of receptors may shape postsynaptic responses to classical neurotransmitters
Adenosine	Transported across cell membranes; not synaptically released	Not released synaptically; mainly has inhibitory effects on neighboring cells
AMINO ACIDS		
Glutamate	The most common excitatory neurotransmitter in the central nervous system	Some people have reactions to the food additive monosodium glutamate because it can affect the nervous system
Glycine Gamma-aminobutyric acid (GABA)	Common inhibitory neurotransmitters	Drugs called benzodiazepines, used to reduce anxiety and produce sedation, mimic the actions of GABA
PEPTIDES		
Endorphins Enkephalins	Modulation of pain pathways	Receptors are activated by narcotic drugs: opium, morphine, heroin, codeine
Substance P	Used by certain sensory nerves, especially in pain pathways	Released by neurons sensitive to heat and pain
GAS		
Nitric oxide	Widely distributed in the nervous system	Not a classic neurotransmitter, it diffuses across membranes rather than being released synaptically. A means whereby a postsynaptic cell can influence a presynaptic cell

(Page 959)

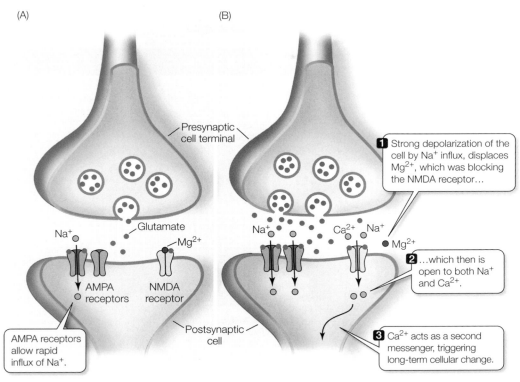

(A) (B)

Presynaptic cell terminal

Na⁺

Glutamate

Mg²⁺

AMPA receptors

NMDA receptor

AMPA receptors allow rapid influx of Na⁺.

Postsynaptic cell

Na⁺ Ca²⁺ Na⁺

Mg²⁺

1 Strong depolarization of the cell by Na⁺ influx, displaces Mg²⁺, which was blocking the NMDA receptor…

2 …which then is open to both Na⁺ and Ca²⁺.

3 Ca²⁺ acts as a second messenger, triggering long-term cellular change.

44.17 Two Ionotropic Glutamate Receptors *(Page 960)*

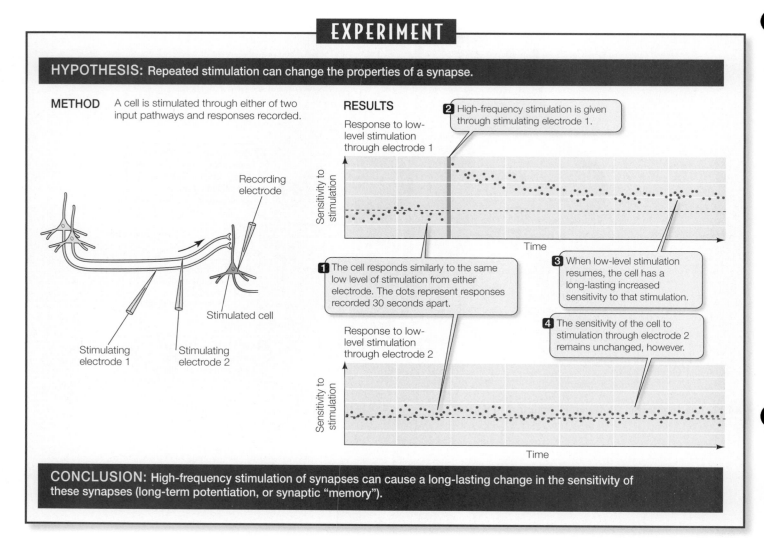

44.18 Repeated Stimulation Can Cause Long-Term Potentiation *(Page 961)*

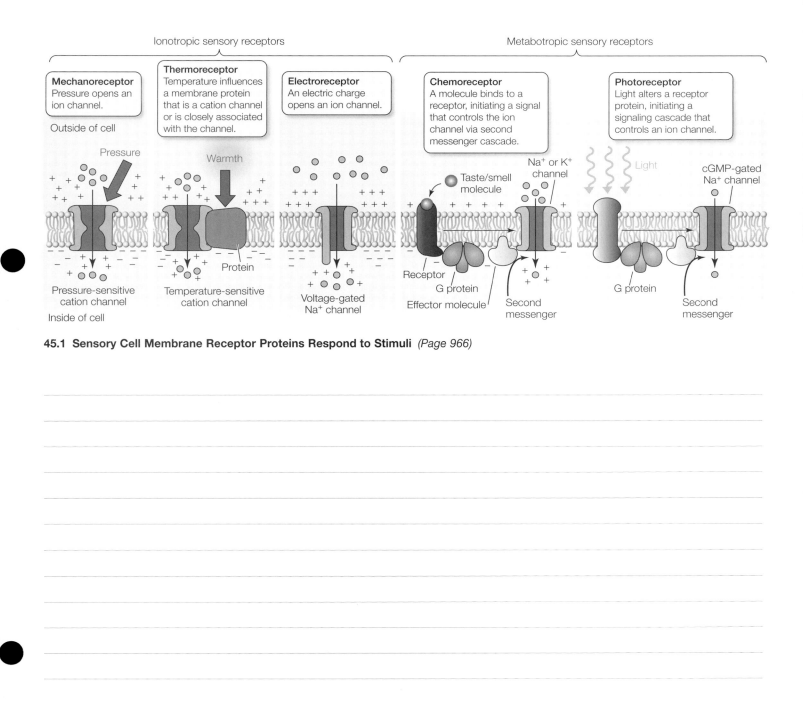

45.1 Sensory Cell Membrane Receptor Proteins Respond to Stimuli *(Page 966)*

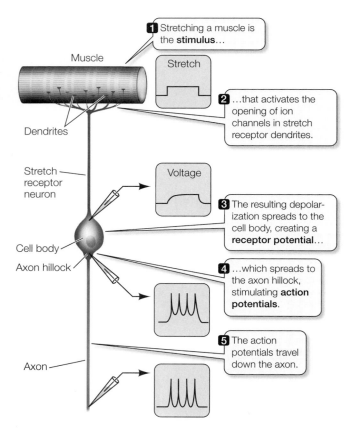

1 Stretching a muscle is the **stimulus**…

Muscle

Stretch

Dendrites

2 …that activates the opening of ion channels in stretch receptor dendrites.

Stretch receptor neuron

Voltage

3 The resulting depolarization spreads to the cell body, creating a **receptor potential**…

Cell body

Axon hillock

4 …which spreads to the axon hillock, stimulating **action potentials**.

5 The action potentials travel down the axon.

Axon

45.2 Stimulating a Sensory Cell Produces a Receptor Potential (Page 966)

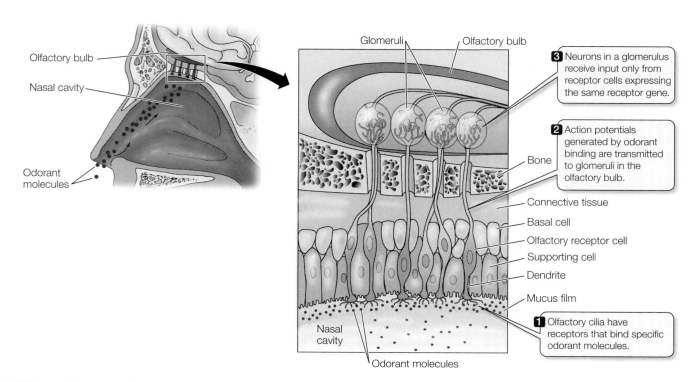

Olfactory bulb

Nasal cavity

Odorant molecules

Glomeruli

Olfactory bulb

3 Neurons in a glomerulus receive input only from receptor cells expressing the same receptor gene.

2 Action potentials generated by odorant binding are transmitted to glomeruli in the olfactory bulb.

Bone

Connective tissue

Basal cell

Olfactory receptor cell

Supporting cell

Dendrite

Mucus film

1 Olfactory cilia have receptors that bind specific odorant molecules.

Nasal cavity

Odorant molecules

45.4 Olfactory Receptors Communicate Directly with the Brain (Page 969)

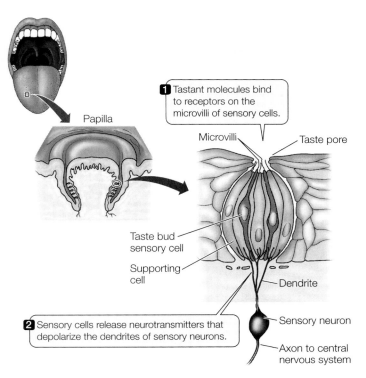

1 Tastant molecules bind to receptors on the microvilli of sensory cells.

Papilla

Microvilli

Taste pore

Taste bud sensory cell

Supporting cell

Dendrite

Sensory neuron

2 Sensory cells release neurotransmitters that depolarize the dendrites of sensory neurons.

Axon to central nervous system

45.5 Taste Buds Are Clusters of Sensory Cells *(Page 970)*

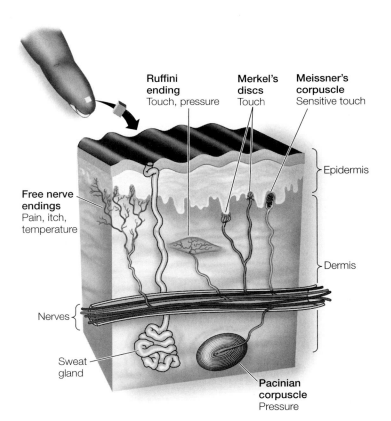

Ruffini ending Touch, pressure

Merkel's discs Touch

Meissner's corpuscle Sensitive touch

Epidermis

Free nerve endings Pain, itch, temperature

Dermis

Nerves

Sweat gland

Pacinian corpuscle Pressure

45.6 The Skin Feels Many Sensations *(Page 971)*

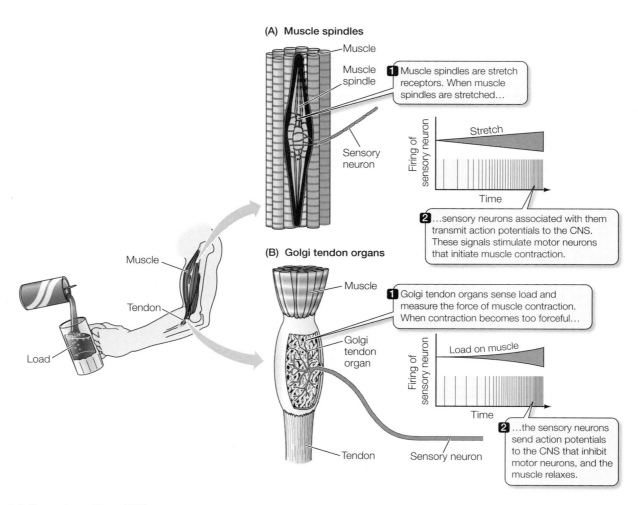

(A) Muscle spindles

Muscle

Muscle spindle

1 Muscle spindles are stretch receptors. When muscle spindles are stretched...

Sensory neuron

Firing of sensory neuron

Stretch

Time

2 ...sensory neurons associated with them transmit action potentials to the CNS. These signals stimulate motor neurons that initiate muscle contraction.

(B) Golgi tendon organs

Muscle

1 Golgi tendon organs sense load and measure the force of muscle contraction. When contraction becomes too forceful...

Golgi tendon organ

Firing of sensory neuron

Load on muscle

Time

2 ...the sensory neurons send action potentials to the CNS that inhibit motor neurons, and the muscle relaxes.

Tendon   Sensory neuron

Muscle

Tendon

Load

45.7 Stretch Receptors *(Page 971)*

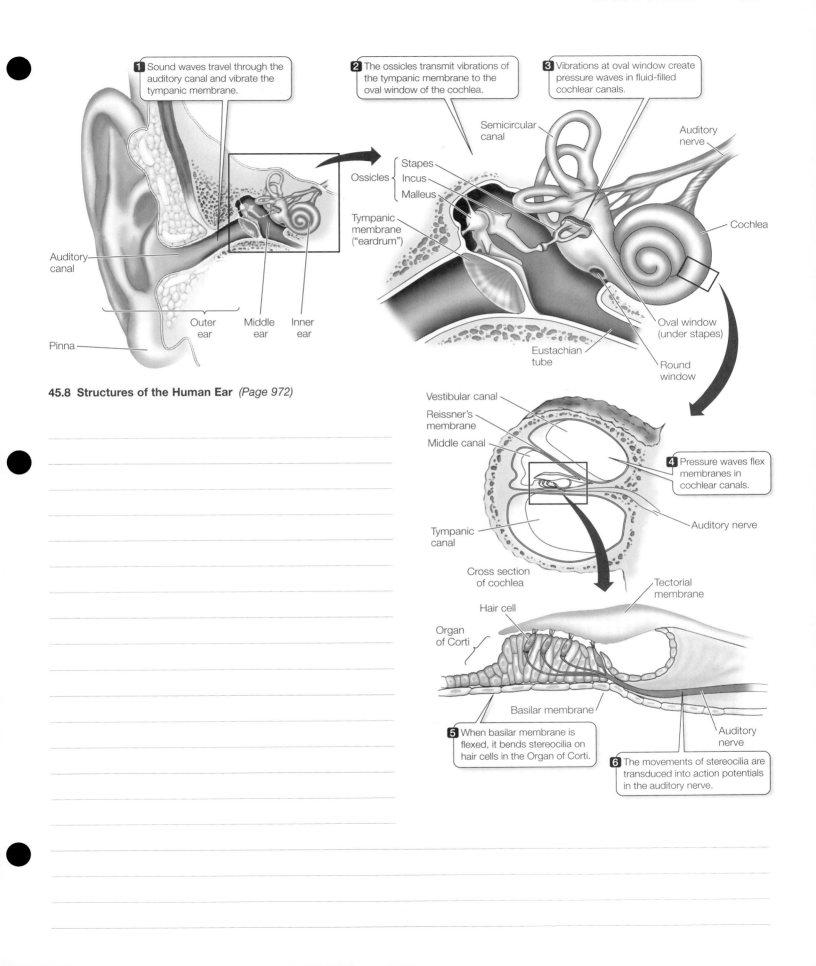

1 Sound waves travel through the auditory canal and vibrate the tympanic membrane.

2 The ossicles transmit vibrations of the tympanic membrane to the oval window of the cochlea.

3 Vibrations at oval window create pressure waves in fluid-filled cochlear canals.

Semicircular canal

Auditory nerve

Ossicles — Stapes / Incus / Malleus

Tympanic membrane ("eardrum")

Cochlea

Auditory canal

Outer ear

Middle ear

Inner ear

Pinna

Oval window (under stapes)

Eustachian tube

Round window

45.8 Structures of the Human Ear *(Page 972)*

Vestibular canal

Reissner's membrane

Middle canal

4 Pressure waves flex membranes in cochlear canals.

Tympanic canal

Auditory nerve

Cross section of cochlea

Tectorial membrane

Hair cell

Organ of Corti

Basilar membrane

Auditory nerve

5 When basilar membrane is flexed, it bends stereocilia on hair cells in the Organ of Corti.

6 The movements of stereocilia are transduced into action potentials in the auditory nerve.

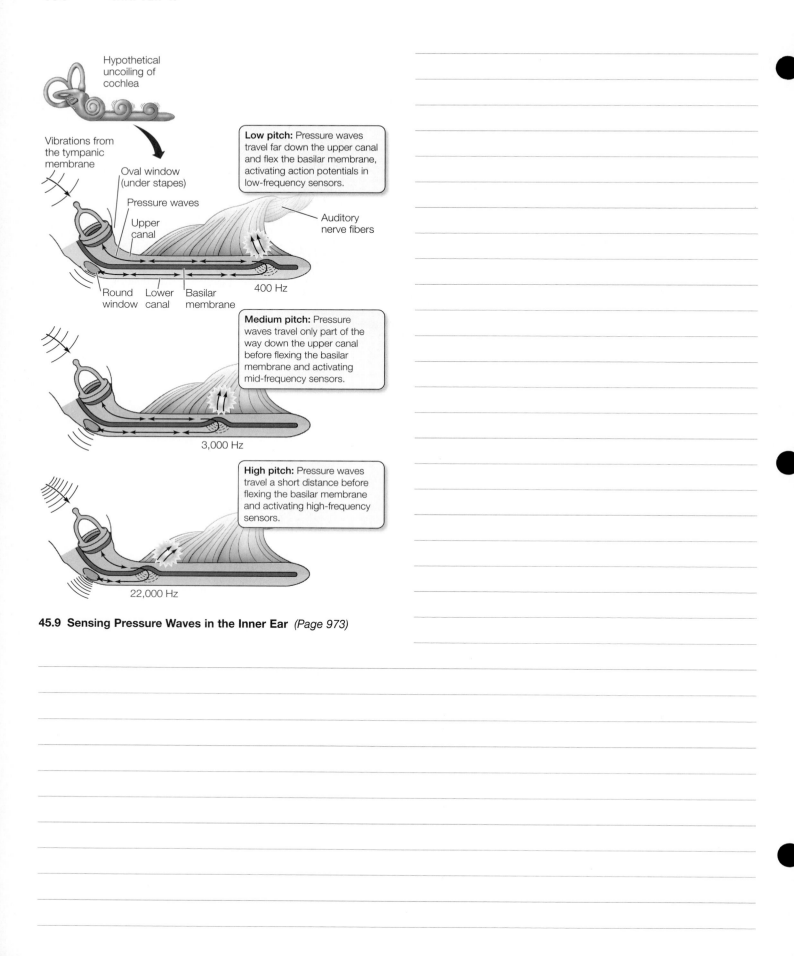

Hypothetical uncoiling of cochlea

Vibrations from the tympanic membrane

Oval window (under stapes)

Pressure waves

Upper canal

Low pitch: Pressure waves travel far down the upper canal and flex the basilar membrane, activating action potentials in low-frequency sensors.

Auditory nerve fibers

Round window   Lower canal   Basilar membrane

400 Hz

Medium pitch: Pressure waves travel only part of the way down the upper canal before flexing the basilar membrane and activating mid-frequency sensors.

3,000 Hz

High pitch: Pressure waves travel a short distance before flexing the basilar membrane and activating high-frequency sensors.

22,000 Hz

45.9 Sensing Pressure Waves in the Inner Ear *(Page 973)*

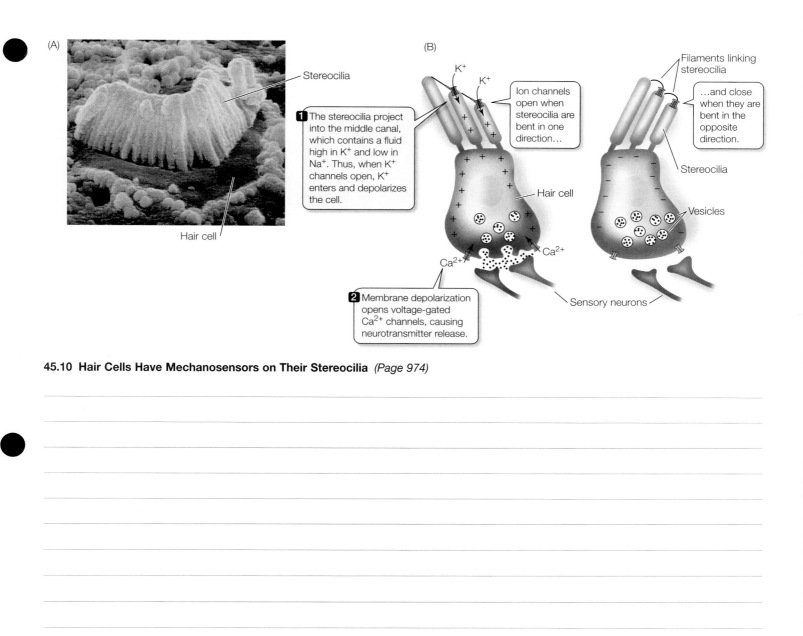

(A)

Stereocilia

Hair cell

(B)

K⁺ K⁺

Ion channels open when stereocilia are bent in one direction…

1 The stereocilia project into the middle canal, which contains a fluid high in K⁺ and low in Na⁺. Thus, when K⁺ channels open, K⁺ enters and depolarizes the cell.

Hair cell

Ca²⁺ Ca²⁺

2 Membrane depolarization opens voltage-gated Ca²⁺ channels, causing neurotransmitter release.

Filaments linking stereocilia

…and close when they are bent in the opposite direction.

Stereocilia

Vesicles

Sensory neurons

45.10 Hair Cells Have Mechanosensors on Their Stereocilia *(Page 974)*

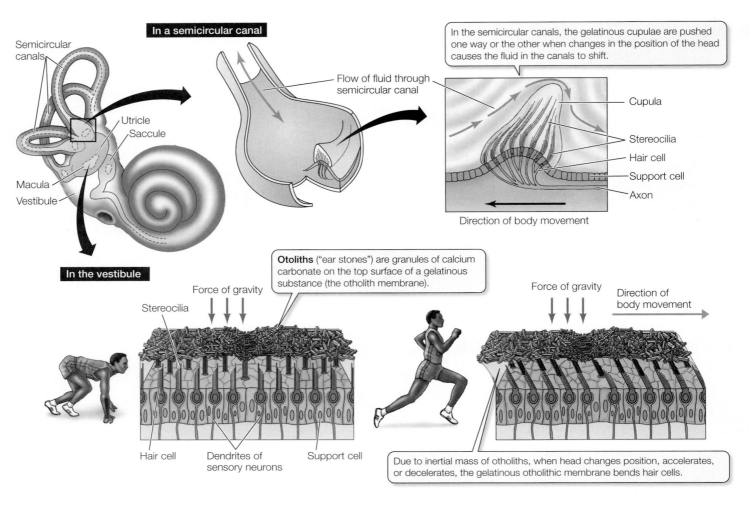

In a semicircular canal

Semicircular canals

Utricle
Saccule

Macula
Vestibule

In the vestibule

Flow of fluid through semicircular canal

In the semicircular canals, the gelatinous cupulae are pushed one way or the other when changes in the position of the head causes the fluid in the canals to shift.

Cupula

Stereocilia

Hair cell

Support cell

Axon

Direction of body movement

Stereocilia

Force of gravity

Otoliths ("ear stones") are granules of calcium carbonate on the top surface of a gelatinous substance (the otolith membrane).

Force of gravity

Direction of body movement

Hair cell   Dendrites of sensory neurons   Support cell

Due to inertial mass of otoliths, when head changes position, accelerates, or decelerates, the gelatinous otolithic membrane bends hair cells.

45.11 Organs of Equilibrium *(Page 975)*

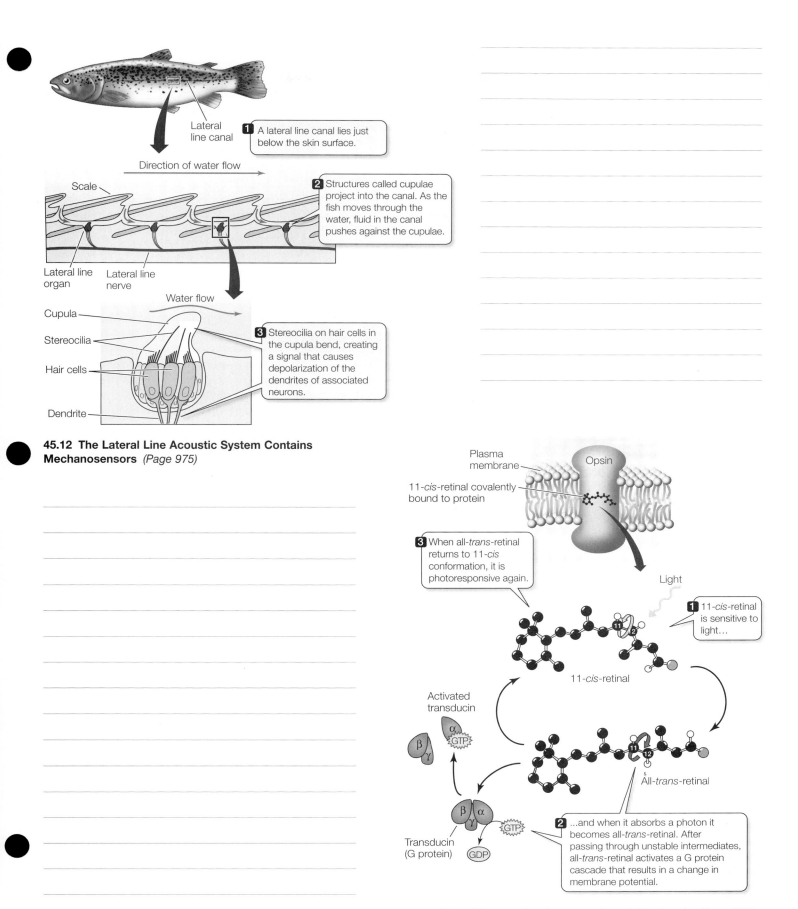

Lateral line canal

1 A lateral line canal lies just below the skin surface.

Direction of water flow

Scale

2 Structures called cupulae project into the canal. As the fish moves through the water, fluid in the canal pushes against the cupulae.

Lateral line organ

Lateral line nerve

Water flow

Cupula

Stereocilia

Hair cells

3 Stereocilia on hair cells in the cupula bend, creating a signal that causes depolarization of the dendrites of associated neurons.

Dendrite

45.12 The Lateral Line Acoustic System Contains Mechanosensors *(Page 975)*

Plasma membrane

Opsin

11-*cis*-retinal covalently bound to protein

3 When all-*trans*-retinal returns to 11-*cis* conformation, it is photoresponsive again.

Light

1 11-*cis*-retinal is sensitive to light…

11-*cis*-retinal

Activated transducin

α
GTP

β
γ

All-*trans*-retinal

β
γ
α

GTP

Transducin (G protein)

GDP

2 …and when it absorbs a photon it becomes all-*trans*-retinal. After passing through unstable intermediates, all-*trans*-retinal activates a G protein cascade that results in a change in membrane potential.

45.13 Light Changes the Conformation of Rhodopsin *(Page 976)*

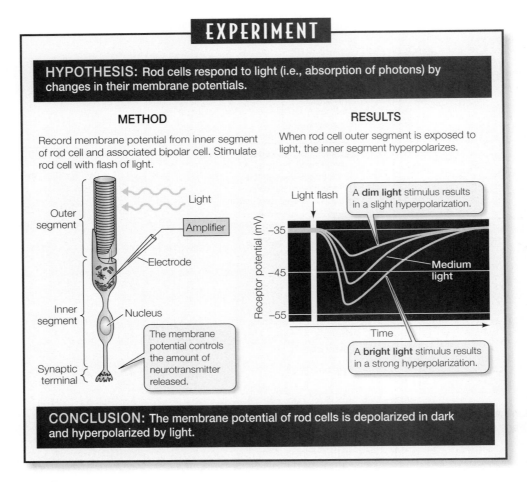

EXPERIMENT

HYPOTHESIS: Rod cells respond to light (i.e., absorption of photons) by changes in their membrane potentials.

METHOD

Record membrane potential from inner segment of rod cell and associated bipolar cell. Stimulate rod cell with flash of light.

Light

Outer segment

Amplifier

Electrode

Inner segment

Nucleus

Synaptic terminal

The membrane potential controls the amount of neurotransmitter released.

RESULTS

When rod cell outer segment is exposed to light, the inner segment hyperpolarizes.

Light flash

A **dim light** stimulus results in a slight hyperpolarization.

Receptor potential (mV)

−35

−45

−55

Medium light

Time

A **bright light** stimulus results in a strong hyperpolarization.

CONCLUSION: The membrane potential of rod cells is depolarized in dark and hyperpolarized by light.

45.14 A Rod Cell Responds to Light *(Page 977)*

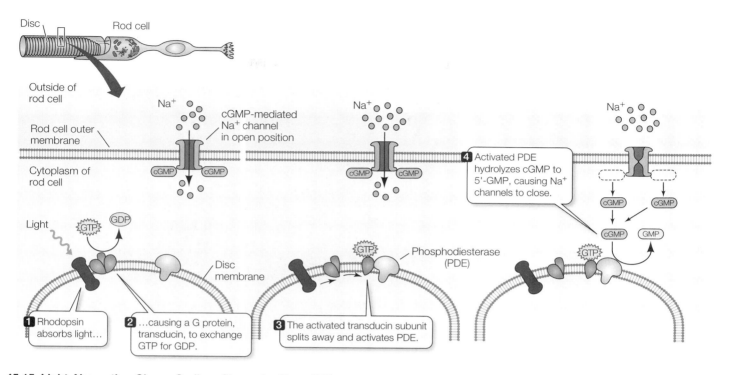

45.15 Light Absorption Closes Sodium Channels *(Page 978)*

Outside of rod cell

Rod cell outer membrane

Cytoplasm of rod cell

Disc   Rod cell

cGMP-mediated Na⁺ channel in open position

4 Activated PDE hydrolyzes cGMP to 5'-GMP, causing Na⁺ channels to close.

Light

1 Rhodopsin absorbs light…

2 …causing a G protein, transducin, to exchange GTP for GDP.

3 The activated transducin subunit splits away and activates PDE.

Disc membrane

Phosphodiesterase (PDE)

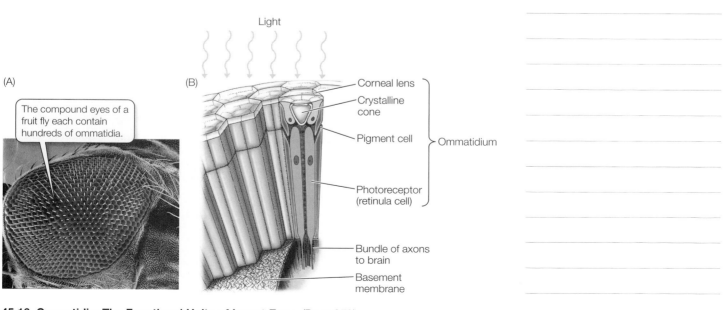

45.16 Ommatidia: The Functional Units of Insect Eyes *(Page 978)*

(A) The compound eyes of a fruit fly each contain hundreds of ommatidia.

(B) Light

Corneal lens
Crystalline cone
Pigment cell
Photoreceptor (retinula cell)
Ommatidium

Bundle of axons to brain
Basement membrane

(A) Human

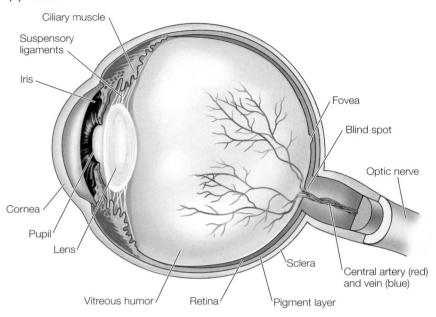

Ciliary muscle

Suspensory ligaments

Iris

Cornea

Pupil

Lens

Vitreous humor

Retina

Sclera

Pigment layer

Fovea

Blind spot

Optic nerve

Central artery (red) and vein (blue)

45.17 Eyes Like Cameras *(Page 979)*

(B) Octopus

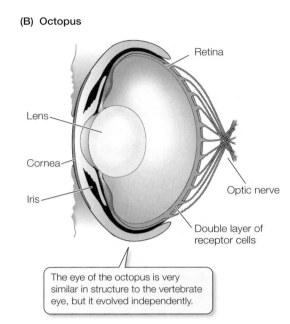

Retina

Lens

Cornea

Iris

Optic nerve

Double layer of receptor cells

The eye of the octopus is very similar in structure to the vertebrate eye, but it evolved independently.

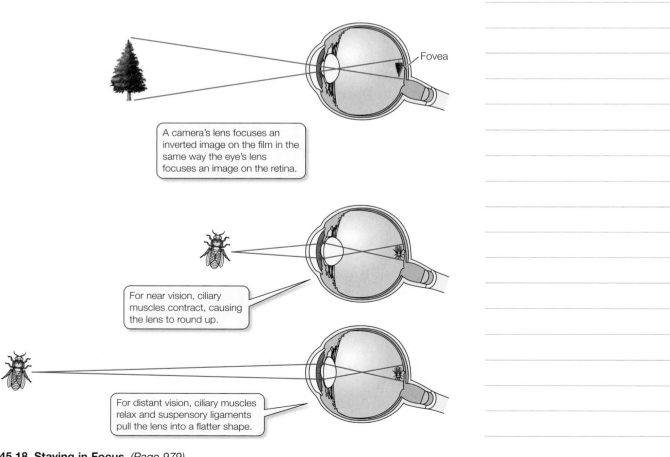

Fovea

A camera's lens focuses an inverted image on the film in the same way the eye's lens focuses an image on the retina.

For near vision, ciliary muscles contract, causing the lens to round up.

For distant vision, ciliary muscles relax and suspensory ligaments pull the lens into a flatter shape.

45.18 Staying in Focus *(Page 979)*

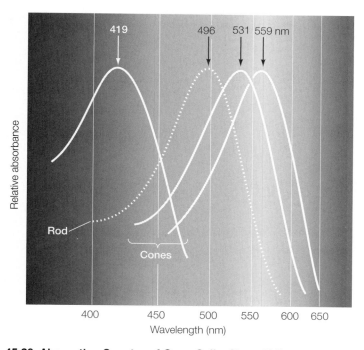

45.20 Absorption Spectra of Cone Cells *(Page 981)*

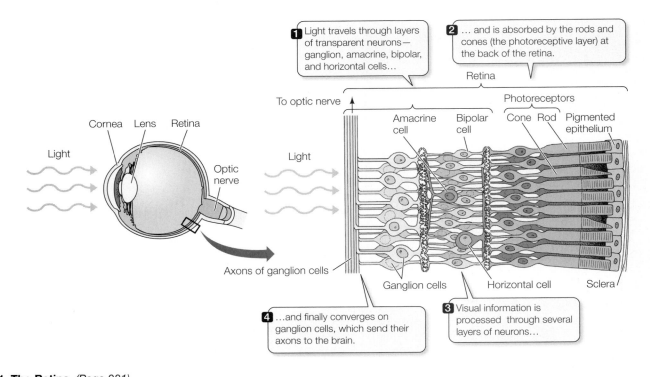

1 Light travels through layers of transparent neurons— ganglion, amacrine, bipolar, and horizontal cells…

2 … and is absorbed by the rods and cones (the photoreceptive layer) at the back of the retina.

Retina

To optic nerve

Photoreceptors

Amacrine cell

Bipolar cell

Cone Rod Pigmented epithelium

Light

Axons of ganglion cells

Ganglion cells

Horizontal cell

Sclera

4 …and finally converges on ganglion cells, which send their axons to the brain.

3 Visual information is processed through several layers of neurons…

Cornea Lens Retina

Light

Optic nerve

45.21 The Retina *(Page 981)*

CHAPTER 46 The Mammalian Nervous System: Structure and Higher Function

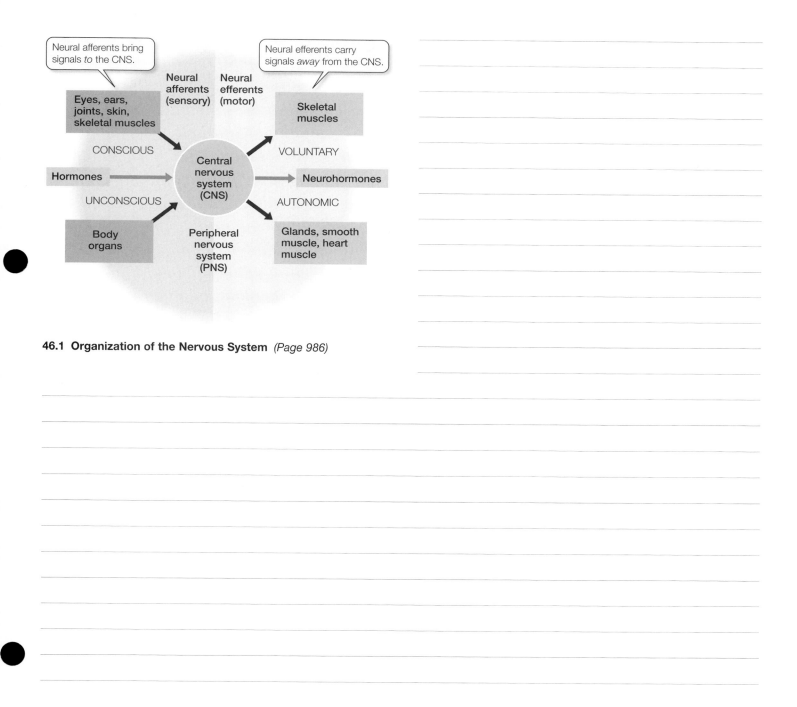

46.1 Organization of the Nervous System *(Page 986)*

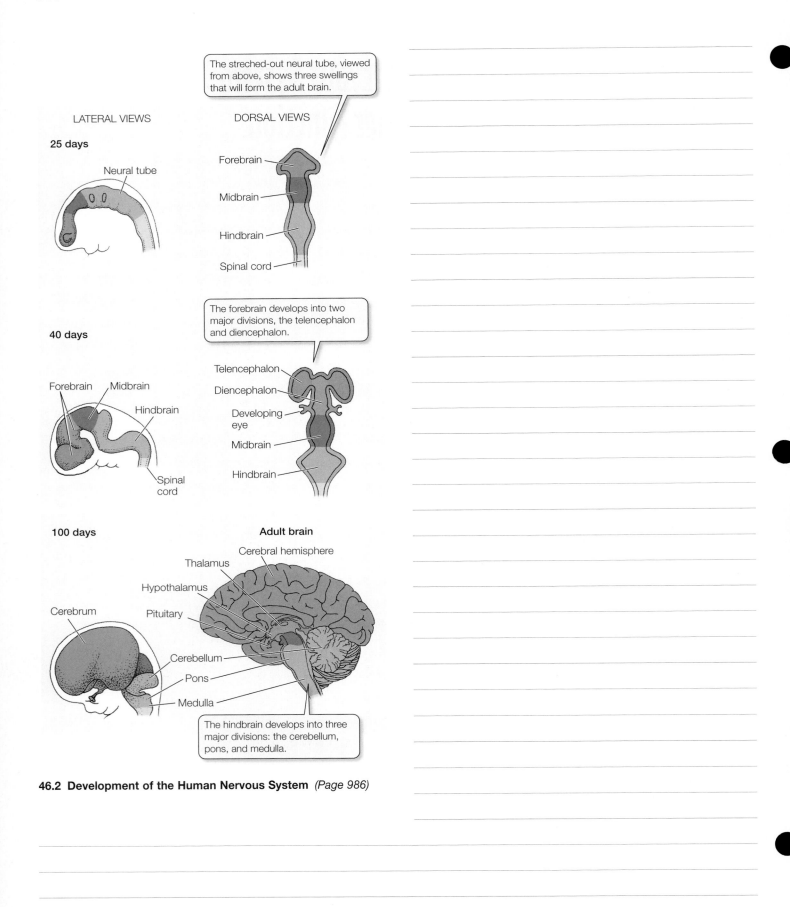

LATERAL VIEWS

DORSAL VIEWS

The streched-out neural tube, viewed from above, shows three swellings that will form the adult brain.

25 days

Neural tube

Forebrain

Midbrain

Hindbrain

Spinal cord

The forebrain develops into two major divisions, the telencephalon and diencephalon.

40 days

Forebrain   Midbrain

Hindbrain

Spinal cord

Telencephalon

Diencephalon

Developing eye

Midbrain

Hindbrain

100 days

Adult brain

Cerebral hemisphere

Thalamus

Hypothalamus

Cerebrum

Pituitary

Cerebellum

Pons

Medulla

The hindbrain develops into three major divisions: the cerebellum, pons, and medulla.

46.2 Development of the Human Nervous System *(Page 986)*

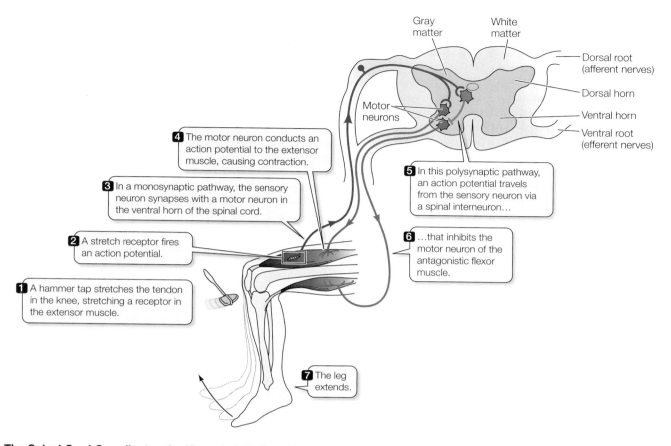

4 The motor neuron conducts an action potential to the extensor muscle, causing contraction.

3 In a monosynaptic pathway, the sensory neuron synapses with a motor neuron in the ventral horn of the spinal cord.

2 A stretch receptor fires an action potential.

1 A hammer tap stretches the tendon in the knee, stretching a receptor in the extensor muscle.

Gray matter

White matter

Dorsal root (afferent nerves)

Dorsal horn

Ventral horn

Ventral root (efferent nerves)

Motor neurons

5 In this polysynaptic pathway, an action potential travels from the sensory neuron via a spinal interneuron…

6 …that inhibits the motor neuron of the antagonistic flexor muscle.

7 The leg extends.

46.3 The Spinal Cord Coordinates the Knee-Jerk Reflex *(Page 987)*

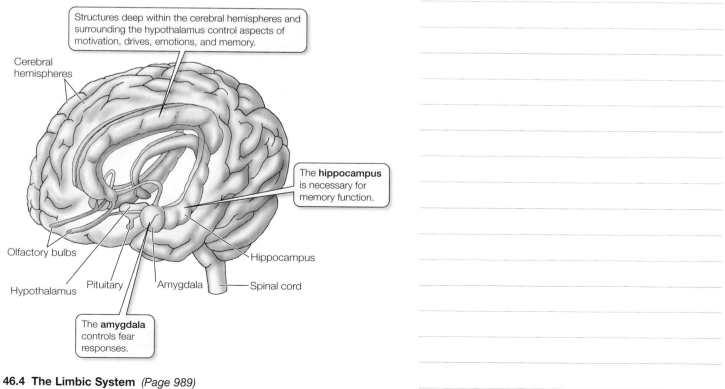

Structures deep within the cerebral hemispheres and surrounding the hypothalamus control aspects of motivation, drives, emotions, and memory.

Cerebral hemispheres

The **hippocampus** is necessary for memory function.

Olfactory bulbs

Hippocampus

Hypothalamus   Pituitary   Amygdala   Spinal cord

The **amygdala** controls fear responses.

46.4 The Limbic System *(Page 989)*

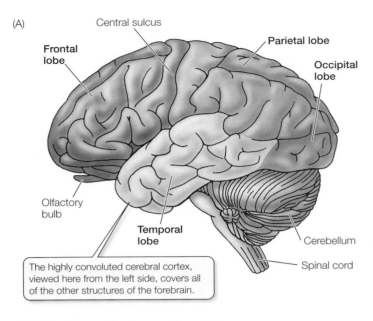

(A)

Central sulcus

Frontal lobe

Parietal lobe

Occipital lobe

Olfactory bulb

Temporal lobe

The highly convoluted cerebral cortex, viewed here from the left side, covers all of the other structures of the forebrain.

Cerebellum

Spinal cord

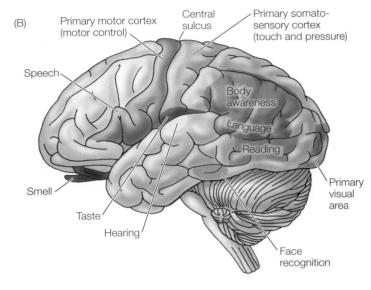

(B)

Primary motor cortex (motor control)

Central sulcus

Primary somato-sensory cortex (touch and pressure)

Speech

Body awareness

Language

Reading

Smell

Taste

Hearing

Primary visual area

Face recognition

46.5 The Human Cerebrum *(Page 989)*

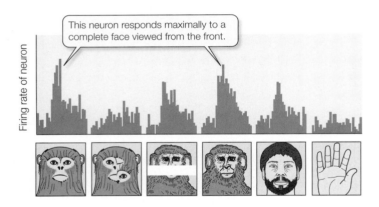

This neuron responds maximally to a complete face viewed from the front.

Firing rate of neuron

46.6 Neurons in One Region of the Temporal Lobe Respond to Faces *(Page 990)*

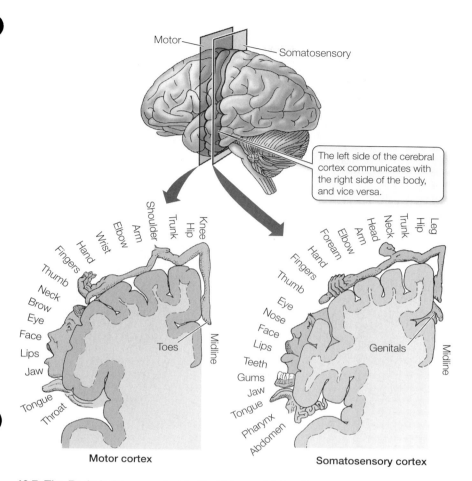

46.7 The Body Is Represented in the Primary Motor Cortex and the Primary Somatosensory Cortex *(Page 990)*

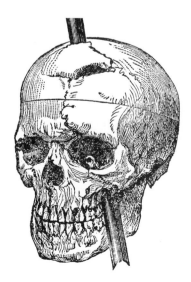

46.8 A Mind-Altering Experience *(Page 991)*

46.9 Contralateral Neglect Syndrome *(Page 991)*

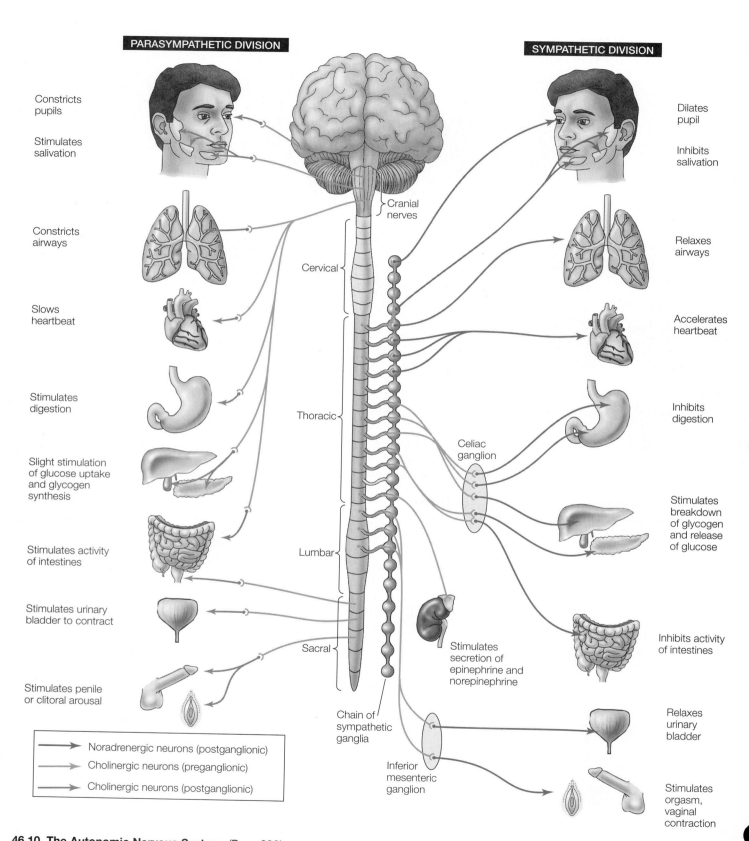

PARASYMPATHETIC DIVISION

SYMPATHETIC DIVISION

Constricts
pupils

Stimulates
salivation

Constricts
airways

Slows
heartbeat

Stimulates
digestion

Slight stimulation
of glucose uptake
and glycogen
synthesis

Stimulates activity
of intestines

Stimulates urinary
bladder to contract

Stimulates penile
or clitoral arousal

Cranial
nerves

Cervical

Thoracic

Lumbar

Sacral

Celiac
ganglion

Stimulates
secretion of
epinephrine and
norepinephrine

Chain of
sympathetic
ganglia

Inferior
mesenteric
ganglion

Dilates
pupil

Inhibits
salivation

Relaxes
airways

Accelerates
heartbeat

Inhibits
digestion

Stimulates
breakdown
of glycogen
and release
of glucose

Inhibits activity
of intestines

Relaxes
urinary
bladder

Stimulates
orgasm,
vaginal
contraction

⟶ Noradrenergic neurons (postganglionic)
⟶ Cholinergic neurons (preganglionic)
⟶ Cholinergic neurons (postganglionic)

46.10 The Autonomic Nervous System *(Page 993)*

EXPERIMENT

HYPOTHESIS: Retinal ganglion cells are excited or inhibited by light and dark stimuli falling on local areas of the retina.

METHOD

Place electrodes in the optic nerve to record from the axons of ganglion cells while stimulating the retina with different combinations of light and dark stimuli. The stimuli are moved around the retina to find the area of sensitivity for a particular ganglion cell.

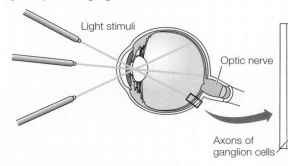

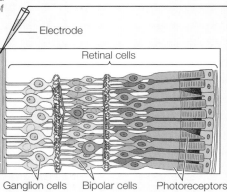

Electrode

Retinal cells

Recording of action potentials in axon of ganglion cell

Light stimuli

Optic nerve

Axons of ganglion cells

Ganglion cells   Bipolar cells   Photoreceptors

RESULTS

Some ganglion cells are maximally stimulated by light falling on the center of their receptive fields. Others are maximally stimulated by light falling on the surround of their receptive fields.

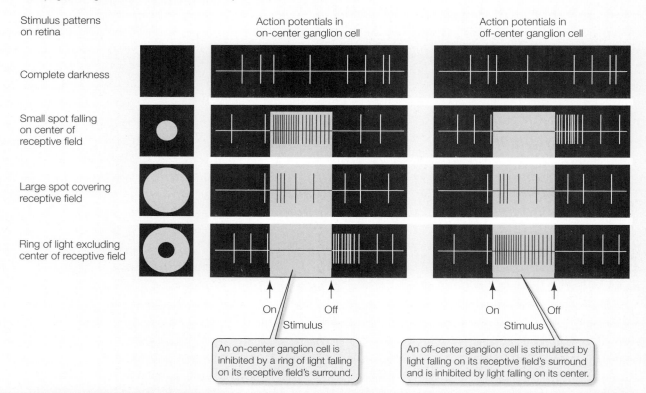

Stimulus patterns on retina

Action potentials in on-center ganglion cell

Action potentials in off-center ganglion cell

Complete darkness

Small spot falling on center of receptive field

Large spot covering receptive field

Ring of light excluding center of receptive field

On   Off
Stimulus

On   Off
Stimulus

An on-center ganglion cell is inhibited by a ring of light falling on its receptive field's surround.

An off-center ganglion cell is stimulated by light falling on its receptive field's surround and is inhibited by light falling on its center.

CONCLUSION: Ganglion cells use a center-surround dichotomy to encode patterns of contrast between light and dark.

46.11 What Does the Eye Tell the Brain? *(Page 994)*

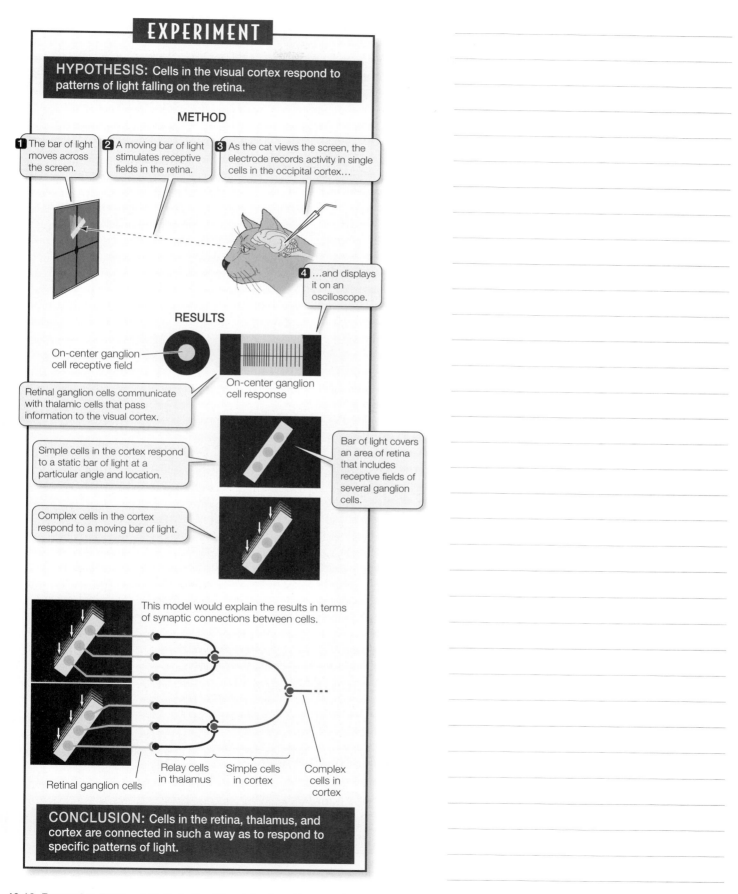

EXPERIMENT

HYPOTHESIS: Cells in the visual cortex respond to patterns of light falling on the retina.

METHOD

1 The bar of light moves across the screen.

2 A moving bar of light stimulates receptive fields in the retina.

3 As the cat views the screen, the electrode records activity in single cells in the occipital cortex…

4 …and displays it on an oscilloscope.

RESULTS

On-center ganglion cell receptive field

On-center ganglion cell response

Retinal ganglion cells communicate with thalamic cells that pass information to the visual cortex.

Simple cells in the cortex respond to a static bar of light at a particular angle and location.

Bar of light covers an area of retina that includes receptive fields of several ganglion cells.

Complex cells in the cortex respond to a moving bar of light.

This model would explain the results in terms of synaptic connections between cells.

Retinal ganglion cells

Relay cells in thalamus

Simple cells in cortex

Complex cells in cortex

CONCLUSION: Cells in the retina, thalamus, and cortex are connected in such a way as to respond to specific patterns of light.

46.12 Receptive Fields of Cells in the Visual Cortex *(Page 996)*

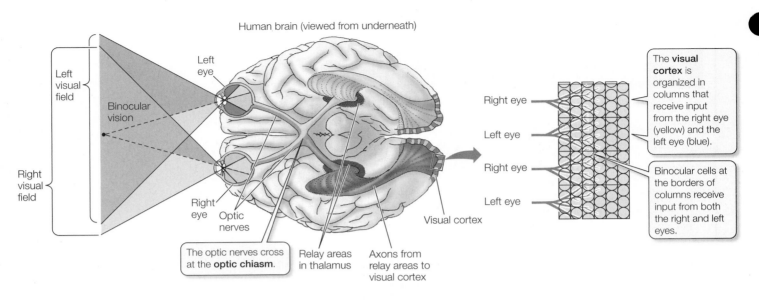

46.13 The Anatomy of Binocular Vision *(Page 997)*

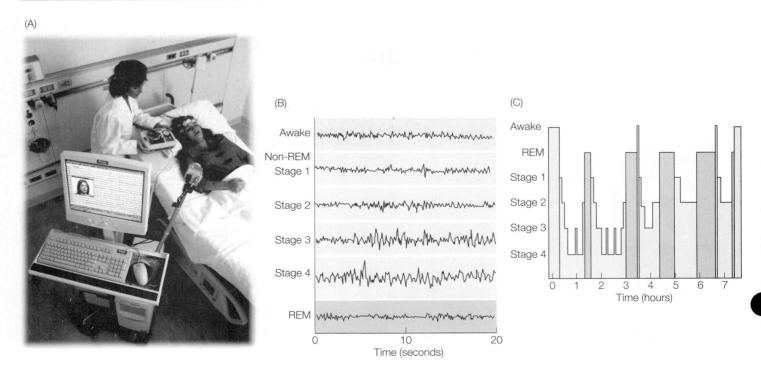

46.14 Patterns of Electrical Activity in the Cerebral Cortex Characterize Stages of Sleep *(Page 998)*

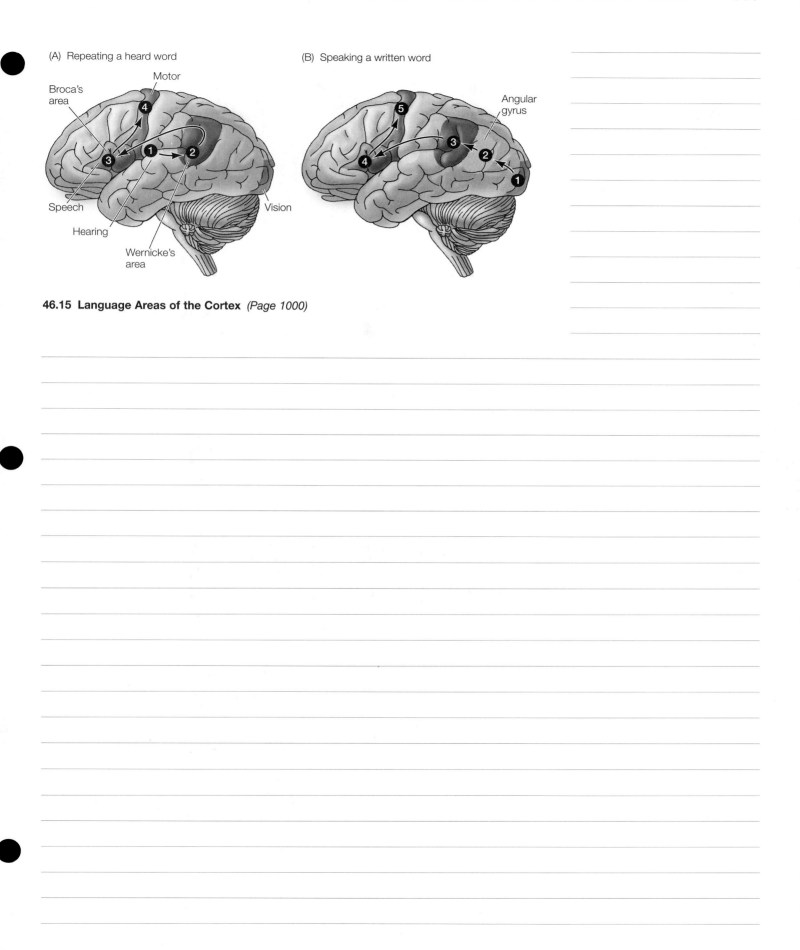

(A) Repeating a heard word

Broca's area

Motor

Speech

Hearing

Wernicke's area

Vision

(B) Speaking a written word

Angular gyrus

46.15 Language Areas of the Cortex *(Page 1000)*

CHAPTER 47 Effectors: How Animals Get Things Done

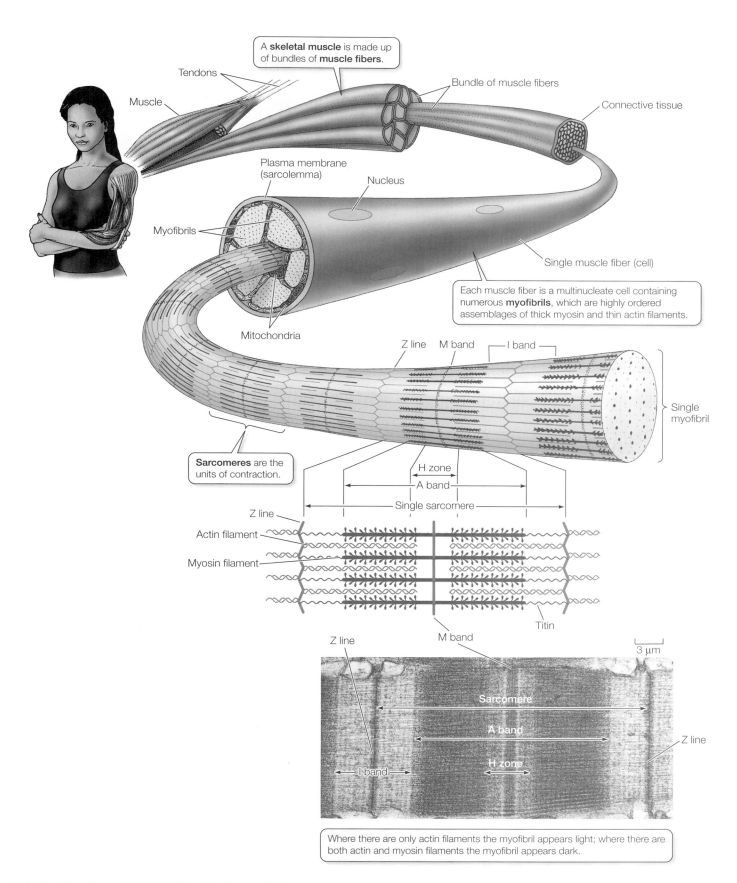

A **skeletal muscle** is made up of bundles of **muscle fibers**.

Tendons

Muscle

Bundle of muscle fibers

Connective tissue

Plasma membrane (sarcolemma)

Nucleus

Myofibrils

Mitochondria

Single muscle fiber (cell)

Each muscle fiber is a multinucleate cell containing numerous **myofibrils**, which are highly ordered assemblages of thick myosin and thin actin filaments.

Z line   M band   I band

Single myofibril

Sarcomeres are the units of contraction.

H zone

A band

Single sarcomere

Z line

Actin filament

Myosin filament

Titin

M band

Z line

3 μm

Sarcomere

A band

H zone

I band

Z line

Where there are only actin filaments the myofibril appears light; where there are both actin and myosin filaments the myofibril appears dark.

47.1 The Structure of Skeletal Muscle *(Page 1006)*

Muscle relaxed

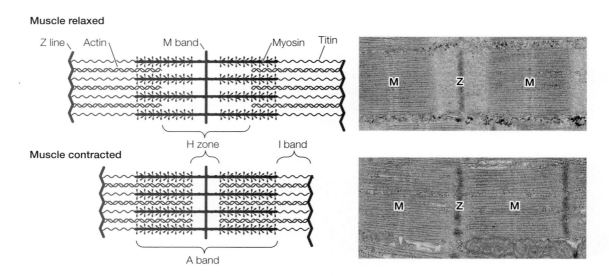

47.2 Sliding Filaments *(Page 1007)*

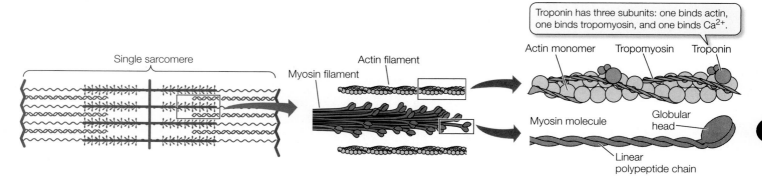

47.3 Actin and Myosin Filaments Overlap to Form Myofibrils *(Page 1007)*

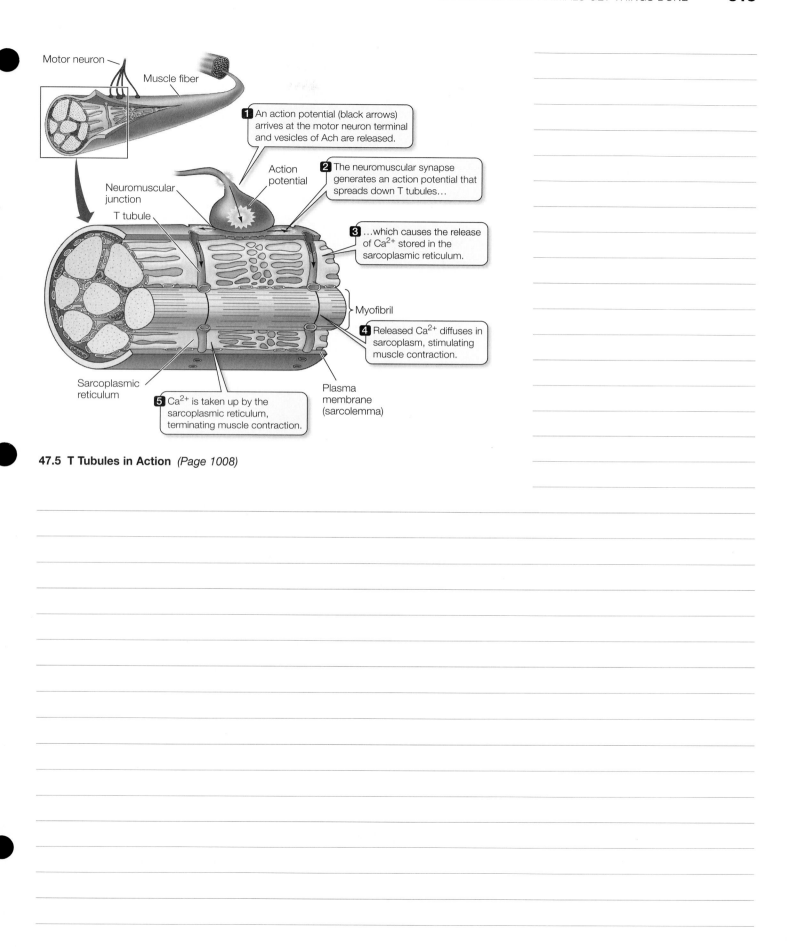

Motor neuron

Muscle fiber

1 An action potential (black arrows) arrives at the motor neuron terminal and vesicles of Ach are released.

Action potential

Neuromuscular junction

2 The neuromuscular synapse generates an action potential that spreads down T tubules…

T tubule

3 …which causes the release of Ca^{2+} stored in the sarcoplasmic reticulum.

Myofibril

4 Released Ca^{2+} diffuses in sarcoplasm, stimulating muscle contraction.

Sarcoplasmic reticulum

5 Ca^{2+} is taken up by the sarcoplasmic reticulum, terminating muscle contraction.

Plasma membrane (sarcolemma)

47.5 T Tubules in Action *(Page 1008)*

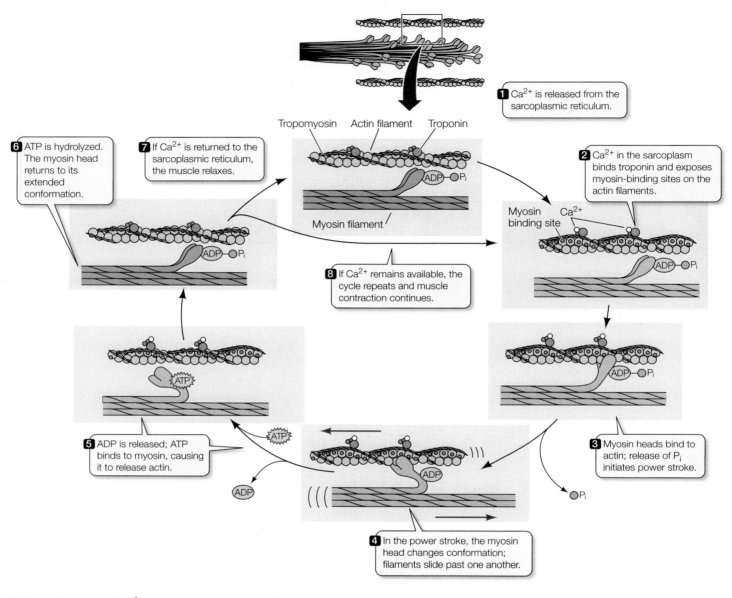

47.6 The Release of Ca²⁺ from the Sarcoplasmic Reticulum Triggers Muscle Contraction *(Page 1009)*

EXPERIMENT

HYPOTHESIS: Stretch and parasympathetic stimulation induce contraction in gut smooth muscle.

METHOD

Incubate a strip of smooth (intestinal) muscle in a saline bath. Measure action potentials and force of contraction.

Experiment 1 Stretch intestinal muscle and analyze response.

In Experiment 2 a pipette drips acetylcholine or norepinephrine onto strip.

2 An electrode detects action potentials in a muscle cell.

3 Muscle membrane potential and action potentials are recorded.

1 The muscle is anchored to a device that applies force to stretch the muscle.

Measuring electrode

Chart recorder

Amplifier

Reference electrode
(outside cell)

Force transducer

Measures muscle contractions

Intestinal muscle   Saline bath

4 The force of contraction of the muscle is measured by a force transducer.

RESULTS Stretching depolarizes the smooth muscle membrane. The depolarization causes action potentials that activate the contractile mechanism.

Experiment 2 Response of muscle strip to neurotransmitters of the autonomic nervous system.

When acetylcholine is dripped onto the muscle, the cells depolarize, fire action potentials more rapidly, and increase their force of contraction.

Norepinephrine, on the other hand, causes the cells to hyperpolarize, decreasing their rate of firing, and decreasing their force of contraction.

Apply acetylcholine   Wash out acetylcholine   Apply norepinephrine   Wash out norepinephrine

Membrane potential (mV)

+25
0
−25
−50

Muscle contracts

Muscle relaxes

Force

RESULTS Autonomic neurotransmitters alter membrane resting potential and thereby determine the rate that smooth muscle cells fire action potentials.

CONCLUSION: Gut smooth muscle contraction is stimulated by stretch and by the neurotransmitter acetylcholine.

47.8 Mechanisms of Smooth Muscle Activation *(Page 1011)*

(A)

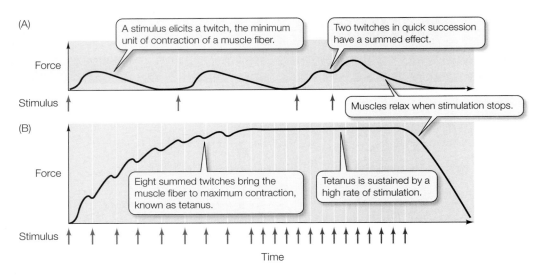

Force

A stimulus elicits a twitch, the minimum unit of contraction of a muscle fiber.

Two twitches in quick succession have a summed effect.

Stimulus

Muscles relax when stimulation stops.

(B)

Force

Eight summed twitches bring the muscle fiber to maximum contraction, known as tetanus.

Tetanus is sustained by a high rate of stimulation.

Stimulus

Time

47.9 Twitches and Tetanus *(Page 1012)*

Distance cyclist

Sprinter

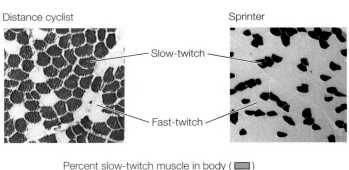

Slow-twitch

Fast-twitch

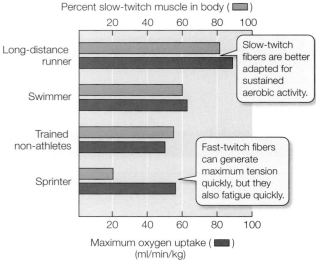

Percent slow-twitch muscle in body ( )

20   40   60   80   100

Long-distance runner

Slow-twitch fibers are better adapted for sustained aerobic activity.

Swimmer

Trained non-athletes

Fast-twitch fibers can generate maximum tension quickly, but they also fatigue quickly.

Sprinter

20   40   60   80   100

Maximum oxygen uptake ( )
(ml/min/kg)

47.10 Slow-and Fast-Twitch Muscle Fibers *(Page 1013)*

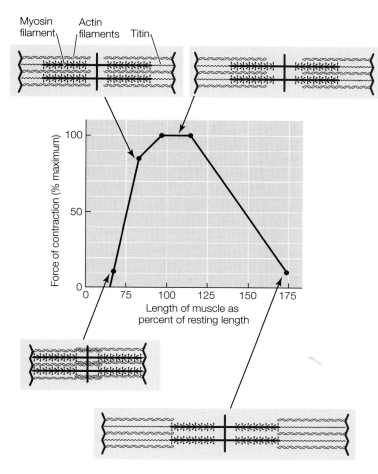

47.11 Strength and Length *(Page 1014)*

(A)

1 Pre-formed ATP and CP are immediately available but rapidly exhausted.

2 Glycolysis comes on line within seconds but lacks sustained efficiency.

3 Sustained ATP production by oxidative metabolism kicks in after about 1 minute.

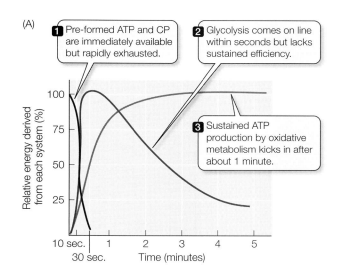

(B)

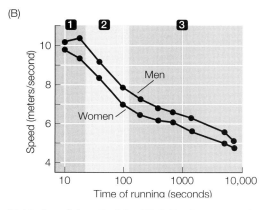

47.12 Supplying Fuel for High Performance *(Page 1015)*

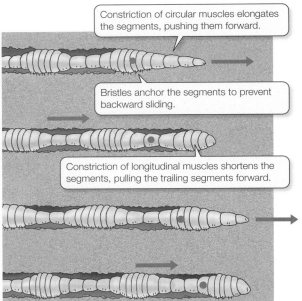

47.13 A Hydrostatic Skeleton *(Page 1016)*

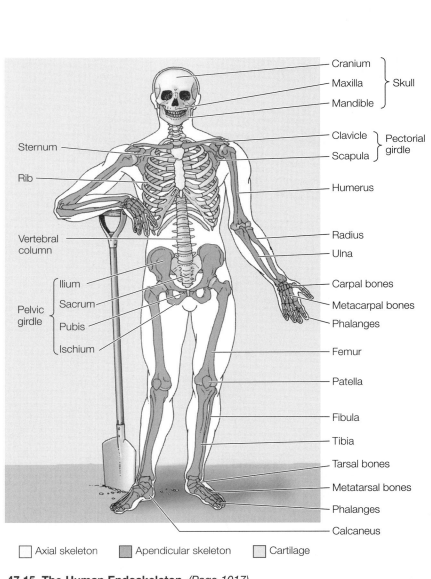

47.15 The Human Endoskeleton *(Page 1017)*

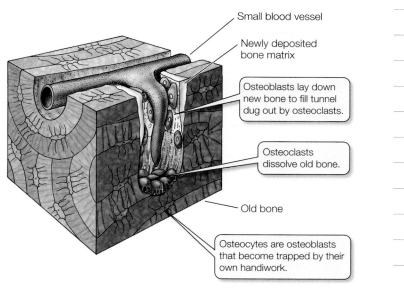

- Small blood vessel
- Newly deposited bone matrix
- Osteoblasts lay down new bone to fill tunnel dug out by osteoclasts.
- Osteoclasts dissolve old bone.
- Old bone
- Osteocytes are osteoblasts that become trapped by their own handiwork.

47.16 Renovating Bone *(Page 1018)*

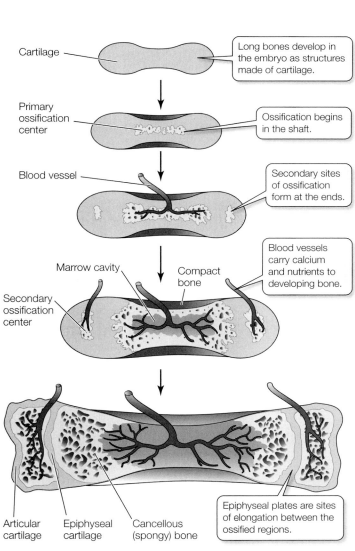

Cartilage — Long bones develop in the embryo as structures made of cartilage.

Primary ossification center — Ossification begins in the shaft.

Blood vessel — Secondary sites of ossification form at the ends.

Marrow cavity — Compact bone — Blood vessels carry calcium and nutrients to developing bone.

Secondary ossification center

Articular cartilage | Epiphyseal cartilage | Cancellous (spongy) bone — Epiphyseal plates are sites of elongation between the ossified regions.

47.17 The Growth of Long Bones *(Page 1018)*

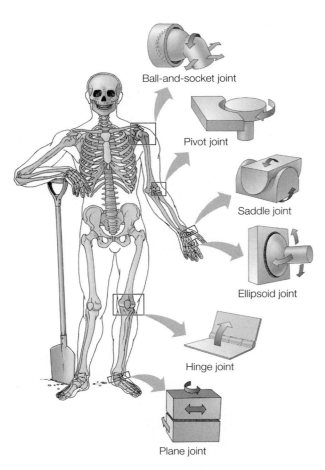

47.19 Types of Joints *(Page 1019)*

- Ball-and-socket joint
- Pivot joint
- Saddle joint
- Ellipsoid joint
- Hinge joint
- Plane joint

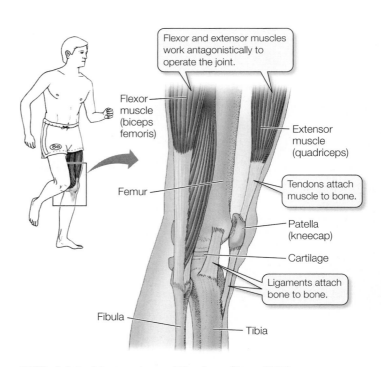

Flexor and extensor muscles work antagonistically to operate the joint.

Flexor muscle (biceps femoris)

Extensor muscle (quadriceps)

Tendons attach muscle to bone.

Femur

Patella (kneecap)

Cartilage

Ligaments attach bone to bone.

Fibula

Tibia

47.20 Joints, Ligaments, and Tendons *(Page 1019)*

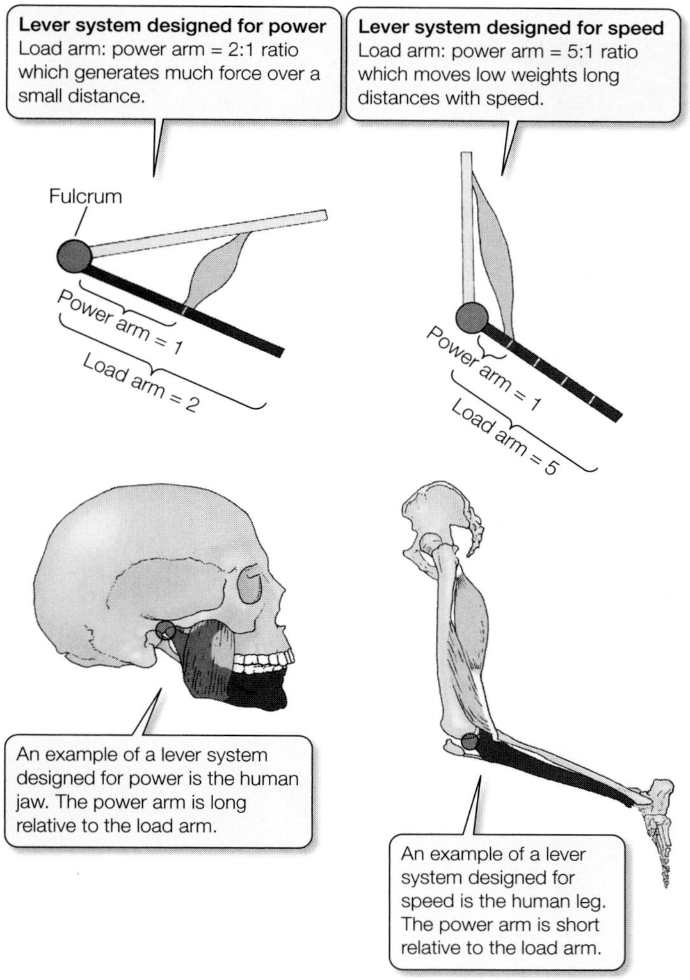

Lever system designed for power
Load arm: power arm = 2:1 ratio which generates much force over a small distance.

Lever system designed for speed
Load arm: power arm = 5:1 ratio which moves low weights long distances with speed.

Fulcrum

Power arm = 1

Load arm = 2

Power arm = 1

Load arm = 5

An example of a lever system designed for power is the human jaw. The power arm is long relative to the load arm.

An example of a lever system designed for speed is the human leg. The power arm is short relative to the load arm.

47.21 Bones and Joints Work Like Systems of Levers *(Page 1020)*

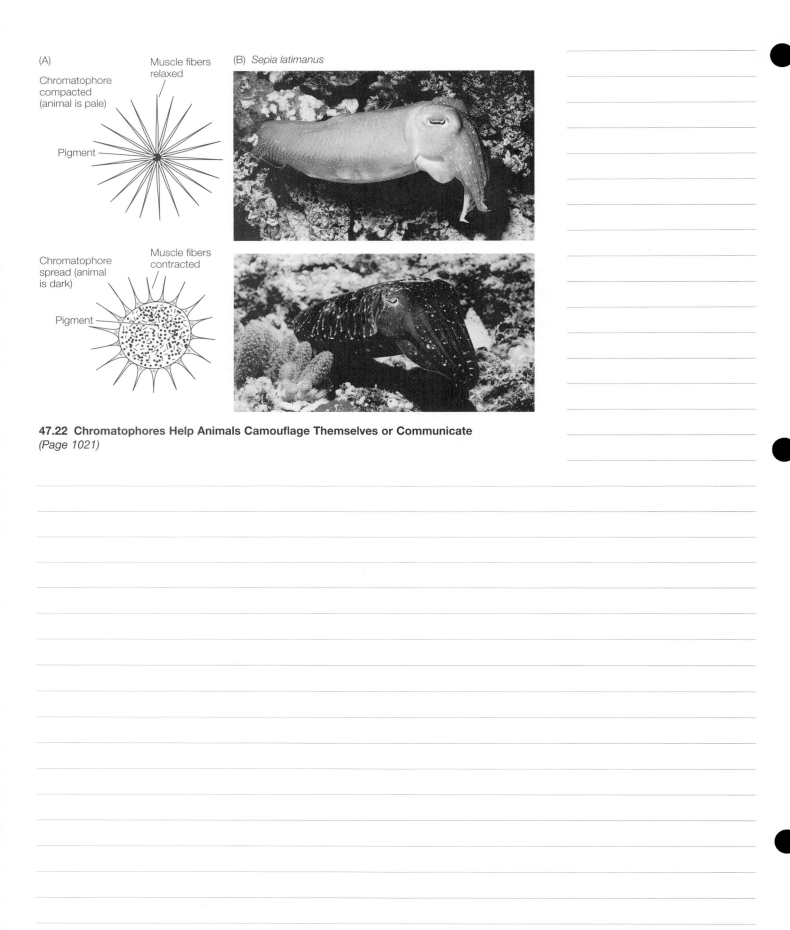

47.22 Chromatophores Help Animals Camouflage Themselves or Communicate
(Page 1021)

CHAPTER 48 Gas Exchange in Animals

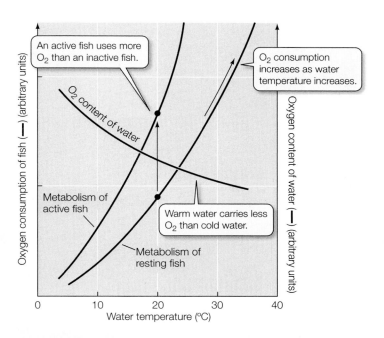

48.2 The Double Bind of Water Breathers *(Page 1027)*

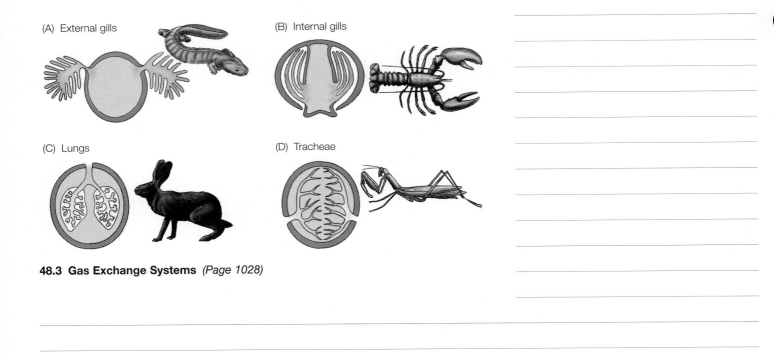

(A) External gills

(B) Internal gills

(C) Lungs

(D) Tracheae

48.3 Gas Exchange Systems *(Page 1028)*

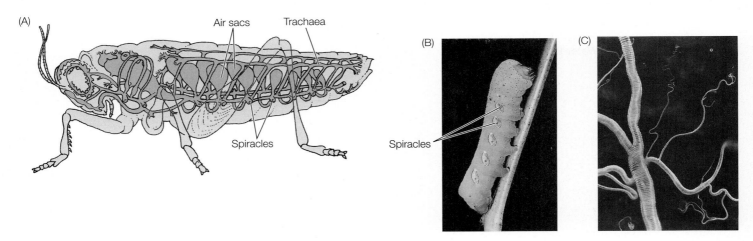

(A)

Air sacs   Trachaea

Spiracles

(B)

Spiracles

(C)

48.4 The Tracheal Gas Exchange System of Insects *(Page 1029)*

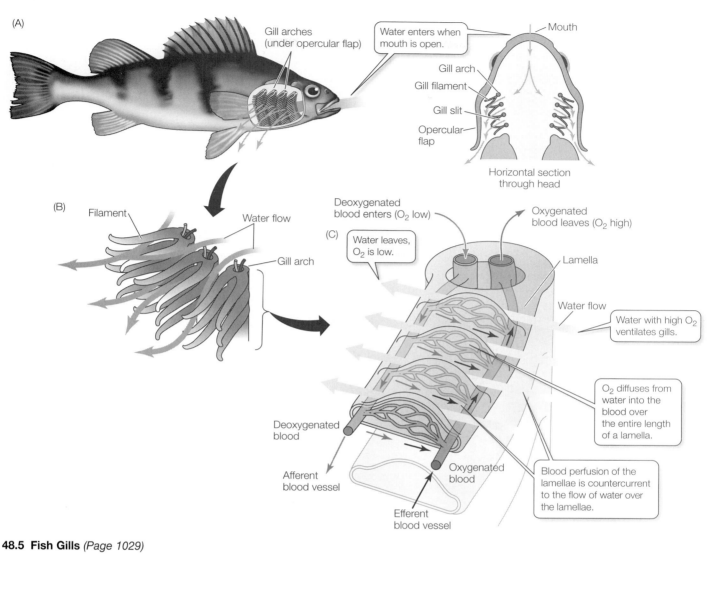

(A)

Gill arches
(under opercular flap)

Water enters when
mouth is open.

Mouth

Gill arch

Gill filament

Gill slit

Opercular
flap

Horizontal section
through head

(B)

Filament

Water flow

Gill arch

(C)

Deoxygenated
blood enters (O₂ low)

Oxygenated
blood leaves (O₂ high)

Water leaves,
O₂ is low.

Lamella

Water flow

Water with high O₂
ventilates gills.

O₂ diffuses from
water into the
blood over
the entire length
of a lamella.

Deoxygenated
blood

Afferent
blood vessel

Oxygenated
blood

Efferent
blood vessel

Blood perfusion of the
lamellae is countercurrent
to the flow of water over
the lamellae.

48.5 Fish Gills *(Page 1029)*

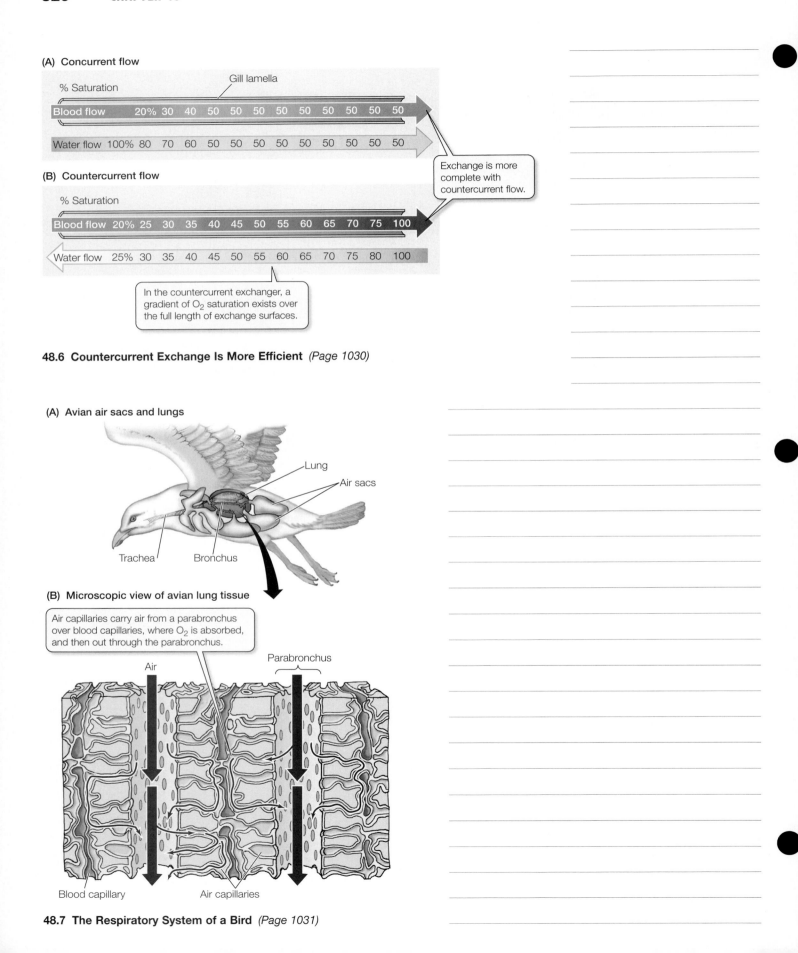

(A) Concurrent flow

Gill lamella

% Saturation

| Blood flow | 20% | 30 | 40 | 50 | 50 | 50 | 50 | 50 | 50 | 50 | 50 | 50 |

| Water flow | 100% | 80 | 70 | 60 | 50 | 50 | 50 | 50 | 50 | 50 | 50 | 50 |

(B) Countercurrent flow

% Saturation

| Blood flow | 20% | 25 | 30 | 35 | 40 | 45 | 50 | 55 | 60 | 65 | 70 | 75 | 100 |

| Water flow | 25% | 30 | 35 | 40 | 45 | 50 | 55 | 60 | 65 | 70 | 75 | 80 | 100 |

Exchange is more complete with countercurrent flow.

In the countercurrent exchanger, a gradient of O_2 saturation exists over the full length of exchange surfaces.

48.6 Countercurrent Exchange Is More Efficient *(Page 1030)*

(A) Avian air sacs and lungs

Lung

Air sacs

Trachea

Bronchus

(B) Microscopic view of avian lung tissue

Air capillaries carry air from a parabronchus over blood capillaries, where O_2 is absorbed, and then out through the parabronchus.

Air

Parabronchus

Blood capillary

Air capillaries

48.7 The Respiratory System of a Bird *(Page 1031)*

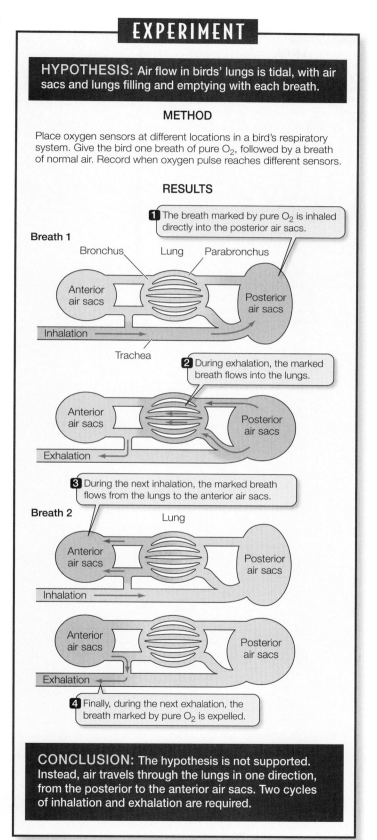

EXPERIMENT

HYPOTHESIS: Air flow in birds' lungs is tidal, with air sacs and lungs filling and emptying with each breath.

METHOD

Place oxygen sensors at different locations in a bird's respiratory system. Give the bird one breath of pure O_2, followed by a breath of normal air. Record when oxygen pulse reaches different sensors.

RESULTS

1 The breath marked by pure O_2 is inhaled directly into the posterior air sacs.

Breath 1

Bronchus   Lung   Parabronchus

Anterior air sacs   Posterior air sacs

Inhalation

Trachea

2 During exhalation, the marked breath flows into the lungs.

Anterior air sacs   Posterior air sacs

Exhalation

3 During the next inhalation, the marked breath flows from the lungs to the anterior air sacs.

Breath 2

Lung

Anterior air sacs   Posterior air sacs

Inhalation

Anterior air sacs   Posterior air sacs

Exhalation

4 Finally, during the next exhalation, the breath marked by pure O_2 is expelled.

CONCLUSION: The hypothesis is not supported. Instead, air travels through the lungs in one direction, from the posterior to the anterior air sacs. Two cycles of inhalation and exhalation are required.

48.8 The Path of Air Flow through Bird Lungs *(Page 1031)*

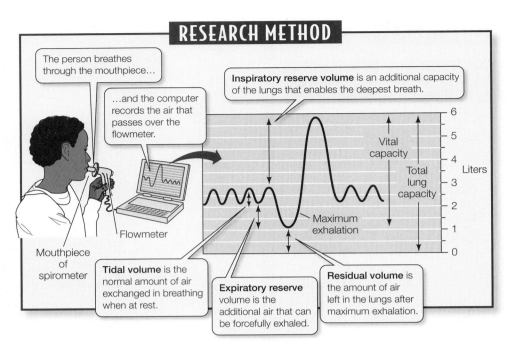

48.9 Measuring Lung Ventilation *(Page 1032)*

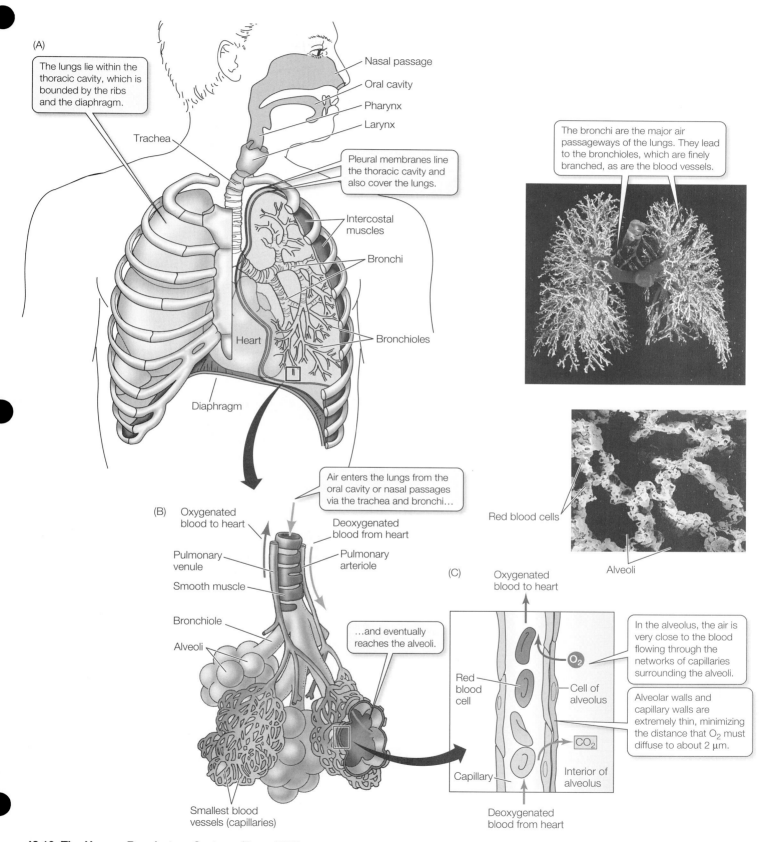

(A)

The lungs lie within the thoracic cavity, which is bounded by the ribs and the diaphragm.

Nasal passage

Oral cavity

Pharynx

Larynx

Trachea

Pleural membranes line the thoracic cavity and also cover the lungs.

Intercostal muscles

Bronchi

Bronchioles

Heart

Diaphragm

The bronchi are the major air passageways of the lungs. They lead to the bronchioles, which are finely branched, as are the blood vessels.

Red blood cells

Alveoli

Air enters the lungs from the oral cavity or nasal passages via the trachea and bronchi...

(B) Oxygenated blood to heart

Deoxygenated blood from heart

Pulmonary venule

Pulmonary arteriole

Smooth muscle

Bronchiole

Alveoli

...and eventually reaches the alveoli.

(C) Oxygenated blood to heart

Red blood cell

O_2

In the alveolus, the air is very close to the blood flowing through the networks of capillaries surrounding the alveoli.

Cell of alveolus

Alveolar walls and capillary walls are extremely thin, minimizing the distance that O_2 must diffuse to about 2 μm.

CO_2

Capillary

Interior of alveolus

Smallest blood vessels (capillaries)

Deoxygenated blood from heart

48.10 The Human Respiratory System *(Page 1033)*

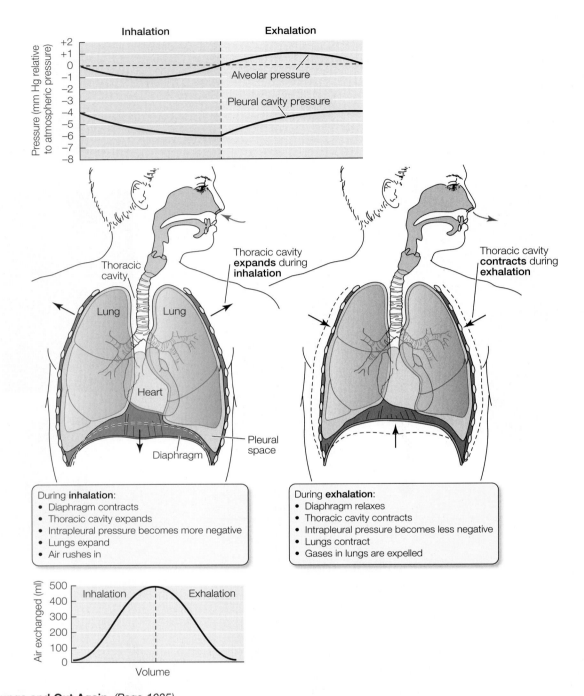

During **inhalation**:
• Diaphragm contracts
• Thoracic cavity expands
• Intrapleural pressure becomes more negative
• Lungs expand
• Air rushes in

During **exhalation**:
• Diaphragm relaxes
• Thoracic cavity contracts
• Intrapleural pressure becomes less negative
• Lungs contract
• Gases in lungs are expelled

48.11 Into the Lungs and Out Again *(Page 1035)*

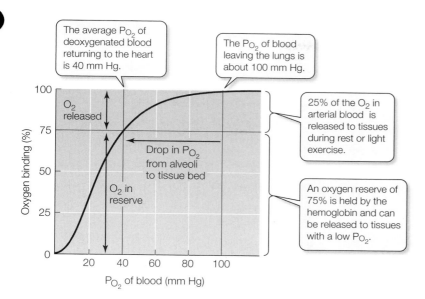

The average P_{O_2} of deoxygenated blood returning to the heart is 40 mm Hg.

The P_{O_2} of blood leaving the lungs is about 100 mm Hg.

25% of the O_2 in arterial blood is released to tissues during rest or light exercise.

An oxygen reserve of 75% is held by the hemoglobin and can be released to tissues with a low P_{O_2}.

O_2 released

Drop in P_{O_2} from alveoli to tissue bed

O_2 in reserve

Oxygen binding (%)

P_{O_2} of blood (mm Hg)

48.12 The Binding of O$_2$ to Hemoglobin Depends on P$_{O_2}$ *(Page 1036)*

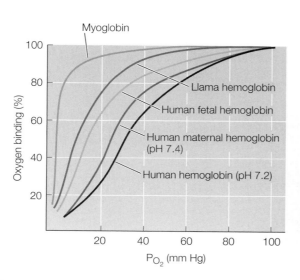

Myoglobin

Llama hemoglobin

Human fetal hemoglobin

Human maternal hemoglobin (pH 7.4)

Human hemoglobin (pH 7.2)

Oxygen binding (%)

P_{O_2} (mm Hg)

Llama guanicoe

48.13 Oxygen-Binding Adaptation *(Page 1037)*

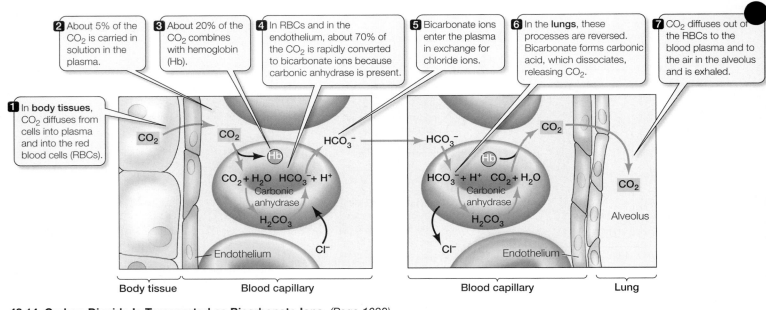

2 About 5% of the CO_2 is carried in solution in the plasma.

3 About 20% of the CO_2 combines with hemoglobin (Hb).

4 In RBCs and in the endothelium, about 70% of the CO_2 is rapidly converted to bicarbonate ions because carbonic anhydrase is present.

5 Bicarbonate ions enter the plasma in exchange for chloride ions.

6 In the **lungs**, these processes are reversed. Bicarbonate forms carbonic acid, which dissociates, releasing CO_2.

7 CO_2 diffuses out of the RBCs to the blood plasma and to the air in the alveolus and is exhaled.

1 In **body tissues**, CO_2 diffuses from cells into plasma and into the red blood cells (RBCs).

CO_2

CO_2

Hb

HCO_3^-

HCO_3^-

CO_2

Hb

CO_2

$CO_2 + H_2O \quad HCO_3^- + H^+$
Carbonic anhydrase

$HCO_3^- + H^+ \quad CO_2 + H_2O$
Carbonic anhydrase

H_2CO_3

H_2CO_3

Cl^-

Cl^-

Endothelium

Alveolus

Endothelium

Body tissue | Blood capillary

Blood capillary | Lung

48.14 Carbon Dioxide Is Transported as Bicarbonate Ions *(Page 1038)*

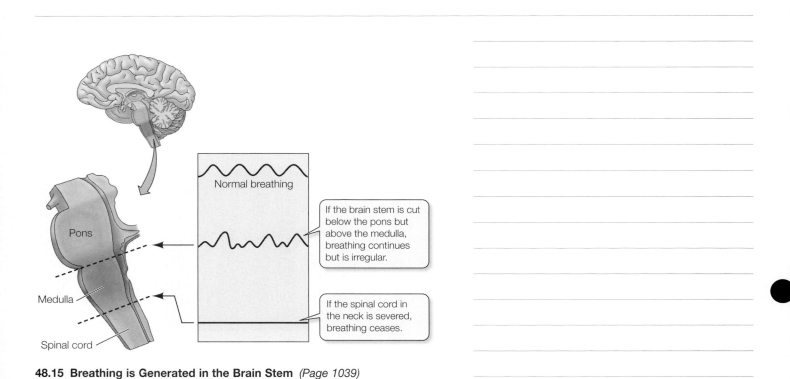

Normal breathing

Pons

If the brain stem is cut below the pons but above the medulla, breathing continues but is irregular.

Medulla

If the spinal cord in the neck is severed, breathing ceases.

Spinal cord

48.15 Breathing is Generated in the Brain Stem *(Page 1039)*

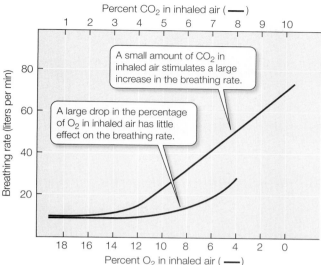

A small amount of CO_2 in inhaled air stimulates a large increase in the breathing rate.

A large drop in the percentage of O_2 in inhaled air has little effect on the breathing rate.

48.16 Carbon Dioxide Affects Breathing Rate *(Page 1039)*

HYPOTHESIS: The rising level of CO_2 in the blood during exercise is the feedback signal that stimulates the increase in respiratory rate.

METHOD

Dogs are equipped for measuring their respiratory rate (L/min) and they are trained to run on a treadmill. Blood samples are taken for measurement of CO_2 levels.

Catheter for taking blood samples

To flowmeter and respiratory analyzer

Experiment 1

Dog runs with the treadmill set at different speeds. The respiratory rate is plotted as a function of arterial CO_2 concentration.

RESULTS

Respiratory rate changes with running speed, but arterial levels of CO_2 do not. The hypothesis is not supported.

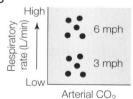

Experiment 2

Dog runs on the treadmill at the same speed, but the slope of the treadmill is elevated, increasing the work load.

RESULTS

Respiratory rate and arterial levels of CO_2 both rise as work load increases. The hypothesis is supported.

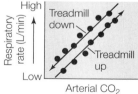

CONCLUSION: Arterial levels of CO_2 are the metabolic feedback signal that regulates respiration in response to work load—an example of feedback control. Information from sensors in joints and muscles that signal rate of limb movement can change the sensitivity of the system to CO_2—an example of feedforward control.

48.17 The Sensitivity of the Respiratory Control System Changes with Exercise *(Page 1040)*

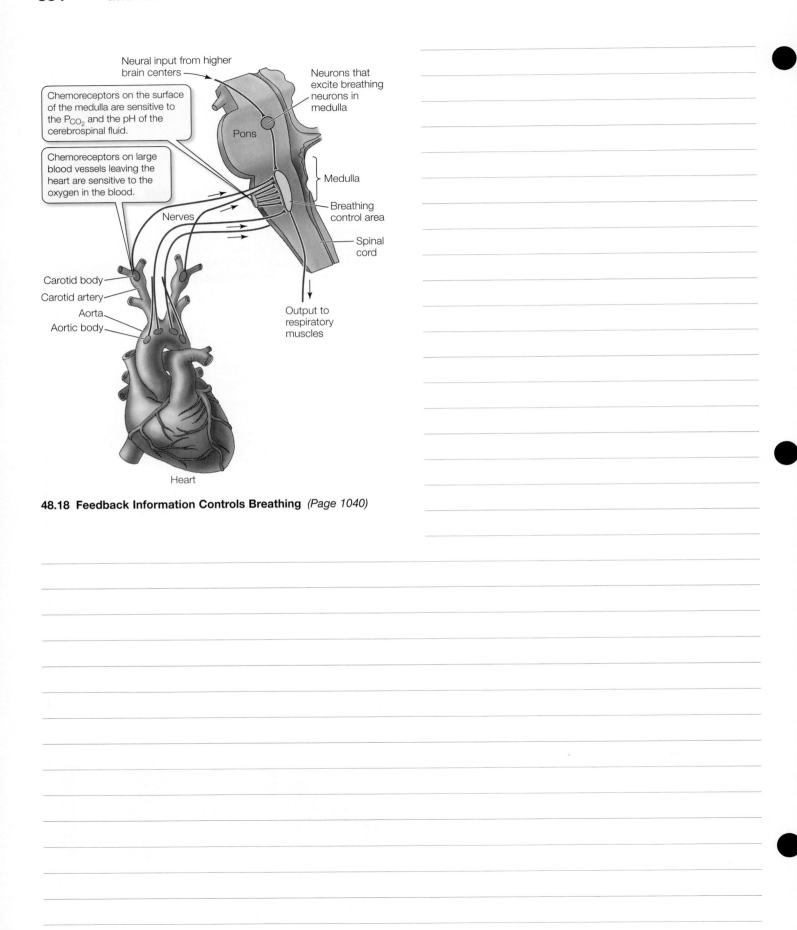

Neural input from higher brain centers

Chemoreceptors on the surface of the medulla are sensitive to the P_{CO_2} and the pH of the cerebrospinal fluid.

Chemoreceptors on large blood vessels leaving the heart are sensitive to the oxygen in the blood.

Neurons that excite breathing neurons in medulla

Pons

Medulla

Breathing control area

Spinal cord

Nerves

Carotid body

Carotid artery

Aorta

Aortic body

Output to respiratory muscles

Heart

48.18 Feedback Information Controls Breathing *(Page 1040)*

(A) Arthropod

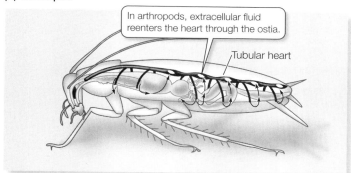

In arthropods, extracellular fluid reenters the heart through the ostia.

Tubular heart

(B) Mollusk

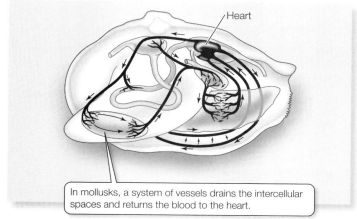

Heart

In mollusks, a system of vessels drains the intercellular spaces and returns the blood to the heart.

(C) Annelid worm

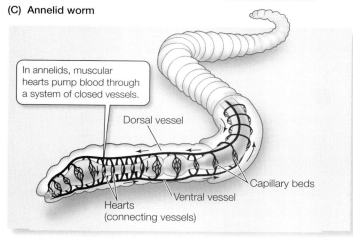

In annelids, muscular hearts pump blood through a system of closed vessels.

Dorsal vessel

Capillary beds

Ventral vessel

Hearts (connecting vessels)

49.1 Circulatory Systems *(Page 1046)*

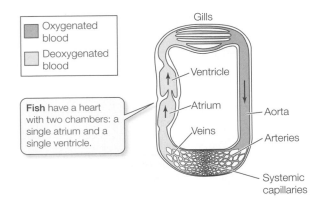

Oxygenated blood

Deoxygenated blood

Gills

Ventricle

Atrium

Veins

Aorta

Arteries

Systemic capillaries

Fish have a heart with two chambers: a single atrium and a single ventricle.

Page 1047 In-Text Art

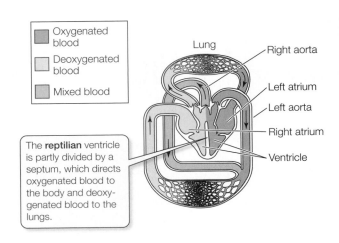

Oxygenated blood

Deoxygenated blood

Mixed blood

Lung

Right aorta

Left atrium

Left aorta

Right atrium

Ventricle

The **reptilian** ventricle is partly divided by a septum, which directs oxygenated blood to the body and deoxygenated blood to the lungs.

Page 1048 In-Text Art (3)

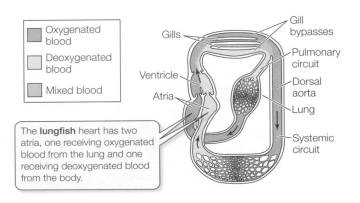

Oxygenated blood

Deoxygenated blood

Mixed blood

Gills

Gill bypasses

Pulmonary circuit

Ventricle

Dorsal aorta

Atria

Lung

Systemic circuit

The **lungfish** heart has two atria, one receiving oxygenated blood from the lung and one receiving deoxygenated blood from the body.

Page 1048 In-Text Art (1)

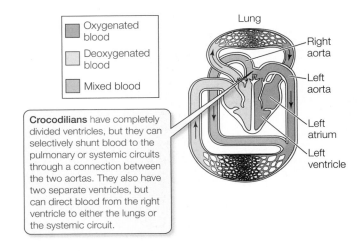

Oxygenated blood

Deoxygenated blood

Mixed blood

Lung

Right aorta

Left aorta

Left atrium

Left ventricle

Crocodilians have completely divided ventricles, but they can selectively shunt blood to the pulmonary or systemic circuits through a connection between the two aortas. They also have two separate ventricles, but can direct blood from the right ventricle to either the lungs or the systemic circuit.

Page 1049 In-Text Art (1)

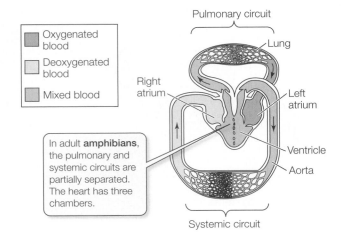

Pulmonary circuit

Oxygenated blood

Deoxygenated blood

Mixed blood

Right atrium

Lung

Left atrium

Ventricle

Aorta

In adult **amphibians**, the pulmonary and systemic circuits are partially separated. The heart has three chambers.

Systemic circuit

Page 1048 In-Text Art (2)

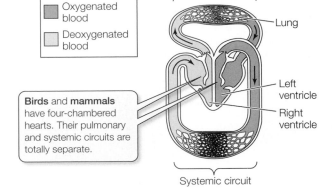

Pulmonary circuit

Oxygenated blood

Deoxygenated blood

Lung

Left ventricle

Right ventricle

Birds and **mammals** have four-chambered hearts. Their pulmonary and systemic circuits are totally separate.

Systemic circuit

Page 1049 In-Text Art (2)

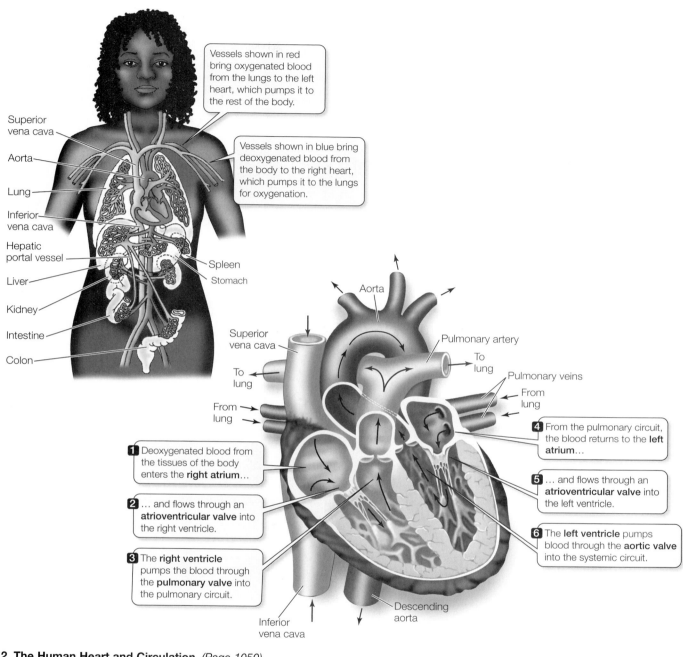

49.2 The Human Heart and Circulation *(Page 1050)*

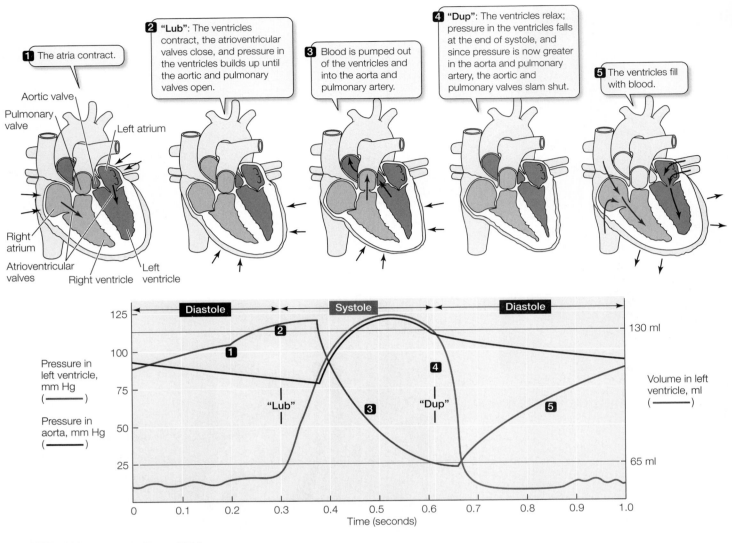

1 The atria contract.

2 "Lub": The ventricles contract, the atrioventricular valves close, and pressure in the ventricles builds up until the aortic and pulmonary valves open.

3 Blood is pumped out of the ventricles and into the aorta and pulmonary artery.

4 "Dup": The ventricles relax; pressure in the ventricles falls at the end of systole, and since pressure is now greater in the aorta and pulmonary artery, the aortic and pulmonary valves slam shut.

5 The ventricles fill with blood.

Aortic valve
Pulmonary valve
Left atrium
Right atrium
Atrioventricular valves
Right ventricle
Left ventricle

Diastole   Systole   Diastole

Pressure in left ventricle, mm Hg (————)

Pressure in aorta, mm Hg (————)

Volume in left ventricle, ml (————)

130 ml
65 ml

"Lub"   "Dup"

Time (seconds)

49.3 The Cardiac Cycle *(Page 1051)*

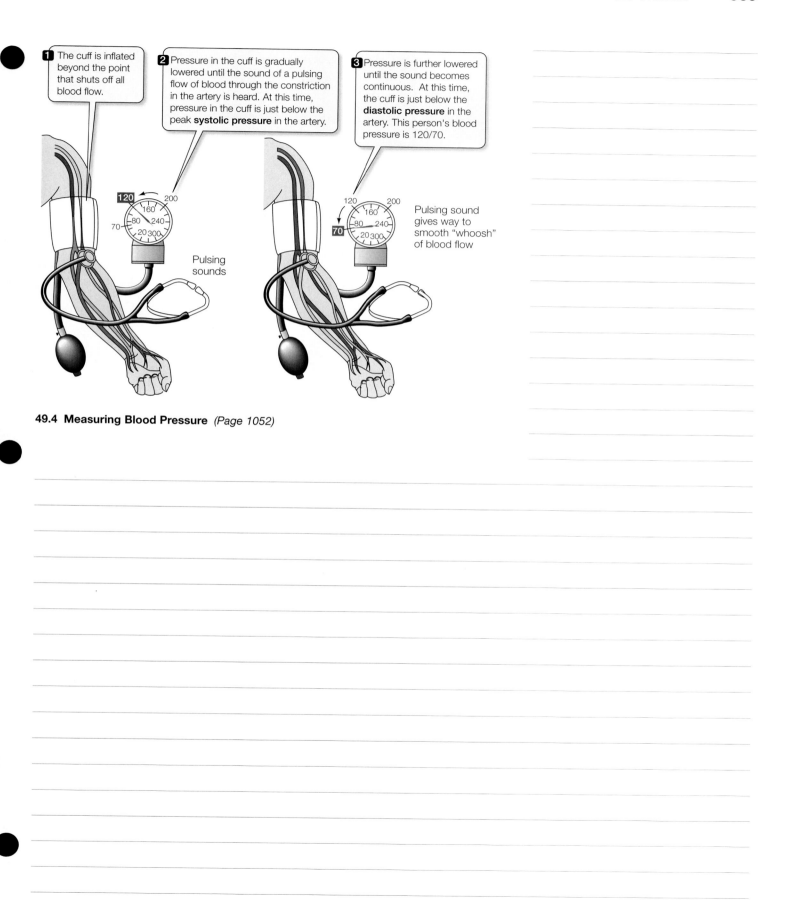

49.4 Measuring Blood Pressure *(Page 1052)*

EXPERIMENT

HYPOTHESIS: Sympathetic and parasympathetic neurotransmitters influence the resting potentials of pacemaker cells.

METHOD

Culture a sinoatrial node in a dish and use an intracellular recording electrode to measure the membrane potential of pacemaker cells while applying acetylcholine and norepinephrine to the node.

RESULTS

Control recording (purple) shows that the resting potential of these cells gradually depolarizes after an action potential is fired.

When norepinephrine is applied to the node, the rate of depolarization of the resting potential increases. Time between action potentials decreases and the heart rate increases.

When acetylcholine is applied to the node, the resting potential is more negative following an action potential and the rate of depolarization is slower. Time between action potentials increases and the heart rate declines.

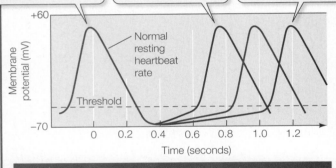

Normal resting heartbeat rate

Threshold

Membrane potential (mV)

+60

−70

0   0.2   0.4   0.6   0.8   1.0   1.2

Time (seconds)

CONCLUSION: The sympathetic and parasympathetic neurotransmitters control the heart rate by influencing the ion channels that determine the resting potential of pacemaker cells.

49.5 The Autonomic Nervous System Controls Heart Rate *(Page 1053)*

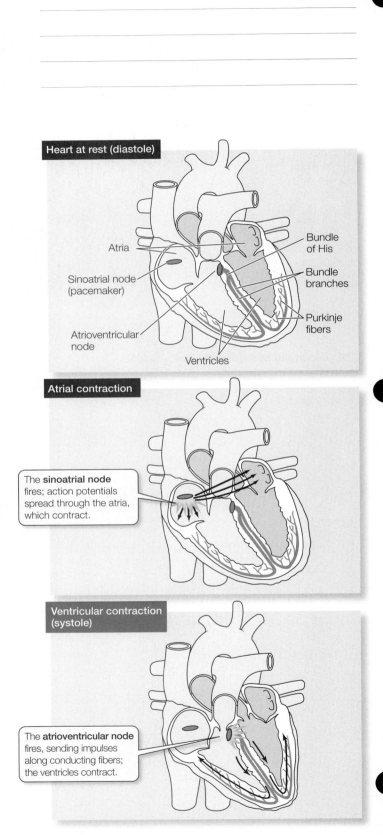

Heart at rest (diastole)

Atria

Sinoatrial node (pacemaker)

Atrioventricular node

Ventricles

Bundle of His

Bundle branches

Purkinje fibers

Atrial contraction

The **sinoatrial node** fires; action potentials spread through the atria, which contract.

Ventricular contraction (systole)

The **atrioventricular node** fires, sending impulses along conducting fibers; the ventricles contract.

49.6 The Heartbeat *(Page 1053)*

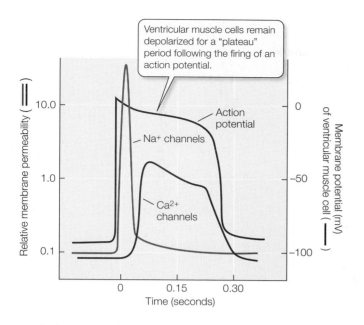

Ventricular muscle cells remain depolarized for a "plateau" period following the firing of an action potential.

Action potential

Na+ channels

Ca2+ channels

49.7 The Action Potential of Ventricular Muscle Fibers
(Page 1054)

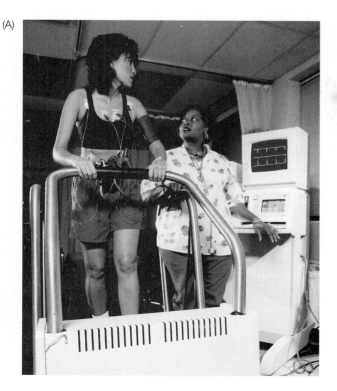

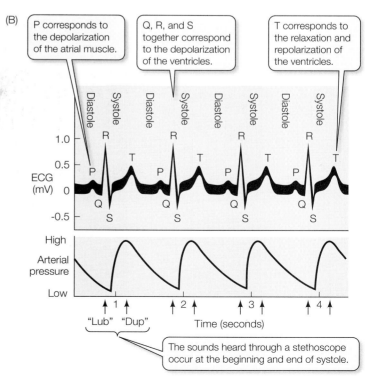

(B)

P corresponds to the depolarization of the atrial muscle.

Q, R, and S together correspond to the depolarization of the ventricles.

T corresponds to the relaxation and repolarization of the ventricles.

The sounds heard through a stethoscope occur at the beginning and end of systole.

49.8 The Electrocardiogram *(Page 1054)*

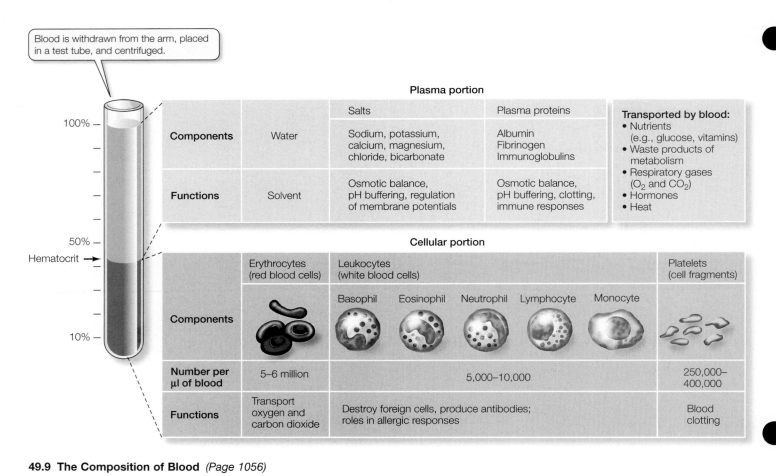

49.9 The Composition of Blood *(Page 1056)*

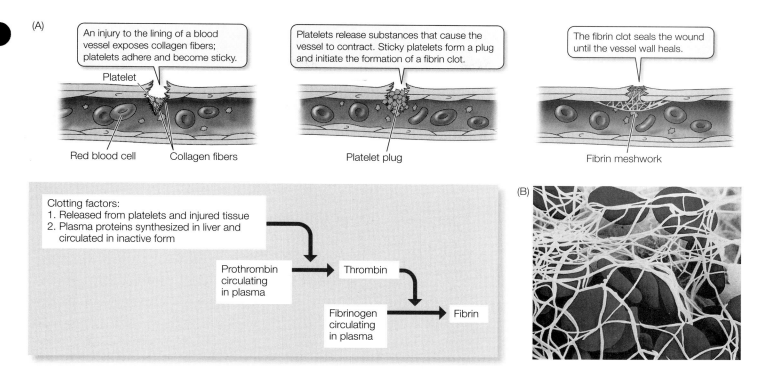

(A)

An injury to the lining of a blood vessel exposes collagen fibers; platelets adhere and become sticky.

Platelets release substances that cause the vessel to contract. Sticky platelets form a plug and initiate the formation of a fibrin clot.

The fibrin clot seals the wound until the vessel wall heals.

Platelet

Red blood cell   Collagen fibers   Platelet plug   Fibrin meshwork

Clotting factors:
1. Released from platelets and injured tissue
2. Plasma proteins synthesized in liver and circulated in inactive form

Prothrombin circulating in plasma → Thrombin

Fibrinogen circulating in plasma → Fibrin

(B)

49.10 Blood Clotting *(Page 1057)*

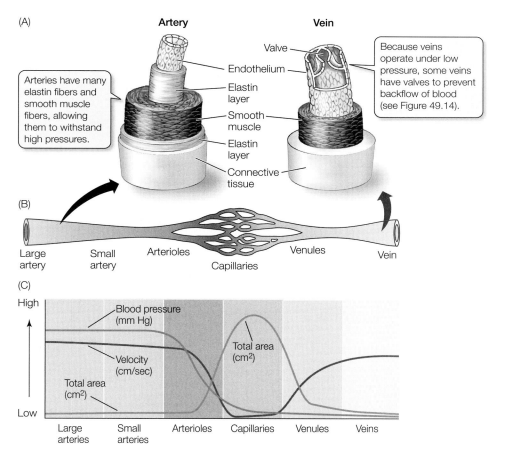

(A)

Artery   **Vein**

Arteries have many elastin fibers and smooth muscle fibers, allowing them to withstand high pressures.

Valve

Endothelium

Elastin layer

Smooth muscle

Elastin layer

Connective tissue

Because veins operate under low pressure, some veins have valves to prevent backflow of blood (see Figure 49.14).

(B)

Large artery   Small artery   Arterioles   Capillaries   Venules   Vein

(C)

High

Blood pressure (mm Hg)

Velocity (cm/sec)

Total area (cm2)

Total area (cm2)

Low

Large arteries   Small arteries   Arterioles   Capillaries   Venules   Veins

49.11 Anatomy of Blood Vessels *(Page 1058)*

(A)

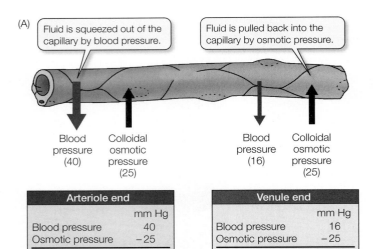

Fluid is squeezed out of the capillary by blood pressure.

Fluid is pulled back into the capillary by osmotic pressure.

Blood pressure (40)

Colloidal osmotic pressure (25)

Blood pressure (16)

Colloidal osmotic pressure (25)

Arteriole end	
	mm Hg
Blood pressure	40
Osmotic pressure	−25
Net outward force	15

Venule end	
	mm Hg
Blood pressure	16
Osmotic pressure	−25
Net inward force	−9

(B)

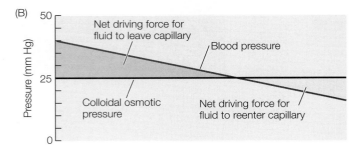

Net driving force for fluid to leave capillary

Blood pressure

Colloidal osmotic pressure

Net driving force for fluid to reenter capillary

49.13 Starling's Forces *(Page 1059)*

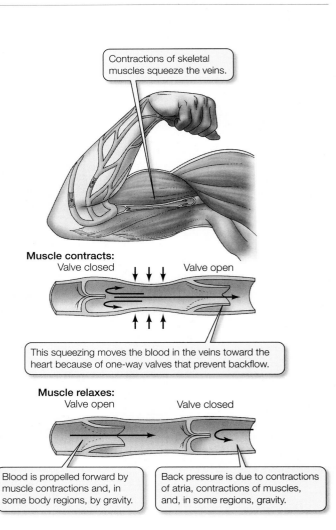

Contractions of skeletal muscles squeeze the veins.

Muscle contracts:
Valve closed   Valve open

This squeezing moves the blood in the veins toward the heart because of one-way valves that prevent backflow.

Muscle relaxes:
Valve open   Valve closed

Blood is propelled forward by muscle contractions and, in some body regions, by gravity.

Back pressure is due to contractions of atria, contractions of muscles, and, in some regions, gravity.

49.14 One-Way Flow *(Page 1060)*

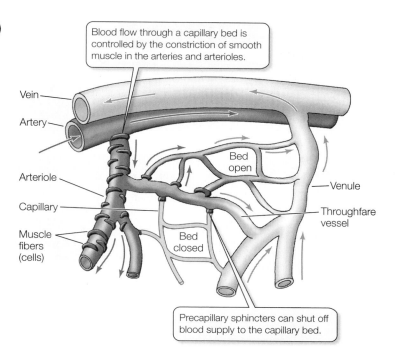

Blood flow through a capillary bed is controlled by the constriction of smooth muscle in the arteries and arterioles.

Vein

Artery

Arteriole

Capillary

Muscle fibers (cells)

Bed open

Venule

Throughfare vessel

Bed closed

Precapillary sphincters can shut off blood supply to the capillary bed.

49.16 Local Control of Blood Flow *(Page 1062)*

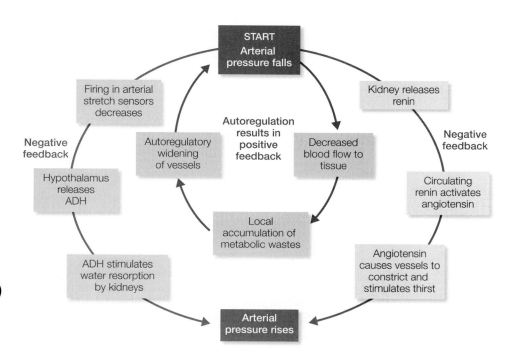

START
Arterial pressure falls

Firing in arterial stretch sensors decreases

Kidney releases renin

Negative feedback

Autoregulation results in positive feedback

Autoregulatory widening of vessels

Decreased blood flow to tissue

Negative feedback

Hypothalamus releases ADH

Circulating renin activates angiotensin

Local accumulation of metabolic wastes

ADH stimulates water resorption by kidneys

Angiotensin causes vessels to constrict and stimulates thirst

Arterial pressure rises

49.17 Control of Blood Pressure through Vascular Resistance *(Page 1062)*

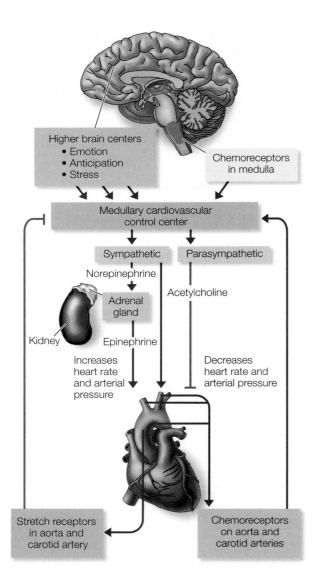

49.18 Regulating Blood Pressure *(Page 1063)*

EXPERIMENT

HYPOTHESIS: Northern elephant seals spend extended periods of time underwater. A slowed heart rate is part of the physiological reflex that makes extended dives possible.

Mirounga angustirostris

METHOD

While seals are on land for the breeding season, attach electronic tracking and recording devices to individual animals to obtain time and location data, along with readings of heart rate, water pressure (indicates depth of dive), and breathing pattern.

RESULTS

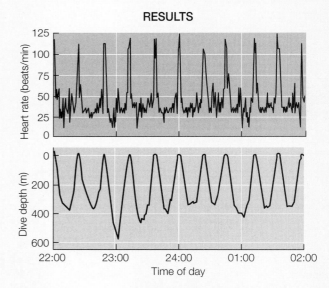

During many days at sea, seals spent more than 80 percent of their time under water, making repeated dives to depths of 300–600 meters. Each dive lasted an average 20 minutes; surface time averaged less than 3 minutes per dive. Heart rate averaged 110 beats per minute at the surface but only 30 bpm while diving, with heart rate dropping as low as 3–4 bpm during parts of the dive.

CONCLUSION: Northern elephant seals spend more than 80 percent of their time at sea underwater. Reduced heart rate is part of the diving reflex that makes this possible.

49.19 Elephant Seal Diving Ability *(Page 1063)*

CHAPTER 50 Nutrition, Digestion, and Absorption

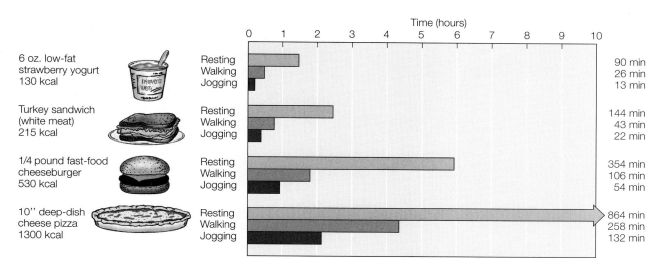

		Time (hours)	
6 oz. low-fat strawberry yogurt 130 kcal	Resting Walking Jogging		90 min 26 min 13 min
Turkey sandwich (white meat) 215 kcal	Resting Walking Jogging		144 min 43 min 22 min
1/4 pound fast-food cheeseburger 530 kcal	Resting Walking Jogging		354 min 106 min 54 min
10'' deep-dish cheese pizza 1300 kcal	Resting Walking Jogging		864 min 258 min 132 min

50.2 Food Energy and How We Use It *(Page 1070)*

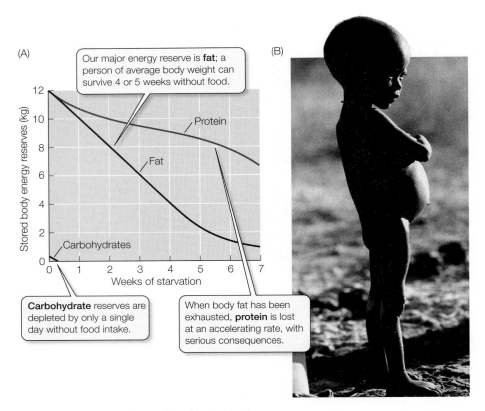

(A)

Our major energy reserve is **fat**; a person of average body weight can survive 4 or 5 weeks without food.

Protein

Fat

Carbohydrates

Carbohydrate reserves are depleted by only a single day without food intake.

When body fat has been exhausted, **protein** is lost at an accelerating rate, with serious consequences.

(B)

50.3 The Course of Starvation *(Page 1071)*

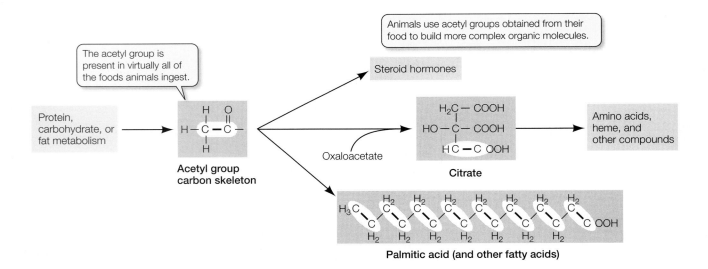

The acetyl group is present in virtually all of the foods animals ingest.

Animals use acetyl groups obtained from their food to build more complex organic molecules.

Steroid hormones

Protein, carbohydrate, or fat metabolism

Acetyl group carbon skeleton

Oxaloacetate

Citrate

Amino acids, heme, and other compounds

Palmitic acid (and other fatty acids)

50.4 The Acetyl Group Is an Acquired Carbon Skeleton *(Page 1072)*

Eight essential amino acids for humans

Tryptophan

Methionine

Valine

Threonine

Phenylalanine

Leucine

Isoleucine

Lysine

Grains
(corn in tortilla chips)

Legumes
(beans in bean dip)

50.5 A Strategy for Vegetarians *(Page 1072)*

TABLE 50.1

Mineral Elements Required by Animals

ELEMENT	SOURCE IN HUMAN DIET	MAJOR FUNCTIONS
MACRONUTRIENTS		
Calcium (Ca)	Dairy foods, eggs, green leafy vegetables, whole grains, legumes, nuts, meat	Found in bones and teeth; blood clotting; nerve and muscle action; enzyme activation
Chlorine (Cl)	Table salt (NaCl), meat, eggs, vegetables, dairy foods	Water balance; digestion (as HCl); principal negative ion in extracellular fluid
Magnesium (Mg)	Green vegetables, meat, whole grains, nuts, milk, legumes	Required by many enzymes; found in bones and teeth
Phosphorus (P)	Dairy, eggs, meat, whole grains, legumes, nuts	Found in nucleic acids, ATP, and phospholipids; bone formation; buffers; metabolism of sugars
Potassium (K)	Meat, whole grains, fruits, vegetables	Nerve and muscle action; protein synthesis; principal positive ion in cells
Sodium (Na)	Table salt, dairy foods, meat, eggs	Nerve and muscle action; water balance; principal positive ion in extracellular fluid
Sulfur (S)	Meat, eggs, dairy foods, nuts, legumes	Found in proteins and coenzymes; detoxification of harmful substances
MICRONUTRIENTS		
Chromium (Cr)	Meat, dairy, whole grains, legumes, yeast	Glucose metabolism
Cobalt (Co)	Meat, tap water	Found in vitamin B12; formation of red blood cells
Copper (Cu)	Liver, meat, fish, shellfish, legumes, whole grains, nuts	Found in active site of many redox enzymes and electron carriers; production of hemoglobin; bone formation
Fluorine (F)	Most water supplies	Found in teeth; helps prevent decay
Iodine (I)	Fish, shellfish, iodized salt	Found in thyroid hormones
Iron (Fe)	Liver, meat, green vegetables, eggs, whole grains, legumes, nuts	Found in active sites of many redox enzymes and electron carriers, hemoglobin, and myoglobin
Manganese (Mn)	Organ meats, whole grains, legumes, nuts, tea, coffee	Activates many enzymes
Molybdenum (Mo)	Organ meats, dairy, whole grains, green vegetables, legumes	Found in some enzymes
Selenium (Se)	Meat, seafood, whole grains, eggs, milk, garlic	Fat metabolism
Zinc (Zn)	Liver, fish, shellfish, and many other foods	Found in some enzymes and some transcription factors; insulin physiology

TABLE 50.2

Vitamins in the Human Diet

VITAMIN	SOURCE	FUNCTION	DEFICIENCY SYMPTOMS
WATER-SOLUBLE			
B₁ (thiamin)	Liver, legumes, whole grains,	Coenzyme in cellular respiration	Beriberi, loss of appetite, fatigue
B₂ (riboflavin)	Dairy, meat, eggs, green leafy vegetables	Coenzyme in FAD	Lesions in corners of mouth, eye irritation, skin disorders
Niacin	Meat, fowl, liver, yeast	Coenzyme in NAD and NADP	Pellagra, skin disorders, diarrhea, mental disorders
B6 (pyridoxine)	Liver, whole grains, dairy foods	Coenzyme in amino acid metabolism	Anemia, slow growth, skin problems, convulsions
Pantothenic acid	Liver, eggs, yeast	Found in acetyl CoA	Adrenal problems, reproductive problems
Biotin	Liver, yeast, bacteria in gut	Found in coenzymes	Skin problems, loss of hair
B₁₂ (cobalamin)	Liver, meat, dairy foods, eggs	Formation of nucleic acids, proteins, and red blood cells	Pernicious anemia
Folic acid	Vegetables, eggs, liver, whole grains	Coenzyme in formation of heme and nucleotides	Anemia
C (ascorbic acid)	Citrus fruits, tomatoes, potatoes	Formation of connective tissues; antioxidant	Scurvy, slow healing, poor bone growth
FAT-SOLUBLE			
A (retinol)	Fruits, vegetables, liver, dairy	Found in visual pigments	Night blindness
D (cholecalciferol)	Fortified milk, fish oils, sunshine	Absorption of calcium and phosphate	Rickets
E (tocopherol)	Meat, dairy foods, whole grains	Muscle maintenance, antioxidant	Anemia
K (menadione)	Intestinal bacteria, liver	Blood clotting	Blood-clotting problems

(Page 1074)

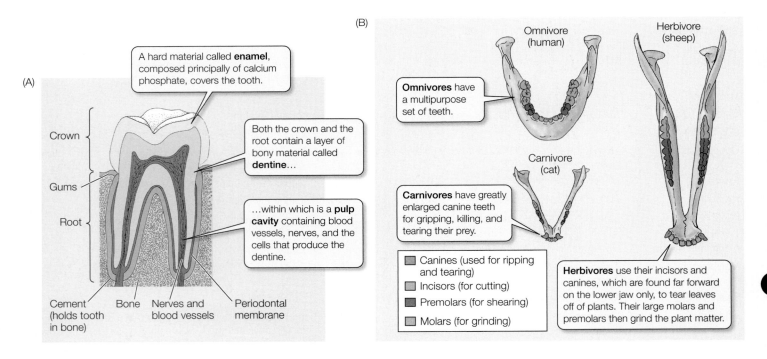

50.7 Mammalian Teeth *(Page 1076)*

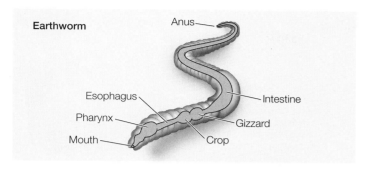

Earthworm

Anus

Esophagus

Pharynx

Mouth

Intestine

Gizzard

Crop

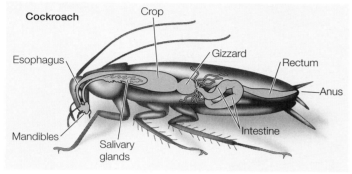

Cockroach

Crop

Esophagus

Gizzard

Rectum

Anus

Mandibles

Salivary glands

Intestine

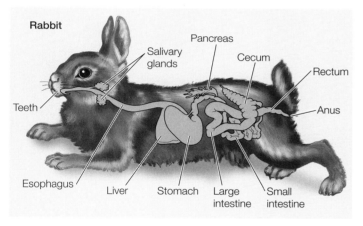

Rabbit

Pancreas

Salivary glands

Cecum

Rectum

Teeth

Anus

Esophagus

Liver

Stomach

Large intestine

Small intestine

50.8 Compartments for Digestion and Absorption *(Page 1077)*

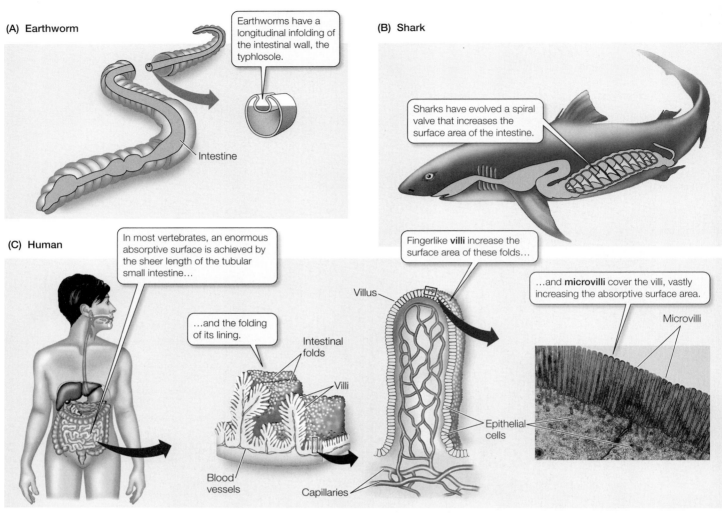

(A) Earthworm

Earthworms have a longitudinal infolding of the intestinal wall, the typhlosole.

Intestine

(B) Shark

Sharks have evolved a spiral valve that increases the surface area of the intestine.

(C) Human

In most vertebrates, an enormous absorptive surface is achieved by the sheer length of the tubular small intestine…

…and the folding of its lining.

Intestinal folds

Villi

Blood vessels

Capillaries

Fingerlike **villi** increase the surface area of these folds…

Villus

Epithelial cells

…and **microvilli** cover the villi, vastly increasing the absorptive surface area.

Microvilli

50.9 Greater Intestinal Surface Area Means More Nutrient Absorption *(Page 1078)*

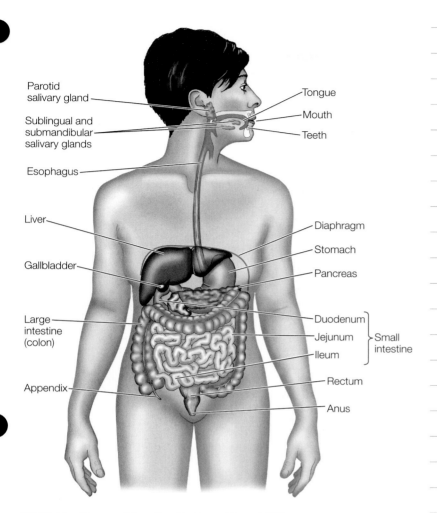

50.10 The Human Digestive System *(Page 1079)*

Parotid salivary gland

Sublingual and submandibular salivary glands

Esophagus

Liver

Gallbladder

Large intestine (colon)

Appendix

Tongue

Mouth

Teeth

Diaphragm

Stomach

Pancreas

Duodenum

Jejunum

Ileum

Rectum

Anus

Small intestine

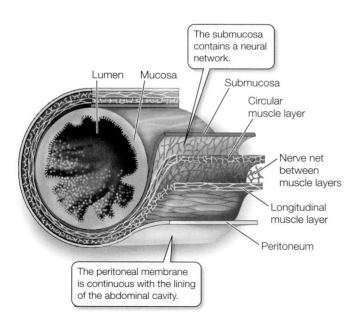

50.11 Tissue Layers of the Vertebrate Gut *(Page 1079)*

The submucosa contains a neural network.

Lumen   Mucosa

Submucosa

Circular muscle layer

Nerve net between muscle layers

Longitudinal muscle layer

Peritoneum

The peritoneal membrane is continuous with the lining of the abdominal cavity.

(A) Swallowing

(B) Peristalsis

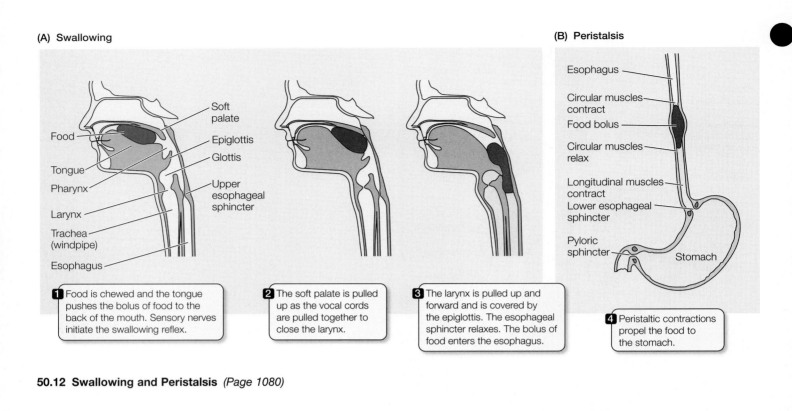

Food

Tongue

Pharynx

Larynx

Trachea
(windpipe)

Esophagus

Soft
palate

Epiglottis

Glottis

Upper
esophageal
sphincter

Esophagus

Circular muscles
contract

Food bolus

Circular muscles
relax

Longitudinal muscles
contract

Lower esophageal
sphincter

Pyloric
sphincter

Stomach

1 Food is chewed and the tongue pushes the bolus of food to the back of the mouth. Sensory nerves initiate the swallowing reflex.

2 The soft palate is pulled up as the vocal cords are pulled together to close the larynx.

3 The larynx is pulled up and forward and is covered by the epiglottis. The esophageal sphincter relaxes. The bolus of food enters the esophagus.

4 Peristaltic contractions propel the food to the stomach.

50.12 Swallowing and Peristalsis *(Page 1080)*

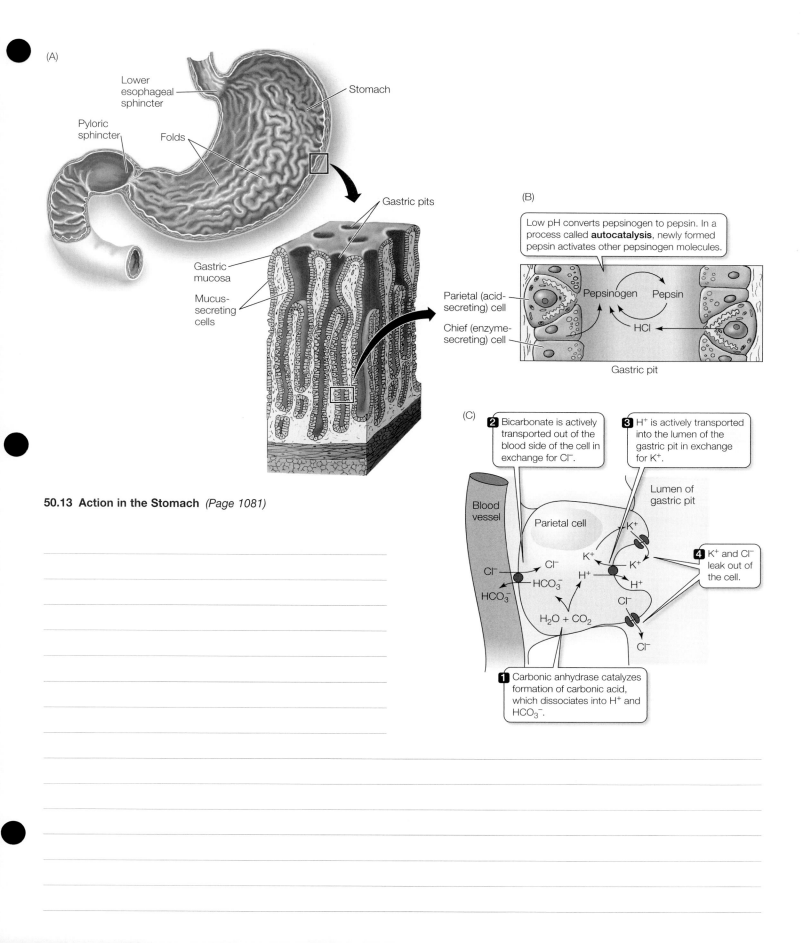

(A)

Lower esophageal sphincter

Stomach

Pyloric sphincter

Folds

Gastric pits

Gastric mucosa

Mucus-secreting cells

(B)

Low pH converts pepsinogen to pepsin. In a process called **autocatalysis**, newly formed pepsin activates other pepsinogen molecules.

Parietal (acid-secreting) cell

Chief (enzyme-secreting) cell

Pepsinogen → Pepsin

HCl

Gastric pit

50.13 Action in the Stomach *(Page 1081)*

(C)

2 Bicarbonate is actively transported out of the blood side of the cell in exchange for Cl⁻.

3 H⁺ is actively transported into the lumen of the gastric pit in exchange for K⁺.

Lumen of gastric pit

Blood vessel

Parietal cell

K^+

K^+

K^+

Cl^- Cl^-

HCO_3^-

HCO_3^-

H^+

H^+

Cl^-

$H_2O + CO_2$

Cl^-

4 K⁺ and Cl⁻ leak out of the cell.

1 Carbonic anhydrase catalyzes formation of carbonic acid, which dissociates into H⁺ and HCO₃⁻.

Marshall and Warren set out to satisfy Koch's postulates:

Test 1

The microorganism must be present in every case of the disease.

Results: Biopsies from the stomachs of many patients revealed that the bacterium was always present if the stomach was inflamed or ulcerated.

Test 2

The microorganism must be cultured from a sick host.

Results: The bacterium was isolated from biopsy material and eventually grown in culture media in the laboratory.

Test 3

The isolated and cultured bacteria must be able to induce the disease.

Results: Marshall was examined and found to be free of bacteria and inflammation in his stomach. After drinking a pure culture of the bacterium, he developed stomach inflammation (gastritis).

Test 4

The bacteria must be recoverable from the infected volunteers.

Results: Biopsy of Marshall's stomach 2 weeks after he ingested the bacteria revealed the presence of the bacterium, now christened *Helicobacter pylori*, in the inflamed tissue.

Conclusion

Antibiotic treatment eliminated the bacteria and the inflammation in Marshall. The experiment was repeated on healthy volunteers, and many patients with gastric ulcers were cured with antibiotics. Thus Marshall and Warren demonstrated that the stomach inflammation leading to ulcers is caused by *H. pylori* infections in the stomach.

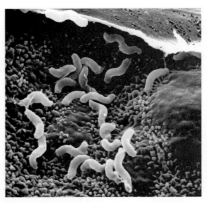

Helicobacter pylori

50.14 Satisfying Koch's Postulates *(Page 1082)*

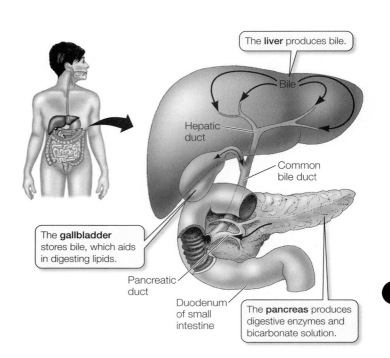

The **liver** produces bile.

Bile

Hepatic duct

Common bile duct

The **gallbladder** stores bile, which aids in digesting lipids.

Pancreatic duct

Duodenum of small intestine

The **pancreas** produces digestive enzymes and bicarbonate solution.

50.15 Ducts of the Gallbladder and Pancreas *(Page 1083)*

(A) Digestion of fats

Large lipid droplet

1 Dietary fats are emulsified into tiny droplets called micelles through the action of bile salts in the intestinal lumen.

Bile salts

Micelles

2 Pancreatic lipase hydrolyzes fats in the micelles to produce fatty acids and monoglycerides.

Monoglycerides

Fatty acids

(B) Absorption of fats

3 Fatty acids and monoglycerides enter the cell by diffusion. They are resynthesized into triglycerides in the endoplasmic reticulum.

4 Triglycerides are packaged with cholesterol and phospholipids in protein-coated chylomicrons.

Endoplasmic reticulum

Intestinal epithelial cell

5 Chylomicrons are enclosed in vesicles and leave the cell by exocytosis.

Lymphatic vessel

50.16 Digestion and Absorption of Fats *(Page 1083)*

Major Digestive Enzymes of Humans

SOURCE/ENZYME	ACTION
SALIVARY GLANDS	
Salivary amylase	Starch → Maltose
STOMACH	
Pepsin	Proteins → Peptides; autocatalysis
PANCREAS	
Pancreatic amylase	Starch → Maltose
Lipase	Fats → Fatty acids and glycerol
Nuclease	Nucleic acids → Nucleotides
Trypsin	Proteins → Peptides; zymogen activation
Chymotrypsin	Proteins → Peptides
Carboxypeptidase	Peptides → Shorter peptides and amino acids
SMALL INTESTINE	
Aminopeptidase	Peptides → Shorter peptides and amino acids
Dipeptidase	Dipeptides → Amino acids
Enterokinase	Trypsinogen → Trypsin
Nuclease	Nucleic acids → Nucleotides
Maltase	Maltose → Glucose
Lactase	Lactose → Galactose and glucose
Sucrase	Sucrose → Fructose and glucose

(Page 1084)

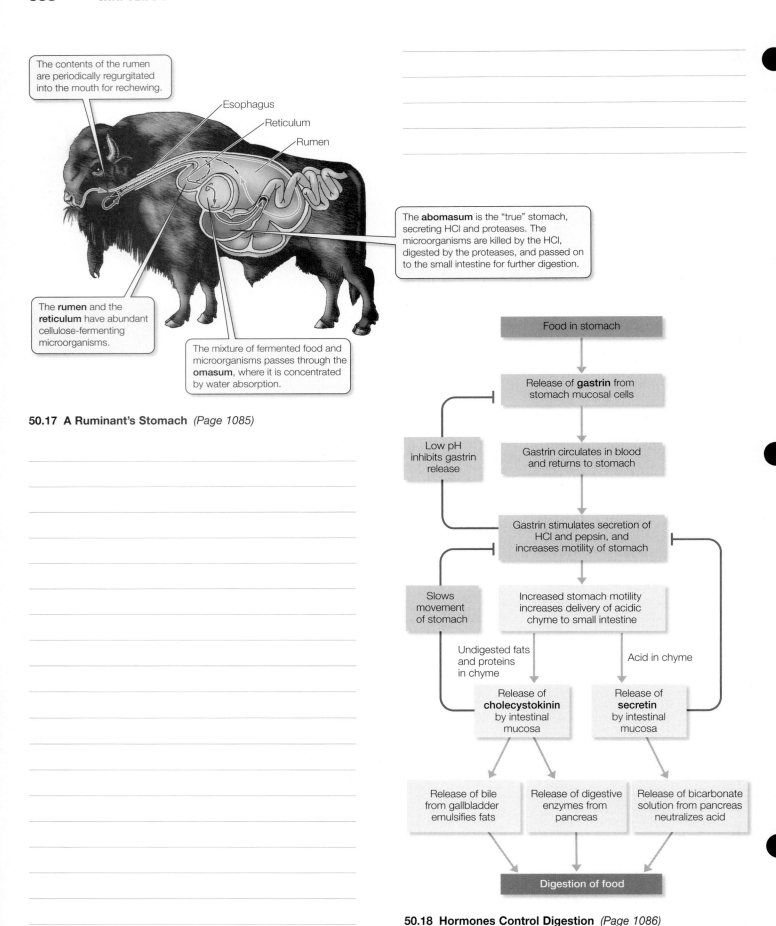

The contents of the rumen are periodically regurgitated into the mouth for rechewing.

Esophagus

Reticulum

Rumen

The **abomasum** is the "true" stomach, secreting HCl and proteases. The microorganisms are killed by the HCl, digested by the proteases, and passed on to the small intestine for further digestion.

The **rumen** and the **reticulum** have abundant cellulose-fermenting microorganisms.

The mixture of fermented food and microorganisms passes through the **omasum**, where it is concentrated by water absorption.

50.17 A Ruminant's Stomach *(Page 1085)*

Food in stomach

Release of **gastrin** from stomach mucosal cells

Low pH inhibits gastrin release

Gastrin circulates in blood and returns to stomach

Gastrin stimulates secretion of HCl and pepsin, and increases motility of stomach

Slows movement of stomach

Increased stomach motility increases delivery of acidic chyme to small intestine

Undigested fats and proteins in chyme

Acid in chyme

Release of **cholecystokinin** by intestinal mucosa

Release of **secretin** by intestinal mucosa

Release of bile from gallbladder emulsifies fats

Release of digestive enzymes from pancreas

Release of bicarbonate solution from pancreas neutralizes acid

Digestion of food

50.18 Hormones Control Digestion *(Page 1086)*

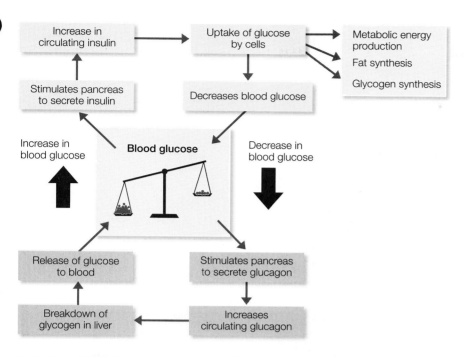

50.19 Regulating Glucose Levels in the Blood *(Page 1087)*

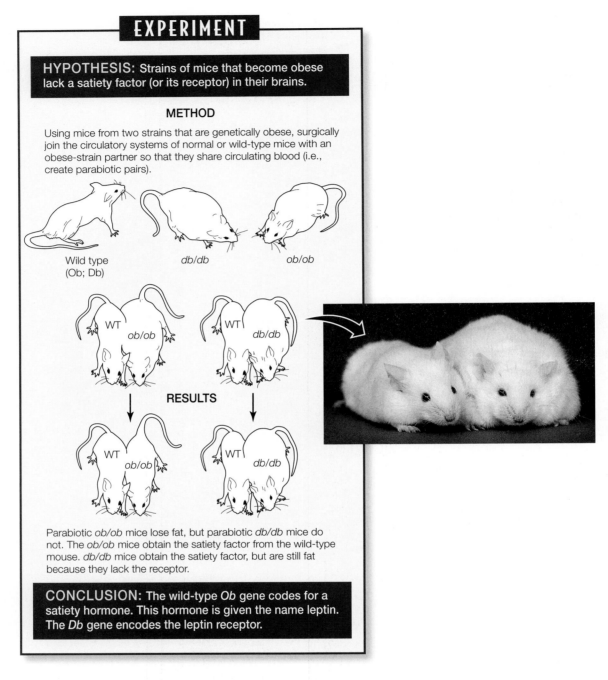

EXPERIMENT

HYPOTHESIS: Strains of mice that become obese lack a satiety factor (or its receptor) in their brains.

METHOD

Using mice from two strains that are genetically obese, surgically join the circulatory systems of normal or wild-type mice with an obese-strain partner so that they share circulating blood (i.e., create parabiotic pairs).

Wild type
(Ob; Db)

db/db

ob/ob

WT *ob/ob*

WT *db/db*

RESULTS

WT *ob/ob*

WT *db/db*

Parabiotic *ob/ob* mice lose fat, but parabiotic *db/db* mice do not. The *ob/ob* mice obtain the satiety factor from the wild-type mouse. *db/db* mice obtain the satiety factor, but are still fat because they lack the receptor.

CONCLUSION: The wild-type *Ob* gene codes for a satiety hormone. This hormone is given the name leptin. The *Db* gene encodes the leptin receptor.

50.20 A Single-Gene Mutation Leads to Obesity in Mice *(Page 1088)*

CHAPTER 51 Salt and Water Balance and Nitrogen Excretion

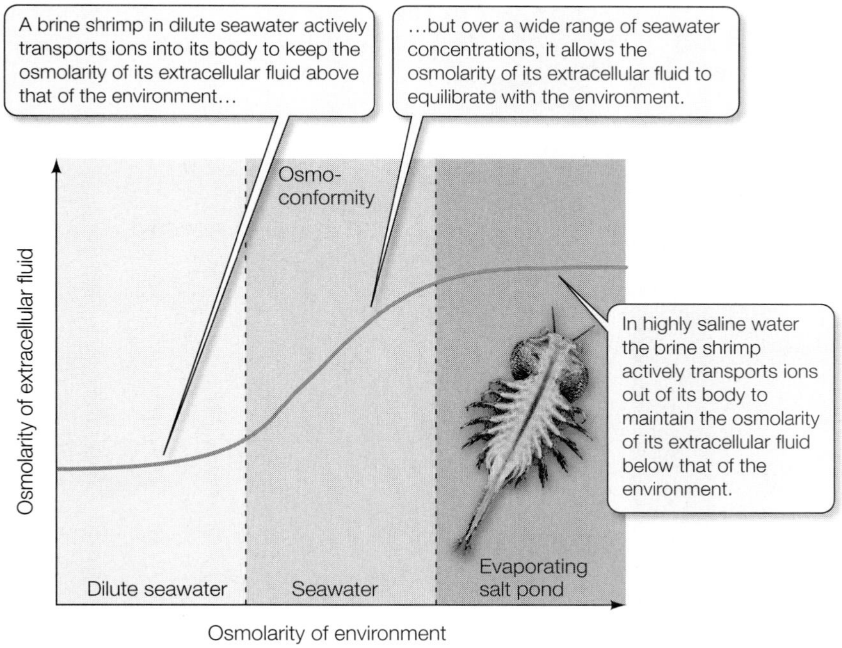

> A brine shrimp in dilute seawater actively transports ions into its body to keep the osmolarity of its extracellular fluid above that of the environment…

> …but over a wide range of seawater concentrations, it allows the osmolarity of its extracellular fluid to equilibrate with the environment.

> In highly saline water the brine shrimp actively transports ions out of its body to maintain the osmolarity of its extracellular fluid below that of the environment.

Osmolarity of extracellular fluid

Osmo-conformity

Dilute seawater

Seawater

Evaporating salt pond

Osmolarity of environment

51.1 Environments Can Vary Greatly in Salt Concentration *(Page 1094)*

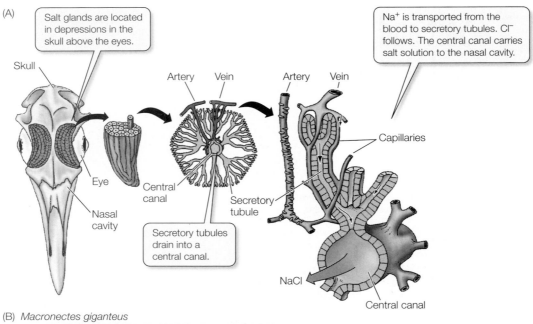

(A)

Salt glands are located in depressions in the skull above the eyes.

Skull

Eye

Nasal cavity

Artery   Vein

Central canal

Secretory tubules drain into a central canal.

Artery   Vein

Na+ is transported from the blood to secretory tubules. Cl− follows. The central canal carries salt solution to the nasal cavity.

Capillaries

Secretory tubule

NaCl

Central canal

(B) *Macronectes giganteus*

Note the secretion at the tip of the bird's beak.

51.2 Nasal Salt Glands Excrete Excess Salt *(Page 1095)*

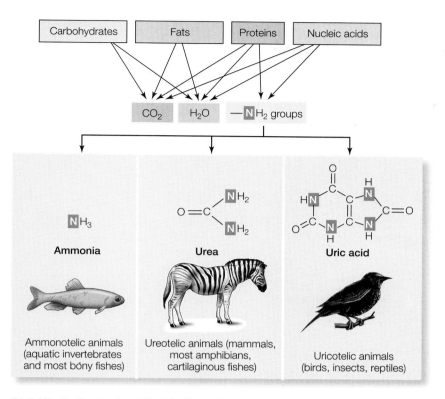

51.3 Waste Products of Metabolism *(Page 1096)*

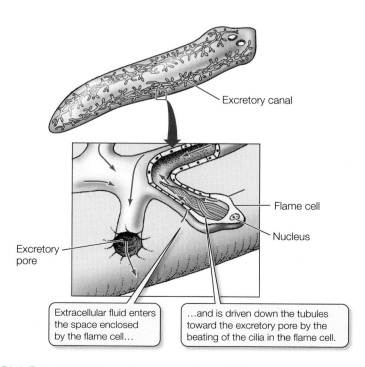

51.4 Protonephridia in Flatworms *(Page 1097)*

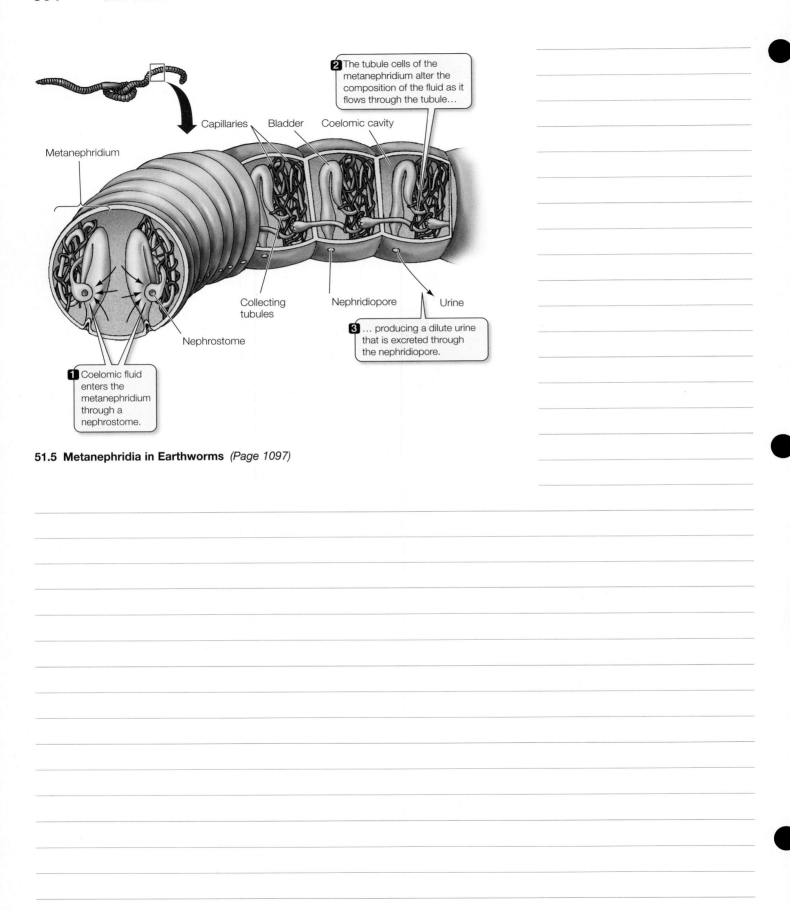

2 The tubule cells of the metanephridium alter the composition of the fluid as it flows through the tubule…

Capillaries Bladder Coelomic cavity

Metanephridium

Collecting tubules

Nephrostome

Nephridiopore Urine

1 Coelomic fluid enters the metanephridium through a nephrostome.

3 … producing a dilute urine that is excreted through the nephridiopore.

51.5 Metanephridia in Earthworms *(Page 1097)*

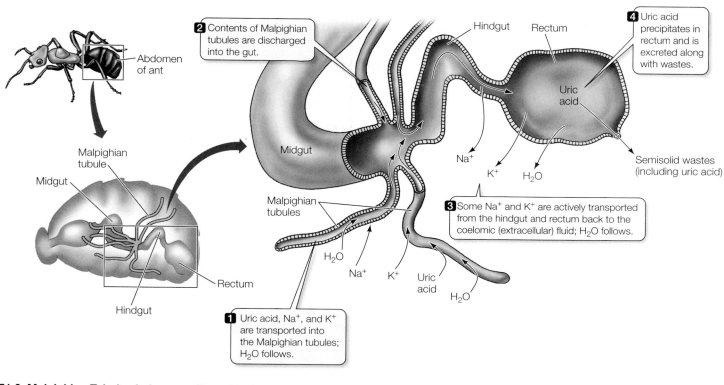

Abdomen of ant

Malpighian tubule

Midgut

Rectum

Hindgut

2 Contents of Malpighian tubules are discharged into the gut.

Midgut

Malpighian tubules

H_2O

Na^+

K^+

Uric acid

H_2O

1 Uric acid, Na^+, and K^+ are transported into the Malpighian tubules; H_2O follows.

Hindgut

Rectum

4 Uric acid precipitates in rectum and is excreted along with wastes.

Uric acid

Na^+

K^+

H_2O

Semisolid wastes (including uric acid)

3 Some Na^+ and K^+ are actively transported from the hindgut and rectum back to the coelomic (extracellular) fluid; H_2O follows.

51.6 Malpighian Tubules in Insects *(Page 1098)*

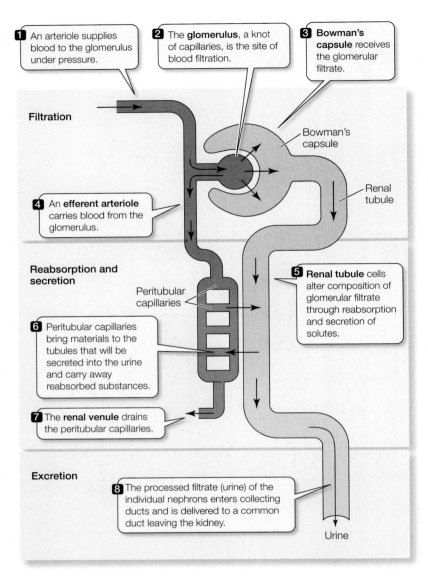

1 An arteriole supplies blood to the glomerulus under pressure.

2 The **glomerulus**, a knot of capillaries, is the site of blood filtration.

3 **Bowman's capsule** receives the glomerular filtrate.

Filtration

Bowman's capsule

Renal tubule

4 An **efferent arteriole** carries blood from the glomerulus.

Reabsorption and secretion

Peritubular capillaries

5 **Renal tubule** cells alter composition of glomerular filtrate through reabsorption and secretion of solutes.

6 Peritubular capillaries bring materials to the tubules that will be secreted into the urine and carry away reabsorbed substances.

7 The **renal venule** drains the peritubular capillaries.

Excretion

8 The processed filtrate (urine) of the individual nephrons enters collecting ducts and is delivered to a common duct leaving the kidney.

Urine

51.7 The Vertebrate Nephron *(Page 1100)*

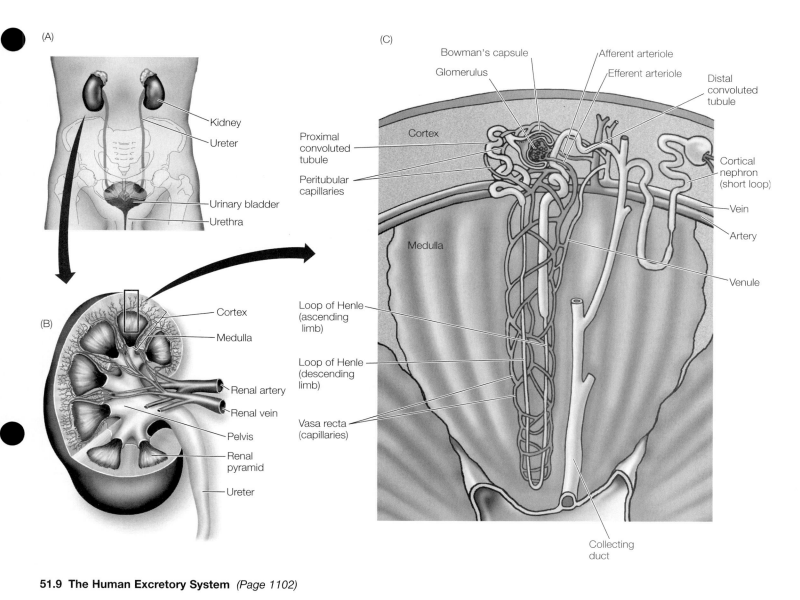

(A)

Kidney

Ureter

Urinary bladder

Urethra

(B)

Cortex

Medulla

Renal artery

Renal vein

Pelvis

Renal pyramid

Ureter

(C)

Bowman's capsule

Glomerulus

Afferent arteriole

Efferent arteriole

Distal convoluted tubule

Proximal convoluted tubule

Cortex

Peritubular capillaries

Cortical nephron (short loop)

Vein

Medulla

Artery

Venule

Loop of Henle (ascending limb)

Loop of Henle (descending limb)

Vasa recta (capillaries)

Collecting duct

51.9 The Human Excretory System *(Page 1102)*

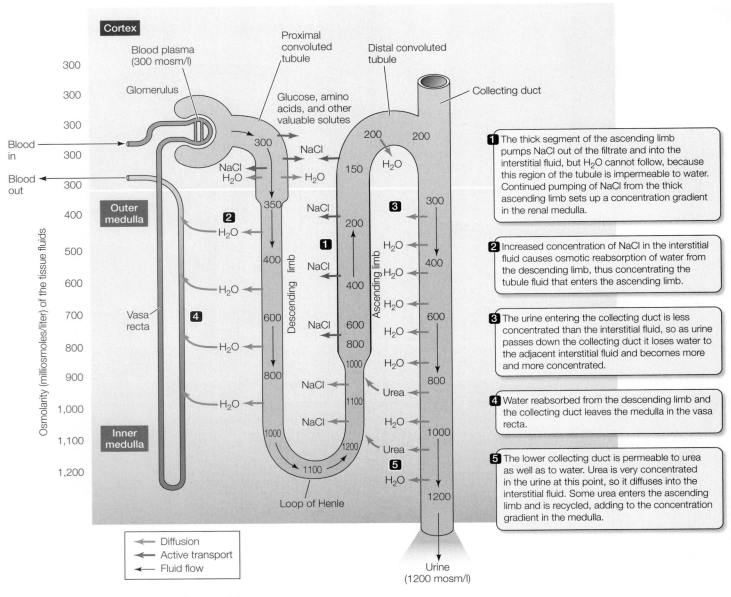

Cortex

Blood plasma
(300 mosm/l)

Glomerulus

Proximal convoluted tubule

Glucose, amino acids, and other valuable solutes

Distal convoluted tubule

Collecting duct

Blood in

Blood out

Outer medulla

Osmolarity (milliosmoles/liter) of the tissue fluids

300
300
300
300
300
400
500
600
700
800
900
1,000
1,100
1,200

Descending limb

Ascending limb

Vasa recta

Inner medulla

Loop of Henle

Diffusion
Active transport
Fluid flow

Urine
(1200 mosm/l)

1 The thick segment of the ascending limb pumps NaCl out of the filtrate and into the interstitial fluid, but H_2O cannot follow, because this region of the tubule is impermeable to water. Continued pumping of NaCl from the thick ascending limb sets up a concentration gradient in the renal medulla.

2 Increased concentration of NaCl in the interstitial fluid causes osmotic reabsorption of water from the descending limb, thus concentrating the tubule fluid that enters the ascending limb.

3 The urine entering the collecting duct is less concentrated than the interstitial fluid, so as urine passes down the collecting duct it loses water to the adjacent interstitial fluid and becomes more and more concentrated.

4 Water reabsorbed from the descending limb and the collecting duct leaves the medulla in the vasa recta.

5 The lower collecting duct is permeable to urea as well as to water. Urea is very concentrated in the urine at this point, so it diffuses into the interstitial fluid. Some urea enters the ascending limb and is recycled, adding to the concentration gradient in the medulla.

51.10 Concentrating the Urine *(Page 1104)*

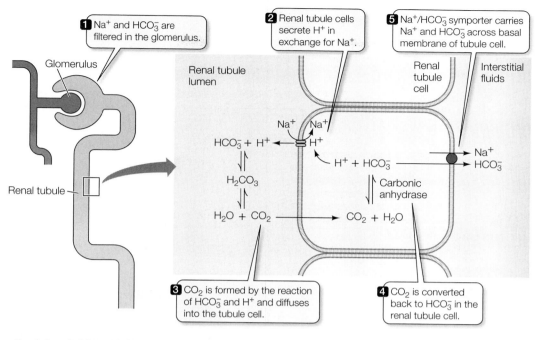

1 Na^+ and HCO_3^- are filtered in the glomerulus.

2 Renal tubule cells secrete H^+ in exchange for Na^+.

5 Na^+/HCO_3^- symporter carries Na^+ and HCO_3^- across basal membrane of tubule cell.

Glomerulus

Renal tubule lumen

Renal tubule cell

Interstitial fluids

Na^+

Na^+

$HCO_3^- + H^+ \longleftarrow$ H^+

Na^+

$H^+ + HCO_3^-$ $\longrightarrow$ HCO_3^-

H_2CO_3

Carbonic anhydrase

Renal tubule

$H_2O + CO_2$ $\longrightarrow$ $CO_2 + H_2O$

3 CO_2 is formed by the reaction of HCO_3^- and H^+ and diffuses into the tubule cell.

4 CO_2 is converted back to HCO_3^- in the renal tubule cell.

51.12 The Kidney Excretes Acids and Conserves Bases *(Page 1106)*

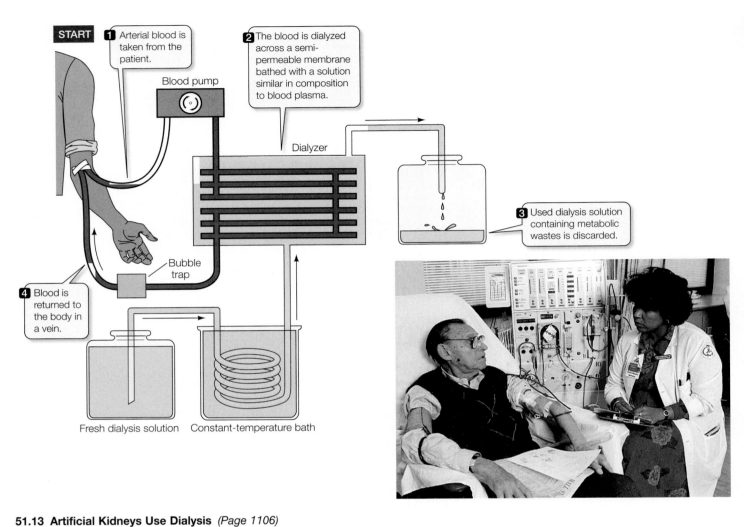

START
1 Arterial blood is taken from the patient.

Blood pump

2 The blood is dialyzed across a semi-permeable membrane bathed with a solution similar in composition to blood plasma.

Dialyzer

3 Used dialysis solution containing metabolic wastes is discarded.

Bubble trap

4 Blood is returned to the body in a vein.

Fresh dialysis solution   Constant-temperature bath

51.13 Artificial Kidneys Use Dialysis *(Page 1106)*

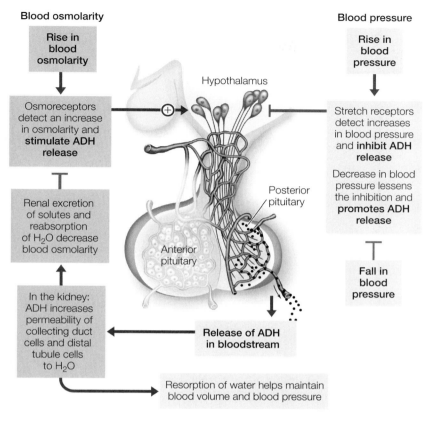

Blood osmolarity

Rise in blood osmolarity

↓

Osmoreceptors detect an increase in osmolarity and **stimulate ADH release**

↓

Renal excretion of solutes and reabsorption of H_2O decrease blood osmolarity

↑

In the kidney: ADH increases permeability of collecting duct cells and distal tubule cells to H_2O

Hypothalamus

⊕

Posterior pituitary

Anterior pituitary

Release of ADH in bloodstream

Blood pressure

Rise in blood pressure

↓

Stretch receptors detect increases in blood pressure and **inhibit ADH release**

Decrease in blood pressure lessens the inhibition and **promotes ADH release**

Fall in blood pressure

Resorption of water helps maintain blood volume and blood pressure

51.14 Antidiuretic Hormone Increases Blood Pressure and Promotes Water Reabsorption *(Page 1107)*

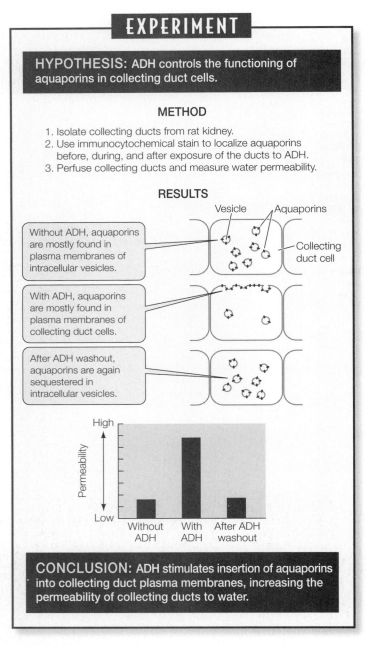

EXPERIMENT

HYPOTHESIS: ADH controls the functioning of aquaporins in collecting duct cells.

METHOD

1. Isolate collecting ducts from rat kidney.
2. Use immunocytochemical stain to localize aquaporins before, during, and after exposure of the ducts to ADH.
3. Perfuse collecting ducts and measure water permeability.

RESULTS

Vesicle   Aquaporins

Without ADH, aquaporins are mostly found in plasma membranes of intracellular vesicles.

Collecting duct cell

With ADH, aquaporins are mostly found in plasma membranes of collecting duct cells.

After ADH washout, aquaporins are again sequestered in intracellular vesicles.

High

Permeability

Low

Without ADH   With ADH   After ADH washout

CONCLUSION: ADH stimulates insertion of aquaporins into collecting duct plasma membranes, increasing the permeability of collecting ducts to water.

51.15 ADH Induces Insertion of Aquaporins into Collecting Duct Cell Plasma Membranes *(Page 1108)*

CHAPTER 52 Ecology and the Distribution of Life

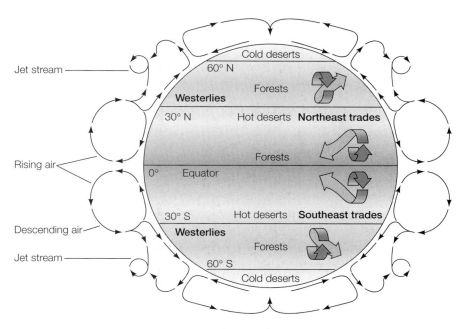

52.2 The Circulation of Earth's Atmosphere *(Page 1115)*

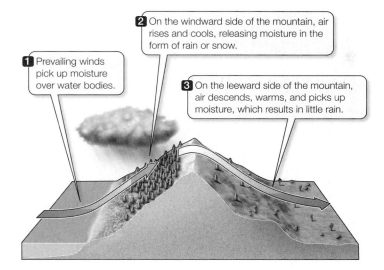

1 Prevailing winds pick up moisture over water bodies.

2 On the windward side of the mountain, air rises and cools, releasing moisture in the form of rain or snow.

3 On the leeward side of the mountain, air descends, warms, and picks up moisture, which results in little rain.

52.3 A Rain Shadow *(Page 1115)*

573

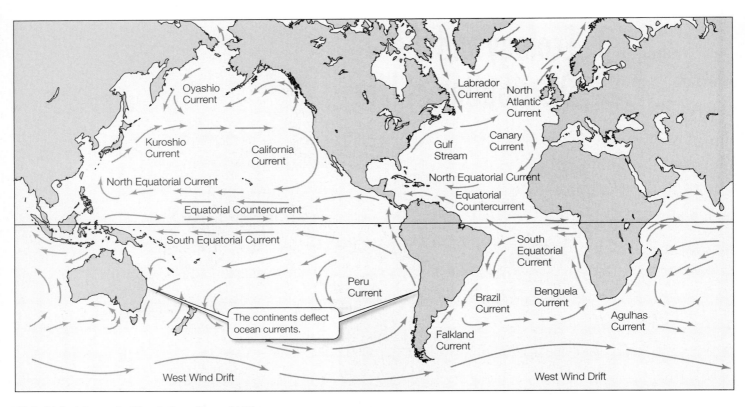

52.4 Global Oceanic Circulation *(Page 1116)*

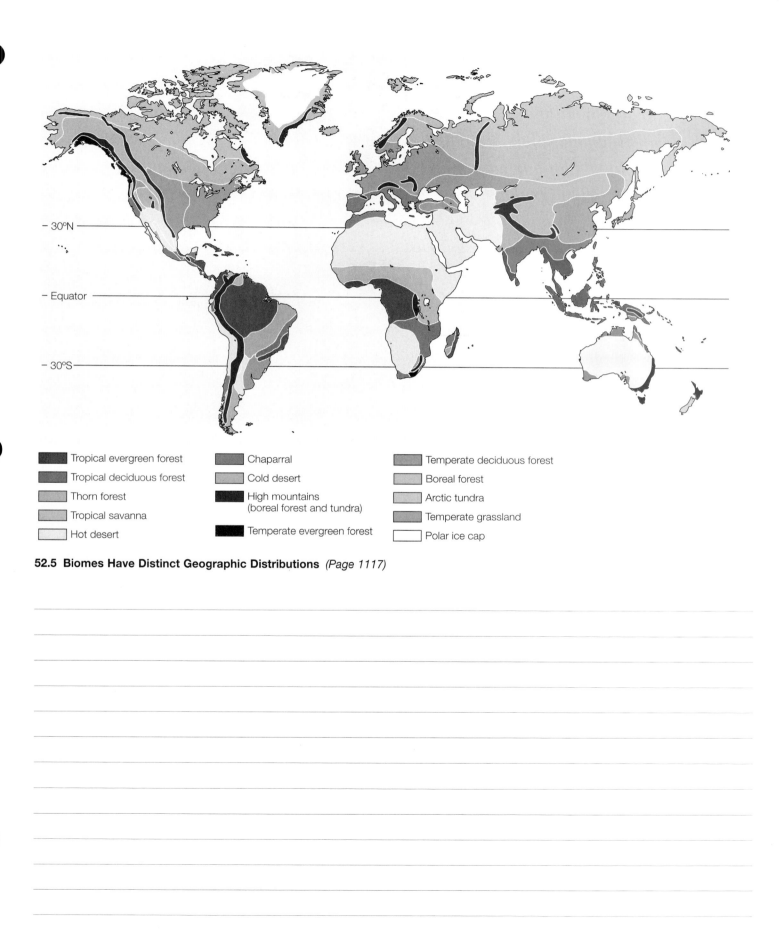

- 30°N
- Equator
- 30°S

	Tropical evergreen forest		Chaparral		Temperate deciduous forest
	Tropical deciduous forest		Cold desert		Boreal forest
	Thorn forest		High mountains (boreal forest and tundra)		Arctic tundra
	Tropical savanna				Temperate grassland
	Hot desert		Temperate evergreen forest		Polar ice cap

52.5 Biomes Have Distinct Geographic Distributions *(Page 1117)*

TUNDRA

Temperature

20°C is a "comfortable" 68°F.

0°C is the freezing point of water (=32°F).

Upernavik, Greenland 73°N

Winter is very cold and long.

Summer is cool and short.

Range 28°C

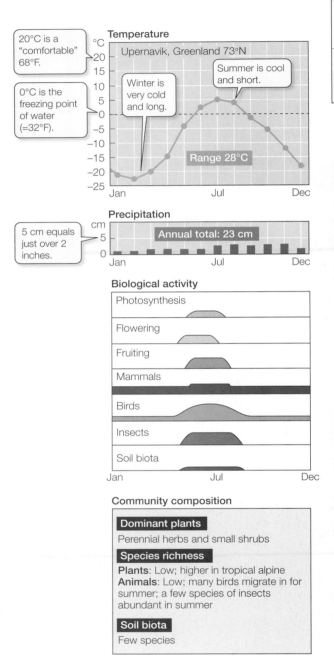

Precipitation

5 cm equals just over 2 inches.

Annual total: 23 cm

Biological activity

Photosynthesis

Flowering

Fruiting

Mammals

Birds

Insects

Soil biota

Community composition

Dominant plants

Perennial herbs and small shrubs

Species richness

Plants: Low; higher in tropical alpine
Animals: Low; many birds migrate in for summer; a few species of insects abundant in summer

Soil biota

Few species

Arctic tundra, Greenland

Tropical alpine tundra, Teleki Valley, Mt. Kenya, Kenya

Figure 52A *(Page 1118)*

BOREAL FOREST and TEMPERATE EVERGREEN FOREST

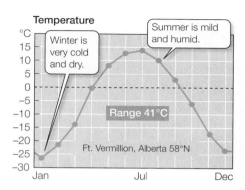

Temperature

°C

Winter is very cold and dry.

Summer is mild and humid.

Range 41°C

Ft. Vermillion, Alberta 58°N

Jan · Jul · Dec

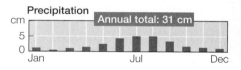

Precipitation

cm · Annual total: 31 cm

Jan · Jul · Dec

Biological activity

| Photosynthesis |
| Flowering |
| Fruiting |
| Mammals |
| Birds |
| Insects |
| Soil biota |

Jan · Jul · Dec

Community composition

Dominant plants

Trees, shrubs, and perennial herbs

Species richness

Plants: Low in trees, higher in understory
Animals: Low, but with summer peaks in migratory birds

Soil biota

Very rich in deep litter layer

Figure 52B *(Page 1119)*

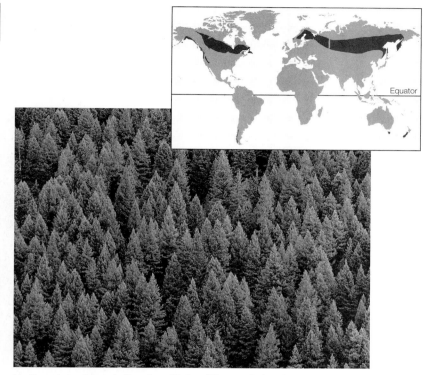

Northern boreal forest, Gunnison National Forest, Colorado

Southern boreal forest, Fiordland National Park, New Zealand

TEMPERATE DECIDUOUS FOREST

A Rhode Island forest in summer and…

…in winter

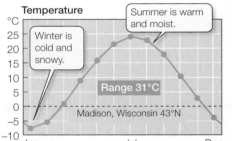

Temperature

Winter is cold and snowy.

Summer is warm and moist.

Range 31°C

Madison, Wisconsin 43°N

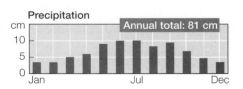

Precipitation

Annual total: 81 cm

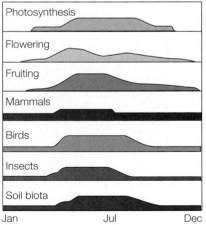

Biological activity

Photosynthesis

Flowering

Fruiting

Mammals

Birds

Insects

Soil biota

Community composition

Dominant plants

Trees and shrubs

Species richness

Plants: Many tree species in southeastern U.S. and eastern Asia, rich shrub layer

Animals: Rich; many migrant birds, richest amphibian communities on Earth, rich summer insect fauna

Soil biota

Rich

Figure 52C *(Page 1120)*

TEMPERATE GRASSLANDS

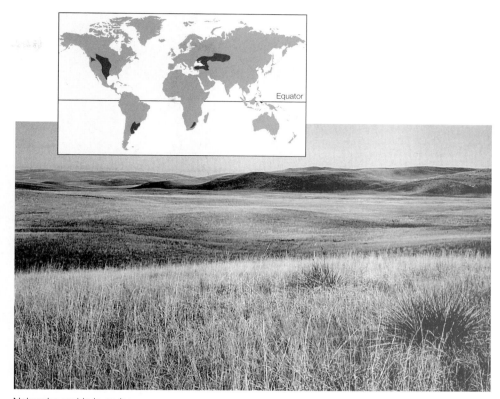

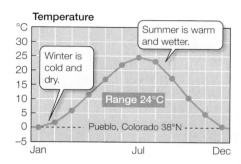

Temperature

Winter is cold and dry.

Summer is warm and wetter.

Range 24°C

Pueblo, Colorado 38°N

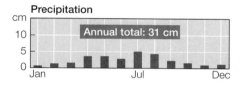

Precipitation

Annual total: 31 cm

Biological activity

Photosynthesis

Flowering

Fruiting

Mammals

Birds

Insects

Soil biota

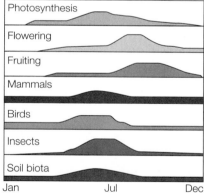

Community composition

Dominant plants
Perennial grasses and forbs

Species richness
Plants: Fairly high
Animals: Relatively few birds because of simple structure; mammals fairly rich

Soil biota
Rich

Nebraska prairie in spring

The Veldt, Natal, South Africa

Figure 52D *(Page 1121)*

COLD DESERT

Temperature

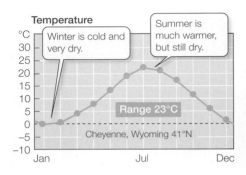

°C
30
25
20
15
10
5
0
−5
−10

Winter is cold and very dry.

Summer is much warmer, but still dry.

Range 23°C

Cheyenne, Wyoming 41°N

Jan — Jul — Dec

Precipitation

cm

Annual total: 38 cm

5

0

Jan — Jul — Dec

Biological activity

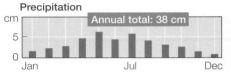

Photosynthesis

Flowering

Fruiting

Mammals

Birds

Insects

Soil biota

Jan — Jul — Dec

Community composition

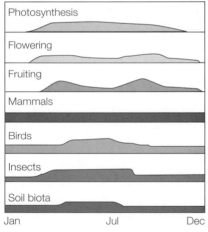

Dominant plants
Low-growing shrubs and herbaceous plants

Species richness
Plants: Few species
Animals: Rich in seed-eating birds, ants, and rodents; low in all other taxa

Soil biota
Poor in species

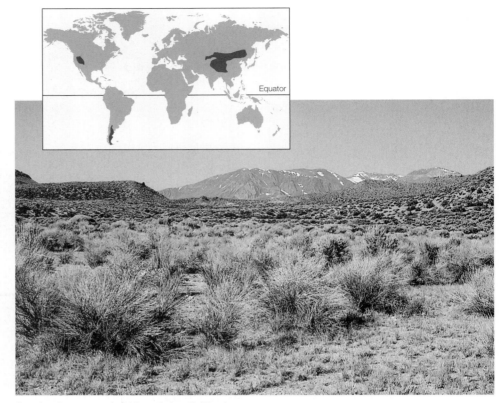

Equator

Sagebrush steppe near Mono Lake, California

Patagonia, Argentina

Figure 52E *(Page 1122)*

HOT DESERT

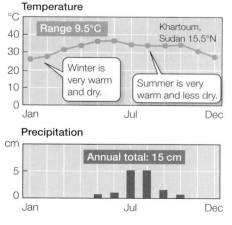

Anza Borrego Desert, California

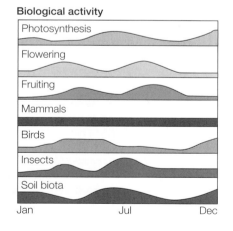

Simpson Desert, Australia, following rain

Temperature

°C

Range 9.5°C   Khartoum,
                     Sudan 15.5°N
40
30
20   Winter is
            very warm
10   and dry.   Summer is very
                            warm and less dry.
0
Jan   Jul   Dec

Precipitation

cm

            Annual total: 15 cm

5

0
Jan   Jul   Dec

Biological activity

Photosynthesis

Flowering

Fruiting

Mammals

Birds

Insects

Soil biota

Jan   Jul   Dec

Community composition

Dominant plants

Many different growth forms

Species richness

Plants: Moderately rich; many annuals
Animals: Very rich in rodents; richest bee
communities on Earth; very rich in reptiles
and butterflies

Soil biota

Poor in species

Figure 52F *(Page 1123)*

CHAPARRAL

Southwest Australia

Santa Barbara County, California

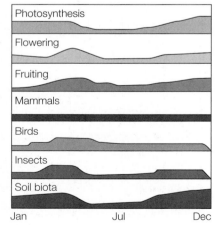

Temperature

°C
Winter is mild and humid.
Summer is mild and very dry.
Range 7°C
Monterey, California 36°N

Precipitation

cm
Annual total: 42 cm

Jan   Jul   Dec

Biological activity

Photosynthesis

Flowering

Fruiting

Mammals

Birds

Insects

Soil biota

Jan   Jul   Dec

Community composition

Dominant plants
Low-growing shrubs and herbaceous plants

Species richness
Plants: Extremely high in South Africa and Australia
Animals: Rich in rodents and reptiles; very rich in insects, especially bees

Soil biota
Moderately rich

Figure 52G *(Page 1124)*

THORN FOREST and TROPICAL SAVANNA

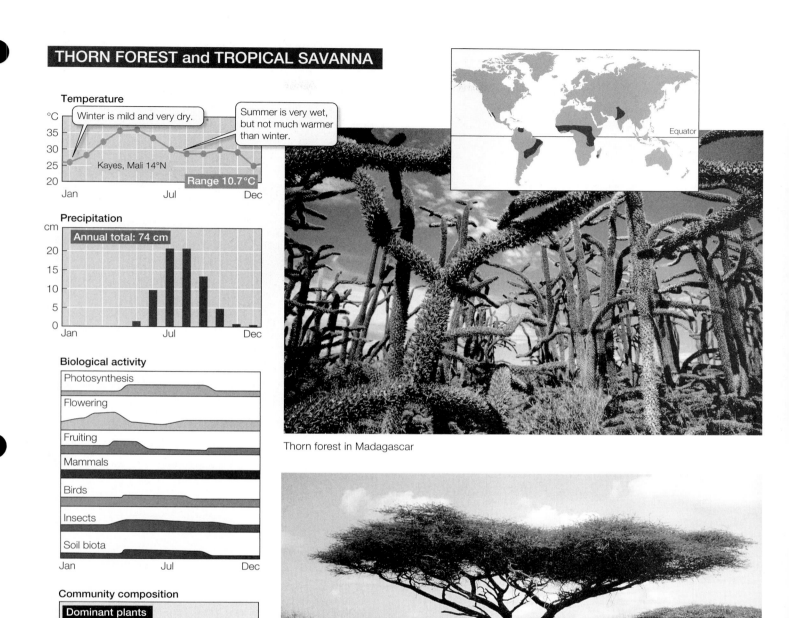

Temperature

°C

Winter is mild and very dry.

Summer is very wet, but not much warmer than winter.

Kayes, Mali 14°N

Range 10.7°C

35
30
25
20

Jan   Jul   Dec

Precipitation

cm

Annual total: 74 cm

20
15
10
5
0

Jan   Jul   Dec

Biological activity

Photosynthesis

Flowering

Fruiting

Mammals

Birds

Insects

Soil biota

Jan   Jul   Dec

Community composition

Dominant plants

Shrubs and small trees; grasses

Species richness

Plants: Moderate in thorn forest; low in savanna
Animals: Rich mammal faunas; moderately rich in birds, reptiles, and insects

Soil biota

Rich

Equator

Thorn forest in Madagascar

KwaZulu-Natal, South Africa

Figure 52H *(Page 1225)*

TROPICAL DECIDUOUS FOREST

Temperature

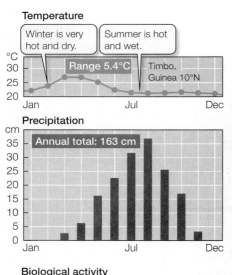

Winter is very hot and dry.

Summer is hot and wet.

Range 5.4°C

Timbo, Guinea 10°N

°C
30
25
20

Jan   Jul   Dec

Precipitation

cm
35
30
25
20
15
10
5
0

Annual total: 163 cm

Jan   Jul   Dec

Biological activity

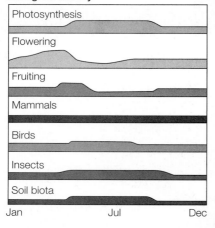

Photosynthesis

Flowering

Fruiting

Mammals

Birds

Insects

Soil biota

Jan   Jul   Dec

Community composition

Dominant plants
Deciduous trees
Species richness
Plants: Moderately rich in tree species
Animals: Rich mammal, bird, reptile, and amphibian communities; rich in insects
Soil biota
Rich, but poorly known

Palo Verde National Park, Costa Rica, in the rainy season…

…and in the dry season

Figure 52I *(Page 1126)*

TROPICAL EVERGREEN FOREST

The exterior of lowland wet forest...

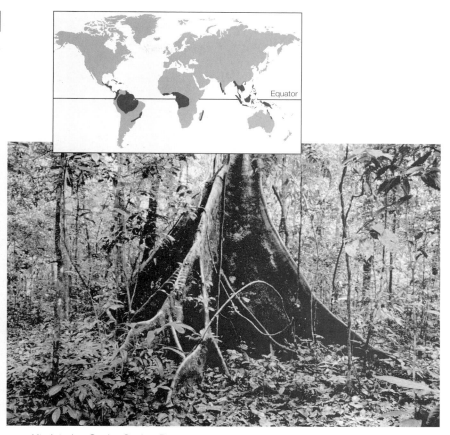

...and its interior, Cocha Cashu, Peru

Equator

Temperature

°C

The weather is warm and rainy all year.

Range 2.2°C   Iquitos, Peru 3°S

Precipitation

cm

Annual total: 262 cm

Jan   Jul   Dec

Biological activity

Photosynthesis

Flowering

Fruiting

Mammals

Birds

Insects

Soil biota

Jan   Jul   Dec

Biological activity is high year round.

Community composition

Dominant plants
Trees and vines

Species richness
Plants: Extremely high
Animals: Extremely high in mammals, birds, amphibians, and arthropods

Soil biota
Very rich but poorly known

Figure 52J *(Page 1127)*

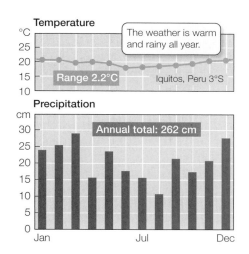

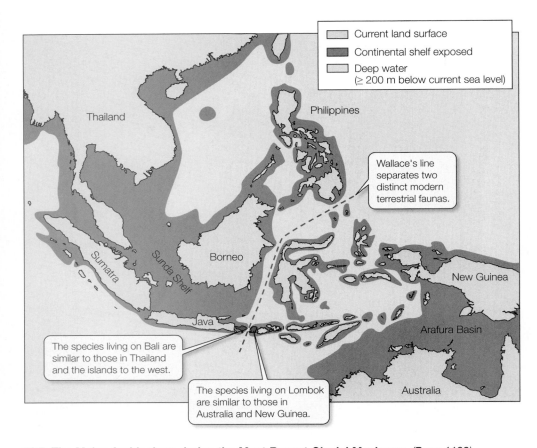

52.7 The Malay Archipelago during the Most Recent Glacial Maximum *(Page 1129)*

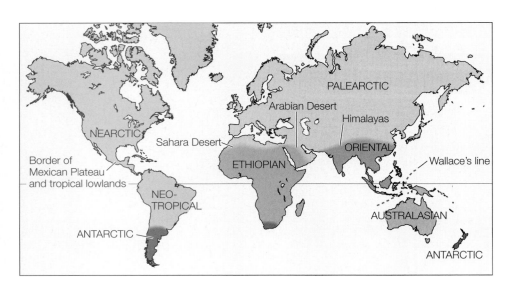

52.8 Major Biogeographic Regions *(Page 1129)*

Taxonomic phylogeny

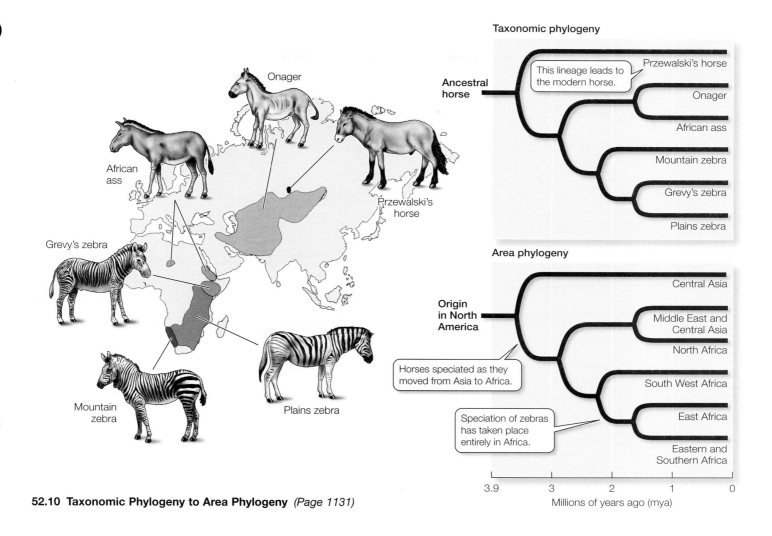

This lineage leads to the modern horse.

Ancestral horse

Przewalski's horse

Onager

African ass

Mountain zebra

Grevy's zebra

Plains zebra

Area phylogeny

Origin in North America

Central Asia

Middle East and Central Asia

North Africa

South West Africa

East Africa

Eastern and Southern Africa

Horses speciated as they moved from Asia to Africa.

Speciation of zebras has taken place entirely in Africa.

3.9 3 2 1 0
Millions of years ago (mya)

Onager

African ass

Grevy's zebra

Przewalski's horse

Mountain zebra

Plains zebra

52.10 Taxonomic Phylogeny to Area Phylogeny *(Page 1131)*

(A)

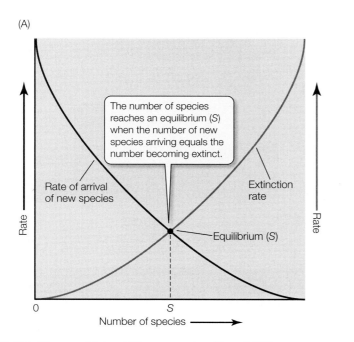

The number of species reaches an equilibrium (*S*) when the number of new species arriving equals the number becoming extinct.

Rate of arrival of new species

Extinction rate

Equilibrium (*S*)

Rate

Rate

0   *S*
Number of species

(B)

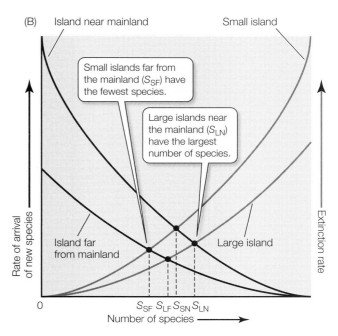

Island near mainland

Small island

Small islands far from the mainland (S_{SF}) have the fewest species.

Large islands near the mainland (S_{LN}) have the largest number of species.

Island far from mainland

Large island

Rate of arrival of new species

Extinction rate

0   S_{SF} S_{LF} S_{SN} S_{LN}
Number of species

52.11 The Theory of Island Biogeography *(Page 1132)*

TABLE 52.1

Number of Species of Resident Land Birds on Krakatau

PERIOD	NUMBER OF SPECIES	EXTINCTIONS	COLONIZATIONS
1908	13		
1908–1919		2	17
1919–1921	28		
1921–1933		3	4
1933–1934	29		
1934–1951		3	7
1951	33		
1952–1984		4	7
1984–1996	36		

(Page 1133)

EXPERIMENT

HYPOTHESIS: Defaunated islands will be rapidly recolonized, eventually achieving about the same number of species that they had prior to defaunation.

METHOD

Erect scaffolding and tent to enclose islets. Fumigate small islets with a chemical (methyl bromide) that kills arthropods but does not harm plants. Periodically monitor recolonizations and extinctions of arthropods on the islands.

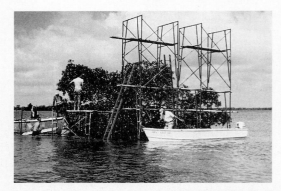

RESULTS

Recolonization was rapid, turnover rates were high, and the rate of recolonization was slowest on the most remote island.

CONCLUSION: An island can support a certain equilibrium number of species.

52.12 Experimental Island Defaunation *(Page 1133)*

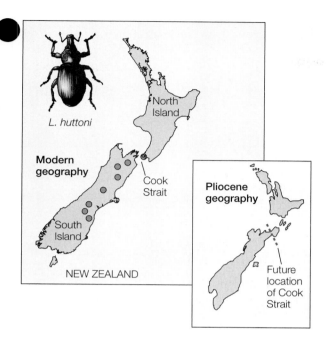

52.14 A Vicariant Distribution Explained *(Page 1135)*

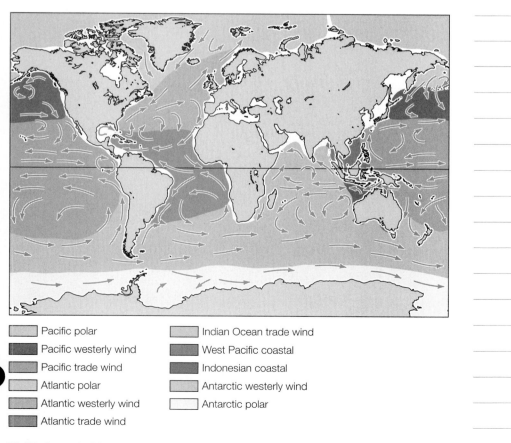

Pacific polar

Pacific westerly wind

Pacific trade wind

Atlantic polar

Atlantic westerly wind

Atlantic trade wind

Indian Ocean trade wind

West Pacific coastal

Indonesian coastal

Antarctic westerly wind

Antarctic polar

52.15 Oceanic Biogeographic Regions Are Determined by Ocean Currents
(Page 1136)

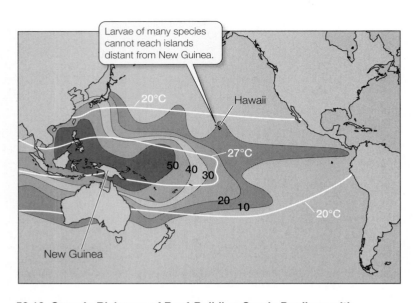

52.16 Generic Richness of Reef-Building Corals Declines with Distance from New Guinea *(Page 1136)*

CHAPTER 53 Behavior and Behavioral Ecology

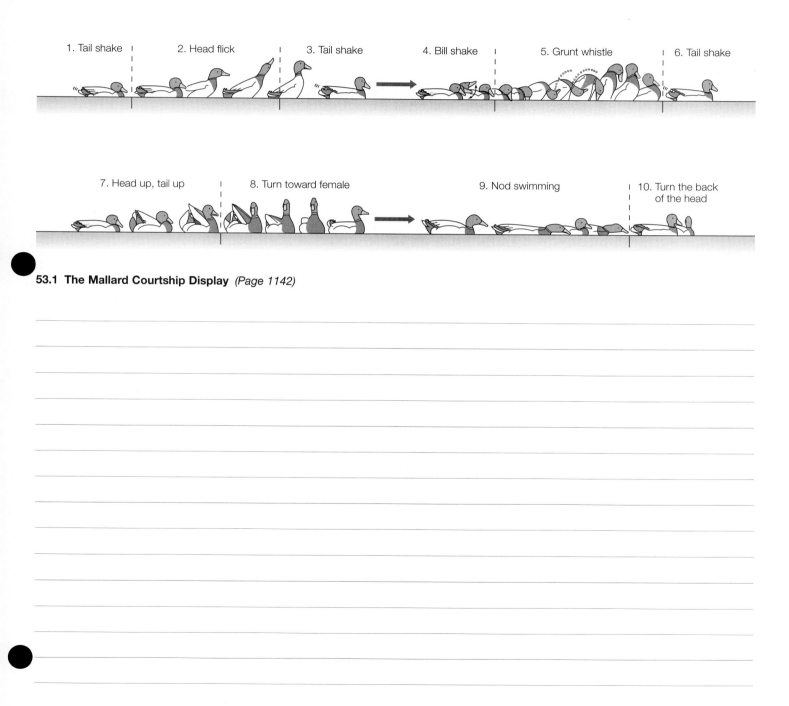

1. Tail shake   2. Head flick   3. Tail shake   4. Bill shake   5. Grunt whistle   6. Tail shake

7. Head up, tail up   8. Turn toward female   9. Nod swimming   10. Turn the back of the head

53.1 The Mallard Courtship Display *(Page 1142)*

(A)

(B)

EXPERIMENT

HYPOTHESIS: Absence of the *fosB* gene changes the maternal behavior of female mice.

METHOD

1. Inactivate ("knock out") the *fosB* gene in a strain of female mice. When mature, allow them to mate and give birth at the same time wild-type females of the same strain do so.

2. Immediately upon birth, separate mouse pups from mothers. Place three newborn pups in opposite corners of a cage with their mother.

3. Count number of pups each mother retrieves within 20 minutes.

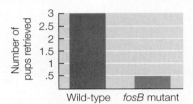

RESULTS

Normal females retrieved all of their pups within 20 minutes, whereas *fosB* mutant females retrieved an average of only 0.5 pups.

CONCLUSION: The protein product of the *fosB* gene is needed in order for female mice to display normal maternal behavior.

53.2 A Single Gene Affects Maternal Behavior in Mice
(Page 1143)

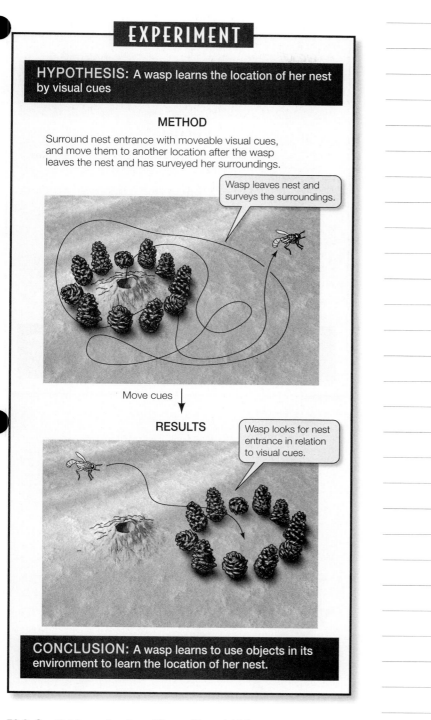

EXPERIMENT

HYPOTHESIS: A wasp learns the location of her nest by visual cues

METHOD

Surround nest entrance with moveable visual cues, and move them to another location after the wasp leaves the nest and has surveyed her surroundings.

Wasp leaves nest and surveys the surroundings.

Move cues ↓

RESULTS

Wasp looks for nest entrance in relation to visual cues.

CONCLUSION: A wasp learns to use objects in its environment to learn the location of her nest.

53.3 Spatial Learning by a Wasp *(Page 1144)*

EXPERIMENT 1

HYPOTHESIS: Learning is essential for song acquisition in white-crowned sparrows.

METHOD

Raise young sparrows in the presence of an adult male sparrow singing. Record song of these control birds when they mature and plot as a sonogram.

RESULTS

Control or wild bird

Frequency (kilocycles per second): 6 5 4 3 2 1

Time (seconds): 0.5 1.0 1.5 2.0

METHOD

Hatch eggs in an incubator and rear the birds in isolation. Record and plot their song. Compare to the control birds' song.

RESULTS

Isolated hand-reared bird

Frequency (kilocycles per second): 6 5 4 3 2 1

Time (seconds): 0.5 1.0 1.5 2.0

CONCLUSION: White-crowned sparrows that do not hear adult song as nestlings do not express their species-specific song when they mature.

53.5 Two Critical Periods for Song Learning *(Page 1146)*

EXPERIMENT 2

HYPOTHESIS: Maturing white-crowned sparrows require auditory feedback to learn to express their species-specific song.

METHOD

Deafen subadult birds that have heard the song of adult males when they were nestlings. Record their song and plot as a sonogram.

RESULTS

Deaf bird

Frequency (kilocycles per second): 6 5 4 3 2 1

Time (seconds): 0.5 1.0 1.5 2.0

CONCLUSION: Even if white-crowned sparrows have formed a song memory, they need auditory feedback to learn to match it.

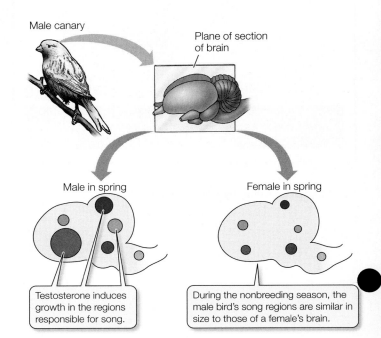

Male canary

Plane of section of brain

Male in spring

Female in spring

Testosterone induces growth in the regions responsible for song.

During the nonbreeding season, the male bird's song regions are similar in size to those of a female's brain.

53.6 Effects of Testosterone on Bird Brains *(Page 1147)*

EXPERIMENT

HYPOTHESIS: Collared flycatchers use information on the reproductive success of already established individuals to decide where to settle.

METHOD

Transfer nestlings from nests in one area to nests in another area to create an area with super-sized broods and another area with under-sized broods. Leave a third area unaltered to serve as a control. Observe settlement the following year.

RESULTS

Immigration was higher in plots with super-sized broods…

…and lower in plots with undersized broods.

Relative immigration rates (year $t + 1$)

High

Low

Broods increased | Controls | Broods decreased

Experimental treatment

CONCLUSION: Collared flycatchers use brood size of already settled individuals to assess how good an area is for breeding.

53.7 Flycatchers Use Neighbors' Success to Assess Habitat Quality *(Page 1148)*

EXPERIMENT

HYPOTHESIS: Testosterone-induced aggressiveness is too costly to Yarrow's spiny lizards to maintain year round.

METHOD

Insert testosterone capsules under skin of males during the summer and observe their behavior and survivorship.

Yarrow's spiny lizard (*Sceloporus jarrovii*)

RESULTS

Testosterone-treated males were more active and displayed more territorial behaviors than untreated males.

Percentage of lizards active

80

60

40

20

0

Implants (treated with testosterone)

No implants (untreated controls)

6 am | 8 am | 10 am | 12 noon | 2 pm | 4 pm

Time of day

Percentage surviving

100

80

60

40

Treated males survived less well than untreated males.

No implants (untreated controls)

Implants (treated with testosterone)

0 | 10 | 20 | 30 | 40 | 50

Time after receiving implant (days)

CONCLUSION: If Yarrow's spiny lizards were territorial during the summer, they would survive less well than they do.

53.8 The Costs of Defending a Territory *(Page 1149)*

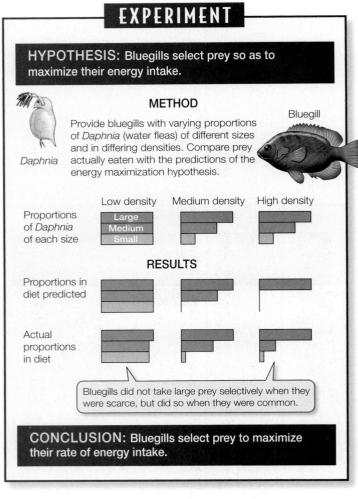

EXPERIMENT

HYPOTHESIS: Bluegills select prey so as to maximize their energy intake.

METHOD

Daphnia

Bluegill

Provide bluegills with varying proportions of *Daphnia* (water fleas) of different sizes and in differing densities. Compare prey actually eaten with the predictions of the energy maximization hypothesis.

Low density   Medium density   High density

Proportions of *Daphnia* of each size

Large
Medium
Small

RESULTS

Proportions in diet predicted

Actual proportions in diet

Bluegills did not take large prey selectively when they were scarce, but did so when they were common.

CONCLUSION: Bluegills select prey to maximize their rate of energy intake.

53.10 Bluegills Are Energy Maximizers *(Page 1150)*

EXPERIMENT

HYPOTHESIS: Spices commonly used in cooking have antibacterial properties.

METHOD

Prepare alcohol extracts of spices and test whether they inhibit growth of food-borne bacteria in culture media.

RESULTS

● Complete inhibition   ● Moderate inhibition   ● No inhibition

Spice	*Staphylococcus aureus*	*Bacillus stearothermophilus*	*Bacillus coagulans*	*Vibrio cholerae*
Mace	●	●	●	●
Bay leaf	●	●	●	●
Nutmeg	●	●	●	●
Garlic	●	●	●	●
Sage	●	●	●	●
Cinnamon	●	●	●	●
Thyme	●	●	●	●
Paprika	●	●	●	●
Oregano	●	●	●	●
Anise	●	●	●	●
Turmeric	●	●	●	●
Cardamom	●	●	●	●
White pepper	●	●	●	●
Black pepper	●	●	●	●
Allspice	●	●	●	●
Rosemary	●	●	●	●

Bacterium

CONCLUSION: Most commonly used spices have strong antibacterial activity against more than one kind of food-borne bacteria.

53.12 Most Spices Have Antimicrobial Activity *(Page 1151)*

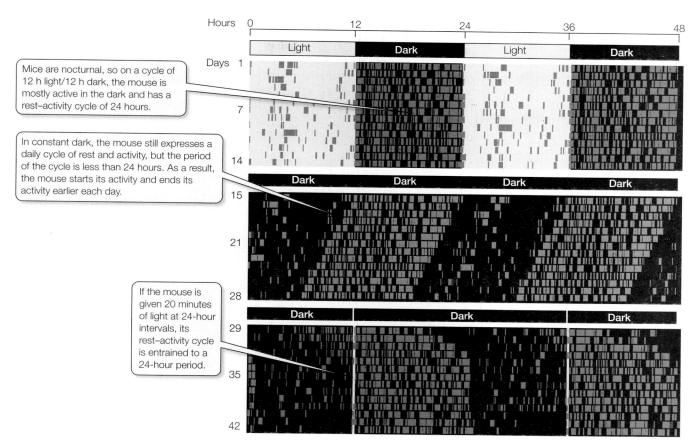

Hours

Mice are nocturnal, so on a cycle of 12 h light/12 h dark, the mouse is mostly active in the dark and has a rest–activity cycle of 24 hours.

In constant dark, the mouse still expresses a daily cycle of rest and activity, but the period of the cycle is less than 24 hours. As a result, the mouse starts its activity and ends its activity earlier each day.

If the mouse is given 20 minutes of light at 24-hour intervals, its rest–activity cycle is entrained to a 24-hour period.

53.14 Circadian Rhythms Are Entrained *(Page 1153)*

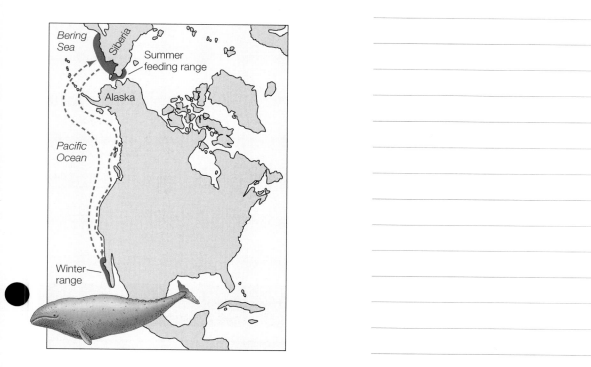

53.15 Piloting *(Page 1154)*

(A)

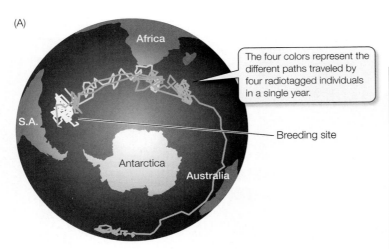

The four colors represent the different paths traveled by four radiotagged individuals in a single year.

Africa

S.A.

Antarctica

Australia

Breeding site

(B) *Diomedea chrysostoma*

53.16 Coming Home *(Page 1155)*

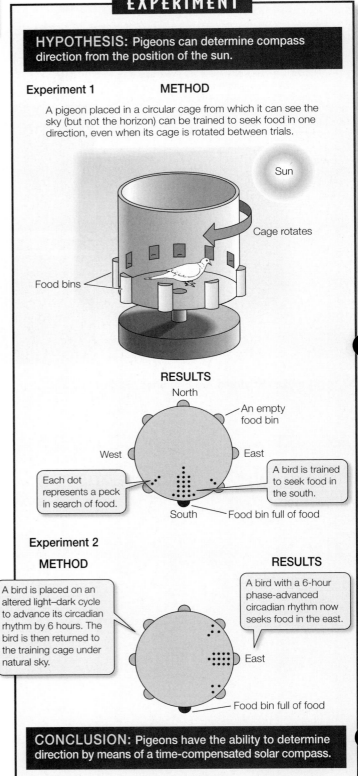

EXPERIMENT

HYPOTHESIS: Pigeons can determine compass direction from the position of the sun.

Experiment 1 **METHOD**

A pigeon placed in a circular cage from which it can see the sky (but not the horizon) can be trained to seek food in one direction, even when its cage is rotated between trials.

Sun

Cage rotates

Food bins

RESULTS

North

An empty food bin

West   East

Each dot represents a peck in search of food.

A bird is trained to seek food in the south.

South   Food bin full of food

Experiment 2

METHOD

A bird is placed on an altered light–dark cycle to advance its circadian rhythm by 6 hours. The bird is then returned to the training cage under natural sky.

RESULTS

A bird with a 6-hour phase-advanced circadian rhythm now seeks food in the east.

East

Food bin full of food

CONCLUSION: Pigeons have the ability to determine direction by means of a time-compensated solar compass.

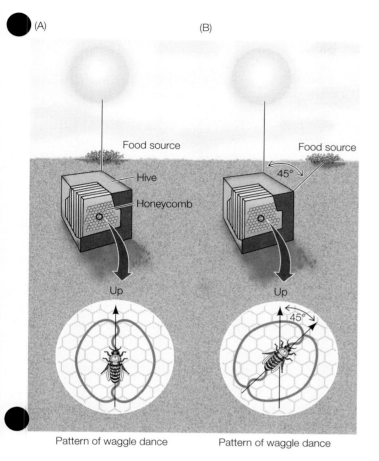

(A)   (B)

Food source

Food source

45°

Hive

Honeycomb

Up   Up

45°

Pattern of waggle dance   Pattern of waggle dance

53.19 The Waggle Dance of the Honeybee *(Page 1158)*

EXPERIMENT

HYPOTHESIS: Flocking helps animals evade predators.

METHOD

Release a hawk near pigeon flocks of different sizes. Observe whether the hawk captures a pigeon.

Goshawk

Wood pigeon

RESULTS

The more pigeons in the flock, the sooner the hawk is spotted…

…and the lower the hawk's attack success.

CONCLUSION: Flocking behavior provides protection against predation.

53.20 Groups Provide Protection from Predators *(Page 1159)*

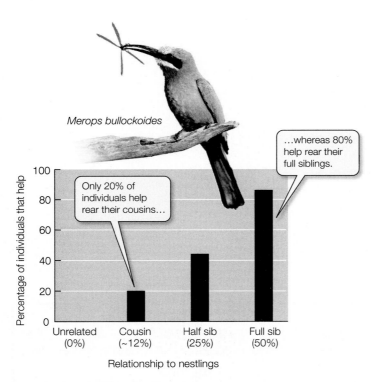

53.21 White-Fronted Bee-Eaters Are Discriminating Altruists
(Page 1160)

CHAPTER 54 Population Ecology

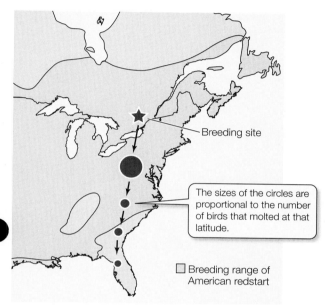

54.2 **Hydrogen Isotopes Tell Where Migratory American Redstarts Molted Their Feathers** *(Page 1169)*

Breeding site

The sizes of the circles are proportional to the number of birds that molted at that latitude.

☐ Breeding range of American redstart

TABLE 54.1

Life Table of the 1978 Cohort of the Cactus Finch (*Geospiza scandens*) on Isla Daphne

AGE IN YEARS (x)	NUMBER ALIVE	SURVIVORSHIP[a]	SURVIVAL RATE[b]	MORTALITY RATE[c]
0	210	1.000	0.434	0.566
1	91	0.434	0.857	0.143
2	78	0.371	0.898	0.102
3	70	0.333	0.928	0.072
4	65	0.309	0.955	0.045
5	62	0.295	0.678	0.322
6	42	0.200	0.548	0.452
7	23	0.109	0.652	0.348
8	15	0.071	0.933	0.067
9	14	0.067	0.786	0.214
10	11	0.052	0.909	0.091
11	10	0.048	0.400	0.600
12	4	0.019	0.750	0.250
13	3	0.014	0.996	

[a]Survivorship = the proportion of newborns who survive to age x.
[b]Survival rate = the proportion of individuals of age x who survive to age $x + 1$.
[c]Mortality rate = the proportion of individuals of age x who die before the age of $x + 1$.

(Page 1170)

601

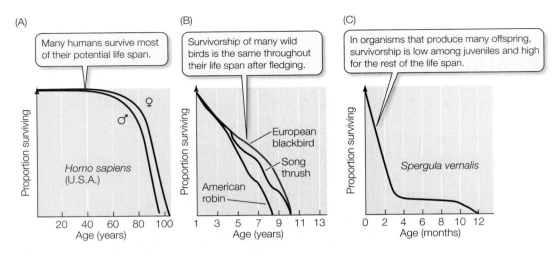

(A) Many humans survive most of their potential life span.

Homo sapiens (U.S.A.)

Proportion surviving

♀
♂

20 40 60 80 100
Age (years)

(B) Survivorship of many wild birds is the same throughout their life span after fledging.

Proportion surviving

European blackbird
Song thrush
American robin

1 3 5 7 9 11 13
Age (years)

(C) In organisms that produce many offspring, survivorship is low among juveniles and high for the rest of the life span.

Proportion surviving

Spergula vernalis

0 2 4 6 8 10 12
Age (months)

54.3 Survivorship Curves *(Page 1170)*

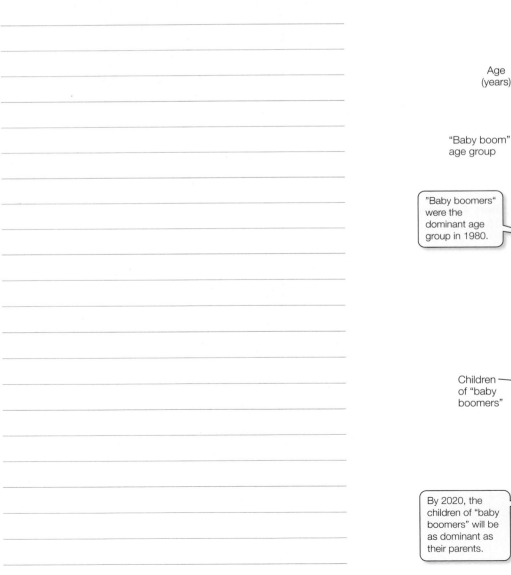

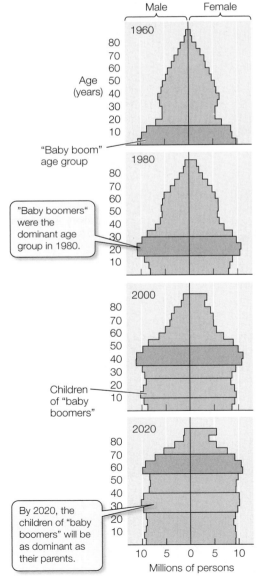

Male   Female

1960
Age (years) 80 70 60 50 40 30 20 10

"Baby boom" age group

1980
80 70 60 50 40 30 20 10

"Baby boomers" were the dominant age group in 1980.

2000
80 70 60 50 40 30 20 10

Children of "baby boomers"

2020
80 70 60 50 40 30 20 10

By 2020, the children of "baby boomers" will be as dominant as their parents.

10 5 0 5 10
Millions of persons

54.4 Age Distributions Change over Time *(Page 1171)*

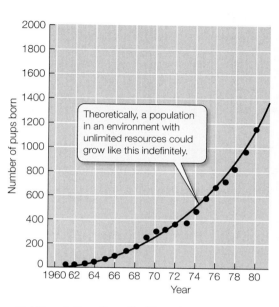

Mirounga angustirostris

54.7 Exponential Population Growth *(Page 1174)*

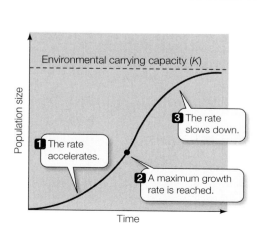

54.8 Logistic Population Growth *(Page 1174)*

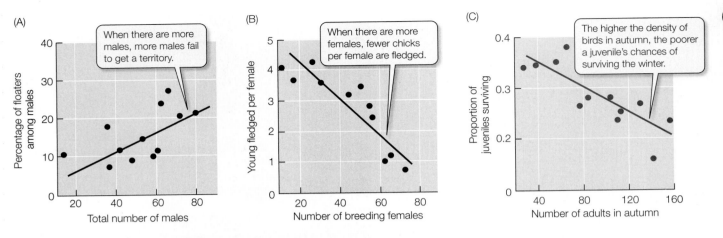

54.9 Regulation of an Island Population of Song Sparrows *(Page 1175)*

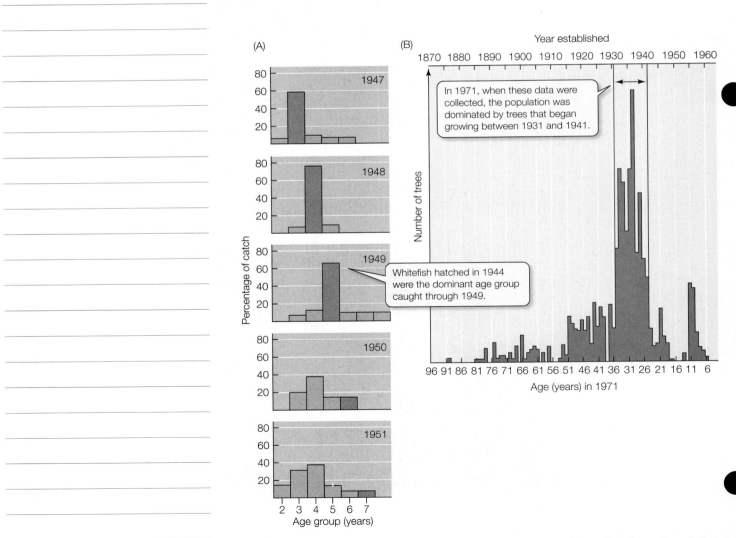

54.10 Individuals Born during Years of Good Reproduction May Dominate Populations
(Page 1176)

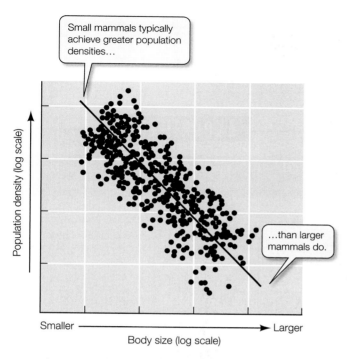

54.11 Population Density Decreases as Body Size Increases *(Page 1176)*

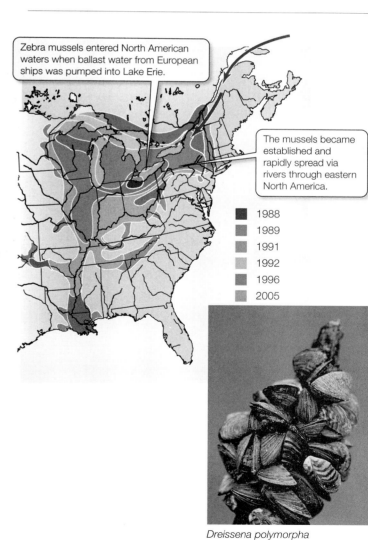

Dreissena polymorpha

54.12 Introduced Zebra Mussels Have Spread Rapidly
(Page 1177)

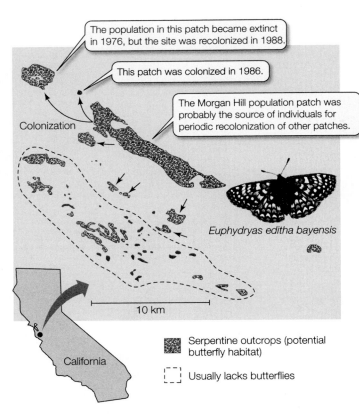

54.14 Metapopulation Dynamics *(Page 1178)*

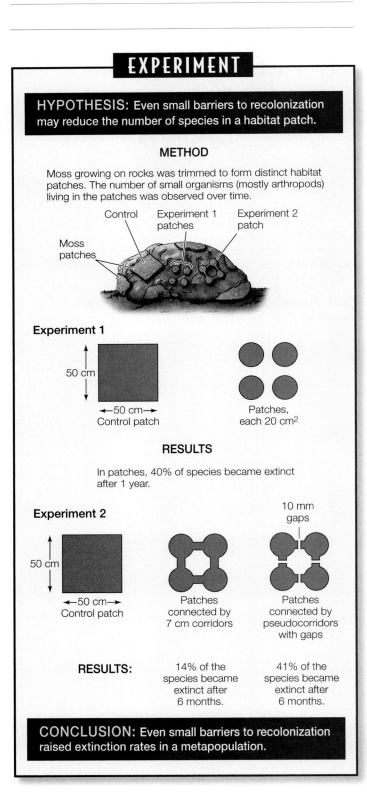

54.15 Narrow Barriers Suffice to Separate Arthropod Subpopulations *(Page 1179)*

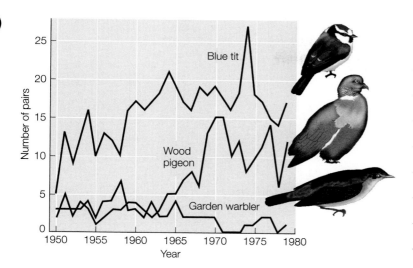

54.16 Populations May Be Influenced by Remote Events
(Page 1179)

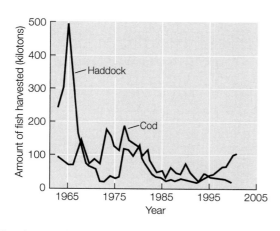

54.17 Overharvesting Can Reduce Fish Populations *(Page 1180)*

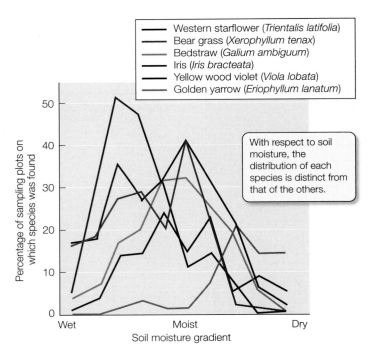

Legend:
— Western starflower (*Trientalis latifolia*)
— Bear grass (*Xerophyllum tenax*)
— Bedstraw (*Galium ambiguum*)
— Iris (*Iris bracteata*)
— Yellow wood violet (*Viola lobata*)
— Golden yarrow (*Eriophyllum lanatum*)

Percentage of sampling plots on which species was found

With respect to soil moisture, the distribution of each species is distinct from that of the others.

Soil moisture gradient: Wet — Moist — Dry

55.1 Plant Distributions along an Environmental Gradient
(Page 1186

TABLE 55.1

The Major Trophic Levels

TROPHIC LEVEL	SOURCE OF ENERGY	EXAMPLES
Photosynthesizers (primary producers)	Solar energy	Green plants, photosynthetic bacteria and protists
Herbivores (primary consumers)	Tissues of primary producers	Termites, grasshoppers, gypsy moth larvae, anchovies, deer, geese, white-footed mice
Primary carnivores (secondary consumers)	Herbivores	Spiders, warblers, wolves, copepods
Secondary carnivores (tertiary consumers)	Primary carnivores	Tuna, falcons, killer whales
Omnivores	Several trophic levels	Humans, opossums, crabs, robins
Detritivores (decomposers)	Dead bodies and waste products of other organisms	Fungi, many bacteria, vultures, earthworms

(Page 1186)

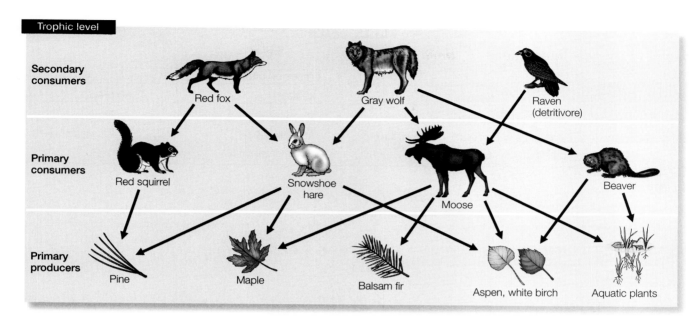

Trophic level

Secondary consumers — Red fox, Gray wolf, Raven (detritivore)

Primary consumers — Red squirrel, Snowshoe hare, Moose, Beaver

Primary producers — Pine, Maple, Balsam fir, Aspen, white birch, Aquatic plants

55.2 Food Webs Show Trophic Interactions in a Community *(Page 1187)*

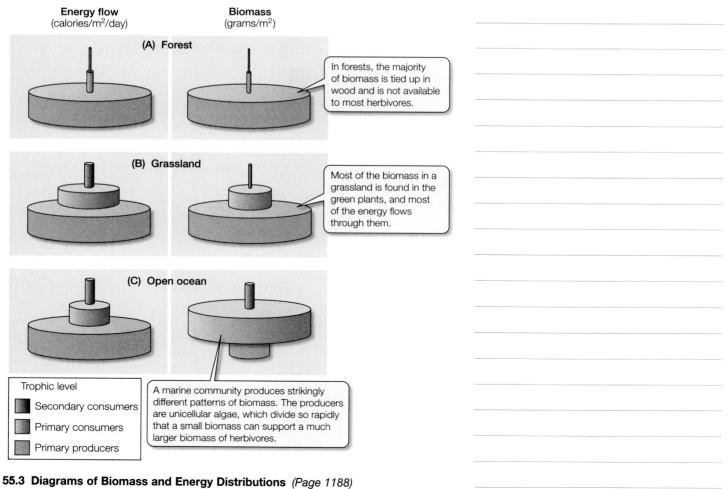

Energy flow
(calories/m²/day)

Biomass
(grams/m²)

(A) Forest

In forests, the majority of biomass is tied up in wood and is not available to most herbivores.

(B) Grassland

Most of the biomass in a grassland is found in the green plants, and most of the energy flows through them.

(C) Open ocean

A marine community produces strikingly different patterns of biomass. The producers are unicellular algae, which divide so rapidly that a small biomass can support a much larger biomass of herbivores.

Trophic level
- Secondary consumers
- Primary consumers
- Primary producers

55.3 Diagrams of Biomass and Energy Distributions *(Page 1188)*

TABLE 55.2

Types of Ecological Interactions

EFFECT OF ORGANISM 1		EFFECT ON ORGANISM 2		
		HARM	BENEFIT	NO EFFECT
	HARM	Competition (–/–)	Predation or parasitism (–/+)	Amensalism (–/0)
	BENEFIT	Predation or parasitism (+/–)	Mutualism (+/+)	Commensalism (+/0)
	NO EFFECT	Amensalism (0/–)	Commensalism (0/+)	—

(Page 1189)

55.4 Hare and Lynx Populations Cycle in Nature *(Page 1189)*

EXPERIMENT

HYPOTHESIS: Population cycles of hares are influenced by both food supply and predators.

METHOD

1. Select 9 1-km^2 blocks of undisturbed coniferous forest.
2. In two of the blocks give the hares supplemental food year-round.
3. Erect an electric fence around two other blocks, with mesh large enough to allow hares, but not lynxes, to pass through.
4. Provide extra food in one of these enclosed blocks.
5. In two other blocks add fertilizer to increase food quality.
6. Use three other blocks as unmanipulated controls.

RESULTS

Food added
Adding food tripled hare density.

Predators excluded
Excluding predators doubled hare density.

Fertilizer added
Fertilizing vegetation to increase its food quality had no significant effect.

Food added and predators excluded
Both adding food and excluding predators increased hare density dramatically.

Ratio of hare density to controls

Control

One hare population cycle (11 years)

CONCLUSION: Population cycles of the snowshoe hare are influenced by their food supply as well as by interactions with their predators.

55.5 Prey Population Cycles May Have Multiple Causes
(Page 1190)

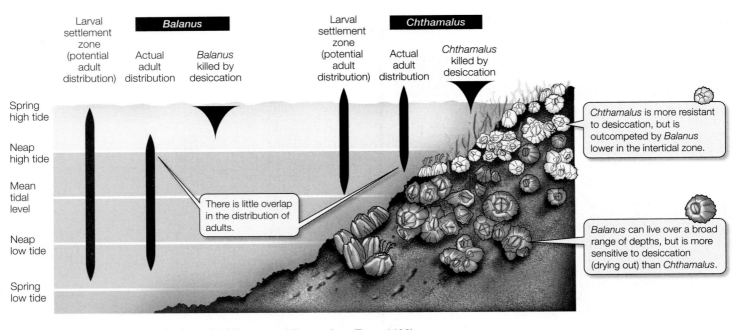

55.9 Competition Restricts the Intertidal Ranges of Barnacles *(Page 1192)*

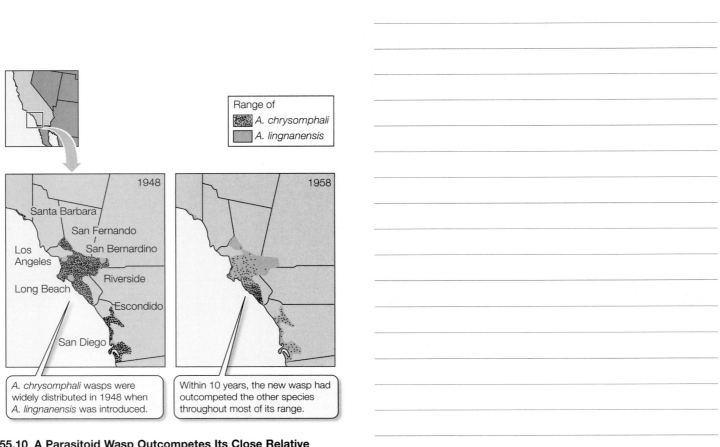

55.10 A Parasitoid Wasp Outcompetes Its Close Relative
(Page 1193)

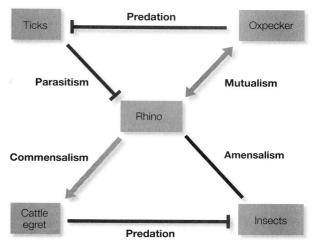

55.11 A Single Small Community Demonstrates Many Interactions *(Page 1193)*

(A)

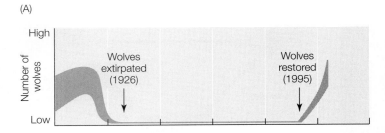

(B)

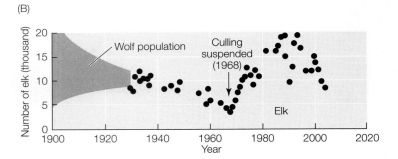

(C)

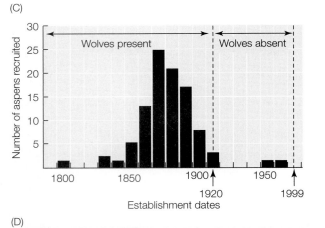

(D)

Populus tremuloides

55.13 Wolves Initiated a Trophic Cascade *(Page 1195)*

(A)

Libellula pulchella

(B)

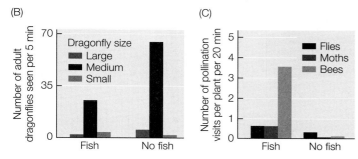

55.14 Trophic Cascades May Cross Habitats *(Page 1196)*

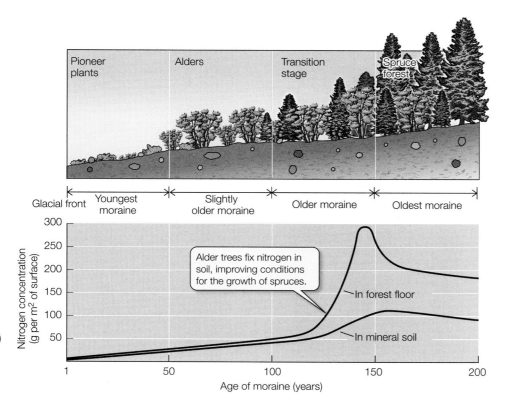

55.17 Primary Succession on a Glacial Moraine *(Page 1198)*

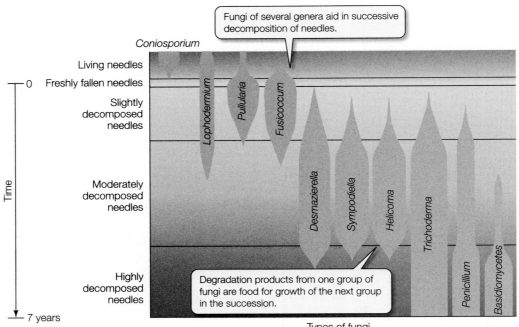

55.18 Secondary Succession on Pine Needles *(Page 1198)*

EXPERIMENT

HYPOTHESIS: Small boulders have fewer species growing on them than larger boulders because they are subjected to high levels of disturbance.

METHOD

Sterilize a number of small boulders. Secure some of them to the natural substratum with glue. Leave other small boulders unsecured to serve as controls. Observe accumulation of species on the boulders over time.

RESULTS

Secured small boulders accumulated many more species than unsecured small boulders.

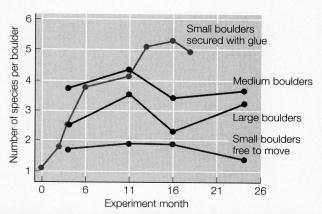

CONCLUSION: Small boulders have fewer species because the higher rates at which they are moved by waves prevent many species from surviving on them, not because they are unsuitable habitat for local species.

55.19 Species Richness Is Greatest at Intermediate Levels of Disturbance *(Page 1199)*

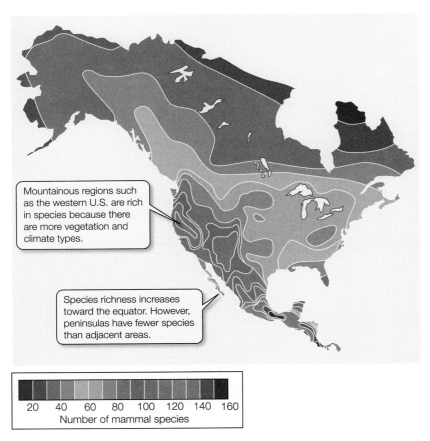

Mountainous regions such as the western U.S. are rich in species because there are more vegetation and climate types.

Species richness increases toward the equator. However, peninsulas have fewer species than adjacent areas.

20 40 60 80 100 120 140 160
Number of mammal species

55.20 The Latitudinal Gradient of Species Richness of North American Mammals *(Page 1200)*

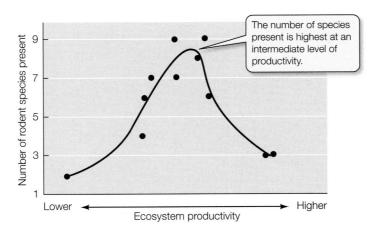

The number of species present is highest at an intermediate level of productivity.

55.21 Species Richness Peaks at Intermediate Productivity *(Page 1201)*

EXPERIMENT

HYPOTHESIS: Communities with many species should have higher productivity and stability than communities with few species.

METHOD

Clear and plant plots with different numbers and mixtures of grass species. Measure productivity and species composition of the plots over 11 years.

RESULTS

(A) Productivity increases with species richness

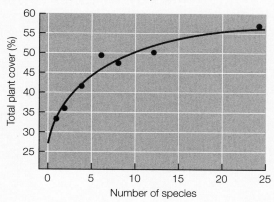

(B) Variation in productivity decreases with species richness

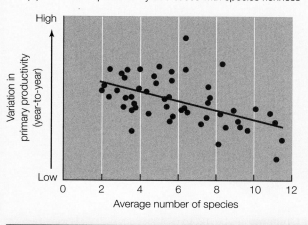

CONCLUSION: Plots with more species were more productive and varied less in productivity.

55.22 Species Richness Enhances Community Productivity
(Page 1201)

CHAPTER 56 Ecosystems and Global Ecology

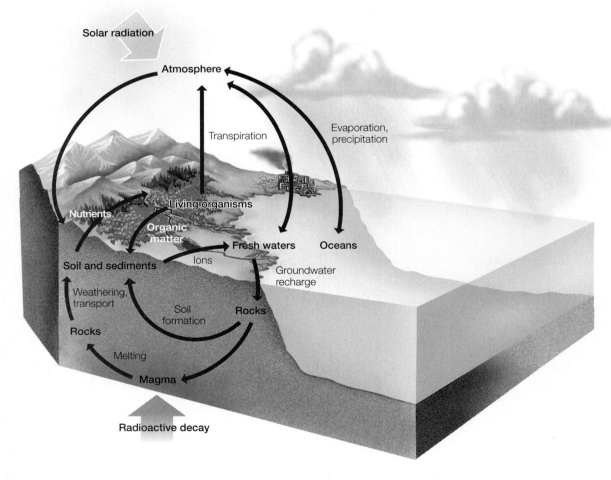

56.1 The Global Ecosystem's Compartments are Connected by the Flow of Elements *(Page 1206)*

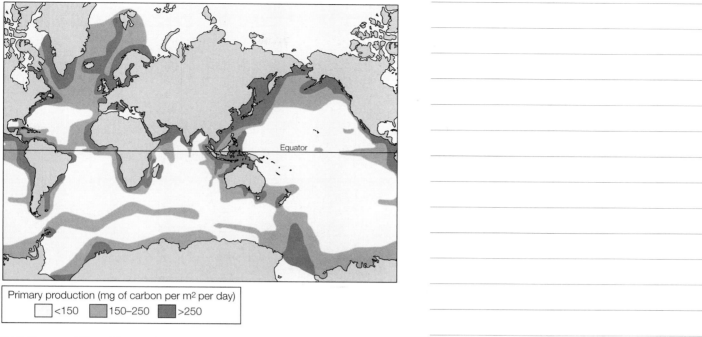

Primary production (mg of carbon per m² per day)

☐ <150 ▨ 150–250 ▧ >250

56.2 Zones of Upwelling are Productive *(Page 1207)*

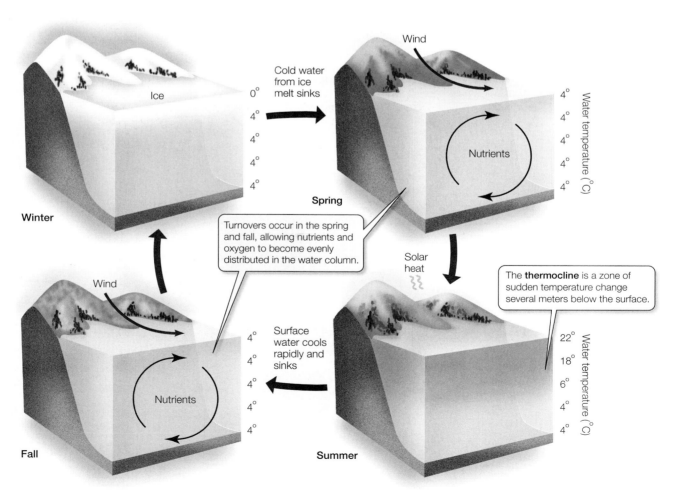

56.3 The Turnover Cycle in a Temperate Lake *(Page 1207)*

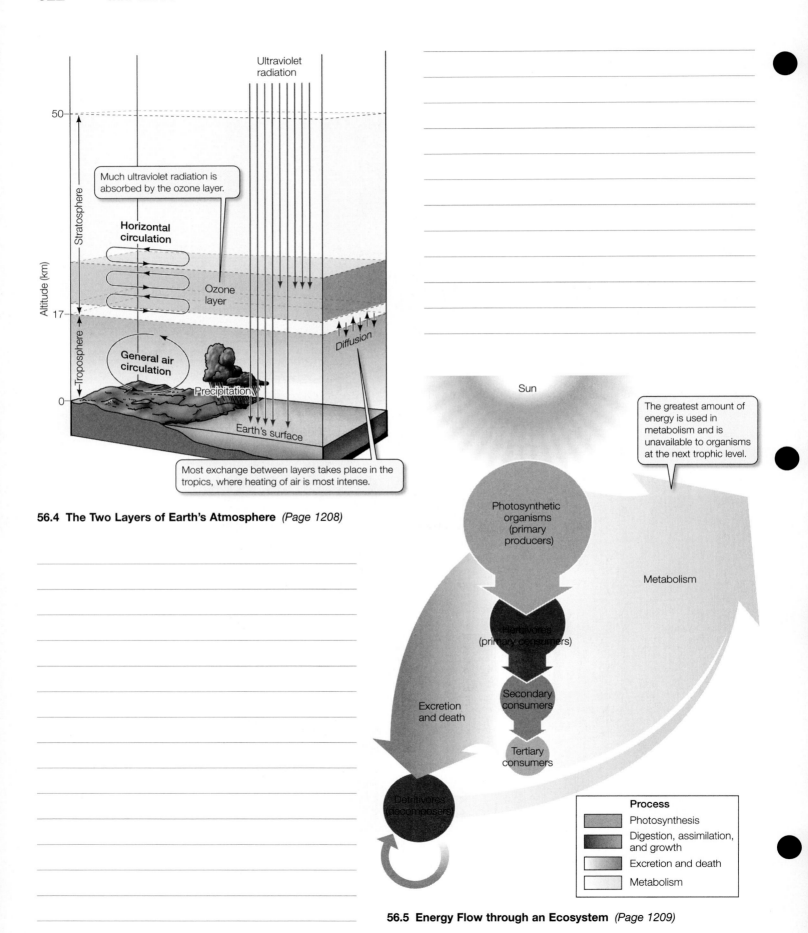

56.4 The Two Layers of Earth's Atmosphere *(Page 1208)*

56.5 Energy Flow through an Ecosystem *(Page 1209)*

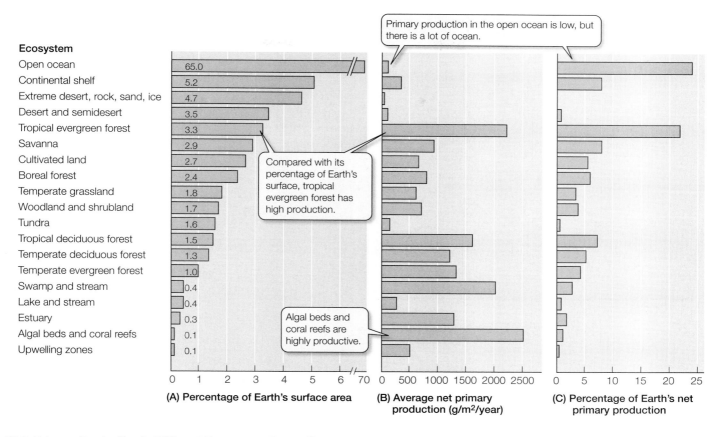

Ecosystem	
Open ocean	65.0
Continental shelf	5.2
Extreme desert, rock, sand, ice	4.7
Desert and semidesert	3.5
Tropical evergreen forest	3.3
Savanna	2.9
Cultivated land	2.7
Boreal forest	2.4
Temperate grassland	1.8
Woodland and shrubland	1.7
Tundra	1.6
Tropical deciduous forest	1.5
Temperate deciduous forest	1.3
Temperate evergreen forest	1.0
Swamp and stream	0.4
Lake and stream	0.4
Estuary	0.3
Algal beds and coral reefs	0.1
Upwelling zones	0.1

Primary production in the open ocean is low, but there is a lot of ocean.

Compared with its percentage of Earth's surface, tropical evergreen forest has high production.

Algal beds and coral reefs are highly productive.

(A) Percentage of Earth's surface area

(B) Average net primary production (g/m^2/year)

(C) Percentage of Earth's net primary production

56.6 Primary Production in Different Ecosystem Types *(Page 1210)*

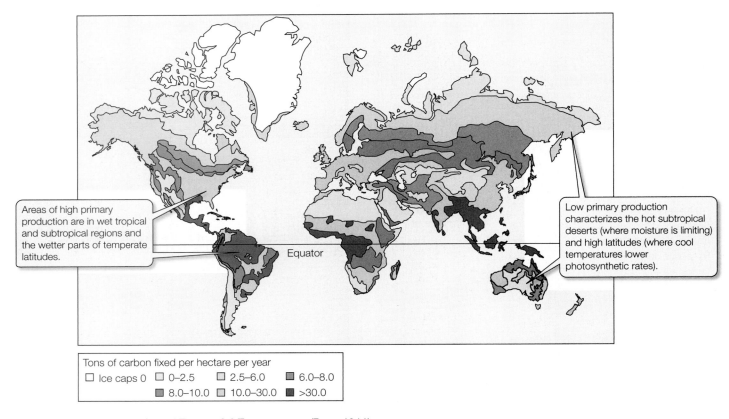

Areas of high primary production are in wet tropical and subtropical regions and the wetter parts of temperate latitudes.

Low primary production characterizes the hot subtropical deserts (where moisture is limiting) and high latitudes (where cool temperatures lower photosynthetic rates).

Equator

Tons of carbon fixed per hectare per year
☐ Ice caps 0 ☐ 0–2.5 ☐ 2.5–6.0 ■ 6.0–8.0
■ 8.0–10.0 ☐ 10.0–30.0 ■ >30.0

56.7 Net Primary Production of Terrestrial Ecosystems *(Page 1211)*

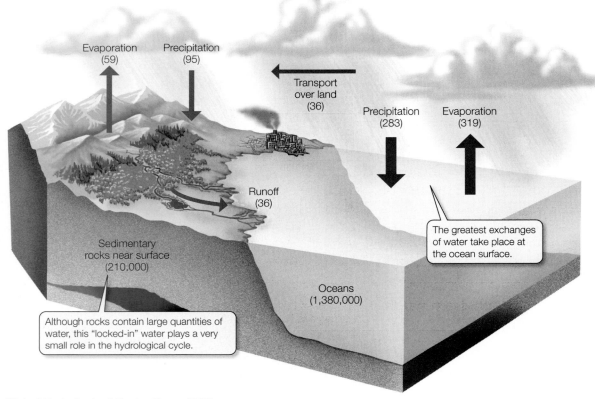

Evaporation (59)

Precipitation (95)

Transport over land (36)

Precipitation (283)

Evaporation (319)

Runoff (36)

The greatest exchanges of water take place at the ocean surface.

Sedimentary rocks near surface (210,000)

Oceans (1,380,000)

Although rocks contain large quantities of water, this "locked-in" water plays a very small role in the hydrological cycle.

56.8 The Global Hydrological Cycle *(Page 1212)*

(A)

(B)

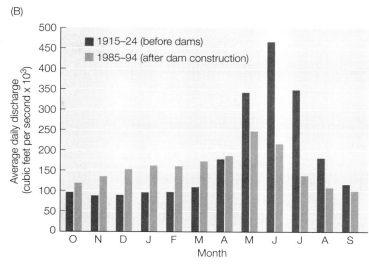

56.9 Columbia River Flows Have Been Massively Altered *(Page 1213)*

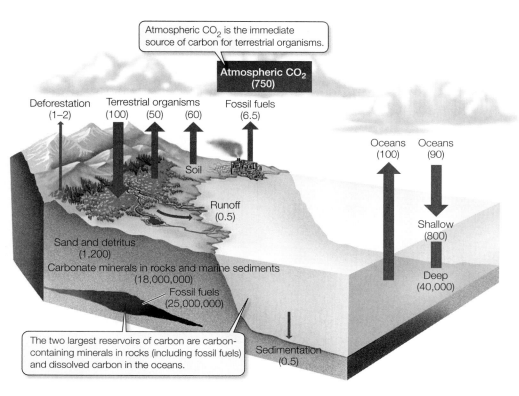

Atmospheric CO_2 is the immediate source of carbon for terrestrial organisms.

Atmospheric CO_2 (750)

Deforestation (1–2)

Terrestrial organisms (100) (50) (60)

Fossil fuels (6.5)

Soil

Runoff (0.5)

Oceans (100)

Oceans (90)

Shallow (800)

Sand and detritus (1,200)

Carbonate minerals in rocks and marine sediments (18,000,000)

Fossil fuels (25,000,000)

Deep (40,000)

The two largest reservoirs of carbon are carbon-containing minerals in rocks (including fossil fuels) and dissolved carbon in the oceans.

Sedimentation (0.5)

56.10 The Global Carbon Cycle *(Page 1214)*

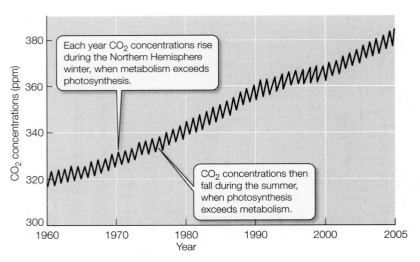

56.11 Atmospheric Carbon Dioxide Concentrations Are Increasing
(Page 1214)

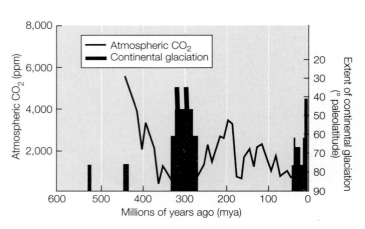

56.12 Higher Atmospheric CO$_2$ Concentrations Correlate with Warmer Temperatures *(Page 1215)*

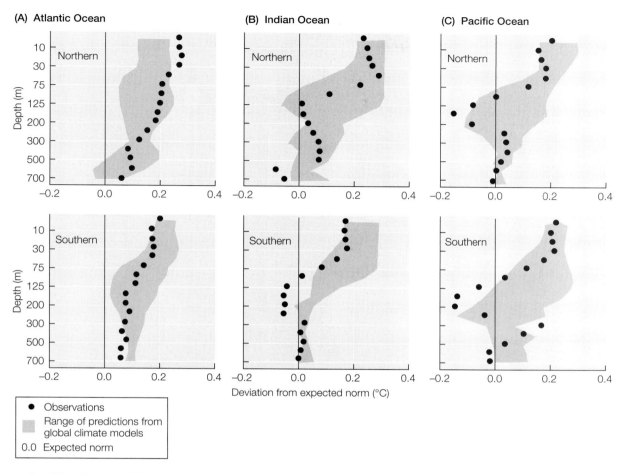

56.13 Oceans Are Warming—and Not Just at the Surface *(Page 1215)*

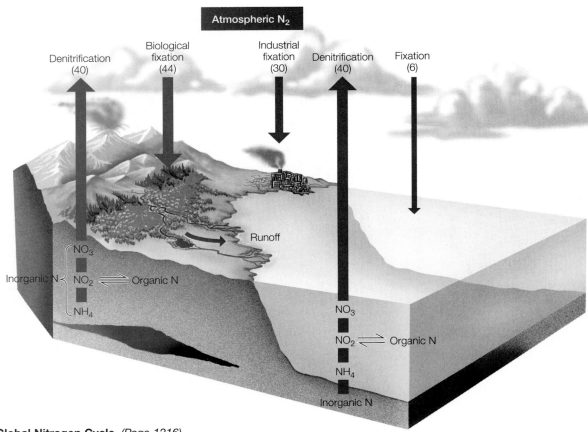

56.14 The Global Nitrogen Cycle *(Page 1216)*

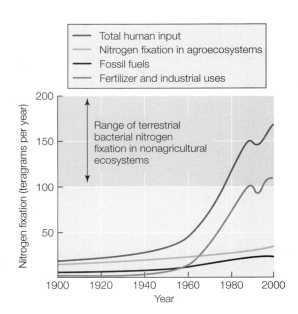

56.15 Human Activities Have Increased Nitrogen Fixation
(Page 1217)

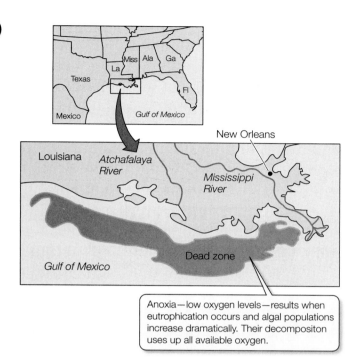

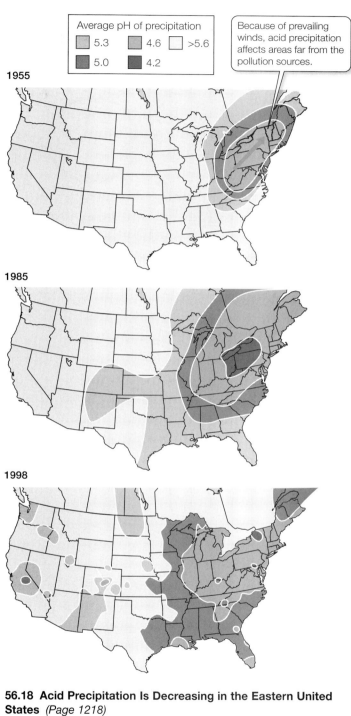

56.16 A "Dead Zone" Near the Mouth of the Mississippi River
(Page 1217)

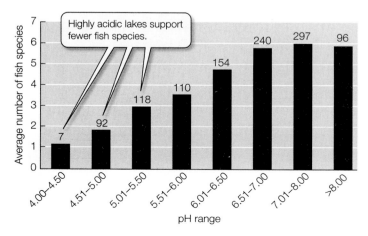

56.17 Acidification of Lakes Exterminates Fish Species
(Page 1218)

56.18 Acid Precipitation Is Decreasing in the Eastern United States *(Page 1218)*

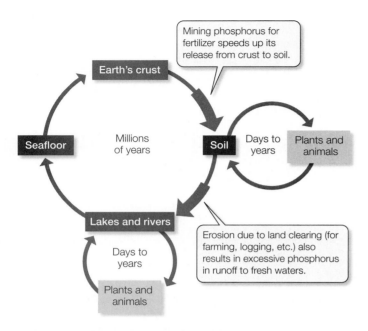

56.19 The Phosphorus Cycle *(Page 1219)*

EXPERIMENT

HYPOTHESIS: Higher CO_2 levels enhance nitrogen fixation.

METHOD

Artificially increase concentration of atmospheric CO_2 around plants. Measure rates of nitrogen fixation over 7 years. Determine concentration of iron and molybdenum in leaves of experimental plants. Maintain normal CO_2 levels around control plants.

RESULTS

Although higher CO_2 levels initially increased the rate of nitrogen fixation, the effect was quickly reversed. By year 4, nitrogen fixation rates were lower than those in the control plants.

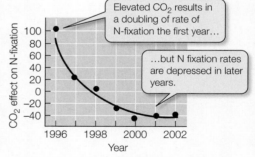

Elevated CO_2 results in a doubling of rate of N-fixation the first year…

…but N fixation rates are depressed in later years.

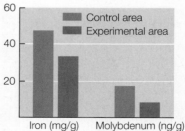

CONCLUSION: To predict the likely effects of CO_2 enrichment we need to investigate the interactions among many factors that might alter the processes we are studying.

56.20 High Atmospheric CO_2 Concentrations Negatively Affect Nitrogen Fixation *(Page 1220)*

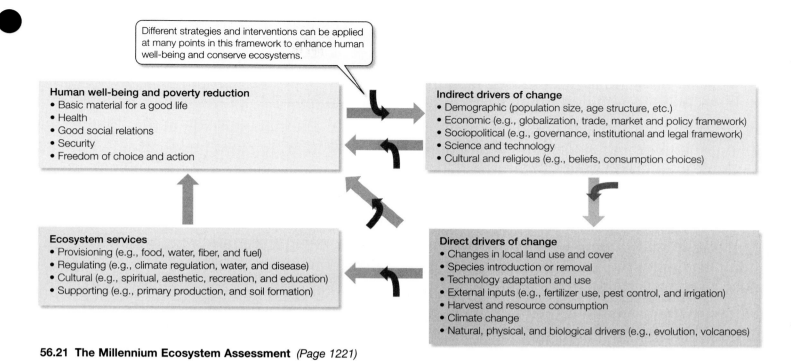

56.21 The Millennium Ecosystem Assessment *(Page 1221)*

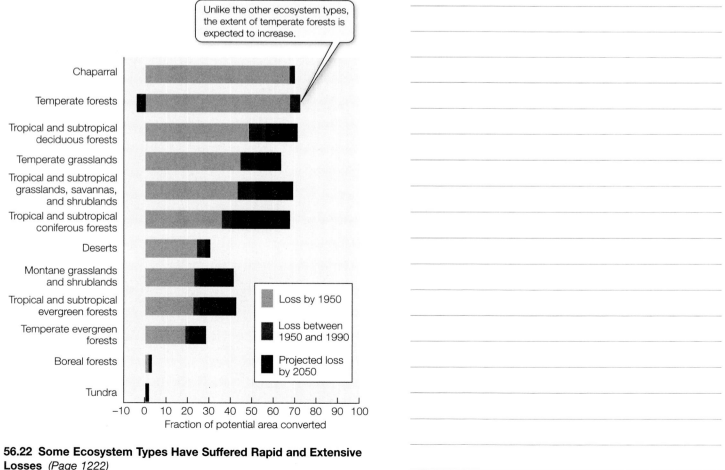

56.22 Some Ecosystem Types Have Suffered Rapid and Extensive Losses *(Page 1222)*

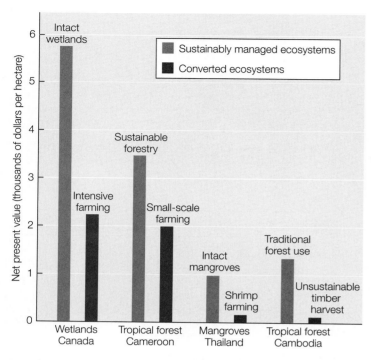

56.23 Economic Values of Sustainably Managed and Converted Ecosystems *(Page 1223)*

CHAPTER 57 Conservation Biology

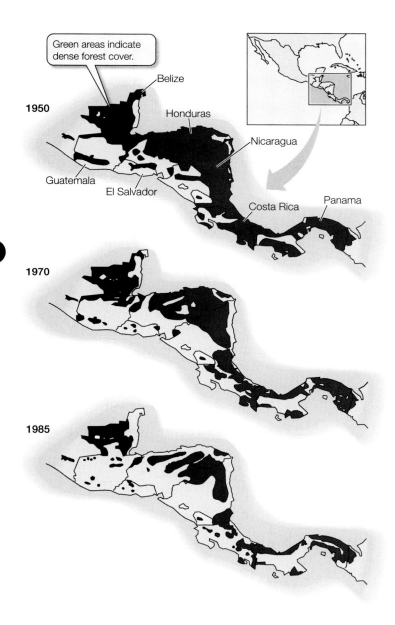

Green areas indicate dense forest cover.

1950

Belize

Honduras

Nicaragua

Guatemala

El Salvador

Costa Rica

Panama

1970

1985

57.3 Deforestation Rates Are High in Tropical Forests *(Page 1230)*

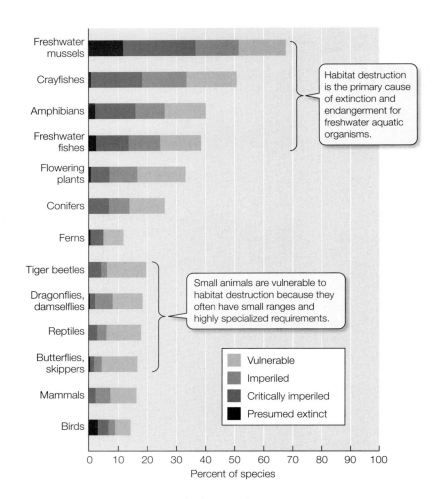

57.4 Proportions of U.S. Species Extinct or Threatened *(Page 1231)*

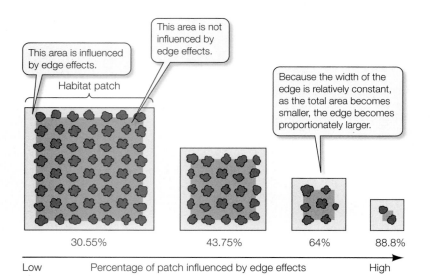

57.5 Edge Effects *(Page 1231)*

EXPERIMENT

HYPOTHESIS: Bluebirds use corridors to move between open patches within a forest.

METHOD

1. Create suitable patches of bluebird habitat in pine forests by cutting down trees to create the open conditions favored by bluebirds. Connect two patches by habitat corridors.

2. In one of the connected patches, plant wax myrtle bushes with fluorescently tagged fruit. Because birds that eat the fruit carry seeds in their guts and later defecate them, fluorescently tagged seeds serve as a tracking device.

3. Observe birds as they fly from wax myrtle bushes until they fly out of sight. Record the direction and distance each bird travels. Check surrounding patches for fluorescent seeds defecated by the birds. Develop a computer model that predicts the movement of seeds based on observations of the birds.

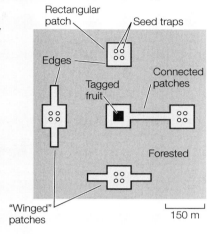

RESULTS

Bluebirds usually remained within a habitat patch when they encountered an edge, but they did move through corridors. Many more fluorescent seeds were moved between patches connected by corridors than between unconnected patches. The computer model correctly predicted the number of seeds that were moved.

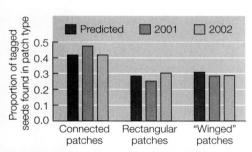

CONCLUSION: Bluebirds use corridors to move between patches of suitable habitat.

57.7 Habitat Corridors Facilitate Movement *(Page 1232)*

(A)

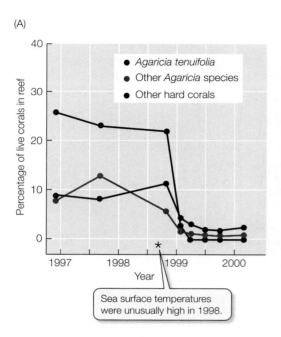

57.10 Global Warming Threatens Corals *(Page 1234)*

Centers of bird species richness

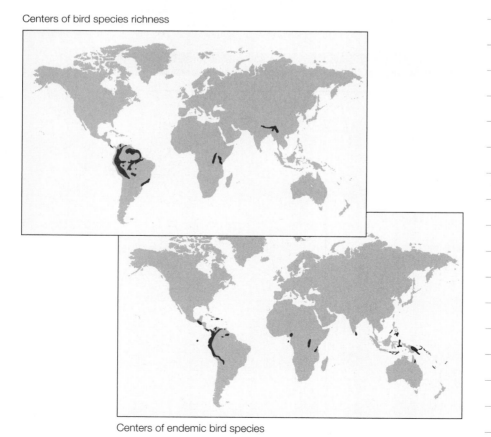

Centers of endemic bird species

57.11 Hotspots of Avian Biodiversity *(Page 1235)*

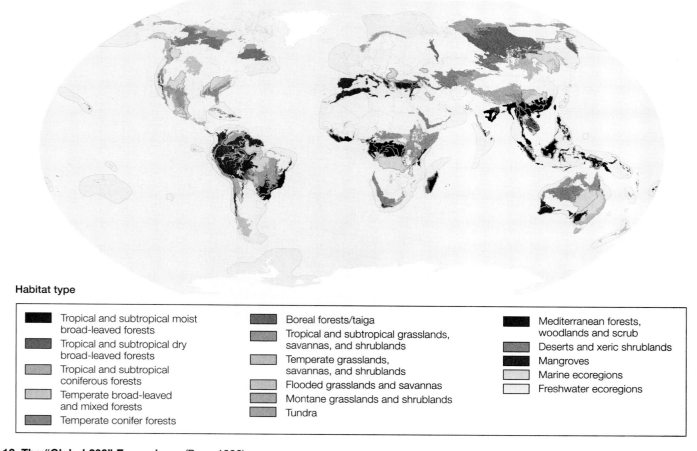

Habitat type

■	Tropical and subtropical moist broad-leaved forests
■	Tropical and subtropical dry broad-leaved forests
■	Tropical and subtropical coniferous forests
■	Temperate broad-leaved and mixed forests
■	Temperate conifer forests
■	Boreal forests/taiga
■	Tropical and subtropical grasslands, savannas, and shrublands
■	Temperate grasslands, savannas, and shrublands
■	Flooded grasslands and savannas
■	Montane grasslands and shrublands
■	Tundra
■	Mediterranean forests, woodlands and scrub
■	Deserts and xeric shrublands
■	Mangroves
■	Marine ecoregions
■	Freshwater ecoregions

57.12 The "Global 200" Ecoregions *(Page 1236)*

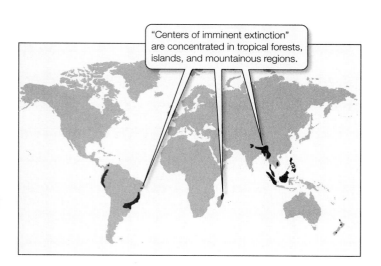

"Centers of imminent extinction" are concentrated in tropical forests, islands, and mountainous regions.

57.13 Centers of Imminent Extinction *(Page 1236)*

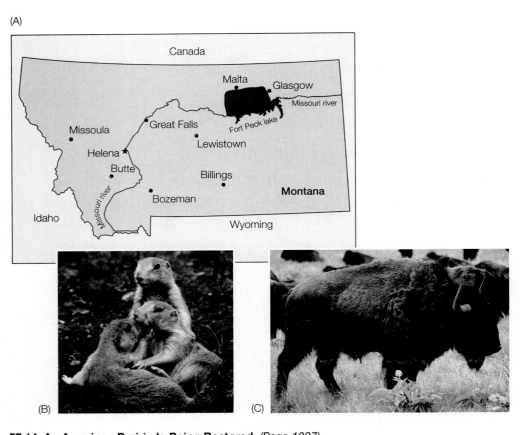

(A)

(B)

(C)

57.14 An American Prairie Is Being Restored *(Page 1237)*

EXPERIMENT

HYPOTHESIS: Wetland plots to be restored that are planted with mixtures of species will approach their original condition more rapidly than single-species plots.

METHOD

Plant some plots with only one of each of the 8 plant species typical of wetlands in the region. Plant other plots with randomly chosen assemblages of 3 and 6 species. Plant the same density of seedlings in all plots. Replant and weed as necessary to compensate for early mortality of seedlings.

■ Plot with 1 species
▲ Plot with 3 species
● Plot with 6 species

RESULTS

CONCLUSION: Vegetation cover, canopy complexity, and nitrogen accumulation are enhanced by species richness. In future wetland restoration attempts, a rich mixture of species should be planted.

57.16 Species Richness Enhances Wetland Restoration *(Page 1239)*

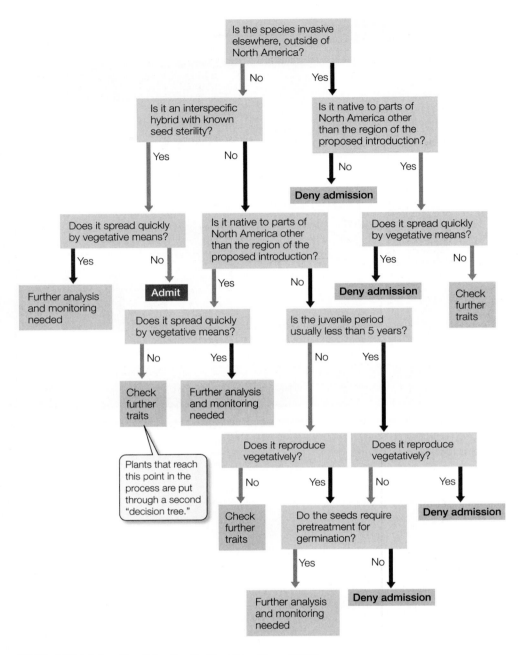

57.17 A "Decision Tree" for Exotic Species *(Page 1240)*

(A)

(B)

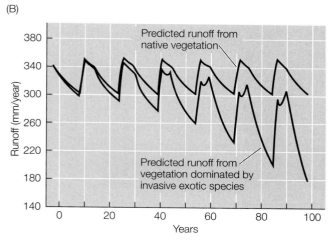

57.18 Invasive Species Disrupt Ecosystem Functioning
(Page 1241)

EXPERIMENT

HYPOTHESIS: Native pollinators (bees) of coffee flowers that live in forests increase both the quantity and quality of coffee seeds produced on adjacent plantations.

METHOD

Establish 12 sites that are near (50 m), intermediate (800 m), and far (1600 m) from forest patches. Measure the quality of seeds produced per plant and the frequency of small, misshapen seeds. To test whether the observed differences are due to the frequency of pollinator visits, hand-pollinate flowers.

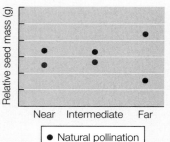
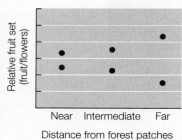
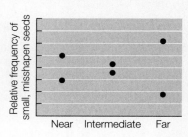

Distance from forest patches

- Natural pollination
- Hand pollination

RESULTS

Forest-dwelling bees increased coffee yields by 20% at sites within 1 kilometer of a forest patch. There were sufficient visits from bees to support full production at the near and intermediate sites. At those sites, hand pollination did not increase seed production. However, at far sites, hand pollination increased the mass of harvested seeds and fruit set. Hand pollination also decreased the proportion of small, misshapen seeds.

CONCLUSION: Coffee plants close to forest patches produced more seeds and had fewer misshapen seeds than coffee plants farther from the forest patches.

57.19 Establishing the Economic Value of Forest Patches *(Page 1242)*

Notes